NETWORK
ANALYSIS
&
CIRCUITS

NETWORK ANALYSIS & CIRCUITS

By

M. Arshad

INFINITY SCIENCE PRESS LLC
Hingham, Massachusetts
New Delhi

INFINITY SCIENCE PRESS LLC
11 Leavitt Street
Hingham, MA 02043
Tel. 877-266-5796 (toll free)
Fax 781-740-1677
info@infinitysciencepress.com
www.infinitysciencepress.com

This book is printed on acid-free paper.

Network Analysis and Circuits by M. Arshad
ISBN: 978-1-934015-19-3

The publisher recognizes and respects all marks used by companies, manufacturers, and developers as a means to distinguish their products. All brand names and product names mentioned in this book are trademarks or service marks of their respective companies. Any omission or misuse (of any kind) of service marks or trademarks, etc. is not an attempt to infringe on the property of others.

Library of Congress Cataloging-in-Publication Data

Arshad, M., b. 1979-
   Network analysis and circuits / M. Arshad.
     p. cm.
   ISBN 978-1-934015-19-3 (hardcover)
   1. Electronic circuits. 2. Electronic circuit design–Data processing.
   3. Electric network analysis–Mathematics. I. Title.
   TK7867.A77 2008
   621.3815–dc22
                                    2008015455
8 9 0 4 3 2 1

Our titles are available for adoption, license or bulk purchase by institutions, corporations, etc. For additional information, please contact the Customer Service Dept. at 877-266-5796 (toll free).

Requests for replacement of a defective CD-ROM must be accompanied by the original disc, your mailing address, telephone number, date of purchase and purchase price. Please state the nature of the problem, and send the information to INFINITY SCIENCE PRESS, 11 Leavitt Street, Hingham, MA 02043.

The sole obligation of INFINITY SCIENCE PRESS to the purchaser is to replace the disc, based on defective materials or faulty workmanship, but not based on the operation or functionality of the product.

Dedicated to my mother
(Late) Mrs. Shahjahan Begum

CONTENTS

Chapter 1

INTRODUCTION

1.1 WHAT IS A NETWORK?

In general terms, a network is a combination of elements. The interconnection of two or more simple elements (active and passive) is called an electrical network.

1.1.1 Active Elements

Elements, which can generate energy, are known as active elements (*e.g.*, battery, generator).

1.1.2 Passive Elements

Elements, which cannot generate energy but can dissipate or store energy are known as passive elements (*e.g.*, resistor, inductor, capacitor).

A network, which contains at least one active element, is known as an active network. A network, which does not contain any active elements, is a passive network.

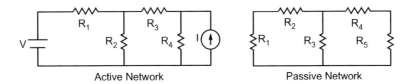

| Active Network | Passive Network |

FIGURE 1.1   *FIGURE 1.2*

If a network contains at least one energized closed path, it is also an electric circuit. Every circuit is a network but not all networks are circuits (*e.g.*, T-networks).

1.1.3 Unilateral or Bilateral Elements

A unilateral network has different relationships with the two possible directions of current between voltage and current (*e.g.*, diodes, SCR).

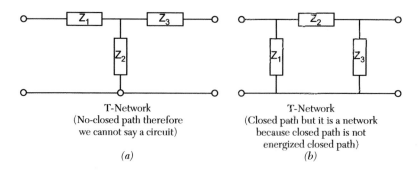

T-Network
(No-closed path therefore
we cannot say a circuit)

(a)

T-Network
(Closed path but it is a network
because closed path is not
energized closed path)

(b)

FIGURE 1.3

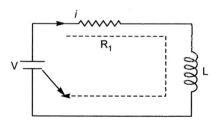

(This is a circuit. There is a energized closed path.)

FIGURE 1.4

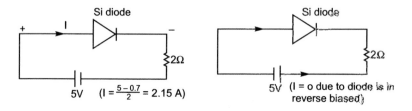

$\left(I = \frac{5 - 0.7}{2} = 2.15 \text{ A}\right)$

(I = o due to diode is in
reverse biased)

FIGURE 1.5   **FIGURE 1.6**

In the above example the relation between voltage and current is different. Therefore, diode is a unilateral element. A bilateral network has the same relationship between voltage and current for the two possible directions of current, *e.g.*, R, L, C (non-polar capacitor only).

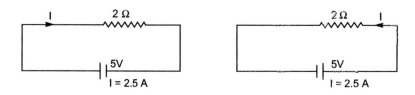

1.1.4 Linear or Non-linear Elements

A circuit element is linear if the principle of superposition holds and the relationship between the current and the voltage involves a constant coefficient, *e.g.*,

$$v = i\text{R}, \quad v = \text{L}\frac{di}{dt}, \quad v = \frac{1}{\text{C}}\int i\,dt.$$

A non-linear element is one in which the principle of superposition fails (*e.g.*, Varactor diode).

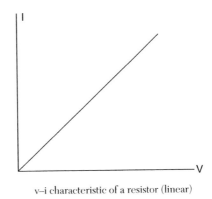

v–i characteristic of a resistor (linear)

FIGURE 1.7

1.1.5 Lumped or Distributed Network

A network with physically separate resistors, capacitors, and inductors is known as a lumped network (*e.g.*, all networks have different physically separate elements).

In the above diagram we can separate any passive element.

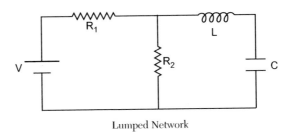

Lumped Network

FIGURE 1.8

A network in which network elements are not physically separable (*e.g.*, transmission line), the transmission line has distributed capacitance, resistance, and inductance along its length. If the length of the transmission line increases or decreases, the value of effective capacitance, resistance, and inductance changes and these cannot be separated physically.

1.2   SYSTEMS

A system is a combination of several components used to perform a desired task. If the system components are good, the system is good. System performance directly depends on the system components.

In this figure on the left-hand side of the box the inputs are represented. Suppose there are two inputs that excite the system to the desired output. Suppose there are $c_1(t)$ and $c_2(t)$.

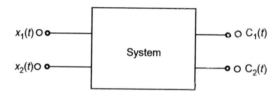

FIGURE 1.9

1.2.1   Classification of Systems

  (i)   Continuous time and discrete-time systems.
 (ii)   Time invariant and time-varying systems.
(iii)   Linear and non-linear systems.

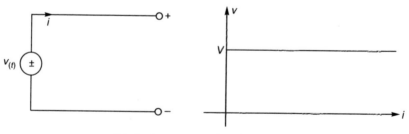

(Ideal voltage source and v-i characteristic)

FIGURE 1.10

(iv) Inverse systems.

(v) Instantaneous and dynamic systems.

(vi) Causal and non-causal systems.

1.2.2 Practical Voltage Source

Due to the power dissipation, the temperature rises and the value of internal resistance increases. Hence, the internal voltage drop rises and terminal voltage decreases.

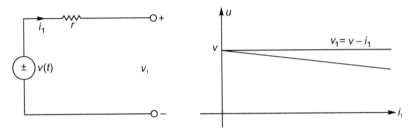

(Practical voltage source and v-i characteristic)

FIGURE 1.11

1.2.3 Independent Current Source

The current source is assumed to deliver energy through its terminal.

Ideal current source Maintains a constant current $i(t)$ independent of the value of voltage through its terminals as shown in Figure 1.12.

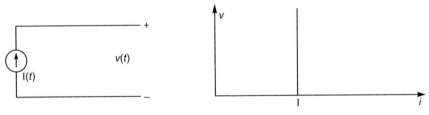

(ideal current source and v-i characteristic)

FIGURE 1.12

Practical current source Current through the terminals of the source keeps falling as the terminal voltage across it increases.

$$i_1 = I - \frac{V_1}{r}$$

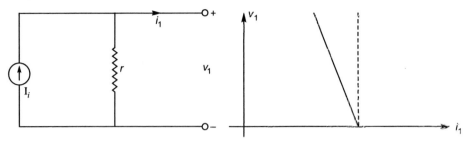

(Practical current source and v-i characteristic)

FIGURE 1.13

1.3   ELECTRICAL ENERGY SOURCES

There are two types of sources of electrical energy, the voltage source and the current source. They are further classified.

1.3.1   Independent Sources

A source in which the voltage is completely independent of the current or the current is completely independent of the voltage.

1.3.2   Independent Voltage Source

The voltage source is assumed to deliver energy with a terminal voltage.

Ideal voltage source Maintains a constant terminal voltage independent of the value of the current through its terminals as shown in Figure 1.10.

1.4   DEPENDENT SOURCE OR CONTROLLED SOURCE

There are four types as shown by the parallelogram.

1.4.1   Voltage Controlled Voltage Source (VCVS)

If the value of V_1 changes, the value of Vd also changes and hence, dependent source voltage (Vd) depends on V_1 and is known as a voltage-dependent source.

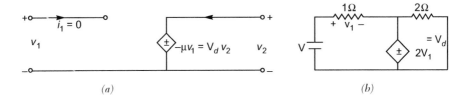

FIGURE 1.14

1.4.2 Current Controlled Voltage Source (CCVS)

The voltage source (Vd) depends on current and this type of source is known as a current-dependent voltage source.

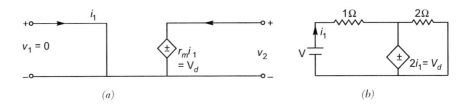

FIGURE 1.15

1.4.3 Current Controlled Current Source (CCCS)

A dependent current source (id) depends on current flowing through the 1Ω resistor. This type of source is known as a current-dependent current source. This is a current source and dependent on the current.

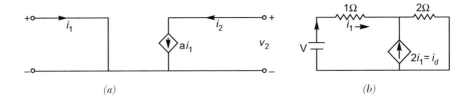

FIGURE 1.16

1.4.4 Voltage Controlled Current Source (VCCS)

A dependent current source current (id) depends on voltage across the 1Ω resistor. This type source is known as a voltage-dependent current source.

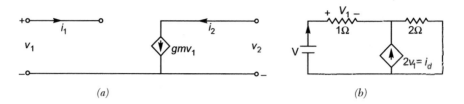

(a)   (b)

FIGURE 1.17

1.5   KIRCHHOFF'S VOLTAGE LAW (KVL)

"The algebraic sum of all voltages around a closed path at any instant is zero."
$\sum v(t) = 0$ "The algebraic sum of the voltage drop is equal to the algebraic sum of the voltage rise."

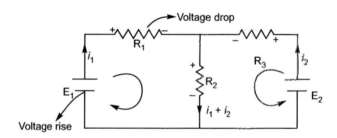

FIGURE 1.18

Steps for Kirchhoff's Voltage Law Equations

1. First show the current direction and value of current.
2. Current entering into a passive element makes a + ve sign at a point where current enters and negative where current leave.
3. Now start to move from any element but remember the element sign should be consecutive and take the sign which is first.

Now the equations. Let us start from source E_1 sign -ve,

$$-E_1 + i_1 R_1 + (i_1 + i_2)R_2 = 0.$$

Similarly, $- E_2 + i_2 R_3 + (i_1 + i_2)R_2 = 0.$

Example 1.1. *Find the KVL equation for the following circuits:*

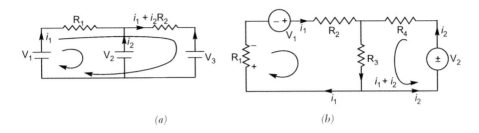

(a)   (b)

FIGURE 1.19

Ans.

(a) $-V_1 + i_1 R_1 + V_2 = 0;\ -V_1 + i_1 R_1 + (i_1 + i_2)R_2 + V_3 = 0$

(b) $+i_1 R_1 - V_1 + i_1 R_2 + (i_1 + i_2)R_3 = 0;\ -V_2 + i_2 R_4 + (i_1 + i_2)R_3 = 0$

KVL follows the law of conservation of energy.

1.6 KIRCHHOFF'S CURRENT LAW (KCL)

"At any instant of time, the algebraic sum of currents at a node is zero."

$$\sum_{\text{node } n} i(t) = 0$$

"The algebraic sum of currents entering at a node is equal to the algebraic sum of currents leaving."

If the current entering a node is assigned a positive sign, *i.e.*,

$$-i_1 + i_2 - i_3 + i_4 + i_5 = 0,$$

$$\underbrace{i_2 + i_4 + i_5}_{\substack{\text{currents}\\ \text{entering}\\ \text{or}\\ \text{incoming}\\ \text{currents}}} = \underbrace{i_1 + i_3}_{\substack{\text{currents}\\ \text{leaving}\\ \text{or}\\ \text{outgoing}\\ \text{currents}}}.$$

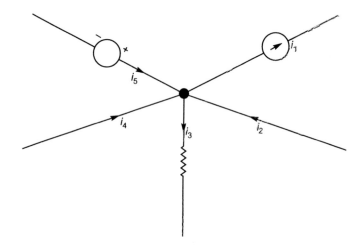

FIGURE 1.20

Example 1.2. *Find the KCL equations for the following diagrams at given node A:*

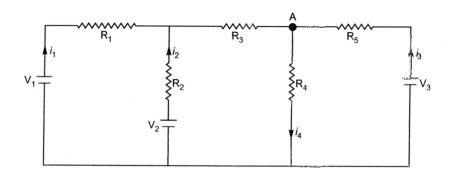

FIGURE 1.21

Ans. $i_1 + i_2 + i_3 - i_4 = 0$

1.7 MESH ANALYSIS

The mesh or loop method of analysis is illustrated by the following circuit. In an individual mesh, assume a current. There are three mesh in the figure and three currents I_1, I_2, and I_3, respectively.

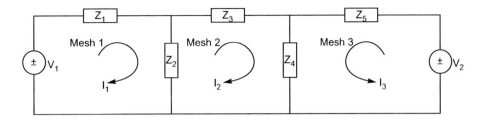

FIGURE 1.22

Step 1. First show the direction of the mesh current.

Step 2. Apply KVL and think what is the effective value of the current in each element. (*i.e.*, the effective value of current in Z_2 is $I_1 - I_2$ for mesh 1).

Step 3. In every mesh preference is given to its mesh current (*i.e.*, suppose we are applying mesh analysis in mesh 1 and 2. the effective current in mesh 1 at Z_2 is $I_1 - I_2$ and $I_2 - I_1$ for mesh 2. I_1 dominates in mesh 1 and I_2 dominates in mesh 2, which is due to polarity).

Now Mesh Analysis

$$-V_1 + Z_1 I_1 + (I_1 - I_2)Z_2 = 0 \tag{1.1}$$

$$+(I_2 - I_1)Z_2 + I_2 Z_3 + (I_2 + I_3)Z_4 = 0 \tag{1.2}$$

$$-V_2 + I_3 Z_5 + Z_4(I_3 + I_2) = 0 \tag{1.3}$$

If any current value is negative, it shows the current direction is opposite as we assume.

Example 1.3. *Find using mesh analysis as shown in Figure 1.23.*

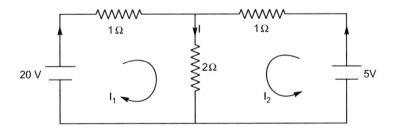

FIGURE 1.23

Solution: Mesh equations for mesh 1,

$$-20 + I_1 + 2(I_1 + I_2) = 0$$

$$3I_1 + 2I_2 = 20. \tag{1}$$

For mesh 2,

$$-5 + I_2 + (I_2 + I_1)2 = 0$$
$$2I_1 + 3I_2 = 5. \tag{2}$$

By Equations (1) and (2),

$$I_1 = 10A, I_2 = -5A$$
$$I = I_1 + I_2 = 10 - 5 = 5A.$$

NOTE I_2 negative sign shows the opposite direction of the current whatever we assume. It means a 5 V battery is not supplying current, it is charging as shown in Figure 1.24.

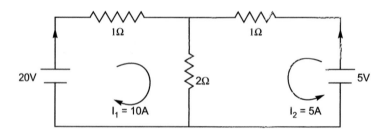

FIGURE 1.24

Example 1.4. *Find the current in each passive element by mesh analysis (i.e., I_1, I_2, I_3), etc.*

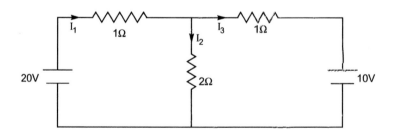

FIGURE 1.25

Ans. $I_1 = 8A, I_2 = 6A, I_3 = 2A$

1.8 NODAL ANALYSIS

Nodal analysis is a very useful method if the number of parallel branches is more in a given network.

Step 1. In this method some convenient junction between elements is chosen as a reference or Datum node.

Step 2. Show the direction of current and keep in mind that current flows from higher potential to lower potential.

Step 3. Write down the KCL equation at the given node.

Now, it is explained by an example.

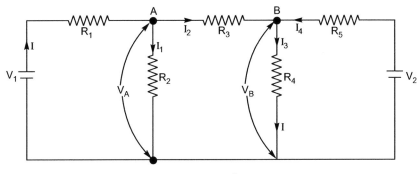

Datum node

FIGURE 1.26

By KCL at node A,

$$I = I_1 + I_2 \text{ and } I = \frac{V_1 - V_A}{R_1}, I_1 = \frac{V_A}{R_2} \qquad \text{(By Ohm's law)}$$

$$I_2 = \frac{V_A - V_B}{R_3}$$

Now,

$$\frac{V_1 - V_A}{R_1} = \frac{V_A}{R_2} + \frac{V_A - V_B}{R_3}. \qquad (1.4)$$

Similarly, at node B

$$I_2 + I_4 = I_3$$

$$\frac{V_A - V_B}{R_3} + \frac{V_2 - V_B}{R_5} = \frac{V_B}{R_4}. \qquad (1.5)$$

V_A and V_B are variables.

Example 1.5. *Using the Nodal method, find the current through* r_2.

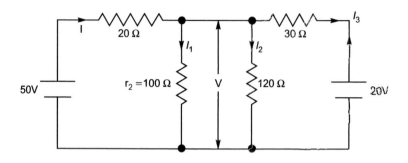

FIGURE 1.27

Solution: By KCL,

$$I = I_1 + I_2 + I_3$$

$$\frac{50 - V}{20} = \frac{V}{100} + \frac{V}{120} + \frac{V - 20}{30}.$$

After solving, $V = 31.14\,\text{V}$ $I_1 = \frac{V}{r_2} = \frac{31.14}{100} = 311.4\,\text{mA}$.

Example 1.6. *Find the current in the* 1Ω *resistor using Nodal analysis.*

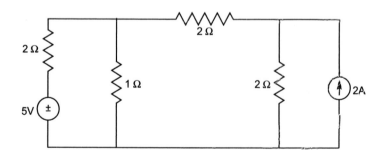

FIGURE 1.28

Ans. 2 Amp.

1.9   PASSIVE ELEMENTS

There are three passive elements.

1.9.1 Resistance

Resistance is the property of a material which offers opposition to the current and dissipates energy. In a conductor the resistance is due to the collision between moving electrons and fixed ions, heat is generated due to the collision, and the resistance of a conductor depends on its length, cross-sectional area, material, and temperature and is denoted R.

$$R = \rho \frac{l}{A}$$

Internal resistance is shown by r and it is battery own resistance,

$$R_t = R_0(1 + \alpha \Delta t).$$

α is + ve for the conductor and −ve for the insulator and the semiconductor, which means, if the temperature increases, the resistance of the conductor increases but the resistance of the insulator and semiconductor decreases.

There are certain alloys such as constantan, nickel, chromium, and magnin where practically the temperature coefficients of resistance are negligibly small. Such alloys are used in the manufacture of standard resistors and heating elements.

In a series, the equivalent resistance is the sum of resistances.

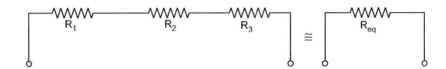

FIGURE 1.29

Suppose there are three resistances connected in a series,

$$R_{eq} = R_1 + R_2 + R_3. \tag{1.6}$$

If there are n resistances connected in a series,

$$R_{eq} = R_1 + R_2 + R_3 + \cdots + R_n. \tag{1.7}$$

In a series equivalent, resistance is always greater than the maximum resistance in the circuit.

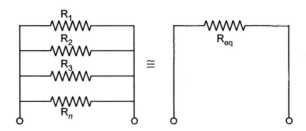

FIGURE 1.30

In parallel,

$$\frac{1}{R_{eq}} = \frac{1}{R_1} + \frac{1}{R_2} + \frac{1}{R_3} + \cdots + \frac{1}{R_n}. \qquad (1.8)$$

For two resistances connected in parallel,

$$R_{eq} = \frac{R_1 \cdot R_2}{R_1 + R_2}.$$

In parallel, equivalent resistance (R_{eq}) is less than the minimum resistance in the circuit.

1.9.2 Inductance

Inductance is the property of a circuit element in which it opposes any change in current, and due to the current changing, a voltage is induced. It is known as the induced voltage. When the voltage is induced in the element by the current changing in its own circuit the property is called self inductance (L).

$$V_L = L\frac{di}{dt} \text{ Volts.} \qquad (1.9)$$

The property of inductance comes into play only when the current in the circuit changes. When the circuit current changes inductance opposes this change. If the circuit current is constant, $\frac{di}{dt} = 0$, there is no induced voltage.

$$\text{Energy stored in an inductor} = \frac{1}{2} LI^2 \text{ Joule.} \qquad (1.10)$$

When two circuits are so placed that a portion of the magnetic flux produced by one links with the turns of both the circuits, they are said to be mutually coupled. Mutual inductance is a property of two circuits when a voltage is induced in one circuit by a change of current in the other circuit

induced voltage. In coil 2, $V_2 = M\frac{di_1}{dt}$ by the changing current in coil 1.

$$M \leq \sqrt{L_1 L_2}, \tag{1.11}$$

where M is mutual inductance and L_1 and L_2 are self inductance of coil 1 and coil 2.

In a series equivalent, inductance of a circuit is equal to the total sum of individual inductance.

$$L_{eq} = L_1 + L_2 + L_3 + \cdots + L_n.$$

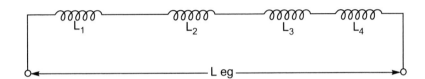

FIGURE 1.31

Similarly, inductances are connected in parallel,

$$\frac{1}{L_{eq}} = \frac{1}{L_1} + \frac{1}{L_2} + \cdots + \frac{1}{L_n}. \tag{1.12}$$

If two inductances are connected in parallel,

$$L_{eq} = \frac{L_1 \cdot L_2}{L_1 + L_2}. \tag{1.13}$$

1.9.3 Capacitance

Capacitance is the property of a circuit element or device which is capable of storing charge or energy in an electric field. A device which has this property is known as a capacitor. A capacitor whose characteristic does not change with time is called a time-invariant capacitor. If the characteristic changes with time, the capacitor is called a time-varying capacitor.

$$c = \frac{q}{v}. \tag{1.14}$$

q is charge stored in a capacitor and v is voltage applied across the capacitor.

Capacitance of a parallel plate capacitor is

$$C = \frac{\in A}{d}, \tag{1.15}$$

where A = area of plates and d = distance between two plates.

If q is measured in coulombs and v in volts, the unit of C is the farad (in honor of Michael Faraday),

$$q = i \quad t = CV. \tag{1.16}$$

Current and voltage are related by the equation

$$i = \frac{dq}{dt} = \frac{d}{dt}[CV] = C\frac{dV}{dt}. \tag{1.17}$$

C is assumed to be constant. If C is not constant then,

$$i = C\frac{dV}{dt} + v\frac{dC}{dt}. \tag{1.18}$$

Capacitors are also variable and fixed and according to polarity, capacitors are polar and non-polar. If we are using polar capacitors or electrolytic capacitors, polarity must be observed when connecting it into a circuit. Capacitors other than electrolytics can be connected into a circuit without concern for proper polarity.

If capacitors are connected in a series, equivalent capacitance of a circuit will be

$$\frac{1}{C_{eq}} = \frac{1}{C_1} + \frac{1}{C_2} + \frac{1}{C_3} + \cdots + \frac{1}{C_n}. \tag{1.19}$$

If capacitors are connected in parallel equivalent capacitance,

$$C_{eq} = C_1 + C_2 + C_3 + \cdots + C_n. \tag{1.20}$$

1.9.3.1 Source transformation.

Any practical voltage source is converted into a current source.

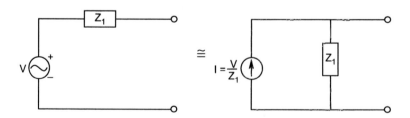

FIGURE 1.32

Similarly, any practical current source can be converted into a voltage source.

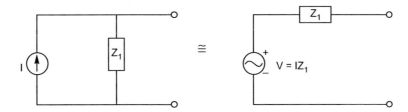

FIGURE 1.33

Example 1.7. *Convert the following current source into a voltage source.*

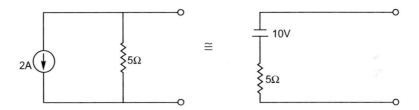

FIGURE 1.34

Example 1.8. *Convert the following voltage source into a current source.*

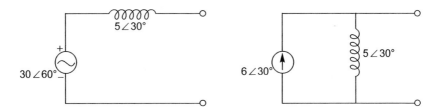

FIGURE 1.35

1.10 CURRENT DIVISION FORMULA

Suppose two resistances R_1 and R_2 are connected in parallel as shown in Figure 1.36.

$$V = IR_{eq} = I_1R_1 = I_2R_2$$

and

$$R_{eq} = \frac{R_1 \cdot R_2}{R_1 + R_2}$$

$$\therefore \quad I_1 = \frac{R_{eq}}{R_1} \text{ (by the above relation)}$$

$$I_1 = \left(\frac{R_2}{R_1 + R_2}\right)I. \qquad (1.21)$$

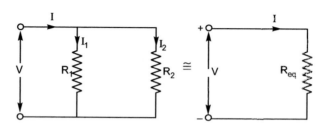

FIGURE 1.36

Similarly,

$$I_2 = \left(\frac{R_1}{R_1 + R_2}\right)I. \qquad (1.22)$$

Thus, the current in one of the two parallel resistors is found by multiplying the total current by the other resistance and dividing by the sum of the two resistances.

For impedance,

$$I_1 = \left(\frac{Z_2}{Z_1 + Z_2}\right)I$$

$$I_2 = \left(\frac{Z_1}{Z_1 + Z_2}\right)I.$$

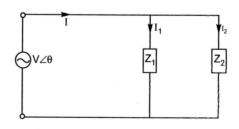

FIGURE 1.37

Example 1.9. *Find I_1 and I_2 as shown in Figure 1.38.*

Solution: $I = 100$ A (given). By using formula $I_1 = (\frac{R_2}{R_1 + R_2})$, $I = \frac{10}{25} \times 100 = 40$ A.

Similarly,

$$I_2 = \frac{15}{25} \times 100 = 60 \text{ A.}$$

Example 1.10. *Find I_1 and I_2 as shown in Figure 1.39.*

Ans. $15\angle93.75°$, $5\angle48.73°$

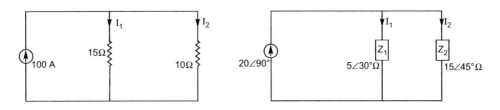

FIGURE 1.38   FIGURE 1.39

1.11   VOLTAGE DIVISION FORMULA FOR TWO RESISTORS

If there are two resistors connected in a series,

$$I = \frac{V}{R_1 + R_2}$$
$$V_1 = IR_1$$
$$V_1 = \left(\frac{R_1}{R_1 + R_2}\right)V$$

and

$$V_2 = IR_2$$
$$V_2 = \left(\frac{R_2}{R_1 + R_2}\right)V.$$

Thus, the voltage on either of the series resistors is equal to the applied voltage times the ratio of that resistance to the total resistance.

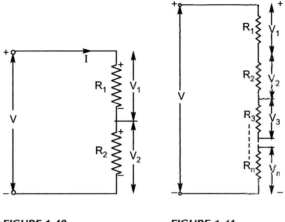

FIGURE 1.40   FIGURE 1.41

For n-resistors connected in a series,

$$V_n = \left(\frac{R_n}{R_1 + R_2 + R_3 + \cdots + R_n} \right) V.$$

Similarly, for impedance,

$$V_n = \left(\frac{Z_n}{Z_1 + Z_2 + Z_3 + \cdots + Z_n} \right) V. \qquad (1.23)$$

Example 1.11. *Find the voltage drop across 5Ω resistor as shown in Figure 1.42.*

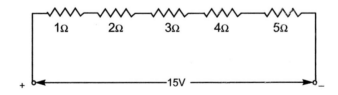

FIGURE 1.42

Ans.

$$V_5 = \left(\frac{R_5}{R_1 + R_2 + R_3 + R_4 + R_5} \right) \times V = \frac{5}{15} \times 15 = 5V$$

Example 1.12. *Find the voltage drop across* Z_1 *as shown in Figure 1.43.*

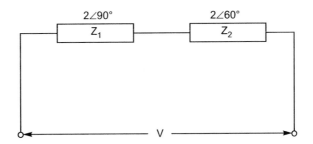

FIGURE 1.43

Ans. Some important formulas for complex numbers.

(i) $M\angle = M(\cos\theta + J\sin\theta)$ (1.24)

(ii) $M_1\angle\theta_1 \times M_2\angle\theta_2 = M_1 M_2\angle\theta_1 + \theta_2$ (1.25)

(iii) $\frac{M_1\angle\theta_1}{M_2\angle\theta_2} = \frac{M_1}{M_2}\angle\theta_1 - \theta_2$ (1.26)

(iv) $M_1\angle\theta_1 + M_2\angle\theta_2 = (M_1\cos\theta_1 + M_2\cos\theta_2)$
$\qquad + J(M_1\sin\theta_1 + M_2\sin\theta_2)$ (1.27)

Example 1.13. *Find* $Z_1 + Z_2, Z_1 Z_2, \frac{Z_1}{Z_2}$ *and* $Z_1 - Z_2$, *where*

$$Z_1 = 10\angle 30^\circ, Z_2 = 5\angle 30^\circ.$$

Ans. $15\angle 30^\circ, 50\angle 60^\circ, 2\angle 0^\circ, 5\angle 30^\circ$

1.12   POWER AND RMS VALUES

The power p converted in a resistor (*i.e.*, the rate of conversion of electrical energy to heat) is $p(t) = iv = v^2/R = i^2 R$.

We use lowercase $p(t)$ because this is the expression for the instantaneous power at time t. Usually, we are interested in the mean power delivered, which is normally written P. P is the total energy converted in one cycle, divided by the period T of the cycle

$$P = \frac{1}{T}\int_0^t vi\,dt. \qquad (1.28)$$

This set of equations is useful because they are exactly those normally used for a resistor in D.C. electricity. However, one must remember that P is the average power, and $V = V_m/2^{1/2}$ and $I = I_m/2^{1/2}$. Looking at the integral above, and dividing by R, we see that I is equal to the square root of the mean value of i^2, so I is called the root-mean-square or **RMS value**. Similarly, $V = V_m/2^{1/2} \sim 0.71 {*} V_m$ is the RMS value of the voltage.

When talking of A.C. RMS values are so commonly used that, unless otherwise stated, you may assume that RMS values are intended.

Power in a resistor In a resistor R, the peak power (achieved instantaneously 100 times per second for 50 Hz AC) is $V_m^2/R = i_m^{2} {*} R$. As discussed above, the voltage, current, and the power pass through zero volts 100 times per second, so the average power is less than this. The average is exactly as shown above: $P = V_m^2/2R = V^2/R$.

Power in inductors and capacitors In ideal inductors and capacitors, a sinusoidal current produces voltages that are respectively 90° ahead and behind the phase of the current. So if $i = I_m \sin \omega t$, the voltages across the inductor and capacitor are $V_m \cos \omega t$ and $-V_m \cos \omega t$, respectively. Now the integral of $\cos {*} \sin$ over a whole number of cycles is zero. Consequently, ideal inductors and capacitors do not take power from the circuit.

$$\text{Active power} \quad P = VI \cos \phi \qquad (1.29)$$

$$\text{Reactive power} \quad Q = VI \sin \phi \qquad (1.30)$$

$$S^2 = P^2 + Q^2 \qquad (1.31)$$

where ϕ is known as a power factor angle and $\cos \phi$ is known as a power factor.

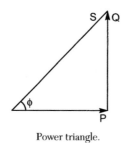

Power triangle.

FIGURE 1.44

SOLVED PROBLEMS

Problem 1.1. *In the network system shown in Figure 1.45, find the current through Z_3 using the Nodal method. The values of voltages are given in volts and the impedances are given in ohms.*

Solution:

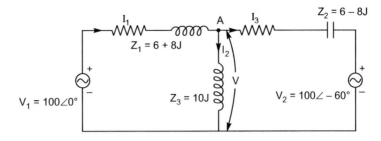

FIGURE 1.45

By KCL at node A,

$$I_1 = I_2 + I_3$$
$$\frac{100\angle 0° - V}{6 + 8J} = \frac{V}{10J} + \frac{V - 100\angle - 60°}{6 - 8J}.$$

After solving,

$$V = 110.83\angle - 43.32°,$$

the current through

$$Z_3 = I = \frac{V}{10J} = \frac{110.83\angle - 43.32°}{10\angle 90°} = 11.089\angle - 133.32°.$$

Problem 1.2. *Determine the current in the 1 Ω resistor of the network shown in Figure 1.46.*

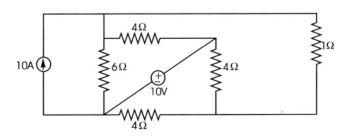

FIGURE 1.46

Solution: The $10\,A$ current source and 6Ω resistor are connected in parallel, using the source transformation formula. Now apply the mesh analysis formula. In this problem there are three meshes and three unknown variables (I_1, I_2, and I_3) as shown in Figure 1.47.

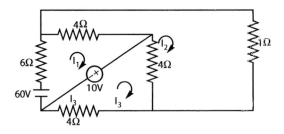

FIGURE 1.47

For mesh 1,

$$60 = 6I_1 + 4(I_1 - I_2) + 10 \Rightarrow 5I_1 - 2I_2 = 25. \tag{1}$$

For mesh 2,

$$4(I_2 - I_3) + 4(I_2 - I_1) + I_2 = 0$$

$$-4I_1 + 9I_2 - 4I_3 = 0 \Rightarrow I_3 = \frac{-4I_1 + 9I_2}{4}. \tag{2}$$

For mesh 3,

$$-10 + 4(I_3 - I_2) + 4I_3 = 0$$
$$-4I_2 + 8I_3 = 10 \Rightarrow -2I_2 + 4I_3 = 5. \tag{3}$$

By Equations (2) and (3),

$$-2I_2 + (-4I_1 + 9I_2) = 5 \Rightarrow -4I_1 + 7I_2 = 5. \tag{4}$$

By Equations (1) and (4), and $I_2 = 4.63A$ hence, current flowing through resistor 1Ω is

$$I_2 = 4.63A.$$

Problem 1.3. *Find i and the power absorbed by the 30Ω resistor as shown in Figure 1.48.*

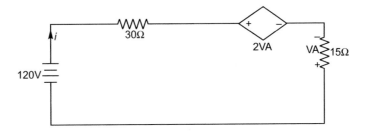

FIGURE 1.48

Solution: Applying KVL,

$$120 = 30i + 2V_A - V_A$$
$$120 = 30i + V_A \tag{1}$$
$$V_A = -15i. \tag{2}$$

Now by Equations (1) and (2),

$$15i = 120$$
$$i = 8 \text{ A}$$
$$P_{30} = i^2 \times 30 = 64 \times 30 = 1920 \text{ W}.$$

Problem 1.4. *Find i_2 and i_3 as shown in Figure 1.49.*

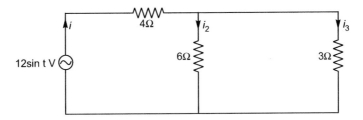

FIGURE 1.49

Solution:

$$i = \frac{12\sin t}{4 + 6113}$$
$$i = 2\sin t$$
$$i_2 = \left(\frac{3}{6+3}\right) 2\sin t = \frac{2}{3}\sin t$$
$$i_3 = \left(\frac{6}{6+3}\right) 2 \text{ in } t = \frac{4}{3}\sin t$$

Problem 1.5. *Find the power absorbed by each element as shown in Figure 1.50.*

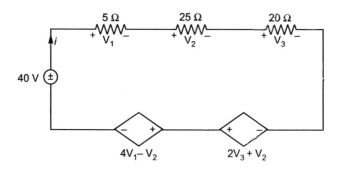

FIGURE 1.50

Solution: Applying KVL,

$$40 = V_1 + V_2 + V_3 - (2V_3 + V_2) + (4V_1 - V_2)$$
$$40 = 5V_1 - V_2 - V_3$$
$$V_1 = 5i, V_2 = 25i, V_3 = 20i$$
$$40 = 25i - 25i - 20i$$
$$i = -2\text{A} \quad (-\text{ve sign shows } i \text{ is flowing in the opposite direction})$$
$$P_{40\,V} = 40 \times 2 = 80\,\text{W(absorbing energy, charging mode)}$$
$$P_{5\omega} = 20\,\text{W}$$
$$P_{25\omega} = 4 \times 25 = 100\,\text{W}$$
$$P_{20\omega} = (2)^2 \times 20 = 80\,\text{W}$$
$$\left.\begin{array}{l} P_{2v_3+v_2} = -260\,\text{W} \\ P_{4v_1-v_2} = -20\,\text{W} \end{array}\right\} \quad \text{Not absorbing they are supplying energy.}$$

Problem 1.6. *Find v_1, v_2, and v_3 using Nodal analysis.*

Solution: By applying KCL,

$$8 = 4 + i_1 \Rightarrow i_1 = 4_\text{A}$$

and

$$i_1 = i_2 + i_3 \Rightarrow i_2 + i_3 = 4 \tag{1}$$

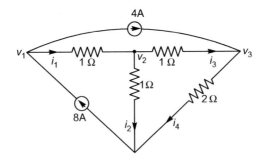

FIGURE 1.51

and

$$4 + i_3 = i_4. \tag{2}$$

Now

$$i_2 = \frac{v_2}{1} \quad [\text{Using } V = IR] = v_2$$

$$i_3 = \frac{v_2 - v_3}{1} = v_2 - v_3$$

$$i_4 = \frac{v_3}{2}.$$

Putting the values of i_2, i_3, and i_4 into Equations (1) and (2),

$$v_2 + v_2 - v_3 = 4 \Rightarrow 2v_2 - v_3 = 4 \tag{3}$$

$$4 + v_2 - v_3 = \frac{v_3}{2} \Rightarrow 2v_2 - 3v_3 = -8. \tag{4}$$

By Equations (3) and (4),

$$v_3 = 6 \quad \text{and} \quad v_2 = 5,$$

which means the i_3 direction is reverse (from v_3 to v_2) because the value of v_3 is higher than v_2.

$$i_1 = 4\text{A} = \frac{v_1 - v_2}{1}$$

$$v_1 = 4 + 5 = 9\,\text{V}$$

Problem 1.7. *Find* i_1, i_2, *and* i_3 *using Mesh analysis as shown in Figure 1.52.*

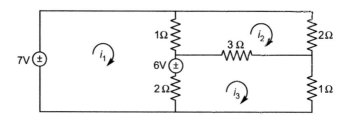

FIGURE 1.52

Solution: Mesh equations

At mesh (1),

$$-7 + 1(i_1 - i_2) + 6 + 2(i_1 - i_3) = 0$$
$$3i_1 - i_2 - 2i_3 = 1. \tag{1}$$

At mesh (2),

$$1(i_2 - i_1) + 2(2 + 3(i_2 - i_3)) = 0$$
$$-i_1 + 6i_2 - 3i_3 = 0. \tag{2}$$

At mesh (3),

$$2(i_3 - i_1) - 6 + 3(i_3 - i_2) + 1i_3 = 0$$
$$-2i_1 - 3i_2 + 6i_3 = 6. \tag{3}$$

After solving Equations (1), (2), and (3),

$$i_1 = 3\text{A}, i_2 = 2\text{A}, i_3 = 3\text{A}.$$

Problem 1.8. *Find I using Mesh and Nodal analysis as shown in Figure 1.53.*

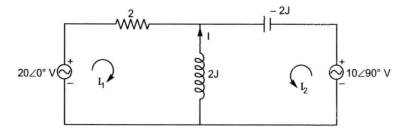

FIGURE 1.53

Solution: By using Mesh analysis,

$$-20° \angle 0 + 2I_1 + 2J(I_1 + I_2) = 0$$
$$10 = (1 + J)I_1 + JI_2 \qquad (1)$$
$$-10\angle 90° - 2JI_2 + (I_1 + I_2)2J = 0$$
$$10J = 2JI_1 + 2JI_2 - 2JI_2$$
$$I_1 = 5.$$

Putting the value of I_1 into Equation (1), $10 - 5 - 5J = JI_2$

$$I_2 = \frac{5(1 - J)}{J} \times \frac{J}{J} = -5(J + 1)$$
$$I = I_1 + I_2 = 5 - 5J - 5 = -5J = 5\angle -90°A.$$

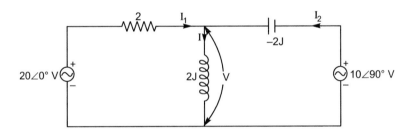

FIGURE 1.54

By Nodal analysis, applying KCL,

$$I_1 + I_2 = I \qquad\qquad [10 < 90° = 10J]$$
$$\frac{20 - V}{2} + \frac{10\angle 90° - V}{-2J} = \frac{V}{2J}$$
$$V = 10\,V$$
$$I = \frac{V}{2J} = \frac{10}{2J} = -5J = 5\angle -90°.$$

Problem 1.9. *Find V_A using nodal analysis as shown in Figure 1.55.*

Solution: Applying KCL at node N,

$$I_1 + 10\angle 0° = I_2 \qquad (1)$$

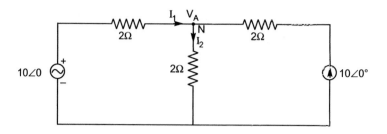

FIGURE 1.55

and

$$I_1 = \frac{10 - V_A}{3}$$

$$I_2 = \frac{V_A}{2}$$

Putting these values into Equation (1),

$$\frac{10 - V_A}{2} + 10 = \frac{V_A}{2}$$

$$10 - V_A + 20 = V_A$$

$$2V_A = 30$$

$$V_A = 15\,\text{V}.$$

QUESTIONS FOR DISCUSSION

1. What is a network?
2. What is the difference between a network and a circuit?
3. Define active and passive elements.
4. What is an active and a passive network?
5. Define unilateral and bilateral networks.
6. What is linearity?
7. What is a lumped and a distributed network?
8. Draw a V-I characteristic for an ideal source.
9. Draw a V-I characteristic for an ideal and practical source.
10. What is internal resistance?
11. Define dependent and independent sources.
12. Draw a circuit diagram for a current dependent current source.
13. Define KVL and KCL.

14. Can KVL and KCL apply for a dependent source?
15. What is a Mesh?
16. What is the advantage of Nodal analysis?
17. Define self and mutual inductance.
18. What is a positive temperature coefficient?
19. Does the temperature of a carbon rod increase its resistance or decrease it?
20. Give some examples with negative temperature coefficients.

OBJECTIVE QUESTIONS

1. Two coils having equal resistance but different inductances are connected in a series. The time constant of the series combination is the
   (a) Sum of the time constants of the individual coils.
   (b) Average of the time constants of the individual coils.
   (c) Geometric mean of the time constants of the individual coils.
   (d) Product of the time constants of the individual coils.

2. All the resistances in Figure 1.56 are 1Ω each. The value of current I is

   (a) $\dfrac{1}{15}$ A   (b) $\dfrac{2}{15}$ A   (c) $\dfrac{4}{15}$ A   (d) $\dfrac{8}{15}$ A.

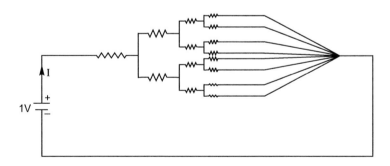

FIGURE 1.56

3. An ideal voltage source will charge an ideal capacitor
   (a) In infinite time   (b) Exponentially
   (c) Instantaneously   (d) None of the above.

4. A practical current source is usually represented by
   (a) A resistance in series with an ideal current source
   (b) A resistance in parallel with an ideal current source
   (c) A resistance in parallel with an ideal voltage source
   (d) None of the above.

5. If the length of a wire of resistance R is uniformly stretched to n times its original value its new resistance is
   (a) nR   (b) $\dfrac{R}{n}$ (c) n^2R   (d) $\dfrac{R}{n^2}$.

6. When the plate area of a parallel plate capacitor is increased keeping the capacitor voltage constant, the force between the plates
   (a) Increases
   (b) Decreases
   (c) Remains constant
   (d) May increase or decrease depending on the metal making up the plates.

7. The voltage phaser of a circuit is $10\angle15°$ V and the current phaser is $2\angle-45°$A. The active and the reactive powers in the circuit are:
   (a) 10 W and 17.34 V Ar
   (b) 5 W and 8.66 V Ar
   (c) 20 W and 60 V Ar
   (d) $20\sqrt{2}$ W and $10\sqrt{2}$ V Ar.

8. The circuit shown in Figure 1.57 is equivalent to a load of
   (a) $\dfrac{4}{3}\Omega$ (b) $\dfrac{8}{3}\Omega$ (c) 4Ω (d) 2Ω.

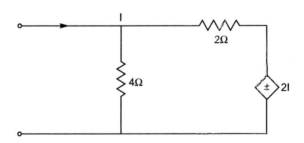

FIGURE 1.57

9. In the circuit shown, the V(t) and i(t) will be
   (a) 1V, 1A   (b) 1V, 6A   (c) 5V, 5A   (d) None of the above.

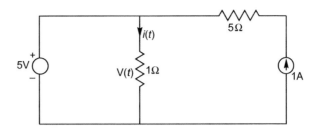

FIGURE 1.58

10. Two wires A and B of the same material and length of L and 2L have radius r and 2r, respectively. The ratio of their specific resistance will be
    (a) 1:1   (b) 1:2   (c) 1:4   (d) 1:8.
11. A unit of capacitance is
    (a) $\dfrac{\text{Coulomb}}{\text{Volt}}$ (b) $\dfrac{\text{Volt}}{\text{Coulomb}}$ (c) Coulomb $\times$ Volt   (d) Daraf.
12. The rate of rise of the voltage in a capacitor $(2\mu F)$ is $20\,\mu/\text{sec}$. The value of current passing through the capacitor is
    (a) 40 A   (b) 20 A   (c) 10 A   (d) 5 A.
13. The function $\dfrac{df V}{dV}$ is called an incremental
    (a) Resistance   (b) Capacitor   (c) Constant   (d) Inductor.
14. An element having a characteristic in current (i) and flux (ϕ) plane is called a
    (a) Resistor   (b) Capacitor   (c) Inductor   (d) Current flux.
15. Two voltage sources can be connected in parallel if they have equal
    (a) Magnitude   (b) Frequency   (c) Phase   (d) All of these.
16. The number of independent equations used to solve a network is equal to the number of
    (a) Branches   (b) Nodes   (c) (Nodes -1)   (d) Chords.
17. If two equal impedance $10\angle 60°$ are connected in parallel their equivalent impedance will be
    (a) $20\angle 60°$ (b) $10\angle 120°$ (c) $15\angle 120°$ (d) $5\angle 60°$.
18. Two series connected batteries supply 2A, and each carries a current of
    (a) 1 A   (b) 2 A   (c) 4 A   (d) 8 A.
19. A 40Ω capacitor is connected in series with a 30Ω resistor across a $100\,V$ single phase supply. The power dissipated in the circuit will be
    (a) 30   (b) 60   (c) 100   (d) 120.
20. The voltage source transformation of a current source between AB in Figure 1.59 will be a series combination of
    (a) $10\,V, 3\Omega$ (b) $15\,V, 2\Omega$ (c) $15\,V, 3\Omega$ (d) $-15\,V, 3\Omega$.

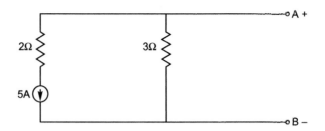

FIGURE 1.59

UNSOLVED PROBLEMS

1. Determine i_1, i_2, and i_3 using Mesh analysis as shown in Figure 1.60.

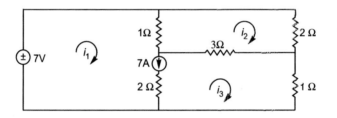

FIGURE 1.60

2. Determine i_1, i_2, and i_3 using Mesh analysis as shown in Figure 1.61.

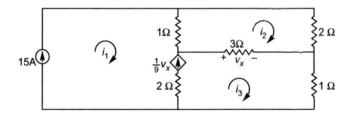

FIGURE 1.61

3. Find I using Nodal and Mesh analysis as shown in Figure 1.62.
4. Find I using Nodal and Mesh analysis as shown in Figure 1.63.
5. Determine the voltage drop across the 2Ω resistor using Nodal analysis as shown in Figure 1.64.
6. Find the voltage drop across the 2Ω resistor using Mesh analysis (Figure 1.64).

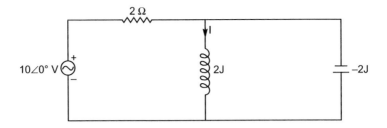

FIGURE 1.62

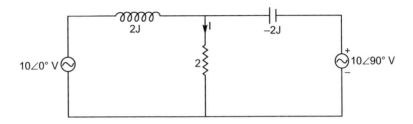

FIGURE 1.63

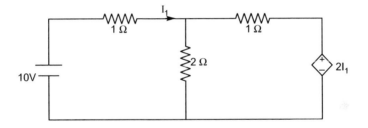

FIGURE 1.64

7. Find the voltage drop across the 2Ω resistor as shown in Figure 1.65 by using Nodal analysis.

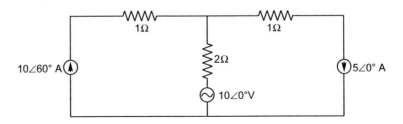

FIGURE 1.65

8. Find I_1, I_2 using Mesh analysis as shown in Figure 1.66.

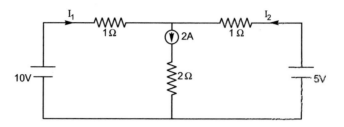

FIGURE 1.66

9. Determine I_1 using Nodal analysis as shown in Figure. 1.67.

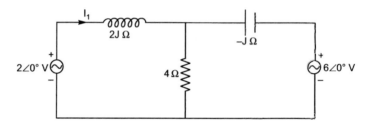

FIGURE 1.67

10. Determine I_1 using Nodal and Mesh analysis as shown in Figure 1.68.

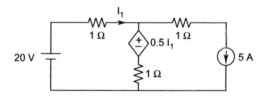

FIGURE 1.68

11. Determine the current in the 2Ω resistor as shown in Figure 1.69 using Nodal and Mesh analysis.

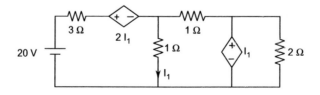

FIGURE 1.69

KIRCHHOFF'S
LAWS AND THEIR
APPLICATIONS

Chapter **2**

2.1 INTRODUCTION

Generally speaking, *network analysis* is any structured technique used to mathematically analyze a circuit (a "network" of interconnected components). This chapter presents a few techniques useful in analyzing such complex circuits.

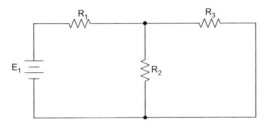

FIGURE 2.1

To analyze the above circuit, we would first find the equivalent of R_2 and R_3 in parallel, then add R_1 in series to arrive at a total resistance. Then, taking the voltage of battery E_1 with the total circuit resistance, the total current could be calculated through the use of Ohm's law ($I = E/R$), then that current figure is used to calculate voltage drops in the circuit. All in all, a fairly simple procedure.

However, the addition of just one more battery could change all of that:

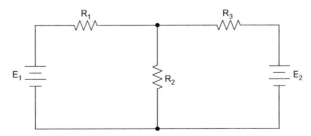

FIGURE 2.2

Resistors R_2 and R_3 are no longer in parallel with each other, because E_2 has been inserted into R_3's branch of the circuit. Upon closer inspection, it appears there are *no* two resistors in this circuit directly in series or parallel with each other. This is the crux of our problem: in series-parallel analysis, we started off by identifying sets of resistors that were directly in series or parallel with each other, and then reduced them to single, equivalent resistances. If there are no resistors in a simple series or parallel configuration with each other, then what can we do?

It should be clear that this seemingly simple circuit, with only three resistors, is impossible to reduce as a combination of simple series and simple parallel sections. It is something different altogether.

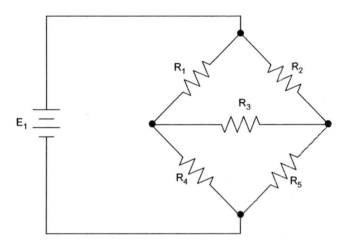

FIGURE 2.3

Here we have a bridge circuit, and for the sake of the example we will suppose that it is not balanced (ratio R_1/R_4 not equal to ratio R_2/R_5). If it were balanced, there would be zero current through R_3, and it could be approached as a series/parallel combination circuit (R_1–R_4 // R_2–R_5). However, any current through R_3 makes a series/parallel analysis impossible. R_1 is not in series with R_4 because there's another path for electrons to flow through R_3. Neither is R_2 in series with R_5 for the same reason. Likewise, R_1 is not in parallel with R_2 because R_3 is separating their bottom leads. Neither is R_4 in parallel with R_5.

Example 2.1. *Find the value of current in resistance R_3 shown in Figure 2.3, where $R_1 = 1\,\Omega$, $R_2 = 2\,\Omega$, $R_4 = 3\,\Omega$, $R_5 = 6\,\Omega$, and $R_3 = 2\,\Omega$, $E_1 = 10\,V$.*

Solution:

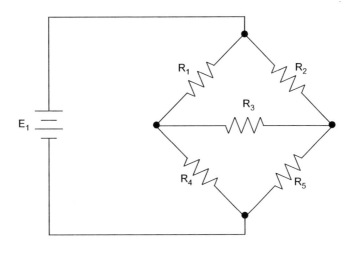

$$\frac{R_1}{R_4} = \frac{R_2}{R_5}$$

therefore, current flowing through $R_3 = 0$ A (balanced bridge circuit).

Example 2.2. *Find the value of current in R_1 as shown in Figure 2.3. Other parameters are the same as used in Example 2.1.*

Ans. $I = 2.5$ A

Although it might not be apparent at this point, the heart of the problem is the existence of multiple unknown quantities. At least in a series/parallel combination circuit, there was a way to find total resistance and total voltage, leaving total current as a single unknown value to calculate (and then that current was used to satisfy previously unknown variables in the reduction process until the entire circuit could be analyzed). With these problems, more than one parameter (variable) is unknown at the most basic level of circuit simplification.

With the two-battery circuit, there is no way to arrive at a value for "total resistance," because there are *two* sources of power to provide voltage and current (we would need *two* "total" resistances in order to proceed with any Ohm's Law calculations). With the unbalanced bridge circuit, there is such a thing as total resistance across the one battery (paving the way for a calculation of total current), but that total current immediately splits up into unknown proportions at each end of the bridge, so no further Ohm's Law calculations for voltage ($E = IR$) can be carried out.

So what can we do when we are faced with multiple unknowns in a circuit? The answer is initially found in a mathematical process known as *simultaneous equations or systems of equations*, whereby multiple unknown variables are solved by relating them to each other in multiple equations. In a scenario with only one unknown (such as every Ohm's Law equation we have dealt with thus far), there only needs to be a single equation to solve for the single unknown.

$$E = IR \text{ (E is unknown; I and R are known)} \tag{2.1}$$

or

$$I = \frac{E}{R} \text{ (I is unknown; E and R are known)} \tag{2.2}$$

or

$$R = \frac{E}{I} \text{ (R is unknown; E and I are known)} \tag{2.3}$$

However, when we are solving for multiple unknown values, we need to have the same number of equations as we have unknowns in order to reach a solution. There are several methods of solving simultaneous equations, all rather intimidiating and all too complex for explanation in this chapter.

Later on we'll see that some clever people have found tricks to avoid having to use simultaneous equations on these types of circuits. We call these tricks *network theorems*, and we will explore a few later in Chapter 8.

- Some circuit configurations ("networks") cannot be solved by reduction according to series/parallel circuit rules, due to multiple unknown values.
- Mathematical techniques to solve for multiple unknowns (called "simultaneous equations" or "systems") can be applied to basic laws of circuits to solve networks.

2.2   BRANCH CURRENT METHOD

The first and most straight-forward network analysis technique is called the *Branch Current Method*. In this method, we assume directions of currents in a network, then write equations describing their relationships to each other through Kirchhoff's and Ohm's Laws. Once we have one equation for every unknown current, we can solve the simultaneous equations and determine all currents, and therefore all voltage drops in the network.

Let's use this circuit to illustrate the method.

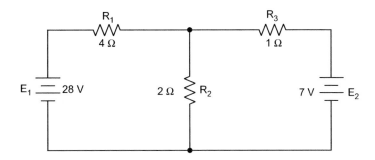

FIGURE 2.4

The first step is to choose a node (junction of wires) in the circuit to use as a point of reference for our unknown currents. We'll choose the node joining the right of R_1, the top of R_2, and the left of R_3.

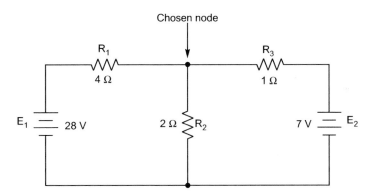

FIGURE 2.5

At this node, guess which directions the three wires' currents take, labeling the three currents as I_1, I_2, and I_3, respectively. Bear in mind that these directions of current are speculative at this point. Fortunately, if it turns out that any of our guesses were wrong, we will know when we mathematically solve for the currents (any "wrong" current directions will show up as negative numbers in our solution).

Kirchhoff's Current Law (KCL) tells us that the algebraic sum of currents entering and exiting a node must equal zero, so we can relate these three

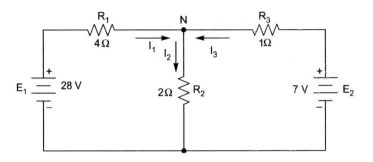

FIGURE 2.6

currents (I_1, I_2, and I_3) to each other in a single equation. For the sake of convention, we'll denote any current *entering* the node as positive in sign, and any current *exiting* the node as negative in sign.

According to KCL, the algebraic sum of currents entering the node = the algebraic sum of currents exiting the node.

Kirchhoff's Current Law (KCL) applied to currents at node (N),

$$I_1 + I_3 - I_2 = 0. \tag{2.4}$$

The next step is to label all voltage drop polarities across resistors according to the assumed directions of the currents. Remember that the "upstream" end of a resistor will always be negative, and the "downstream" end of a resistor will be positive with respect to each other, since electrons are negatively charged:

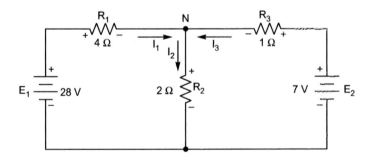

FIGURE 2.7

The battery polarities, of course, remain as they were according to their symbology (short end negative, long end positive). It is okay if the polarity of a resistor's voltage drop doesn't match with the polarity of the nearest battery, so long as the resistor voltage polarity is correctly based on the assumed direction of current through it. In some cases we may discover that current will be forced *backwards* through a battery, causing this very effect. The important thing to remember here is to base all your resistor polarities and subsequent calculations on the directions of current(s) initially assumed. As stated earlier, if your assumption happens to be incorrect, it will be apparent once the equations have been solved (by means of a negative solution). The magnitude of the solution, however, will still be correct.

Kirchhoff's Voltage Law (KVL) tells us that the algebraic sum of all voltages (drop or rise) in a loop must equal zero, so we can create more equations with current terms (I_1, I_2, and I_3) for our simultaneous equations. To obtain a KVL equation, we must tally voltage drops in a loop of the circuit, as though we were measuring with a real voltmeter. We'll choose to trace the left loop of this circuit first, starting from the upper-left corner and moving counter-clockwise (the choice of starting points and directions is arbitrary). The result will look like this:

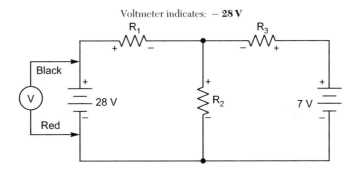

FIGURE 2.8

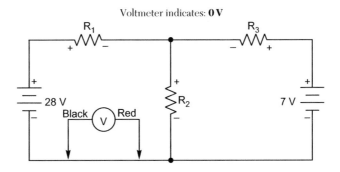

FIGURE 2.9

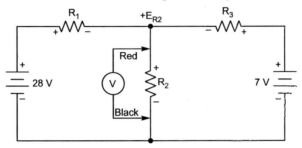

FIGURE 2.10

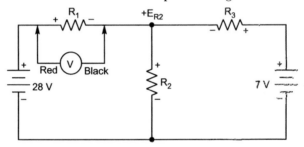

FIGURE 2.11

Having completed our trace of the left loop, we add these voltage indications together for a sum of zero:

Kirchhoff's Voltage Law (KVL) applied to voltage drops in the left loop

$$-28 + 0 + E_{R2} + E_{R1} = 0. \qquad (2.5)$$

Of course, we don't yet know what the voltage is across R_1 or R_2, so we can't insert those values into the equation as numerical figures at this point. However, we *do* know that all three voltages must algebraically add to zero, so the equation is true. We can go a step further and express the unknown voltages as the product of the corresponding unknown currents (I_1 and I_2) and their respective resistors, following Ohm's Law ($E = IR$), as well as eliminate the 0 term:

$$-28 + E_{R2} + E_{R1} = 0 \qquad (2.6)$$

Ohm's Law: $\qquad\qquad\qquad\qquad\qquad E = IR.$

Substituting IR for E in the KVL equation,

$$-28 + I_2 R_2 + I_1 R_1 = 0. \qquad (2.7)$$

Since we know what the values of all the resistors are in ohms, we can just substitute those figures into the equation to simplify things a bit:

$$-28 + 2I_2 + 4I_1 = 0. \tag{2.8}$$

You might be wondering why we went through all the trouble of manipulating this equation from its initial form ($-28 + E_{R2} + E_{R1}$). After all, the last two terms are still unknown, so what advantage is there to expressing them in terms of unknown voltages or as unknown currents (multiplied by resistances)? The purpose in doing this is to get the KVL equation expressed using the *same unknown variables* as the KCL equation, for this is a necessary requirement for any simultaneous equation solution method. To solve for three unknown currents ($I_1, I_2,$ and I_3), we must have three equations relating these three currents (*not voltages!*) together.

Applying the same steps to the right loop of the circuit (starting at the chosen node and moving counter-clockwise), we get another KVL equation:

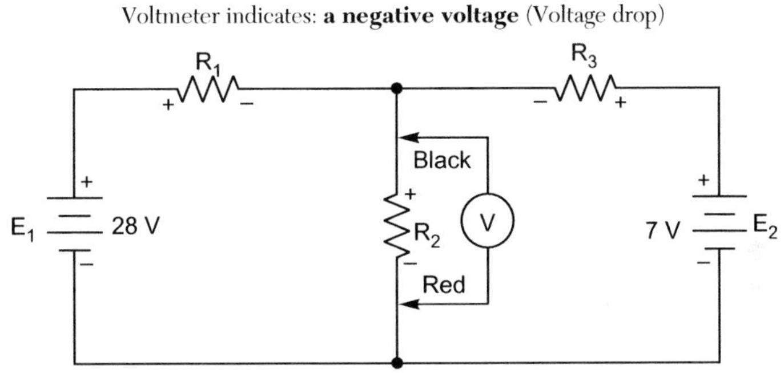

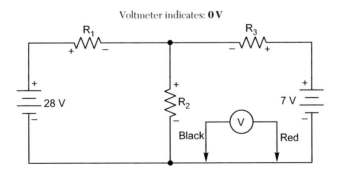

FIGURE 2.12

FIGURE 2.13

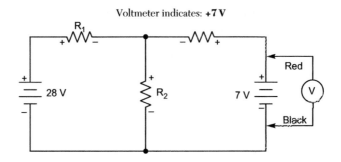

FIGURE 2.14

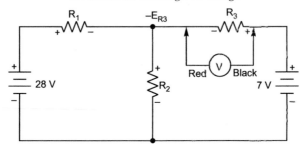

FIGURE 2.15

Kirchhoff's Voltage Law (KVL) applied to voltage drops in the right loop,

$$-E_{R2} + 0 + 7 - E_{R3} = 0. \tag{2.9}$$

Knowing now that the voltage across each resistor can be and *should be* expressed as the product of the corresponding current and the (known) resistance of each resistor, we can re-write the equation as such:

$$-2I_2 + 7 - 1I_3 = 0. \tag{2.10} \quad [E_{R2} = I_2R_2 = 2I_2]$$

Now by Equation 2.4, 2.8, and 2.10,

$$I_1 - I_2 + I_3 = 0 \quad \text{Kirchhoff's Current Law}$$
$$4I_1 + 2I_2 + 0I_3 = 28 \quad \text{Kirchhoff's Voltage Law}$$
$$0I_1 - 2I_2 - 1I_3 = -7. \quad \text{Kirchhoff's Voltage Law}$$

All three variables represented in all three equations.

Using whatever solution techniques are available to us, we should arrive at a solution for the three unknown current values:

Solutions:

$$I_1 = 5\,A$$
$$I_2 = 4\,A$$
$$I_3 = -1\,A.$$

So, I_1 is 5 amps, I_2 is 4 amps, and I_3 is a negative 1 amp. But what does "negative" current mean? In this case, it means that our *assumed* direction for I_3 was the opposite of its *real* direction. Going back to our original circuit, we can re-draw the current arrow for I_3 (and re-draw the polarity of R_3's voltage drop to match).

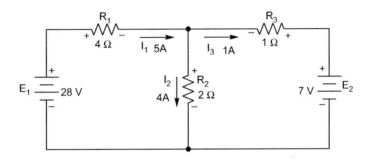

FIGURE 2.16

Now that we know the magnitude of all currents in this circuit, we can calculate voltage drops across all resistors with Ohm's Law ($E = IR$):

$$E_{R1} = I_1 R_1 = (5\,A)(4\,\Omega) = 20\,V$$
$$E_{R2} = I_2 R_2 = (4\,A)(2\,\Omega) = 8\,V$$
$$E_{R3} = I_3 R_3 = (1\,A)(1\,\Omega) = 1\,V.$$

Example 2.3. *Find the value of* I_3 *as shown in Figure 2.17.*

Solution: By applying KCL the algebraic sum of current entering at node N = the algebraic sum of current exiting at node N

$$I_2 + I_4 = I_1 + I_3$$
$$5 + e^{-2t} = 2 + I_3$$
$$I_3 = (3 + e^{-2t}).$$

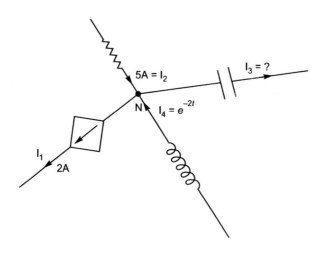

FIGURE 2.17

Example 2.4. *Find V_2 and I_2 with the help of Kirchhoff's Current Law (KCL).*

Ans.

$$V_3 = e^{-2t} \quad \text{and} \quad I_1 = 2e^{-2t}$$
$$I_2 = 10.5(e^{-2t} - 1),$$
$$V_2 = 21(1 - t - e^{-2t}).$$

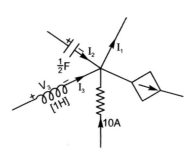

FIGURE 2.18

NOTE ▸ Steps to follow for the "Branch Current" method of analysis:

1. Choose a node and assume directions of currents. (Current entering into a passive element is assumed to be positive and leaving as to be negative.) [—+WWW–—]

2. Write a KCL equation relating currents at the node.
3. Write KVL equations for each loop of the circuit, substituting the product IR for E in each resistor term of the equations.
4. Solve for unknown branch currents (simultaneous equations).
5. If any solution is negative, then the assumed direction of current for that solution is wrong!
6. Solve for voltage drops across all resistors (E = IR).

Example 2.5. *Find the value of current in* R_3 *using Kirchhoff's Laws.*

Solution: By applying KCL at node N,

$$I_1 + I_2 = I_3. \tag{1}$$

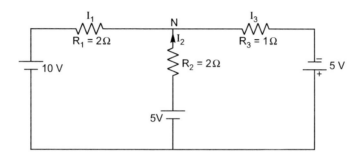

In this case there are three closed loops $(10\,\text{V}, R_1, R_2, 5\,\text{V}]$, $[5\,\text{V}, R_2, R_3, 5\,\text{V}]$, and $[10\,\text{V}, R_1, R_3, 5\,\text{V}]$.

Now applying KVL in the first and second loop,

$$-10 + 2I_1 - 2I_2 + 5 = 0 \implies 2I_1 - 2I_2 = 5 \tag{2}$$
$$-5 + 2I_2 + I_3 - 5 = 0 \implies 2I_2 + I_3 = 10. \tag{3}$$

By Equations (1) and (3),

$$2I_2 + I_1 + I_2 = 10 \implies I_1 + 3I_2 = 10. \tag{4}$$

By Equations (2) and (4),

$$I_2 = \frac{15}{8} \quad \text{and} \quad I_1 = \frac{35}{8}.$$

Current flowing in

$$R_3 = I_3 = I_1 + I_2 = \frac{50}{8} = 6.25\,\text{Amp.}$$

Example 2.6. *Find the value of current in R3 using Kirchhoff's Laws.*

Ans. 3.75 A

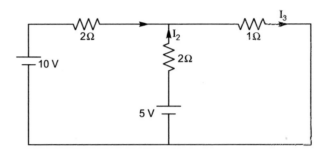

Mesh Current Method

The *Mesh Current Method* is quite similar to the branch current method in that it uses simultaneous equations, Kirchhoff's Voltage Law, and Ohm's Law to determine unknown currents in a network. It differs from the branch current method in that it does *not* use Kirchhoff's Current Law, and it is usually able to solve a circuit with less unknown variables and less simultaneous equations, which is especially nice if you are forced to solve without a calculator.

Let's see how this method works on the same example problem:

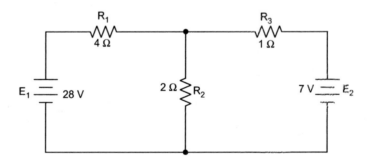

FIGURE 2.19

The first step in the mesh current method is to identify "loops" within the circuit encompassing all components. In our example circuit, the loop formed by E_1, R_1, and R_2 will be the first while the loop formed by E_2, R_2, and R_3 will be the second. The strangest part of the mesh current method is envisioning circulating currents in each of the loops.

The choice of each current's direction is entirely arbitrary, just as in the branch current method, but the resulting equations are easier to solve if the

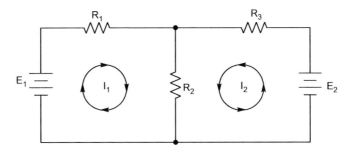

FIGURE 2.20

currents are going the same direction through intersecting components (note how currents I_1 and I_2 are both going "up" through resistor R_2, where they "mesh," or intersect). If the assumed direction of a mesh current is wrong, the answer for that current will have a negative value.

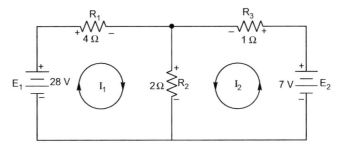

FIGURE 2.21

Using Kirchhoff's Voltage Law, we can now step around each of these loops, generating equations representative of the component voltage drops and polarities. As with the branch current method, we will denote a resistor's voltage drop as the product of the resistance (in ohms) and its respective mesh current (that quantity being unknown at this point). Where two currents mesh together, we will write that term in the equation with the resistor current being the *sum* of the two meshing currents.

$$-28 + 2(I_1 + I_2) + 4I_1 = 0 \qquad (2.10) \quad \text{[For mesh 1]}$$

Notice that the middle term of the equation uses the sum of mesh currents I_1 and I_2 as the current through resistor R_2. This is because mesh currents I_1 and I_2 are going the same direction through R_2, and thus complement each other. Distributing the coefficient of 2 to the I_1 and I_2 terms, and then combining I_1 terms in the equation, we can simplify as such:

$$-28 + 2(I_1 + I_2) + 4I_1 = 0.$$

The original form of the equation distributing to terms within parentheses

$$-28 + 2I_1 + 2I_2 + 4I_1 = 0$$

and combining like terms

$$-28 + 6I_1 + 2I_2 = 0 \implies 3I_1 + I_2 = 14. \tag{2.11}$$

Simplified Form of Equation

At this time we have one equation with two unknowns. To be able to solve for two unknown mesh currents, we must have two equations. If we trace the other loop of the circuit, we can obtain another KVL equation and have enough data to solve for the two currents.

$$-7 + I_2 + 2(I_1 + I_2) = 0. \tag{2.12}$$

Simplifying the equation as before, we end up with:

$$2I_1 + 3I_2 = 7. \tag{2.13}$$

Now, with two equations, we can use one of several methods to mathematically solve for the unknown currents I_1 and I_2. Now by Equations (2.12) and (2.14),

Solutions:

$$I_1 = 5\,\text{A}$$
$$I_2 = -1\,\text{A}.$$

Knowing that these solutions are values for *mesh* currents, not *branch* currents, we must go back to our diagram to see how they fit together to give currents through all components:

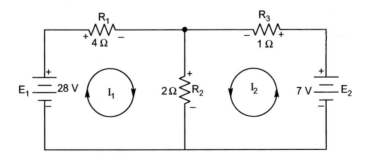

FIGURE 2.22

The solution of -1 amp for I_2 means that our initially assumed direction of current was incorrect. In actuality, I_2 is flowing in a clockwise direction at a value of (positive) 1 amp:

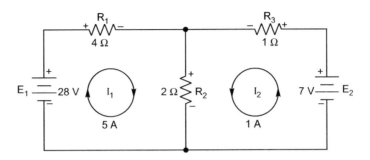

FIGURE 2.23

This change of current direction from what was first assumed will alter the polarity of the voltage drops across R_2 and R_3 due to current I_2. From here, we can say that the current through R_1 is 5 amps, with the voltage drop across R_1 being the product of current and resistance ($E = IR$), 20 volts (positive on the left and negative on the right). Also, we can safely say that the current through R_3 is 1 amp, with a voltage drop of 1 volt ($E = IR$), positive on the left and negative on the right. But what is happening at R_2?

Mesh current I_1 is going "down" through R_2, while mesh current I_2 is going "up" through R_2. To determine the actual current through R_2, we must see how mesh currents I_1 and I_2 interact (in this case they're in opposition), and algebraically add them to arrive at a final value. Since I_1 is going "down" at 5 amps, and I_2 is going "up" at 1 amp, the *real* current through R_2 must be a value of 4 amps, going "down."

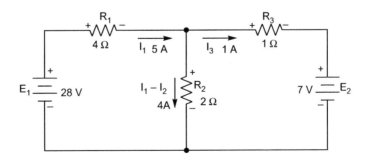

FIGURE 2.24

A current of 4 amps through R_2's resistance of 2 Ω gives us a voltage drop of 8 volts ($E = IR$), positive on the top and negative on the bottom.

The primary advantage of mesh current analysis is that it generally allows for the solution of a large network with fewer unknown values and fewer simultaneous equations. Our example problem took three equations to solve the branch current method and only two equations using the mesh current method. This advantage is much greater as networks increase in complexity:

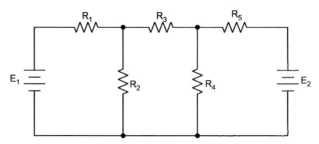

FIGURE 2.25

To solve this network using branch currents, we would have to establish five variables to account for each and every unique current in the circuit (I_1 through I_5). This would require five equations for the solution, in the form of two KCL equations and three KVL equations (two equations for KCL at the nodes, and three equations for KVL in each loop):

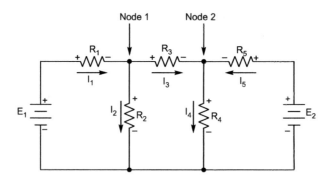

FIGURE 2.26

$$I_1 - I_2 - I_3 = 0 \quad (2.15) \qquad \text{Kirchhoff's Current Law at node 1}$$
$$I_3 + I_5 - I_4 = 0 \quad (2.16) \qquad \text{Kirchhoff's Current Law at node 2}$$
$$-E_1 + I_2R_2 + I_1R_1 = 0 \quad (2.17) \qquad \text{Kirchhoff's Voltage Law in the left loop}$$
$$-I_2R_2 + I_4R_4 + I_3R_3 = 0 \quad (2.18) \qquad \text{Kirchhoff's Voltage Law in the middle loop}$$
$$-I_4R_4 + E_2 - I_5R_5 = 0 \quad (2.19) \qquad \text{Kirchhoff's Voltage Law in the right loop}$$

The mesh current method is easier, requiring only three unknowns and three equations to solve:

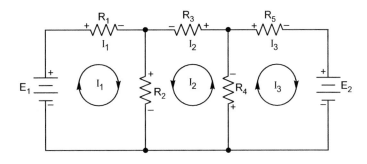

FIGURE 2.27

$$-E_1 + R_2(I_1 + I_2) + I_1 R_1 = 0 \quad (2.20)$$ Kirchhoff's Voltage Law in the left loop

$$-R_2(I_2 + I_1) - R_4(I_2 + I_3) - I_2 R_3 = 0 \quad (2.21)$$ Kirchhoff's Voltage Law in the middle loop

$$R_4(I_3 + I_2) + E_2 + I_3 R_5 = 0 \quad (2.22)$$ Kirchhoff's Voltage Law in the right loop

Fewer equations to work with is a decided advantage, especially when performing simultaneous equation solutions by hand (without a calculator).

Another type of circuit that lends itself well to mesh current is the unbalanced Wheatstone Bridge. Take this circuit.

Since the ratios of R_1/R_4 and R_2/R_5 are unequal, we know that there will be voltage across resistor R_3 and some amount of current passing through it.

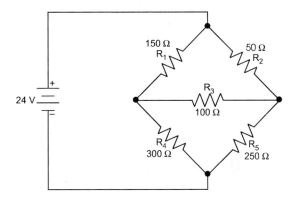

FIGURE 2.28

As discussed at the beginning of this chapter, this type of circuit is irreducible by normal series-parallel analysis, and may only be analyzed by some other method.

We could apply the branch current method to this circuit, but it would require *six* currents (I_1 through I_6), leading to a very large set of simultaneous equations to solve. Using the mesh current method, though, we may solve for all currents and voltages with fewer variables.

The first step in the mesh current method is to draw just enough mesh currents to account for all components in the circuit. Looking at our bridge circuit, it should be obvious where to place two of these currents.

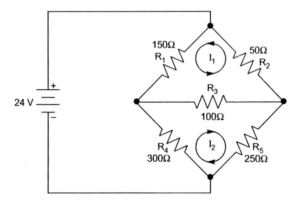

FIGURE 2.29

The directions of these mesh currents, of course, is arbitrary. However, two mesh currents are not enough in this circuit, because neither I_1 nor I_2 goes through the battery. So, we must add a third mesh current, I_3:

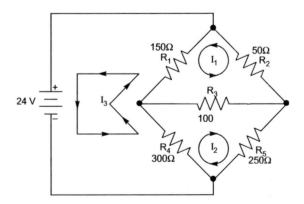

FIGURE 2.30

Here, we have chosen I_3 to loop from the bottom side of the battery, through R_4, through R_1, and back to the top side of the battery. This is not the only path we could have chosen for I_3, but it seems the simplest.

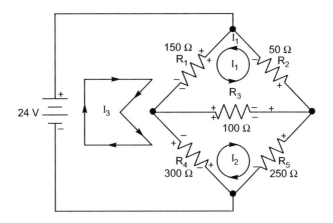

FIGURE 2.31

Notice something very important here: at resistor R_4, the polarities for the respective mesh currents do not agree. This is because those mesh currents (I_2 and I_3) are going through R_4 in different directions. Normally, we try to avoid this when establishing our mesh current directions, but in a bridge circuit it is unavoidable: Two of the mesh currents will inevitably clash through a component.

Generating a KVL equation for the top loop of the bridge, starting from the top node and tracing in an anti-clockwise direction:

$$50I_1 + 100(I_1 + I_2) + 150(I_1 + I_3) = 0.$$
$$(2.22) \quad \text{(Original form of equation)}$$

Distributing to terms within parentheses,

$$50I_1 + 100I_1 + 100I_2 + 150I_1 + 150I_3 = 0 \qquad (2.23)$$

and combining like terms

$$300I_1 + 100I_2 + 150I_3 = 0. \qquad (2.24) \quad \text{(Simplified form of equation)}$$

In this equation, we represent the common directions of currents by their *sums* through common resistors. For example, resistor R_3, with a value of 100 Ω, has its voltage drop represented in the above KVL equation by the expression $100(I_1 + I_2)$, since both currents I_1 and I_2 go through R_3 from right to left. The same may be said for resistor R_1, with its voltage drop expression shown as $150(I_1 + I_3)$, since both I_1 and I_3 go from bottom to top through that resistor, and thus work *together* to generate its voltage drop.

Generating a KVL equation for the bottom loop of the bridge will not be easy, since we have two currents going against each other through resistor R_4. Here is how we do it (starting at the right-hand node, and tracing clockwise):

$$100(I_1 + I_2) + 300(I_2 - I_3) + 250I_2 = 0.$$
$$(2.25) \quad \text{(Original form of equation)}$$

Distributing to terms within parentheses,

$$100I_1 + 100I_2 + 300I_2 - 300I_3 + 250I_2 = 0 \qquad (2.26)$$

and combining like terms

$$100I_1 + 650I_2 - 300I_3 = 0. \qquad (2.27) \quad \text{(Simplified form of equation)}$$

Note how the second term in the equation's original form has resistor R_4's value of 300 Ω multiplied by the *difference* between I_2 and $I_3(I_2 - I_3)$. This is how we represent the combined effect of two mesh currents going in opposite directions through the same component. Choosing the appropriate mathematical signs is very important here: $300(I_2 - I_3)$ does not mean the same thing as $300(I_3 - I_2)$. We chose to write $300(I_2 - I_3)$ because we were thinking first of I_2's effect (creating a positive voltage drop, measuring with an imaginary voltmeter across R_4, red lead on the bottom and black lead on the top), and secondarily of I_3's effect (creating a negative voltage drop, red lead on the bottom and black lead on the top). If we had thought in terms of I_3's effect first and I_2's effect secondarily, holding imaginary voltmeter leads in the same positions (red on bottom and black on top), the expression would have been $-300(I_3 - I_2)$. Note that this expression is mathematically equivalent to the first one: $+300(I_2 - I_3)$.

Well, that takes care of two equations, but we still need a third equation to complete our simultaneous equation set of three variables, three equations. This third equation must also include the battery's voltage, which up to this point does not appear in either of the previous KVL equations. To generate this equation, we will trace a loop again with our imaginary voltmeter starting from the battery's bottom (negative) terminal, stepping anti-clockwise (again, the direction in which we step is arbitrary, and does not need to be the same as the direction of the mesh current in that loop):

$$24 - 150(I_3 + I_1) - 300(I_3 - I_2) = 0.$$
$$(2.28) \quad \text{(Original form of equation)}$$

Distributing to terms within parentheses,

$$24 - 150I_3 - 150I_1 - 300I_3 + 300I_2 = 0 \qquad (2.29)$$

and combining like terms

$$-150I_1 + 300I_2 - 450I_3 = -24. \quad (2.30) \quad \text{Simplified form of equation}$$

Solving for I_1, I_2, and I_3 using whatever simultaneous equation method we prefer:

$$300I_1 + 100I_2 + 150I_3 = 0$$
$$100I_1 + 650I_2 - 300I_3 = 0$$
$$-150I_1 + 300I_2 - 450I_3 = -24.$$

Solutions: By the above three equations,

$$I_1 = -93.793 \, \text{mA}$$
$$I_2 = 77.241 \, \text{mA}$$
$$I_3 = 136.092 \, \text{mA}.$$

The negative value arrived at for I_1 tells us that the assumed direction for that mesh current was incorrect. Thus, the actual current values through each resistor is such:

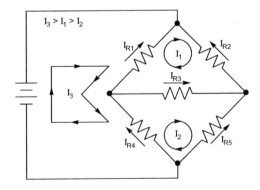

FIGURE 2.32

$$I_{R1} = I_3 - I_1 = 136.092 \, \text{mA} - 93.793 \, \text{mA} = 42.299 \, \text{mA}$$
$$I_{R2} = I_1 = 93.793 \, \text{mA}$$
$$I_{R3} = I_1 - I_2 = 93.793 \, \text{mA} - 77.241 \, \text{mA} = 16.552 \, \text{mA}$$
$$I_{R4} = I_3 - I_2 = 136.092 \, \text{mA} - 77.241 \, \text{mA} = 58.851 \, \text{mA}$$
$$I_{R5} = I_2 = 77.241 \, \text{mA}$$

Calculating voltage drops across each resistor:

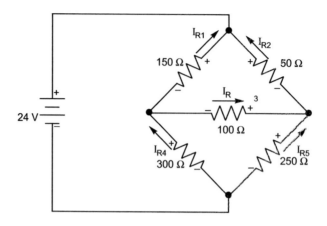

FIGURE 2.33

$$E_{R1} = I_{R1}R_1 = (42.299\,\text{mA})(150\,\Omega) = 6.3448\,\text{V}$$
$$E_{R2} = I_{R2}R_2 = (93.793\,\text{mA})(50\,\Omega) = 4.6897\,\text{V}$$
$$E_{R3} = I_{R3}R_3 = (16.552\,\text{mA})(100\,\Omega) = 1.6552\,\text{V}$$
$$E_{R4} = I_{R4}R_4 = (58.851\,\text{mA})(300\,\Omega) = 17.6552\,\text{V}$$
$$E_{R5} = I_{R5}R_5 = (77.241\,\text{mA})(250\,\Omega) = 19.3103\,\text{V}$$

NOTE ▶ Steps to follow for the "Mesh Current" method of analysis:

1. Draw mesh currents in each loop of circuit, enough to account for all components.
2. Write KVL equations for the loop of the circuit, substituting the product IR for E in each resistor term of the equation. Where two mesh currents intersect through a component, express the current as the algebraic sum of those two mesh currents (*i.e.*, $I_1 + I_2$) if the currents go in the same direction through that component. If not, express the current as the difference (*i.e.*, $I_1 - I_2$).
3. Solve for unknown mesh currents (simultaneous equations).
4. If any solution is negative, then the assumed current direction is wrong!
5. Algebraically add mesh currents to find the current in components sharing multiple mesh currents.
6. Solve for the voltage drop across all resistors (E = IR).

SOLVED PROBLEMS

Problem 2.1. *Find the value of current passing through* $R_1, R_2, R_3, R_4,$ *and* R_5 *using loop current analysis or mesh analysis.*

Solution: Assume the loop currents are I_1 and I_2 as shown in Figure 2.34.

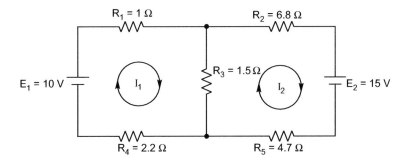

FIGURE 2.34

By applying the KVL in mesh (1) and (2),

$$-10 + I_1 + 1.5(I_1 - I_2) + 2.2I_1 = 0$$
$$\implies 4.7I_1 - 1.5I_2 = 10 \qquad (1)$$
$$1.5(I_2 - I_1) + 6.8I_2 + 15 - 4.7I_2 = 0$$
$$1.5I_1 - 13I_2 = 15. \qquad (2)$$

By Equations (1) and (2),

$$I_1 = -2.59 \, \text{A} \quad \text{and} \quad I_2 = -1.45 \, \text{A}.$$

The negative sign shows the direction of current is opposite as we assumed. The direction of currents I_1 and I_2 are as shown in the figure.

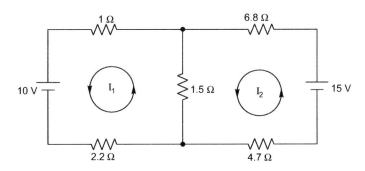

$$\text{Current passing through } R_1 = I_1 = 2.59\,\text{A}$$
$$\text{Current passing through } R_2 = I_2 = 1.45\,\text{A}$$
$$\text{Current passing through } R_3 = I_1 - I_2$$
$$= 2.59 - 1.45$$
$$= 1.14\,\text{A}$$
$$\text{Current passing through } R_4 = I_1 = 2.59\,\text{A}$$
$$\text{Current passing through } R_5 = I_2 = 1.45\,\text{A}$$

Problem 2.2. *Determine the values of* I_1 *and* I_2 *using mesh analysis (Figure 2.35).*

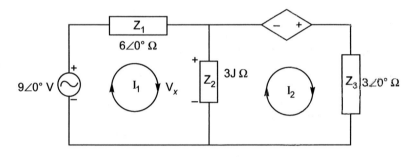

FIGURE 2.35

Solution: By using mesh analysis in mesh (1) and (2) for mesh 1

$$-9\angle 0° + (6\angle 0°)I_1 - 3J(I_1 - I_2) = 0$$

or

$$6\angle 0°I_1 - 3J(I_1 - I_2) = 9\angle 0° \implies (2 - J)I_1 + JI_2 = 3$$

$$-V_x - 2V_x + 3\angle 0°I_2 = 0$$

$$V_x = I_2.$$

But

$$V_x = -3J(I_1 - I_2) = I_2 \tag{1}$$

$$3JI_1 + (1 - 3J)I_2 = 0. \tag{2}$$

By solving Equations (1) and (2),

$$I_1 = 1.3\angle 2.4°\,\text{A} \quad \text{and} \quad I_2 = 1.24\angle - 16°\,\text{A}.$$

Problem 2.3. *Determine* V_1 *using mesh analysis as shown in Figure 2.36.*

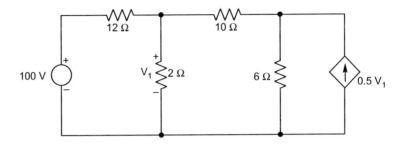

FIGURE 2.36

Solution: Apply the source transformation method and convert the dependent current source with parallel resistance $6\,\Omega$ to the dependent voltage source, as shown in Figure 2.37.

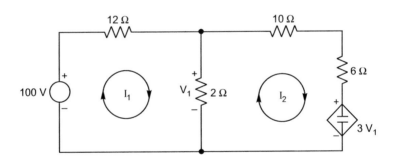

FIGURE 2.37

Now apply mesh analysis for mesh 1,

$$-100 + 12I_1 + 2(I_1 - I_2) = 0$$

or
$$14I_1 - 2I_2 = 100, \qquad (1)$$

for mesh 2,

$$2(I_2 - I_1) + 10I_2 + 6I_2 + 3V_1 = 0 \implies -2I_1 + 18I_2 = -3V_1$$

and

$$V_1 = -2(I_2 - I_1) \text{ or } 2(I_1 - I_2)$$
$$-2I_1 + 18I_2 = -6I_1 + 6I_2 \implies 4I_1 + 12I_2 = 0. \qquad (2)$$

By Equations (1) and (2),

$$I_1 = 6.32\,\text{A}, \quad I_2 = -2.27\,\text{A}. \tag{3}$$

Problem 2.4. *Determine I_1 using mesh analysis as shown in Figure 2.38.*

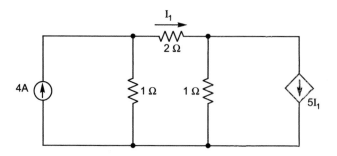

FIGURE 2.38

Solution: Apply the source transformation method and convert the current source into the voltage source. Now there is a single mesh.

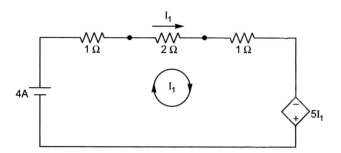

FIGURE 2.39

Applying mesh analysis,

$$-4 + I_1 + 2I_1 + I_1 - 5I_1 = 0$$
$$4 = -I_1$$
$$I_1 = -4\text{A}.$$

The negative sign shows the direction of current is opposite as we assumed.

Problem 2.5. *For the network of the figure below write a set of loop equations using Kirchhoff's Law with loop currents.*

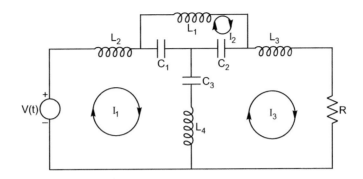

FIGURE 2.40

Solution: Since there are three loops in the given network, there will be three loop equations. If the operating frequency is ω,

$$-V(t) + J\omega L_2 i_1 + (i_1 - i_2)\frac{1}{J\omega C_1} + (i_1 - i_3)\left(\frac{1}{J\omega C_3} - J\omega L_4\right) = 0$$

$$J\omega L_1 i_2 + (i_2 - i_3)\frac{1}{J\omega C_2} + (i_2 - i_1)\frac{1}{J\omega C_1} = 0$$

$$J\omega L_3 i_3 + i_3 R + \left(J\omega L_4 + \frac{1}{J\omega C_3}\right)(i_3 - i_1) + (i_3 - i_2)\frac{1}{J\omega C_2} = 0.$$

These equations can be written in an appropriate form as

$$V(t) = i_1\left(J\omega L_2 + \frac{1}{J\omega C_1} + \frac{1}{J\omega C_3} + J\omega L_4\right)$$

$$- \frac{i_2}{J\omega C_1} - i_3\left(\frac{1}{J\omega C_3} + J\omega L_4\right) = 0 \qquad (1)$$

$$- i_1\frac{1}{J\omega C_1} + i_2\left(J\omega L_1 + \frac{1}{J\omega C_2} + \frac{1}{J\omega C_1}\right) - \frac{i_3}{J\omega C_2} = 0 \qquad (2)$$

$$- i1\left(J\omega L_4 + \frac{1}{J\omega C_3}\right) - \frac{i_2}{J\omega C_2}$$

$$+ i_3\left(J\omega L_3 + R + J\omega L_4 + \frac{1}{J\omega C_3} + \frac{1}{J\omega C_2}\right) = 0. \qquad (3)$$

Problem 2.6. *Determine the voltage V in the circuit of Figure 2.41 using the source transformation technique or any other method.*

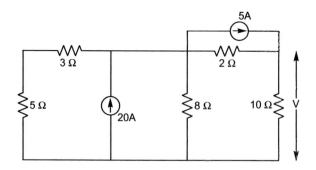

FIGURE 2.41

Solution: The given network can be transformed to the circuit shown in the figure using the source transformation method.

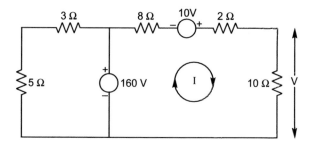

FIGURE 2.42

By using mesh analysis,

$$-160 + 8I - 10 + 2I + 10I = 0$$

$$I = \frac{170}{20} = 8.5 \, \text{Amp.}$$

$$V = I \times 10 = 85 \, \text{V}.$$

Problem 2.7. *Determine the voltage V in the circuit of Figure 2.43 below using nodal analysis.*

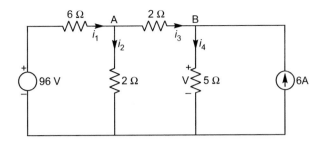

FIGURE 2.43

Solution: Let V_A and V_B be the voltage at Node A and B, respectively and $[V_B = V]$. Applying KCL at A$-$,

$$i_1 = i_2 + i_3 \tag{1}$$

and

$$i_1 = \frac{96 - V_A}{6}; \quad i_2 = \frac{V_A}{2}; \quad i_3 = \frac{V_A - V_B}{2}.$$

Putting the values $i_1, i_2,$ and i_3 into Equation (1),

$$\frac{96 - V_A}{6} = \frac{V_A}{2} + \frac{V_A - V_B}{2}$$

or

$$2(96 - V_A) = 6(2V_A - V_B)$$
$$96 - V_A = 6V_A - 3V_B$$
$$7V_A - 3V_B = 96. \tag{2}$$

Applying KCL at B$-$,

$$i_3 + 6 = i_4 \qquad \left[i_4 = \frac{V_B}{5}\right]$$

$$\frac{V_A - V_B}{2} + 6 = \frac{V_B}{5}$$
$$5[V_A - V_B + 12] = 2V_B \implies 5V_A - 7V_B = -60. \tag{3}$$

By Equations (2) and (3),

$$V_B = 26.47\,\text{V} = \text{V}.$$

Problem 2.8. *Write the mesh equations for the network shown in Figure 2.44 and find the power absorbed by the 3 Ω resistor.*

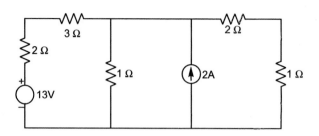

FIGURE 2.44

Solution: By using the source transformation formula,

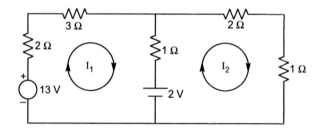

FIGURE 2.45

By applying mesh analysis for mesh 1,

$$-13 + 5I_1 + 1(I_1 - I_2) + 2 = 0$$
$$6I_1 - I_2 = 11. \tag{1}$$

For mesh 2,

$$-2 + 1(I_2 - I_1) + 3I_2 = 0$$
$$-I_1 + 4I_2 = 2 \tag{2}$$
$$\text{or} \quad -6I_1 + 24I_2 = 12. \tag{3}$$

By Equations (1) and (3),

$$23I_2 = 23$$
$$I_2 = 1\,\text{A}$$
$$I_1 = 2\,\text{A}.$$

The current in the $3\,\Omega$ resistor $= I_1 = 2\,\text{A}$.

$$\text{Power} = (2)^2 \times 3 = 12\,\Omega.$$

Problem 2.9. *Convert the network shown in Figure 2.46 into a single voltage source with a suitable resistance.*

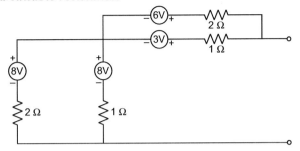

FIGURE 2.46

Solution: First convert all the voltage sources to current sources.

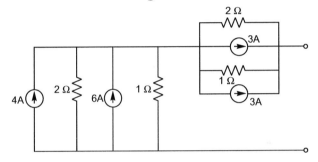

FIGURE 2.47

Combine the two current sources into a single, and the two resistors are connected in parallel.

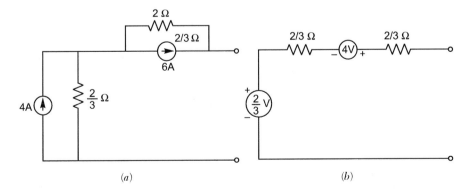

(a)   *(b)*

FIGURE 2.48   (Continued)

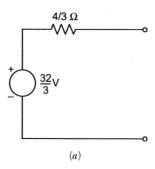

(a)

FIGURE 2.48

QUESTIONS FOR DISCUSSION

1. What is Kirchhoff's Current Law?
2. Explain Kirchhoff's Voltage Law.
3. Explain the Mesh current method with a suitable example.
4. What is battery polarity?

OBJECTIVE QUESTIONS

1. The value of V_{AB} as shown in Figure 2.49, is
   (a) 6 V   (b) 12 V   (c) 18 V   (d) 19 V.

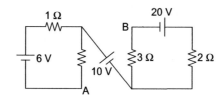

FIGURE 2.49

2. As shown in Figure 2.50, the value of current through the $10\,\Omega$ resistor is
   (a) 0.25 A   (b) 10.5 A   (c) 10.75 A   (d) 1 A.

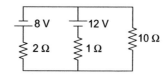

FIGURE 2.50

Chapter 3

GRAPH THEORY

3.1 INTRODUCTION

A combination of active and passive elements are known as an electrical network. Any electrical network can be converted into a graph and this is known as topology. In topology only the geometrical pattern of a network is considered and no distinction is made between the different types of physical elements. The basic elements, according to topology, are branches, nodes, and loops.

All the elements (voltage source, resistor, inductor, capacitor) in a network are replaced by line segments and the junction points by nodes.

3.2 RULES FOR DRAWING A GRAPH

1. Ideal current sources are open circuit and ideal voltage sources are short circuit.
2. If resistance is in series with active (voltage source) and passive (L and C) elements, it is assumed to be the internal resistance of the elements. If the given resistor is an external resistor then this is a separate branch.
3. (a) A passive element in series with a current source is not a branch.

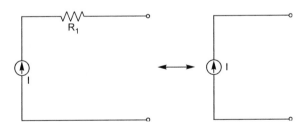

FIGURE 3.1

(b) A passive element in parallel with a voltage source is not a branch.

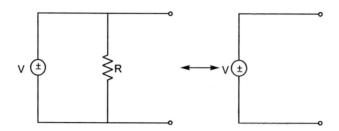

FIGURE 3.2

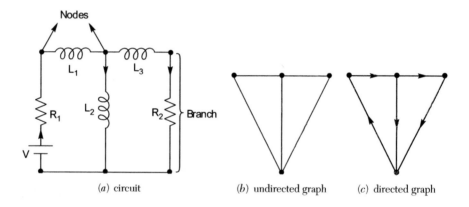

(a) circuit   (b) undirected graph   (c) directed graph

FIGURE 3.3

3.3   SOME USEFUL DEFINITIONS

3.3.1 Connected Graph

A graph is said to be connected if a path can be found between any two nodes of it. Figure 3.3(b) is a connected graph. A graph whose branches are oriented is called a directed graph (Figure 3.3c), otherwise the graph is undirected as shown in Figure 3.3(b).

3.3.2 Linear Graph

A linear graph or simply a graph is a collection of nodes and branches. It shows the geometrical interconnection of the elements of a network.

3.3.3 Planar Circuit

A circuit that may be drawn on a plane surface in such a way that no branch passes over or under any other branch, as shown in Figure 3.3(a), is called a planar circuit.

3.3.4 Non-planar Circuit

Any circuit that is not planar as shown in Figure 3.4.

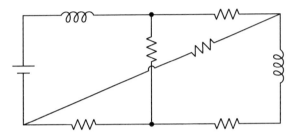

FIGURE 3.4

3.3.5 Node

A node is defined as an end point of a line segment or an isolated point. A node is sometimes also called a vertex or a junction. It is represented by black dots as shown in Figure 3.3(a).

3.3.6 Branch

A branch is a line segment that represents a network element or a combination of elements connected between two points. Each branch joins two distinct nodes as shown in Figure 3.3(a).

3.3.7 Loop

A closed path is known as a loop but an inner closed path is known as a mesh, as shown in Figure 3.5.

3.3.8 Mesh

A mesh is a property of a planar circuit and is not defined for a non-planar circuit. We define a mesh as a loop which does not contain any other loops within it.

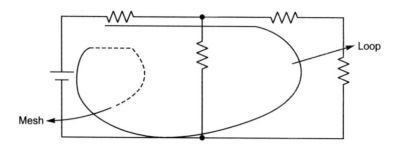

FIGURE 3.5

3.3.9 Tree

It is the subgraph of a connected graph containing all the nodes of the graph, and is itself a connected graph and contains no loops. Many trees can be found for a graph. The branches of the tree are called twigs.

$$N = \text{Number of nodes}$$
$$\text{Number of twigs} = N - 1$$

Properties of a tree:

1. In a tree, there exists one and only one path between any two nodes.
2. It should have all nodes.
3. There should be no closed path.
4. Each tree has $(n - 1)$ branches.
5. The rank of a tree is $(n - 1)$, where n is the number of nodes or vertex.

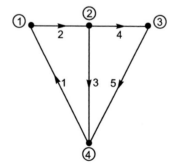

FIGURE 3.6 **Graph G.**

It should be noted that even a simple graph has many different trees. Consider the graph shown in Figure 3.6. It contains 8 trees as shown in Figure 3.7.

3.3.10 Co-tree

Co-tree is the complement of a tree. The branches of a graph not included in the tree are called chords or links. The set of links of a tree constitutes the co-tree.

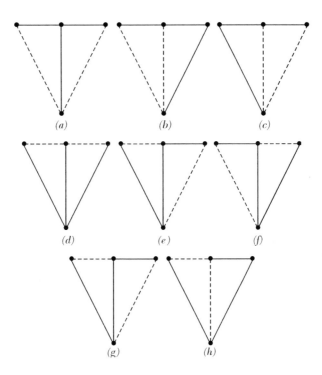

FIGURE 3.7   **Trees of graph G.**

Number of links $= B - (N - 1) = B - N + 1$, where B is the total number of branches of the graph.

Properties of a co-tree:

1. Path may be open or closed.
2. Total number of links should be ($l = b - n + 1$), where b is the number of branches and n is the number of nodes.

3.3.11 Path

If no node was encountered more than once, then the sets of nodes and elements that we have passed through is defined as a path.

There are two paths shown in Figure 3.8. These paths are known as open paths if the node at which we started is the same as the node at which we ended, and then the path is defined as a closed path or a loop.

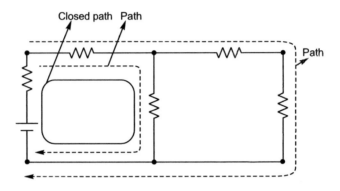

FIGURE 3.8

3.3.12 Walk

In a walk, no edge is covered more than once but a node may appear more than once. If a node appears more than once it is known as a closed walk. Otherwise, it is an open walk or path.

3.3.13 Cut-set

A cut-set is a minimal set of branches that, if it divides a connected graph into two connected subgraphs, it separates the nodes of the graph into two groups.

3.3.14 Subgraph

A portion of a graph is called a subgraph. A graph G_1 is said to be a subgraph of a graph G if every node of G_1 is a node of G and every branch of G_1 is a branch of G.

In other words, a subgraph G_1 of a graph G is a collection of branches and nodes of G such that every branch and node in G_1 is also contained in G. Thus, if a graph G is given, we can find the subgraph G_1 by deleting some branches and nodes from graph G. There is no lower limit on the number

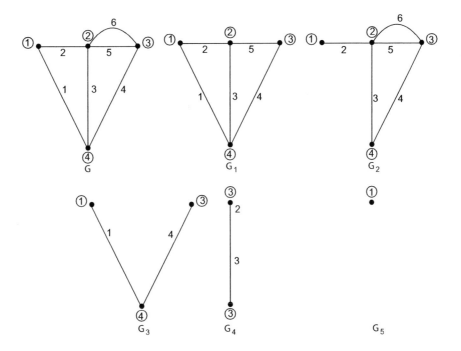

FIGURE 3.9   **Subgraph of graph G.**

of nodes or branches in a subgraph. For example, a subgraph G_5 consists of only one node and is called a degenerate subgraph. A subgraph is said to be a proper subgraph if it does not contain all the branches and nodes of the graph.

3.4   COMPLETE INCIDENCE MATRIX (A_C)

An oriented or directed graph may be described completely in a compact and convenient matrix. This is known as a complete incidence matrix. In a complete incidence matrix, we specify the orientation of each branch in the graph and the nodes at which this branch is incident. Each row of this matrix corresponds to a node of the graph and each column corresponds to a branch. For a graph having n nodes and b branches, the complete incidence matrix A_C is an $n \times b$ rectangular matrix.

The complete incidence matrix is also called the augmented incidence matrix.

The directed graph is shown in Figure 3.10.

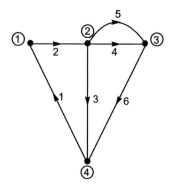

FIGURE 3.10

The incoming current is assumed to be negative and the outgoing current is assumed to be positive.

Nodes	Branches					
	1	2	3	4	5	6
1	−1	1	0	0	0	0
2	0	−1	1	1	1	0
3	0	0	0	−1	−1	1
4	1	0	−1	0	0	−1

There are four nodes and six branches in a given directed graph. Node (1) is connected with branch 1 and 2 and the direction of the current in branch 1 is incoming and branch 2 is outgoing and the other four branch (3, 4, 5, 6) entries are equal to zero.

Similarly for node 2, four branches (2, 3, 4, 5) are connected with node (2) and the direction of the current is incoming in branch 2 and outgoing for branches 3, 4, and 5, and the other two branch (1, 6) entries are equal to zero.

$$[A_C] = \begin{bmatrix} -1 & 1 & 0 & 0 & 0 & 0 \\ 0 & -1 & 1 & 1 & 1 & 0 \\ 0 & 0 & 0 & -1 & -1 & 1 \\ 1 & 0 & -1 & 0 & 0 & -1 \end{bmatrix}$$

A_C is the complete incidence matrix of a given directed graph, as shown in Figure 3.10.

3.4.1 Properties of Complete Incidence Matrix

(i) The algebraic sum of the column entries of an incidence matrix is zero.

(ii) The rank of the complete incidence matrix of a connected graph is $(n - 1)$, where n is the total number of nodes.

(iii) The order of a complete incidence matrix will be $n \times b$, where b is the total number of branches and n is the number of nodes.

(iv) The determinant of the incidence matrix of a closed loop is zero.

3.4.2 Reduced Incidence Matrix

We have seen that any one of the rows of a complete incidence matrix can be obtained from the remaining rows. Thus, it is possible to delete any one row from A_C without losing any information contained in A_C. When one row is deleted from the complete incidence matrix A_C, the remaining matrix is called a reduced incidence matrix.

The reduced incidence matrix is represented by the symbol A. For a graph with n nodes and b branches, the order of the reduced incidence matrix is $(n - 1) \times b$.

The node corresponding to the deleted row is called the reference node or datum node. In Figure 3.10 node (4) is a datum node. Therefore, when node (4) is deleted from the complete incidence matrix we get the reduced incidence matrix [A].

$$[A] = \begin{bmatrix} -1 & +1 & 0 & 0 & 0 & 0 \\ 0 & -1 & 1 & 1 & 1 & 0 \\ 0 & 0 & 0 & -1 & -1 & 1 \end{bmatrix}$$

The number of possible trees of a graph $= \det(AA')$, where A' is the transpose of the reduced incidence matrix A.

3.5 DRAWING A GRAPH FROM AN INCIDENCE MATRIX

We can draw a corresponding graph from any incidence matrix. We know that each column of A_C has exactly two non-zero elements, one of which is $+1$ and the other is -1. First draw all the nodes in a random manner. Connect these nodes according to the above property. Consider the following complete incidence matrix.

$$A_C = \begin{bmatrix} -1 & 1 & 0 & 0 & 0 \\ 0 & -1 & -1 & 1 & 0 \\ 0 & 0 & 0 & -1 & -1 \\ 1 & 0 & 1 & 0 & 1 \end{bmatrix}_{4 \times 5}$$

In the complete incidence matrix, there are 4 nodes and 5 branches.

Mark 4 nodes (1), (2), (3), and (4) arbitrarily as shown in Figure 3.11.

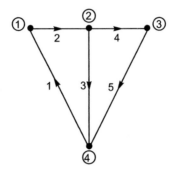

FIGURE 3.11

The first column of A_C corresponds to branch 1. It is seen from this column that branch 1 is incident at nodes (1) and (4). The entries -1 at node (1) and $+1$ at node (4) show that branch 1 enters node (1) and leaves node (4). Therefore, connect node (1) and (4) by branch 1 with the orientation as shown in the figure. Similarly, the second and third column of A_C correspond to branch 2 and 3 and the same procedure is applied at (2), (3), and (4).

3.6   DRAWING A GRAPH FROM A REDUCED INCIDENCE MATRIX

If the reduced incidence matrix A has n nodes, add a datum node and first convert this reduced incidence matrix A into a complete incidence matrix A_C, according to the properties of the complete incidence matrix (the algebraic sum of each column entry of a complete incidence matrix is zero). After converting the reduced incidence matrix A into a complete incidence matrix A_C, the procedure is the same as in the last Section (3.5).

3.7   KIRCHHOFF'S CURRENT LAW (KCL)

Consider the graph of Figure 3.12. It has four nodes. Assume node 4 to be a reference node or a datum node. Let the branch currents be denoted $i_1, i_2, \ldots, i_6$.

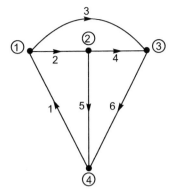

FIGURE 3.12

Applying KCL at node (1), (2), (3), and (4), we get

$$-i_1 + i_2 + i_3 = 0 \tag{1}$$
$$-i_2 + i_4 + i_5 = 0 \tag{2}$$
$$-i_3 - i_4 + i_6 = 0 \tag{3}$$
$$-i_5 - i_6 + i_1 = 0. \tag{4}$$

In matrix form, these equations can be written as

$$\begin{bmatrix} -1 & 0 & 1 & 0 & 0 & 0 \\ 0 & -1 & 0 & 1 & 1 & 0 \\ 0 & 0 & -1 & -1 & 0 & 1 \\ 1 & 0 & 0 & 0 & -1 & -1 \end{bmatrix} \begin{bmatrix} i_1 \\ i_2 \\ i_3 \\ i_4 \\ i_5 \\ i_6 \end{bmatrix} = \begin{bmatrix} 0 \\ 0 \\ 0 \\ 0 \end{bmatrix}$$

or

$$A_C I_b = 0. \tag{3.1}$$

Equation (3.1) gives the matrix representation of KCL. In this equation I_b represents the vector of branch currents. The matrix A_C is called the complete incidence matrix.

3.8   FUNDAMENTAL LOOPS (TIE-SETS) OF A GRAPH

The addition of a link or chord between any two nodes of a tree forms a loop called the fundamental loop. Consider a connected graph G of Figure 3.13.

Let T be a tree of G as shown in Figure 3.13(*b*). The twigs of this tree are branches 2, 4, and 5. The links of this tree are branches 1, 3, and 6. Out of these three links let us replace link 1 in its proper place as shown in Figure 3.13(*c*). It is seen that a loop $\{1, 2, 5\}l_1$ is formed and this loop has one link 1 and a unique set of branches 2 and 5. This loop is called the fundamental loop or *f*-loop of G with respect to a tree. It is also known as a tie-set. The direction of link 1 is assumed to be positive as shown in Figure 3.13(*c*).

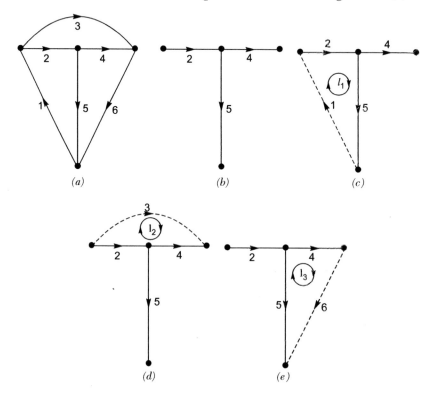

FIGURE 3.13 Formation of *f*-loops.

Similarly, if we replace link 3 in its proper position we find that a loop (3, 4, 2) is formed. This *f*-loop is indicated at loop l_2 in Figure 3.13(*d*). The direction of link 3 is assumed to be a reference.

In a similar manner replacing link 6 in its proper position results in the *f*-loop (6, 5, 4). This loop is shown as l_3 in Figure 3.13(*e*).

Each link of graph G corresponds a fundamental loop, therefore the number of *f*-loops or tie-sets is equal to the number of links of a chosen tree. Thus, for a tree T of a connected graph G with *n* nodes and *b* branches the number of *f*-loops is equal to $b - n + 1$ (the number of links).

3.8.1 Procedure for Obtaining f-loops (tie-sets) of a Graph

1. Select a tree from the given graph.
2. Replace one link in its proper place to form a loop; this is the first f-loop or tie-set.
3. Replace the second link in its proper place to form the second f-loop or tie-set.
4. Repeat the above procedure until each of the remaining links of the chosen tree is replaced in its proper place one at a time.
5. The direction of the link is assumed to be positive.

By the above example, $n = 4$ and $b = 6$.

$$\text{Number of links} = b - n + 1 = 6 - 4 + 1 = 3.$$

Therefore, there are three tie-sets

$$\text{tie-set 1 by loop } l_1[1, 2, 5]$$
$$\text{tie-set 2 by loop } l_2[3, 4, 2]$$
$$\text{tie-set 3 by loop } l_3[6, 5, 4].$$

In order to apply KVL to each fundamental loop, we take the reference direction of the loop which coincides with the reference direction of the link defining the loop.

$$l_1 : \quad v_1 + v_2 + v_5 = 0$$
$$l_2 : -v_2 + v_3 - v_4 = 0$$
$$l_3 : \quad v_4 - v_5 + v_6 = 0$$

In matrix form, we can write

Tie-set	1	2	3	4	5	6
1	1	1	0	0	1	0
2	0	−1	1	−1	0	0
3	0	0	0	1	−1	1

(Branches across columns 1–6)

Or we can also write it in this form

$$\begin{matrix} \text{loop} \\ 1 \\ 2 \\ 3 \end{matrix} \begin{bmatrix} 1 & 1 & 0 & 0 & 1 & 0 \\ 0 & -1 & 1 & -1 & 0 & 0 \\ 0 & 0 & 0 & 1 & -1 & 1 \end{bmatrix} \begin{bmatrix} v_1 \\ v_2 \\ v_3 \\ v_4 \\ v_5 \\ v_6 \end{bmatrix} = \begin{bmatrix} 0 \\ 0 \\ 0 \\ 0 \\ 0 \\ 0 \end{bmatrix}$$

or

$$\boxed{B_f v_b = 0} \quad \text{KVL,} \qquad\qquad (3.2)$$

where the $[B_f]_{l \times b}$ matrix is called the fundamental loop matrix or tie-set matrix.

3.8.2 Rank of the Fundamental Loop Matrix

Let the fundamental loop matrix in the last example be partitioned in the form

Branches →

$$
\begin{array}{c}
\text{loops} \\ \downarrow \\ l_1 \\ l_2 \\ l_3
\end{array}
\begin{array}{cc}
\begin{array}{ccc} \text{twigs} & & \text{links} \\ 2 \;\; 4 \;\; 5 & \;\; 1\;3\;6 \end{array} \\
\left[
\begin{array}{ccc|ccc}
1 & 0 & 1 & 1 & 0 & 0 \\
-1 & -1 & 0 & 0 & 1 & 0 \\
0 & 1 & -1 & 0 & 0 & 1
\end{array}
\right]
\end{array}
$$

$$
B_f =
\left[
\begin{array}{ccc|ccc}
1 & 0 & 1 & 1 & 0 & 0 \\
-1 & -1 & 0 & 0 & 1 & 0 \\
0 & 1 & -1 & 0 & 0 & 1
\end{array}
\right]_{l \times b}
= [B_{ft}, B_{fl}]
$$

$$
[B_{ft}] =
\left[
\begin{array}{ccc}
1 & 0 & 1 \\
-1 & -1 & 0 \\
0 & 1 & -1
\end{array}
\right]_{l \times (n-1)}
\qquad\qquad (3.3)
$$

$$
B_{fl} =
\left[
\begin{array}{ccc}
1 & 0 & 0 \\
0 & 1 & 0 \\
0 & 0 & 1
\end{array}
\right]_{l \times l}
= [U] = \text{Unit matrix}
\qquad\qquad (3.4)
$$

B_{ft} = submatrix of B_f corresponding to the twigs of the chosen tree.
B_{fl} = submatrix of B_f corresponding to the link of the chosen tree.

Rank of B_f is l (number of links) or $(b - n + 1)$ since it has a non-singular submatrix of order $l \times l$ or $(b - n + 1) \times (b - n + 1)$.

3.9 CUT-SET

A connected graph can be separated into two parts by removing certain branches of a graph. This operation is equivalent to cutting a graph into two parts and is known as a cut-set.

Consider the graph shown in Figure 3.14(*a*).

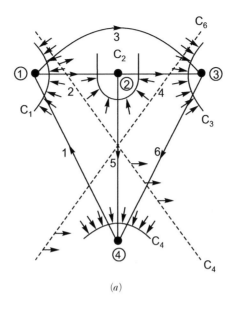

(*a*)

FIGURE 3.14

For cut-set 1 (C_1): the subgraph will be as shown in Figure 3.14(*b*),

$$C_1 = (1, 2, 3).$$

Remove branches 1, 2, and 3 from the graph as well as node 1 and the other three nodes with branches 4, 5, 6.

For cut-set 2 (C_2): the subgraph will be as shown in Figure 3.14(*c*),

$$C_2 = (2, 4, 5).$$

Remove branches 2, 4, and 5 from the graph and divide the graph into two parts (one node (2) and the other three nodes with branches 1, 3, and 6).

Similarly, cut-set 3

$$(C_3) = (3, 4, 6)$$

and cut-set 4

$$C_4 = (1, 5, 6).$$

For cut-set 5, the subgraph will be as shown in Figure 3.14(*d*),

$$C_5 = (2, 3, 5, 6).$$

Remove the cutting branches (2, 3, 5, 6) from the graph.

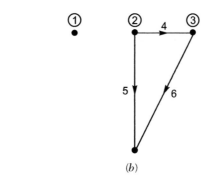

(b)

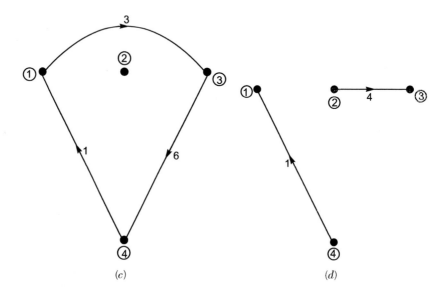

(c) (d)

FIGURE 3.14

For cut-set 6, the subgraph will be as shown in Figure 3.14(e).

$C_6 = (1, 3, 4, 5)$. Remove the cutting branches 1, 3, 4, and 5 from the graph.

A cut-set separates a graph into two parts, the orientation of the cut-set can be selected arbitrarily.

For nodes (1), (2), (3), and (4), the incoming current is assumed to be positive. For cut-set 5 and 6 the reference direction is shown in Figure 3.14(a).

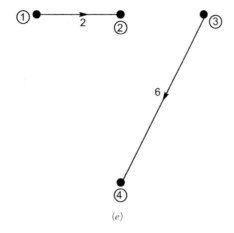

(e)

FIGURE 3.14

3.9.1 Cut-set Matrix

The cut-set matrix is a matrix, between the cut-set versus the total number of branches, by the above example.

$$C_1 = (1, 2, 3), \quad C_2 = (2, 4, 5), \qquad C_3 = (3, 4, 6)$$
$$C_4 = (1, 5, 6), \quad C_5 = (2, 3, 5, 6), \quad C_6 = (1, 3, 4, 5)$$

	Branches					
Cut-sets	1	2	3	4	5	6
1	1	−1	−1	0	0	0
2	0	1	0	−1	−1	0
3	0	0	1	1	0	−1
4	−1	0	0	0	1	1
5	0	1	1	0	−1	−1
6	−1	0	1	1	1	0

3.9.2 Separable and Non-separable Graph

A graph is said to be separable if it can be divided or separated into more than two parts. Otherwise the graph is known as a non-separable graph, and a non-separable graph can be separated only into two parts.

Example 3.1. *Check whether the given graph is separable or not.*

Solution: At node 1, if we make a cut-set, the graph is divided into three parts. Node (1) will be a part of node (2), (3), and (4) and branch 2 and 3 are the second part.

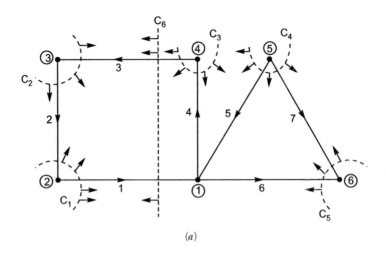

(a)

FIGURE 3.15

Node (5) and (6) and branch 7, create a third part, therefore cut-set (1, 4, 5, 6) is not possible because the graph is divided into more than two parts as shown in Figure 3.15(*b*).

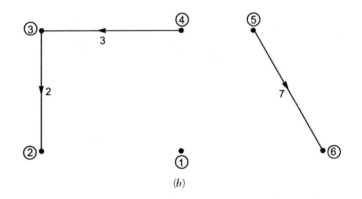

(b)

FIGURE 3.15

Cut-set 1: [1, 2]
Cut-set 2: [2, 3]
Cut-set 3: [3, 4]
Cut-set 4: [5, 7]
Cut-set 5: [6, 7]
Cut-set 6: [1, 3]

Cut-set matrix

Cut-sets ↓	Branches →						
	1	2	3	4	5	6	7
1	1	−1	0	0	0	0	0
2	0	1	−1	0	0	0	0
3	0	0	1	−1	0	0	0
4	0	0	0	0	1	0	−1
5	0	0	0	0	0	−1	1
6	−1	0	1	0	0	0	0

Incidence matrix

Nodes ↓	Branches →						
	1	2	3	4	5	6	7
1	−1	0	0	1	−1	1	0
2	1	−1	0	0	0	0	0
3	0	1	−1	0	0	0	0
4	0	0	1	−1	0	0	0
5	0	0	0	0	1	0	−1
6	0	0	0	0	0	−1	1

For a non-separable graph a complete incidence matrix is contained in the cut-set matrix. But in this case the complete incidence matrix is not a part of the cut-set matrix. Therefore, the given graph is separable graph. It means we can divide this graph into more than two parts.

3.10 FUNDAMENTAL CUT-SETS OR F-CUT-SETS

A fundamental cut-set of a graph G with respect to a tree T is a cut-set that is formed by one twig and a unique set of links (chords). The orientation of a cut-set is chosen to coincide with that of its defining twig. In other words, a fundamental cut-set is assigned the same orientation as that of the twig it contains.

Example 3.2. *Write the fundamental cut-set matrix for the network graph shown in Figure 3.16(a).*

Solution: Step 1: First select a tree T; in this case, we select a tree which has, 1, 4, and 5 branches as shown in Figure 3.16(b), other branches (2, 3, and 6) are links.

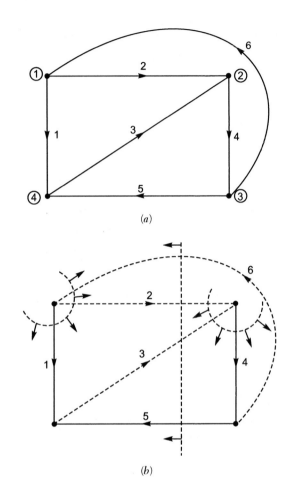

FIGURE 3.16

Step 2: Cut each twig at a time, the link may be more than one.

f-cut-set 1:$(1, 2, 6)$ Only one twig (twig No. 1)
f-cut-set 2:$(2, 3, 4)$ Only one twig (twig No. 4)
f-cut-set 3:$(2, 3, 5, 6)$ Only one twig (twig No. 5)

In f-cut-set 1: Only one, twig 1, and the other two $(2, 6)$ are linked. In f-cut-set 2, only one twig 4 and the other $(2, 3)$ are linked. Similarly, for f-cut-set 3, only one twig 5 and the other $(2, 3, 6)$ are linked.

Step 3: The direction of twig 1, 4, and 5 is assumed to be reference for the f-cut-set 1, 2, and 3.

The f-cut-set matrix is written as follows:

f-cut-sets ↓	Branches → 1	2	3	4	5	6
1	1	1	0	0	0	−1
2	0	−1	−1	1	0	0
3	0	−1	−1	0	1	1

3.10.1 KCL for the f-cut-sets

The basic property of the fundamental cut-sets is that they give linearly independent KCL equations. Since there are $(n-1)$ fundamental cut-sets in a graph of n nodes, to obtain all the linearly independent cut-set equations, it is sufficient to apply KCL to the f-cut-sets of the graph of the figure.

f-cut-set 1: $\qquad i_1 + i_2 - i_6 = 0$
f-cut-set 2: $\qquad -i_2 - i_3 + i_4 = 0$
f-cut-set 3: $\quad -i_2 - i_3 + i_5 + i_6 = 0$

These equations can be written in matrix form as

$$
\begin{array}{c|cccccc}
f\text{-cut-sets} & \multicolumn{6}{c}{\text{Branches} \rightarrow} \\
\downarrow & 1 & 2 & 3 & 4 & 5 & 6 \\
\hline
1 & 1 & 1 & 0 & 0 & 0 & -1 \\
2 & 0 & -1 & -1 & 1 & 0 & 0 \\
3 & 0 & -1 & -1 & 0 & 1 & 1
\end{array}
\begin{bmatrix} i_1 \\ i_2 \\ i_3 \\ i_4 \\ i_5 \\ i_6 \end{bmatrix}
=
\begin{bmatrix} 0 \\ 0 \\ 0 \end{bmatrix}
$$

$$\boxed{Q_f I_b = 0} \; \text{KCL} \qquad\qquad (3.5)$$

Equation 3 is the KCL for the f-cut-sets.

3.10.2 Rank of the f-cut-set Matrix

Since the orientation of the f-cut-sets is the same as the orientation of the tree branches (twigs), Q_f can be partitioned in the form

$$
Q_f = \begin{array}{c} 1 \\ 2 \\ 3 \end{array}
\begin{array}{c}
\overset{\text{twigs}}{\overbrace{1\ 4\ 5}} \quad \overset{\text{links}}{\overbrace{2 \quad 3 \quad 6}} \\
\begin{bmatrix}
1 & 0 & 0 & 1 & 0 & -1 \\
0 & 1 & 0 & -1 & -1 & 0 \\
0 & 0 & 1 & -1 & -1 & 1
\end{bmatrix}
\end{array}
= (Q_{ft}, Q_{fl}]
$$

$$Q_{ft} = \begin{bmatrix} 1 & 0 & 0 \\ 0 & 1 & 0 \\ 0 & 0 & 1 \end{bmatrix}_{(n-1)\times(n-1)} = [U]_{(n-1)\times(n-1)} \tag{3.6}$$

$$Q_{fl} = \begin{bmatrix} 1 & 0 & -1 \\ -1 & -1 & 0 \\ -1 & -1 & 1 \end{bmatrix}_{(n-1)\times(b-n+1)}, \tag{3.7}$$

where

Q_{ft} = submatrix of Q_f corresponding to twigs of chosen tree

Q_{fl} = submatrix of Q_f corresponding to links of a chosen tree

U = unit matrix.

Q_{ft} is an $(n-1) \times (n-1)$ unit matrix corresponding to the twigs of the tree and Q_{fl} is a $(n-1) \times (b-n+1)$ matrix corresponding to the links of the chosen tree. Therefore, the rank of Q_f is $(n-1)$.

3.11   NODAL ANALYSIS BASED ON GRAPH THEORY

The following steps are required for nodal analysis of a given network based on graph theory:

1. Choose an arbitrary node as the datum node and write down the incidence matrix A from the network graph.
2. Determine the branch admittance matrix Y_b.
3. Find the node admittance matrix from

$$Y_n = AY_bA'. \tag{3.8}$$

4. Obtain the voltage and current source vectors V_s and I_s from the network.
5. Determine I_n from the relation

$$I_n = AY_bV_s - AI_s. \tag{3.9}$$

6. Write the node equations of the network from the relation

$$I_n = Y_nV_n. \tag{3.10}$$

7. Use steps 5 and 6 to determine V_n as follows:

$$V_n = Y_n^{-1}AY_bV_s - Y_n^{-1}AI_s \quad \text{or} \quad Y_n^{-1}I_n. \tag{3.11}$$

8. Use the node transformation equation

$$V_b = A'V_n \quad \text{and} \tag{3.12}$$

$$I_b = I_s + Y_b(V_b - V_s). \tag{3.13}$$

3.12 LOOP ANALYSIS BASED ON GRAPH THEORY

The following steps are required for loop analysis of a given network based on graph theory.

1. Choose a tree T of the graph of the network. Determine the fundamental loop matrix B_f corresponding to this tree.
2. Determine the branch impedance matrix Z_b and determine the loop impedance matrix

$$Z_L = B_f Z_b B_f'. \tag{3.14}$$

3. Determine V_L from the relationship

$$V_L = B_f Z_b I_s - B_f V_s. \tag{3.15}$$

4. Write down the matrix loop equation from

$$V_L = Z_L I_L. \tag{3.16}$$

Example 3.3. *Write the matrix loop equation for the network as shown in Figure 3.17(a) using loop analysis.*

Solution: The graph of the given network is shown in Figure 3.17(b). Now select a tree with twigs (2, 3) as shown in Figure 3.17(c).

Tie-set matrix

$$
\begin{array}{c}
\text{tie-sets} \quad \text{Branches} \rightarrow \\
\downarrow \quad 1 \ \ 2 \ 3 \ \ 4 \\
B_f = \begin{array}{c} 1 \\ 2 \end{array}
\begin{bmatrix} 1 & 1 & 0 & 0 \\ 0 & -1 & 1 & 1 \end{bmatrix}
\end{array}
$$

The branch impedance matrix Z_b is given as

$$
Z_b = \begin{bmatrix}
R & 0 & 0 & 0 \\
0 & \frac{1}{C_1 S} & 0 & 0 \\
0 & 0 & \frac{1}{LS} & 0 \\
0 & 0 & 0 & \frac{1}{C_2 S}
\end{bmatrix}.
$$

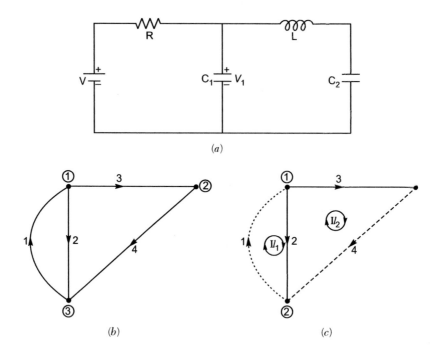

FIGURE 3.17

Now the loop impedance matrix is given as

$$Z_L = B_f Z_b B_f T = \begin{bmatrix} 1 & 1 & 0 & 0 \\ 0 & -1 & 1 & 1 \end{bmatrix} \begin{bmatrix} R & 0 & 0 & 0 \\ 0 & \frac{1}{C_1 S} & 0 & 0 \\ 0 & 0 & \frac{1}{LS} & 0 \\ 0 & 0 & 0 & \frac{1}{C_2 S} \end{bmatrix} \begin{bmatrix} 1 & 0 \\ 1 & -1 \\ 0 & 1 \\ 0 & 1 \end{bmatrix}$$

$$= \begin{bmatrix} 1 & 1 & 0 & 0 \\ 0 & -1 & 1 & 1 \end{bmatrix} \begin{bmatrix} R & 0 \\ \frac{1}{C_1 S} & -\frac{1}{C_1 S} \\ 0 & \frac{1}{LS} \\ 0 & \frac{1}{C_2 S} \end{bmatrix} = \begin{bmatrix} R + \frac{1}{C_1 S} & -\frac{1}{C_1 S} \\ -\frac{1}{C_1 S} & LS + \frac{1}{C_1 S} + \frac{1}{C_2 S} \end{bmatrix}.$$

The voltage and current sources matrices are

$$V_S = \begin{bmatrix} \text{Voltage source in branch 1} \\ \text{Voltage source in branch 2} \\ \text{Voltage source in branch 3} \\ \text{Voltage source in branch 4} \end{bmatrix} = \begin{bmatrix} -v_1 \\ v_1 \\ 0 \\ 0 \end{bmatrix}$$

and

$$I_S = \begin{bmatrix} \text{Current source in branch 1} \\ \text{Current source in branch 2} \\ \text{Current source in branch 3} \\ \text{Current source in branch 4} \end{bmatrix} = \begin{bmatrix} 0 \\ 0 \\ 0 \\ 0 \end{bmatrix}$$

and we know,

$$V_l = B_f Z_b I_s - B_f V_s = -B_f V_s [I_s = 0]$$

$$= -\begin{bmatrix} 1 & 1 & 0 & 0 \\ 0 & -1 & 1 & 1 \end{bmatrix} \begin{bmatrix} -V_1 \\ v_1 \\ 0 \\ 0 \end{bmatrix}.$$

$$= -\begin{bmatrix} -V_1 + v_1 \\ -v_1 \end{bmatrix} = \begin{bmatrix} V_1 - v_1 \\ v_1 \end{bmatrix}$$

Therefore, the matrix loop equation can be written as

$$V_l = Z_l I_l$$

$$\begin{bmatrix} V_1 - v_1 \\ v_1 \end{bmatrix} = \begin{bmatrix} R + \frac{1}{C_1 S} & -\frac{1}{C_1 S} \\ -\frac{1}{C_1 S} & LS + \frac{1}{C_1 S} + \frac{1}{C_2 S} \end{bmatrix} \begin{bmatrix} I_{l_1} \\ I_{l_2} \end{bmatrix}.$$

SOLVED PROBLEMS

Problem 3.1. *Draw the incidence matrix of the following graph as shown in Figure 3.18.*

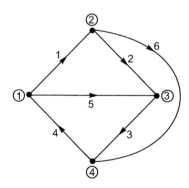

FIGURE 3.18

Solution: The incidence matrix is formed as shown below:

Nodes	Branches					
	1	2	3	4	5	6
1	−1	0	0	+1	−1	0
2	+1	−1	0	0	0	−1
3	0	+1	−1	0	+1	0
4	0	0	+1	−1	0	+1

Problem 3.2. *The complete incidence matrix of a graph is given below. Draw the directed graph.*

Nodes	Branches							
	1	2	3	4	5	6	7	8
1	+1	0	+1	−1	0	0	−1	−1
2	0	+1	0	+1	0	+1	0	0
3	−1	−1	0	0	+1	−	0	+1
4	0	0	−1	0	−1	0	+1	0

Solution: There are 4 nodes and 8 branches. First draw Nodes (1), (2), (3), and (4) arbitrarily. Now it is clear by the complete incidence matrix branch 1 that is connected between nodes (1) and (3). Similarly, branch 2 is connected between nodes (2) and (3) – after drawing all the branches. Now show the orientation as shown in Figure 3.19.

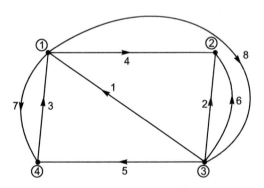

FIGURE 3.19

Problem 3.3. *Develop the graph of the network shown in Figure 3.20(a). Select a tree and obtain the tie-set and f-cut-set matrix.*

Solution: The graph of the given network is shown in Figure 3.20.

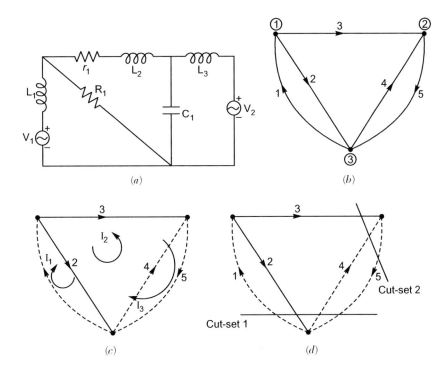

FIGURE 3.20

Assume a tree with fundamental loops that are 1-2, 2-3-4, and 2-3-5. Now the tie-set matrix

	Branches				
	1	2	3	4	5
tie-set 1	1	1	0	0	0
tie-set 4	0	1	−1	1	0
tie-set 5	0	−1	1	0	1

tie-set 1: [1, 2]
tie-set 4: [2, 3, 4]
tie-set 5: [2, 3, 5]

The twig direction is assumed to be the reference.
The f-cut-set matrix:

	Branches				
	1	2	3	4	5
cut-set 1	−1	+1	0	−1	+1
cut-set 2	0	0	+1	+1	−1

Cut-set 1 = [1, 2, 4, 5]
Cut-set 2 = [3, 4, 5]

Problem 3.4. *A graph is shown in Figure 3.21(a). Find the f-cut-set and tie-set matrices.*

Solution: The f-cut-set

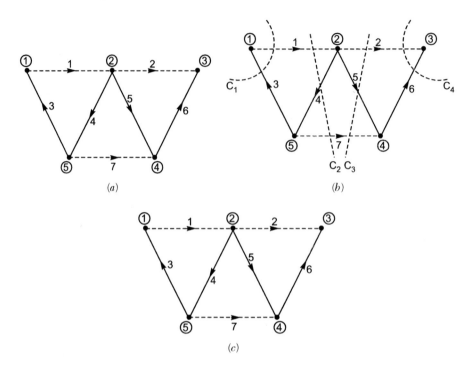

FIGURE 3.21

Cut-set-1

$$(C1) : [1, 3]$$
$$C2 : [1, 4, 7]$$
$$C3 : [2, 5, 7]$$
$$C4 : [2, 6]$$

Now the f-cut-set matrix

			Branches				
	1	2	3	4	5	6	7
C_1	−1	0	+1	0	0	0	0
C_2	−1	0	0	+1	0	0	−1
C_3	0	+1	0	0	+1	0	+1
C_4	0	+1	0	0	0	+1	0

tie-set:

tie-set-1

$$T_1[1,3,4]$$
$$T_2[4,5,7]$$
$$T_3[2,5,6]$$

Now the tie-set matrix

	Branches						
	1	2	3	4	5	6	7
T_1	+1	0	+1	+1	0	0	0
T_2	0	0	0	+1	−1	0	+1
T_3	0	+1	0	0	−1	−1	0

Problem 3.5. *Draw the directed graph for the circuit shown in Figure 3.22(a). Hence, select loop variables and write the network equilibrium equation in matrix form.*

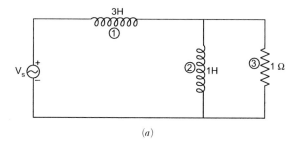

(a)

FIGURE 3.22

Solution: The directed graph of the given network will then be as shown in Figure 3.22(b) and its tree as shown in Figure 3.22(c) where the dotted line shows the branches. These are known as link or chords.

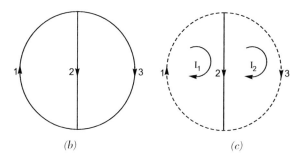

(b)   (c)

FIGURE 3.22

Now tie-set 1: [1, 2] and tie-set 2: [2, 3]

Tie-Set Table

Tie-set	Branches		
	1	2	3
1	+1	+1	0
2	0	−1	1

The tie-set matrix will then be

$$[B] = \begin{bmatrix} 1 & 1 & 0 \\ 0 & -1 & 1 \end{bmatrix}.$$

The transposed tie-set matrix

$$[B]^t = \begin{bmatrix} 1 & 0 \\ 1 & -1 \\ 0 & 1 \end{bmatrix}.$$

The branch impedance matrix

$$[Z_b] = \begin{bmatrix} 3s & \frac{1}{2}s \\ \frac{1}{2}s & s & 0 \\ 0 & 0 & 1 \end{bmatrix}.$$

Now

$$\begin{bmatrix} B & Z_b & B^t \end{bmatrix} = \begin{bmatrix} 1 & 1 & 0 \\ 0 & -1 & 1 \end{bmatrix} \begin{bmatrix} 3s & \frac{1}{2}s & 0 \\ \frac{1}{2}s & s & 0 \\ 0 & 0 & 1 \end{bmatrix} \begin{bmatrix} 1 & 0 \\ 1 & -1 \\ 0 & 1 \end{bmatrix} = \begin{bmatrix} 5s & -\frac{3}{2}s \\ -\frac{3}{2}s & s+1 \end{bmatrix}.$$

Network equilibrium equations

$$[BZ_b B^t][I] = B\{V_S - Z_b I_S\}[\text{In this question } I_S = 0] = BV_S$$

$$\begin{bmatrix} 5s & -\frac{3}{2}s \\ -\frac{3}{2}s & s+1 \end{bmatrix} \begin{bmatrix} I_1 \\ I_2 \end{bmatrix} = \begin{bmatrix} 1 & 1 & 0 \\ 0 & -1 & 1 \end{bmatrix} \begin{bmatrix} V_S \\ 0 \\ 0 \end{bmatrix}.$$

When expanded, this becomes

$$5s\, I_1 - \frac{3}{2}s\, I_2 = V_S$$

$$-\frac{3}{2}s\, I_1 + (s+1)\, I_2 = 0.$$

Problem 3.6. *Find out currents through and voltage across all branches of the network shown in Figure 3.23 with the help of its tie-set schedule.*

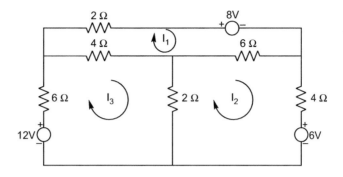

FIGURE 3.23

Solution: The graph of the given network is shown in Figure 3.23(a).

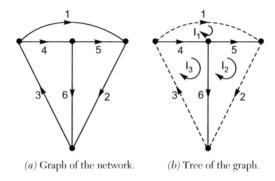

(a) Graph of the network.   (b) Tree of the graph.

FIGURE 3.23

There are three links, hence there are three tie-sets.

$$\text{Tie-set } 1 = [1, 4, 5]$$
$$\text{Tie-set } 2 = [2, 5, 6]$$
$$\text{Tie-set } 3 = [3, 4, 6]$$

The Tie-Set Table

Tie-set	Branch					
	1	2	3	4	5	6
1	1	0	0	−1	−1	0
2	0	1	0	0	1	−1
3	0	0	1	1	0	1

The rows of the tie-set schedule will give the following equations in branch voltages

$$V_1 - V_4 - V_5 = 0 \tag{1}$$
$$V_2 + V_5 - V_6 = 0 \tag{2}$$
$$V_3 + V_4 + V_6 = 0. \tag{3}$$

The columns of the tie-set schedule will give the following equations in branch and loop currents

$$I_1 = i_1$$
$$I_2 = i_2$$
$$I_3 = i_3$$
$$I_4 = i_1 - i_3$$
$$I_5 = i_2 - i_3$$
$$I_6 = i_1 - i_2$$
$$V_1 = 2i_1 + 8$$
$$V_2 = 4i_2 + 6$$
$$V_3 = 6i_3 - 12$$
$$V_4 = 4(I_3 - I_1)$$
$$V_5 = 6(I_2 - I_3)$$
$$V_6 = 2(I_3 - I_2).$$

Putting these values into Equations (1), (2), and (3)

$$2i_1 + 8 - 4I_3 + 4I_1 - 6I_2 + 6I_1 = 0 \tag{4}$$
$$12I_1 - 6I_2 - 4I_3 = -8$$
$$6 + 4i_2 + 6I_2 - 6I_1 - 2I_3 + 2I_2 = 0 \tag{5}$$
$$-6I_1 + 12I_2 - 2I_3 = -6$$
$$6i_3 - 12 + 4I_3 - 4I_1 + 2I_3 - 2I_2 = 0 \tag{6}$$
$$-4I_1 - 2I_2 + 12I_3 = 12.$$

By Equations (4), (5), and (6)

$$I_1 = -0.916 \,\text{Amp.}, \quad I_2 = -0.866 \,\text{Amp.}, \quad I_3 = 0.55 \,\text{Amp.}$$

Now

$$I_4 = I_1 - I_3 = -0.916 - 0.55 = -1.46 \,\text{Amp.}$$
$$I_5 = I_2 - I_3 = -0.866 - 0.55 = 1.416 \,\text{Amp.}$$
$$I_6 = I_1 - I_2 = -0.916 + 0.866 = 0.05 \,\text{Amp.}$$

Problem 3.7. *Draw the graph of the network shown in Figure 3.24, and select a suitable tree to write the tie-set schedule. Then find the three loop currents.*

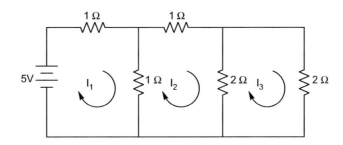

FIGURE 3.24

Solution: The graph of the given network will be as shown in Figure 3.24(*a*) and Figure 3.24(*b*).

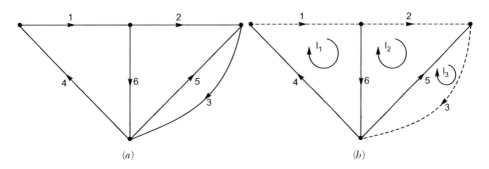

(*a*)   (*b*)

FIGURE 3.24

Now

$$\text{Tie-set } 1 = [1, 4, 6]$$
$$\text{Tie-set } 2 = [2, 5, 6]$$
$$\text{Tie-set } 3 = [3, 5].$$

The tie-set schedule of the graph

Loops Numbers or Tie-Set	Branches					
	1	2	3	4	5	6
1	1	0	0	1	0	1
2	0	1	0	0	−1	−1
3	0	0	1	0	1	0

The rows of the tie-set schedule will give the following equations in branch voltages

$$V_1 + V_4 - V_6 = 0 \qquad (1)$$
$$V_2 - V_5 - V_6 = 0 \qquad (2)$$
$$V_3 + V_5 = 0. \qquad (3)$$

The columns of the tie-set schedule will give branch currents in terms of loop currents

$$I_1 = i_1$$
$$I_2 = i_2$$
$$I_3 = i_3$$
$$I_4 = i_1$$
$$I_5 = -i_2 + i_3$$
$$I_6 = i_1 - i_2.$$

Now

$$V_1 = 1 \times i_1 = i_1$$
$$V_2 = 1 \times i_2 = i2$$
$$V_3 = 2 \times i_3 = 2i_3$$
$$V_4 = -5$$
$$V_5 = 2I_5 = 2(-i_2 + i_3)$$
$$V_6 = 1 \times I_6 = -i_2.$$

Putting these values into Equations (1), (2), and (3)

$$2i_1 - 5 + (-i_2) = 0$$
$$2i_1 - i_2 = +5 \qquad (4)$$
$$-i_1 + i_2 - 2(-i_2 + i_3) + i_2 = 0$$
$$-i_1 + 4i_2 - 2i_3 = 0 \qquad (5)$$
$$2i_3 + 2(-i_2 + i_3) = 0$$
$$-2i_2 + 4i_3 = 0. \qquad (6)$$

By Equations (4), (5), and (6),

$$i_1 = 3\,\text{A}, \quad i_2 = 1\,\text{A}, \quad i_3 = 0.5\,\text{Amp.}$$

Problem 3.8. *Formulate the cut-set matrix of the network shown in Figure 3.25 and hence obtain the node equations.*

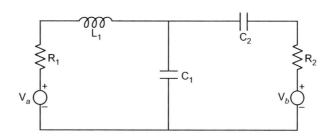

FIGURE 3.25

Solution: Assume a simple tree. This is shown in Figure 3.25(a).

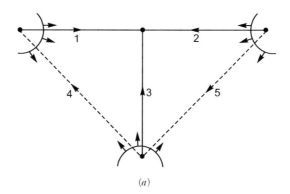

(a)

FIGURE 3.25

Cut-set 1 [1, 4]
Cut-set 2 [2, 5]
Cut-set 3 [3, 4, 5]

The cut-set matrix is given below:

$$
\begin{array}{c}
\text{Cuts} \qquad \text{Branches} \\
\begin{array}{cccccc}
 & 1 & 2 & 3 & 4 & 5
\end{array} \\
Q = \begin{array}{c} 1 \\ 2 \\ 3 \end{array}
\begin{bmatrix}
1 & 0 & 0 & 1 & 0 \\
0 & 1 & 0 & 0 & 1 \\
0 & 0 & 1 & -1 & -1
\end{bmatrix}.
\end{array}
$$

Writing KCL equations from the coeffcients in the rows of matrix Q, we have

$$I_1 + I_4 = 0 \tag{1}$$
$$I_2 + I_5 = 0 \tag{2}$$
$$I_3 - I_4 - I_5 = 0. \tag{3}$$

Column express branch voltages are

$$V_1 = E_1, \quad V_2 = E_2, \quad V_3 = E_3, \quad V_4 = E_1 - E_2, \quad V_5 = E_2 - E_3$$

and

$$V_1 = sL_1I_1, \quad V_2 = \frac{I_2}{sC_2}, \quad V3 = \frac{I_3}{sC_3}, \quad V_4 = \frac{V_a}{s} + R_1I_4$$

$$V_5 = \frac{V_b}{s} + R_2I_5.$$

By using the above equation,

$$I_1 = \frac{E}{sL_1}, \quad I_2 = sC_2E_2, \quad I_3 = sC_1E_3$$

$$I_4 = \frac{1}{R_1}\left[E_1 - E_3 - \frac{V_a}{s}\right]$$

$$I_5 = \frac{1}{R_2}\left[E_2 - E_3 - \frac{V_b}{s}\right]$$

Substituting I_1, I_2, I_3, I_4, and I_5 into Equations (1), (2), and (3)

$$\frac{E}{sL_1} + \frac{1}{R_1}\left[E_1 - E_3 - \frac{V_a}{s}\right] = 0 \tag{4}$$

$$sC_2E_2 + \frac{1}{R_2}\left[E_2 - E_3 - \frac{V_b}{s}\right] = 0 \tag{5}$$

$$sC_1E_3 - \frac{1}{R_1}\left[E_1 - E_3 - \frac{V_s}{s}\right] - \frac{1}{R_2}\left[E_2 - E_3 - \frac{V_b}{s}\right] = 0. \tag{6}$$

Rearranging Equations (4), (5), and (6)

$$\left(\frac{1}{R_1} + \frac{1}{sL_1}\right)E_1 - \frac{1}{R_2}E_3 = \frac{V_a}{sR_1}$$

$$\left(\frac{1}{R_2} + sC_2\right)E_2 - \frac{1}{R_2}E_3 = \frac{V_b}{sR_2}$$

$$-\frac{1}{R_1}E_1 - \frac{1}{R_2}E_2 + \left(\frac{1}{R_1} + \frac{1}{R_2} + sC_1\right)E_3 = -\frac{V_s}{R_1s} - \frac{V_b}{R_2s}.$$

Writing the above equations in matrix form, we have

$$\begin{bmatrix} \left(\frac{1}{R_1} + \frac{1}{sL_1}\right) & 0 & -\frac{1}{R_1} \\ 0 & \left(\frac{1}{R_2} + sC_2\right) & -\frac{1}{R_2} \\ -\frac{1}{R_1} & -\frac{1}{R_2} & \left(\frac{1}{R_1} + \frac{1}{R_2} + sC_1\right) \end{bmatrix} \begin{bmatrix} E_1 \\ E_2 \\ E_3 \end{bmatrix} = \begin{bmatrix} \frac{V_a}{sR_1} \\ \frac{V_b}{sR_2} \\ -\frac{V_a}{R_1 s} - \frac{V_b}{R_2 s} \end{bmatrix}.$$

Problem 3.9. *Determine the total number of trees and draw the trees.*

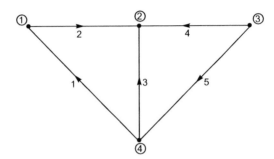

FIGURE 3.26

Solution: Total number of possible trees

$$= \det{(AA')},$$

where A is reduced to an incidence matrix. Node (4) is a reference node or datum node.
    Therefore,

$$A = \begin{bmatrix} -1 & 1 & 0 & 0 & 0 \\ 0 & -1 & 1 & 1 & 0 \\ 0 & 0 & 0 & -1 & 1 \end{bmatrix}$$

$$A' = \begin{bmatrix} -1 & 0 & 0 \\ 1 & -1 & 0 \\ 0 & 1 & 0 \\ 0 & 1 & -1 \\ 0 & 0 & 1 \end{bmatrix}$$

Now

$$AA' = \begin{bmatrix} -1 & 1 & 0 & 0 & 0 \\ 0 & -1 & 1 & 1 & 0 \\ 0 & 0 & 0 & -1 & 1 \end{bmatrix} \begin{bmatrix} -1 & 0 & 0 \\ 1 & -1 & 0 \\ 0 & 1 & 0 \\ 0 & 1 & -1 \\ 0 & 0 & 1 \end{bmatrix}$$

$$= \begin{bmatrix} 2 & -1 & 0 \\ -1 & 3 & -1 \\ 0 & -1 & 2 \end{bmatrix}$$

$$\det(AA') = \begin{vmatrix} 2 & -1 & 0 \\ -1 & 3 & -1 \\ 0 & -1 & 2 \end{vmatrix}.$$

$$= 2(6-1) + 1(-2) = 2 \times 5 - 2$$
$$= 8$$

All 8 trees are shown in Figure 3.27 with a solid line and the cotree is shown with a dotted line.

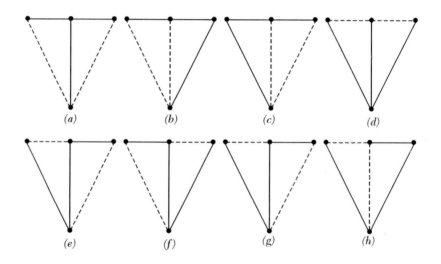

FIGURE 3.27

Problem 3.10. *The reduced incidence matrix of a graph is given below:*

	Branches →					
	1	2	3	4	5	6
Nodes 1	0	+1	0	−1	−1	+1
2	0	−1	+1	0	+1	0
3	+1	0	−1	0	0	0

Draw the oriented graph. Select a tree and write down the f-cut-set matrix.

Solution: The reduced incidence matrix

$$A = \begin{bmatrix} 0 & +1 & 0 & -1 & -1 & +1 \\ 0 & -1 & +1 & 0 & +1 & 0 \\ +1 & 0 & -1 & 0 & 0 & 0 \end{bmatrix}.$$

The complete incidence matrix A_a may be constructed by adding a new row to the matrix A (adding datum node)

$$A_a = \begin{matrix} & \begin{matrix} 1 & 2 & 3 & 4 & 5 & 6 \end{matrix} \\ \begin{matrix} 1 \\ 2 \\ 3 \\ 4 \end{matrix} & \begin{vmatrix} 0 & +1 & 0 & -1 & -1 & +1 \\ 0 & -1 & +1 & 0 & +1 & 0 \\ +1 & 0 & -1 & 0 & 0 & 0 \\ -1 & 0 & 0 & +1 & 0 & -1 \end{vmatrix} \end{matrix}.$$

Now there are four nodes and six branches. First draw all four nodes. Now notice branch (1) is connected between node 3 and 4. Branch 2 is connected between node 1 and 2. Similarly, branch 6 is connected between node 1 and 4.

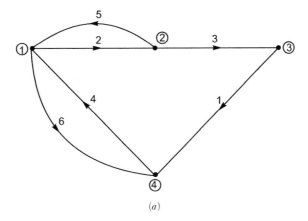

(a)

FIGURE 3.28

Now, show the direction of each branch according to the + and − sign. Suppose the direction of branch 1 is from node 3 to 4, because 3 is at a higher potential or positive with respect to node 4 (−negative).

Fundamental Cut-Set: For the fundamental cut-set assume a tree.

In the fundamental cut-set the twig direction is assumed to be the reference.

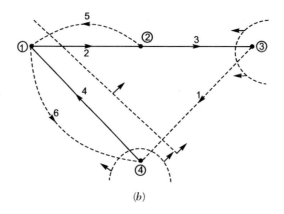

(b)

FIGURE 3.28

Twig Branches	Branches					
	1	2	3	4	5	6
2	−1	1	0	0	−1	0
f-cut-set = 3	−1	1	0	0	−1	0
4	−1	0	1	0	0	0

Problem 3.11. (*i*) *State the properties of the incidence matrix.*

(*ii*) *Draw the reduced incidence matrix of the directed graph shown in Figure 3.29.*

Solution:

Incidence matrix

	Branches					
	1	2	3	4	5	6
Nodes 1	+1	0	0	0	−1	+1
2	0	0	+1	+1	+1	0
3	0	+1	0	−1	0	−1
4	−1	−1	−1	0	0	0

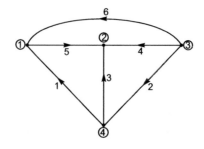

FIGURE 3.29

Node 4 is a datum mode. Remove the datum node from the incidence matrix. It is reduced to the incidence matrix

$$A = \begin{vmatrix} +1 & 0 & 0 & 0 & -1 & +1 \\ 0 & 0 & +1 & +1 & +1 & 0 \\ 0 & +1 & 0 & -1 & 0 & -1 \end{vmatrix}.$$

Problem 3.12. *Draw the oriented graph and the incidence matrix of the following network.*

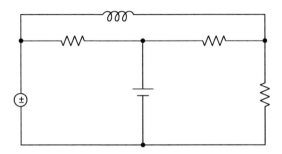

FIGURE 3.30

Solution:

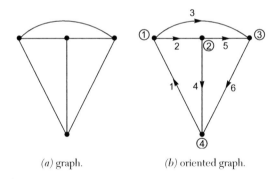

(a) graph.   (b) oriented graph.

FIGURE 3.30

Incidence matrix

Nodes	Branches					
	1	2	3	4	5	6
1	+1	-1	-1	0	0	0
2	0	+1	0	-1	-1	0
3	0	0	+1	0	+1	-1
4	-1	0	0	+1	0	+1

Problem 3.13. *Draw the oriented graph of the following reduced incidence matrix*

$$\begin{bmatrix} -1 & 1 & 0 & 0 & 0 & 0 & -1 & 0 \\ 1 & 0 & 0 & 0 & -1 & 0 & 0 & 0 \\ 0 & -1 & 1 & -1 & 1 & -1 & 0 & 0 \\ 0 & 0 & -1 & 1 & 0 & 0 & 0 & 1 \end{bmatrix}.$$

Solution: The complete incidence matrix of the given reduced incidence matrix

$$A_C = \begin{bmatrix} -1 & 1 & 0 & 0 & 0 & 0 & -1 & 0 \\ 1 & 0 & 0 & 0 & -1 & 0 & 0 & 0 \\ 0 & -1 & 1 & -1 & 1 & -1 & 0 & 0 \\ 0 & 0 & -1 & 1 & 0 & 0 & 0 & 1 \\ 0 & 0 & 0 & 0 & 0 & 1 & 1 & -1 \end{bmatrix}.$$

By the complete incidence matrix, it is clear that the graph has 5 nodes and 8 branches. First mark the given nodes (1), (2), (3), (4), and (5) arbitrarily as shown in Figure 3.31.

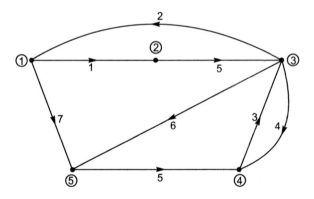

FIGURE 3.31

From the complete incidence matrix, it is clear branch 1 is connected between nodes (1) and (2) and branch 2 is connected between nodes (1) and (3). Similarly, branch 8 is connected between nodes (4) and (5).

After drawing the branches, now show the orientation, according to the complete incidence matrix.

Problem 3.14. *The reduced incidence matrix of a graph is given by*

Nodes ↓	Branches → 1	2	3	4	5	6
a	0	1	0	−1	−1	+1
b	0	−1	1	0	1	0
c	1	0	−1	0	0	0

Draw the oriented graph. Select a tree and find the f-cut-set matrix.

Solution: Now the complete incidence matrix of the given reduced incidence matrix (adding by a datum node according to the properties of complete incidence matrix) is

$$A_C = \begin{bmatrix} 0 & 1 & 0 & -1 & -1 & 1 \\ 0 & -1 & 1 & 0 & 1 & 0 \\ 1 & 0 & -1 & 0 & 0 & 0 \\ -1 & 0 & 0 & 1 & 0 & -1 \end{bmatrix}.$$

The graph of this complete incidence matrix is shown in Figure 3.32.

Assume a tree as shown in Figure 3.32(a). There are 3 twigs hence $3f$-cut-sets

$$C_1 = \{1, 2, 5\}$$
$$C_2 = \{1, 3\}$$
$$C_3 = \{1, 4, 6\}.$$

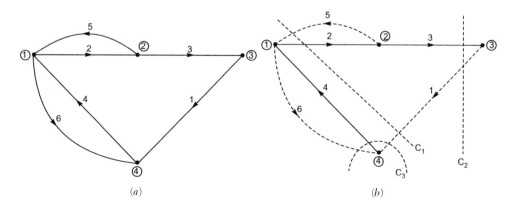

(a)   *(b)*

FIGURE 3.32

Now the f-cut-set matrix

	Branches					
f-cut-sets ↓	1	2	3	4	5	6
1	−1	1	0	0	−1	0
2	−1	0	1	0	0	0
3	−1	0	0	1	0	−1

The f-cut-set matrix

$$Q_f = \begin{bmatrix} -1 & 1 & 0 & 0 & -1 & 0 \\ -1 & 0 & 1 & 0 & 0 & 0 \\ -1 & 0 & 0 & 1 & 0 & -1 \end{bmatrix}.$$

Problem 3.15. *The fundamental cut-set matrix is given as:*

	Branches				Links		
Cut-sets	1	2	3	4	5	6	7
	1	0	0	0	−1	0	0
	0	1	0	0	1	0	1
	0	0	1	0	0	1	1
	0	0	0	1	0	1	0

Draw the oriented graph of the network.

Solution: The graph of the given cut-set matrix is shown in Figure 3.33.

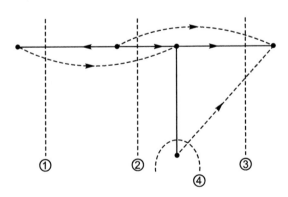

FIGURE 3.33

Problem 3.16. *A network graph has three basic cut-sets and six basic loops. Draw:*

(i) The oriented network graph having all the nodes in one line. (ii) All the six basic loops.

Solution: The number of f-cut-sets = total number of twigs.

$$\text{Number of nodes} = 3 + 1 = 4.$$

The graph is shown in Figure 3.34.

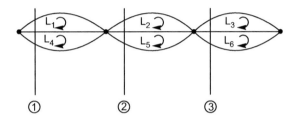

FIGURE 3.34

Problem 3.17. *For the network shown in Figure 3.35(a), draw the network graph, selecting elements 1, 2, and 3 as the tree elements, and obtain the basic cut-sets and write the basic cut-set matrix.*

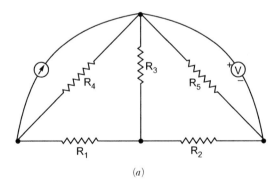

(a)

FIGURE 3.35

Solution: A resistor in parallel with a voltage source may be ignored or omitted entirely from the network representation:

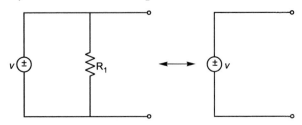

Now the graph. The voltage source is short circuit and the current source is open circuit.

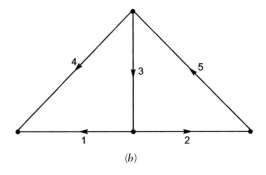

(b)

FIGURE 3.35

Now selecting tree (1, 2, 3) the basic or fundamental cut-sets are :
Cut-set 1 : (1, 4)
Cut-set 2 : (2, 5)
Cut-set 3 : (3, 4, 5)
Therefore, the basic or f-cut-set matrix is given by

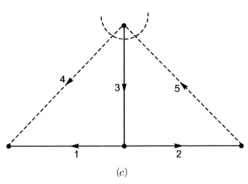

(c)

FIGURE 3.35

Cut-sets	Branches				
	1	2	3	4	5
1	1	0	0	1	0
2	0	1	0	0	−1
3	0	0	1	1	−1

Problem 3.18. *Draw the graph and find a tree of the network shown in Figure 3.36. Consider O as a datum node and assume elements BD and BC as links. Determine the tie-set schedule, branch impedance matrix, and source voltage matrix. Obtain the loop equations using the above matrices.*

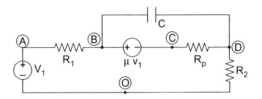

FIGURE 3.36

Solution: The graph of the given network is shown in Figure 3.36(*a*) (considering the voltage source V_1 is short circuited, hence μV_1 is also short circuited).

Now, the tree per the given conditions, *i.e.*, BD and BC are a like, is shown in Figure 3.36(*b*) for the given tree.

Number of tie-sets = number of links = 2
Tie-set 1 = $(1, 2, 4)$ Figure 3.36(*c*)
Tie-set 2 = $(1, 3, 4)$ Figure 3.36(*d*)

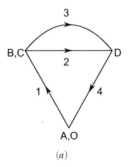

(*a*)

FIGURE 3.36

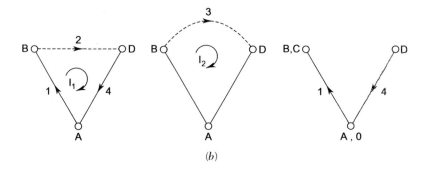

(b)

FIGURE 3.36

The f-loop matrix or tie-set matrix

$$
\begin{array}{cc}
\text{Tie-set} & \text{Branches} \\
 & 1\ 2\ 3\ 4
\end{array}
$$

$$
B_f = \begin{array}{c} 1 \\ 2 \end{array}\begin{bmatrix} 1 & 1 & 0 & 1 \\ 1 & 0 & 1 & 1 \end{bmatrix}
$$

The branch impedance matrix, $V_L = Z_L I_L$,

$$
\begin{bmatrix} V_1 - \mu V_1 \\ V_1 \end{bmatrix} = \begin{bmatrix} R_1 + R_P + R_2 & R_1 + R_2 \\ R_1 + R_2 & R_1 + \frac{1}{CS} + R_2 \end{bmatrix}\begin{bmatrix} I_1 \\ I_2 \end{bmatrix}
$$

$$
Z_b = \begin{bmatrix} R_1 & 0 & 0 & 0 \\ 0 & R_P & 0 & 0 \\ 0 & 0 & \frac{1}{CS} & 0 \\ 0 & 0 & 0 & R_2 \end{bmatrix}.
$$

The source voltage matrix

$$
V_S = \begin{bmatrix} -V_1 \\ \mu V_1 \\ 0 \\ 0 \end{bmatrix}.
$$

The loop voltage matrix per Equation (3.15),

$$
V_L = B_f Z_b I_s - B_f V_s = 0 - \begin{bmatrix} 1 & 1 & 0 & 1 \\ 1 & 0 & 1 & 1 \end{bmatrix}\begin{bmatrix} -V_1 \\ \mu V_1 \\ 0 \\ 0 \end{bmatrix} = \begin{bmatrix} V_1 - \mu V_1 \\ V_1 \end{bmatrix}.
$$

The loop impedance matrix per Equation (3.14),

$$Z_L = B_f Z_b B_f'.$$

$$= \begin{bmatrix} 1 & 1 & 0 & 1 \\ 1 & 0 & 1 & 1 \end{bmatrix} \begin{bmatrix} R_1 & 0 & 0 & 0 \\ 0 & R_P & 0 & 0 \\ 0 & 0 & \frac{1}{CS} & 0 \\ 0 & 0 & 0 & R_2 \end{bmatrix} \begin{bmatrix} 1 & 1 \\ 1 & 0 \\ 0 & 1 \\ 1 & 1 \end{bmatrix}.$$

$$= \begin{bmatrix} 1 & 1 & 0 & 1 \\ 1 & 0 & 1 & 1 \end{bmatrix} \begin{bmatrix} R_1 & R_1 \\ R_P & 0 \\ 0 & \frac{1}{CS} \\ R_2 & R_2 \end{bmatrix} = \begin{bmatrix} R_1 + R_P + R_2 & R_1 + R_2 \\ R_1 + R_2 & R_1 + \frac{1}{CS} + R_2 \end{bmatrix}$$

Now the matrix loop equations,

$$V_L = Z_L I_L.$$

Problem 3.19. *An oriented graph of a certain circuit is shown in Figure 3.37. For the tree [2, 4, 6], find $[Q_f]$ and $[B_f]$ and hence show that $[Q_f][B_f]^T = 0$.*

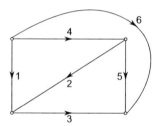

FIGURE 3.37

Solution: For the tree [2, 4, 6] as shown in Figure 3.37(*a*)

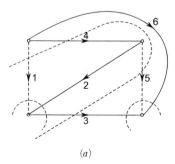

(*a*)

FIGURE 3.37

The fundamental cut-set matrix $[Q_f]$
f-cut-set 1 : (1, 2, 3)
f-cut-set 2 : (3, 5, 6)
f-cut-set 3 : (1, 4, 5, 3)

$$
Q_f = \begin{array}{c} \\ 1 \\ 2 \\ 3 \end{array}
\begin{array}{cccccc}
f\text{-cut-sets} & \multicolumn{3}{c}{\text{Twigs}} & \multicolumn{2}{c}{\text{links}} \\
 & 2\ 6\ 4 & 1 & 5 & 3 \\
\left[\begin{array}{cccccc}
1 & 0 & 0 & 1 & 0 & -1 \\
0 & 1 & 0 & 0 & 1 & 1 \\
0 & 0 & 1 & 1 & -1 & 1
\end{array}\right]
\end{array}
$$

The fundamental tie-set matrix $[B_f]$
Tie-set 1 : (1, 2, 4) Figure 3.37(b)
Tie-set 2 : (4, 5, 6) Figure 3.37(c)
Tie-set 3 : (2, 3, 4, 6) Figure 3.37(d)

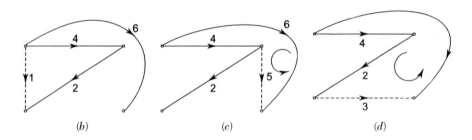

(b)   (c)   (d)

FIGURE 3.37

The f-tie-set matrix or f-loop matrix

Tie-set 1 : (1, 2, 4) Figure 3.37(b)
Tie-set 2 : (4, 5, 6) Figure 3.37(c)
Tie-set 3 : (2, 3, 4) Figure 3.37(d)

$$
B_f = \begin{array}{c} \\ 1 \\ 2 \\ 3 \end{array}
\begin{array}{c}
\text{Tie-sets} \quad \text{Twigs} \quad \text{links} \\
2 \quad 6 \quad 4 \quad 1\ 5\ 3 \\
\left[\begin{array}{cccccc}
-1 & 0 & -1 & 1 & 0 & 0 \\
0 & -1 & 1 & 0 & 1 & 0 \\
1 & -1 & 1 & 0 & 0 & 1
\end{array}\right].
\end{array}
$$

Now

$$Q_f[B_f]^T = \begin{bmatrix} 1 & 0 & 0 & 1 & 0 & -1 \\ 0 & 1 & 0 & 0 & 1 & 1 \\ 0 & 0 & 1 & 1 & -1 & -1 \end{bmatrix} \begin{bmatrix} -1 & 0 & 1 \\ 0 & -1 & -1 \\ -1 & 1 & 1 \\ 1 & 0 & 0 \\ 0 & 1 & 0 \\ 0 & 0 & 1 \end{bmatrix}$$

$$= \begin{bmatrix} 0 & 0 & 0 \\ 0 & 0 & 0 \\ 0 & 0 & 0 \end{bmatrix}$$

$Q_f[B_f]^T = 0.$ **Proved.**

Problem 3.20. *Consider a network as shown in Figure 3.38. Using the loop method of analysis, determine the currents in all the branches, indicating their directions.*

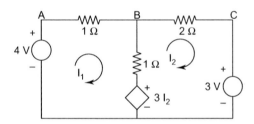

FIGURE 3.38

Solution: The graph of the network is shown in Figure 3.38(a).

$$B_f = \begin{bmatrix} 1 & 0 & -1 \\ 0 & 1 & 1 \end{bmatrix}$$

$$Z_b = \begin{bmatrix} 1 & 0 & 0 \\ 0 & 2 & 0 \\ 0 & 0 & 1 \end{bmatrix}$$

$$Z_L = B_f Z_b B_f'$$

$$= \begin{bmatrix} 1 & 0 & -1 \\ 0 & 1 & 1 \end{bmatrix} \begin{bmatrix} 1 & 0 & 0 \\ 0 & 2 & 0 \\ 0 & 0 & 1 \end{bmatrix} \begin{bmatrix} 1 & 0 \\ 0 & 1 \\ -1 & 1 \end{bmatrix}$$

$$= \begin{bmatrix} 1 & 0 & -1 \\ 0 & 1 & 1 \end{bmatrix} \begin{bmatrix} 1 & 0 \\ 0 & 2 \\ -1 & 1 \end{bmatrix} = \begin{bmatrix} 2 & -1 \\ -1 & 3 \end{bmatrix}$$

$$V_S = \begin{bmatrix} -4 \\ 3 \\ -3I_2 \end{bmatrix}, \quad I_S = \begin{bmatrix} 0 \\ 0 \\ 0 \end{bmatrix}$$

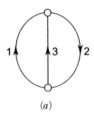

(a)

FIGURE 3.38

Now the loop voltage matrix, $V_L = B_f Z_b I_S - B_f V_S = 0 - B_f V_S$,

$$V_L = - \begin{bmatrix} 1 & 0 & -1 \\ 0 & 1 & 1 \end{bmatrix} \begin{bmatrix} -4 \\ 3 \\ -3I_2 \end{bmatrix} = - \begin{bmatrix} -4 + 3I_2 \\ 3 - 3I_2 \end{bmatrix} = \begin{bmatrix} -4 - 3I_2 \\ 3 + 3I_2 \end{bmatrix}.$$

Now the matrix loop equation, $V_L = Z_L I_L$,

$$\begin{bmatrix} 4 - 3I_2 \\ -3 + 3I_2 \end{bmatrix} = \begin{bmatrix} 2 & -1 \\ -1 & 3 \end{bmatrix} \begin{bmatrix} I_1 \\ I_2 \end{bmatrix}$$

$$4 - 3I_2 = 2I_1 - I_2 \quad \text{or} \quad I_1 + I_2 = 2$$

and

$$-3 + 3I_2 = -I_1 + 3I_2 \quad \text{or} \quad I_1 = 3A.$$

Then

$$I_2 = -1A.$$

Therefore,

the current in Branch 1 = I_1 = 3 A (from A to B).

the current in Branch 2 = I_2 = -1 A (from B to C).

the current in Branch 3 = $I_2 - I_1$ = $-1 - 3 = -4$ A (from 0 to B).

Problem 3.21. *For the network shown in Figure 3.39, draw the network graph, and select 2, 4, 5 as tree branches. Obtain the loop incidence matrix and loop equations.*

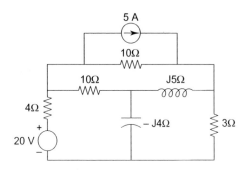

FIGURE 3.39

Solution: The graph of the given network is shown in Figure 3.39(*a*). The voltage source is short circuited and the current source is open circuited, and 2, 4, 5 are twigs. The graph is shown in Figure 3.39(*b*).

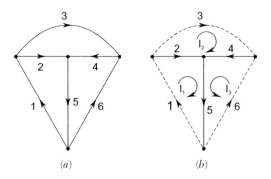

(*a*)   (*b*)

FIGURE 3.39

The f-loop matrix or tie-sets matrix

$$B_f = \begin{bmatrix} 1 & 1 & 0 & 0 & 1 & 0 \\ 0 & -1 & 1 & 1 & 0 & 0 \\ 0 & 0 & 0 & 1 & 1 & 1 \end{bmatrix}.$$

The branch impedance matrix

$$
Z_b = \begin{array}{c} \begin{array}{cccccc} 1 & 2 & 3 & 4 & 5 & 6 \end{array} \\ \begin{bmatrix} 4 & 0 & 0 & 0 & 0 & 0 \\ 0 & 10 & 0 & 0 & 0 & 0 \\ 0 & 0 & 10 & 0 & 0 & 0 \\ 0 & 0 & 0 & J5 & 0 & 0 \\ 0 & 0 & 0 & 0 & -J4 & 0 \\ 0 & 0 & 0 & 0 & 0 & 3 \end{bmatrix} \end{array}.
$$

Now, the loop impedance matrix

$$Z_L = B_f Z_b B_f'$$

$$
= \begin{bmatrix} 1 & 1 & 0 & 0 & 1 & 0 \\ 0 & -1 & 1 & 1 & 0 & 0 \\ 0 & 0 & 0 & 1 & 1 & 1 \end{bmatrix} \begin{bmatrix} 4 & 0 & 0 & 0 & 0 & 0 \\ 0 & 10 & 0 & 0 & 0 & 0 \\ 0 & 0 & 10 & 0 & 0 & 0 \\ 0 & 0 & 0 & J5 & 0 & 0 \\ 0 & 0 & 0 & 0 & -J4 & 0 \\ 0 & 0 & 0 & 0 & 0 & 3 \end{bmatrix} \begin{bmatrix} 1 & 0 & 0 \\ 1 & -1 & 0 \\ 0 & 1 & 0 \\ 0 & 1 & 1 \\ 1 & 0 & 1 \\ 0 & 0 & 1 \end{bmatrix}
$$

$$
= \begin{bmatrix} 4 & 10 & 0 & 0 & -J4 & 0 \\ 0 & -10 & 10 & J5 & 0 & 0 \\ 0 & 0 & 0 & J5 & -J4 & 3 \end{bmatrix} \begin{bmatrix} 1 & 0 & 0 \\ 1 & -1 & 0 \\ 0 & 1 & 0 \\ 0 & 1 & 1 \\ 1 & 0 & 1 \\ 0 & 0 & 1 \end{bmatrix}
$$

$$
= \begin{bmatrix} 14 - J4 & -10 & -J4 \\ -10 & 20 + J5 & J5 \\ -J4 & J5 & 3 + J \end{bmatrix}
$$

$$
V_s = \begin{bmatrix} -20 \\ 0 \\ 0 \\ 0 \\ 0 \\ 0 \end{bmatrix} \quad \text{and} \quad I_s = \begin{bmatrix} 0 \\ 0 \\ 5 \\ 0 \\ 0 \\ 0 \end{bmatrix}.
$$

Hence,

$$V_L = B_f Z_b I_s - B_f V_s$$

$$= \begin{bmatrix} 4 & 10 & 0 & 0 & -J4 & 0 \\ 0 & -10 & 10 & J5 & 0 & 0 \\ 0 & 0 & 0 & J5 & -J4 & 3 \end{bmatrix} \begin{bmatrix} 0 \\ 0 \\ 5 \\ 0 \\ 0 \\ 0 \end{bmatrix}$$

$$- \begin{bmatrix} 1 & 1 & 0 & 0 & 1 & 0 \\ 0 & -1 & 1 & 1 & 0 & 1 \\ 0 & 0 & 0 & 1 & 1 & 1 \end{bmatrix} \begin{bmatrix} -20 \\ 0 \\ 0 \\ 0 \\ 0 \\ 0 \end{bmatrix}.$$

$$= \begin{bmatrix} 0 \\ 50 \\ 0 \end{bmatrix} - \begin{bmatrix} -20 \\ 0 \\ 0 \end{bmatrix} = \begin{bmatrix} 20 \\ 50 \\ 0 \end{bmatrix}$$

Therefore, the matrix loop equation can be written as

$$\begin{bmatrix} 20 \\ 50 \\ 0 \end{bmatrix} = \begin{bmatrix} 14 - J4 & -10 & -J4 \\ -10 & 20 + J5 & J5 \\ -J4 & J5 & 3 + J \end{bmatrix} \begin{bmatrix} I_1 \\ I_2 \\ I_3 \end{bmatrix}.$$

Problem 3.22. *For the graph shown in Figure 3.40,*

(i) Write down the augmented incidence matrix.
(ii) Indicate the number of fundamental loops.

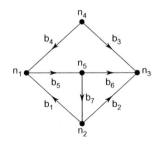

FIGURE 3.40

Solution: (i) The augmented or complete incidence matrix

$$
A_c = \begin{array}{c} \\ n_1 \\ n_2 \\ n_3 \\ n_4 \\ n_5 \end{array}
\begin{array}{c}
\begin{array}{ccccccc} b_1 & b_2 & b_3 & b_4 & b_5 & b_6 & b_7 \end{array} \\
\left[\begin{array}{ccccccc}
-1 & 0 & 0 & +1 & +1 & 0 & 0 \\
+1 & +1 & 0 & 0 & 0 & 0 & -1 \\
0 & -1 & -1 & 0 & 0 & -1 & 0 \\
0 & 0 & +1 & -1 & 0 & 0 & 0 \\
0 & 0 & 0 & 0 & -1 & +1 & +1
\end{array} \right]
\end{array}.
$$

(ii) As per Figure 3.40, the number of nodes $= 5$ and the number of branches $= 7$, therefore, the number of fundamental loops (tie-sets) = the number of links

$$l = b - (n - 1)$$
$$= 7 - (5 - 1) = 3.$$

Problem 3.23. *For the network shown in Figure 3.41, draw the graph, write the tie-set schedule and hence obtain the equilibrium equations on the loop basis. Calculate the values of branch currents and branch voltages.*

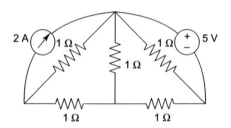

FIGURE 3.41

Solution: The graph of the network is shown in Figure 3.41(a), where the current source is open circuited and the voltage source is short circuited.

Let the twigs be 4, 5, and 6 and the links be 1, 2, 3 as shown in Figure 3.41(b).

The tie-set matrix be per Figure 3.41(b) is

$$
B_f = \begin{array}{c} \\ 1 \\ 2 \\ 3 \end{array}
\begin{array}{c}
\begin{array}{cccccc} 1 & 2 & 3 & 4 & 5 & 6 \end{array} \\
\left[\begin{array}{cccccc}
1 & 0 & 0 & -1 & 1 & -1 \\
0 & 1 & 0 & -1 & 1 & 0 \\
0 & 0 & 1 & 1 & 0 & 0
\end{array} \right]
\end{array}.
$$

The branch impedance matrix is given as

$$
Z_b = \begin{array}{c} \begin{array}{cccccc} 1 & 2 & 3 & 4 & 5 & 6 \end{array} \\ \begin{bmatrix} 1 & 0 & 0 & 0 & 0 & 0 \\ 0 & 1 & 0 & 0 & 0 & 0 \\ 0 & 0 & 1 & 0 & 0 & 0 \\ 0 & 0 & 0 & 0 & 0 & 0 \\ 0 & 0 & 0 & 0 & 1 & 0 \\ 0 & 0 & 0 & 0 & 0 & 1 \end{bmatrix} \end{array}
$$

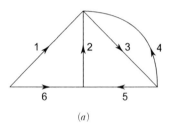

(a)

FIGURE 3.41

$$
V_s = \begin{bmatrix} 0 \\ 0 \\ 0 \\ -5 \\ 0 \\ 0 \end{bmatrix}, \quad I_s = \begin{bmatrix} 2 \\ 0 \\ 0 \\ 0 \\ 0 \\ 0 \end{bmatrix}.
$$

Now, the loop impedance matrix is given as

$$
Z_L = B_f Z_b B_f' = \begin{bmatrix} 1 & 0 & 0 & 0 & 1 & -1 \\ 0 & 1 & 0 & 0 & 1 & 0 \\ 0 & 0 & 1 & 0 & 0 & 0 \end{bmatrix} \begin{bmatrix} 1 & 0 & 0 \\ 0 & 1 & 0 \\ 0 & 0 & 1 \\ -1 & -1 & 1 \\ 1 & 1 & 0 \\ -1 & 0 & 0 \end{bmatrix}
$$

$$
= \begin{bmatrix} 3 & 1 & 0 \\ 1 & 2 & 0 \\ 0 & 0 & 1 \end{bmatrix}
$$

$$V_L = B_f Z_b I_s - B_f V_s$$

$$= \begin{bmatrix} 1 & 0 & 0 & 0 & 1 & -1 \\ 0 & 1 & 0 & 0 & 1 & 0 \\ 0 & 0 & 1 & 0 & 0 & 0 \end{bmatrix} \begin{bmatrix} 2 \\ 0 \\ 0 \\ 0 \\ 0 \\ 0 \end{bmatrix} - \begin{bmatrix} 1 & 0 & 0 & -1 & 1 & -1 \\ 0 & 1 & 0 & -1 & 1 & 0 \\ 0 & 0 & 1 & 1 & 0 & 0 \end{bmatrix} \begin{bmatrix} 0 \\ 0 \\ 0 \\ -5 \\ 0 \\ 0 \end{bmatrix}$$

$$= \begin{bmatrix} -3 \\ -5 \\ 5 \end{bmatrix}.$$

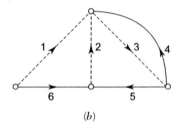

(b)

FIGURE 3.41

The equilibrium equation on the loop basis can be written as

$$V_L = Z_L I_L$$

$$\begin{bmatrix} -3 \\ -5 \\ 5 \end{bmatrix} = \begin{bmatrix} 3 & 1 & 0 \\ 1 & 2 & 0 \\ 0 & 0 & 1 \end{bmatrix} \begin{bmatrix} I_{l_1} \\ I_{l_2} \\ I_{l_3} \end{bmatrix}$$

or

$$I_{l_1} = \frac{\Delta_1}{\Delta}; \quad I_{l_2} = \frac{\Delta_2}{\Delta} \quad \text{and} \quad I_{l_3} = \frac{\Delta_3}{\Delta},$$

where

$$\Delta = |Zl| = 3(2-0) - 1(1-0) + 0 = 5$$
$$\Delta_1 = -3(2-0) - 1(-5-0) + 0 = -1$$
$$\Delta_2 = 3(-5-0) + 3(1-0) + 0 = -12$$
$$\Delta_3 = 3(10-0) - 1(5-0) - 3 \times 0 = 25.$$

Now,

$$I_{l_1} = -\frac{1}{5} \text{ A}, \quad I_{l_2} = -\frac{12}{5} \text{ A} \quad \text{and} \quad I_{l_3} = \frac{25}{5} = 5 \text{ A}.$$

Now, the branch currents can be written as

$$[I_b] = [B_f]'[I_l] - [I_s]$$

$$\begin{bmatrix} i_1 \\ i_2 \\ i_3 \\ i_4 \\ i_5 \\ i_6 \end{bmatrix} = \begin{bmatrix} 1 & 0 & 0 \\ 0 & 1 & 0 \\ 0 & 0 & 1 \\ -1 & -1 & 1 \\ 1 & 1 & 0 \\ -1 & 0 & 0 \end{bmatrix} \begin{bmatrix} -\frac{1}{5} \\ -\frac{12}{5} \\ 5 \end{bmatrix} - \begin{bmatrix} 2 \\ 0 \\ 0 \\ 0 \\ 0 \\ 0 \end{bmatrix} = \begin{bmatrix} -\frac{1}{5} \\ -\frac{12}{5} \\ 5 \\ \frac{38}{5} \\ -\frac{13}{5} \\ \frac{1}{5} \end{bmatrix} - \begin{bmatrix} 2 \\ 0 \\ 0 \\ 0 \\ 0 \\ 0 \end{bmatrix} = \begin{bmatrix} -\frac{11}{5} \\ -\frac{12}{5} \\ 5 \\ \frac{38}{5} \\ -\frac{13}{5} \\ \frac{1}{5} \end{bmatrix}$$

The branch voltage can be written as

$$[V_b] = [Z_b][I_b] + [V_s]$$

$$= \begin{bmatrix} -\frac{11}{5} \\ -\frac{12}{5} \\ 5 \\ 0 \\ -\frac{13}{5} \\ \frac{1}{5} \end{bmatrix} + \begin{bmatrix} 0 \\ 0 \\ 0 \\ -5 \\ 0 \\ 0 \end{bmatrix} = \begin{bmatrix} -\frac{11}{5} \\ -\frac{12}{5} \\ 5 \\ -5 \\ -\frac{13}{5} \\ \frac{1}{5} \end{bmatrix}.$$

QUESTIONS FOR DISCUSSION

1. Define tree, cotree, branch, and node.
2. What are the differences between a tree and cotree?
3. What is a twig?
4. What is a chord or link?
5. Define mesh and loop.
6. Define planar and non-planar graphs.
7. What is the difference between a cut-set and fundamental cut-set?
8. Define walk and path.
9. What is topology?
10. What is mesh analysis?
11. What is the difference between walk and path?
12. What is a connected graph?
13. What is a linear graph?
14. Explain the procedure of f-cut-set matrix.
15. What is the difference between an incidence matrix and a reduced incidence matrix?

OBJECTIVE QUESTIONS

1.   The number of branches in a cotree is equal to
     (a) B − N + 1   (b) B − N − 1   (c) B + N − 1   (d) B + N + 1.
2.   Figure 3.42 shows a resistive network and its graph is drawn aside. A proper tree chosen for analyzing the network will contain the edges,
     (a) ab, bc, ad   (b) ab, bc, ca   (c) ab, bd, cd   (d) ac, bd, ad.

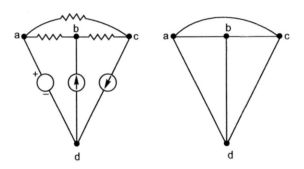

FIGURE 3.42

3.   A connected planar network has 4 nodes and 5 elements. The number of meshes in its dual network is
     (a) 4   (b) 3   (c) 2   (d) 1.
4.   The connection of elements between two terminals is called
     (a) Branch   (b) Loop   (c) Mesh   (d) Node.
5.   The connection of branches forming a common point is
     (a) Joint branch   (b) Loop   (c) Mesh   (d) Node.
6.   A set of branches forming a closed path which passes through a node once only is
     (a) Branch   (b) Node   (c) Loop   (d) Mesh.
7.   A loop which does not contain any other loop within it is a
     (a) Independent loop   (b) Open loop   (c) Closed loop   (d) Mesh.
8.   If a subgraph of (G) does not contain the entire graph (G) then it is a ____ subgraph of G
     (a) Proper   (b) Improper   (c) Spanning   (d) Complementary.
9.   A connected graph has n vertices, and the number of branches in its tree will be
     (a) $\frac{n}{2}$ (b) $(n-1)$ (c) $(n+1)$ (d) $\frac{1}{2}(n-1)$.
10. The order of a complete incidence matrix will be
     (a) $n \times b$ (b) $(n-1) \times b$ (c) $(b-n+1) \times b$ (d) $b \times n$.

11. The number of nodes in a complete incidence matrix is n, and the number of nodes in a reduced incidence matrix will be
    $(a)\, n$ $(b)\, (n+1)$ $(c)\, n-1$ $(d)\, b-n+1$.
12. If A is the reduced incidence matrix and A' is its transpose, then the number of possible trees will be
    $(a)\, \det(A)$ $(b)\, \det(AA')$ $(c)\, \det(A')$ $(d)\, A \times A'$.

UNSOLVED PROBLEMS

1. The reduced incidence matrix of a linear graph is given below:

$$A = \begin{bmatrix} 0 & 0 & 1 & 1 & 1 & 0 & -1 \\ 0 & 1 & 0 & 0 & -1 & 1 & 1 \\ -1 & 0 & -1 & 0 & 0 & -1 & 0 \end{bmatrix}.$$

(a) Draw the graph. The branches $(2, 3, 4)$ constitute a tree. Determine the f-cut-set matrix (b). Determine the number of trees in the graph.

2. In the graphs of Figure 3.37(a), (b) determine the number of trees.

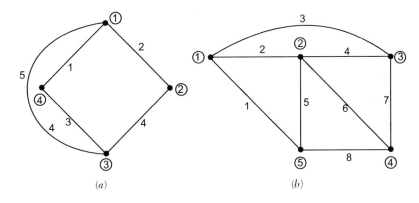

(a) (b)

FIGURE 3.43

3. Differentiate between the following terms related to network graph theory:
    (i) Tree and cotree
    (ii) Planar graph and non-planar graph
    (iii) f-cut-set and f-tie-set
    (iv) Graph and oriented graph
    (v) Link and twig.

4. Draw a graph, select a suitable tree, and obtain the tie-set matrix and f-cut-set matrix.

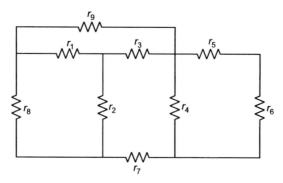

FIGURE 3.44

5. Draw an incidence matrix, select a suitable tree, and obtain the tie-set matrix and f-cut-set matrix.

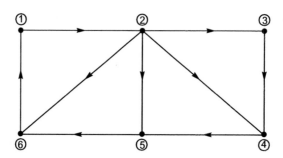

FIGURE 3.45

6. Determine the f-cut-set matrix of the given graph as shown in Figure 3.46. Assume a tree [1, 2, 3, and 6].

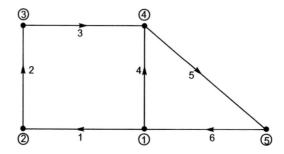

FIGURE 3.46

7. Determine the tie-set matrix of a given graph as shown in Figure 3.47.

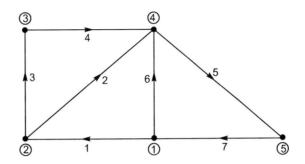

FIGURE 3.47

8. Determine the matrix loop equation for the network as shown in Figure 3.48 using loop analysis.

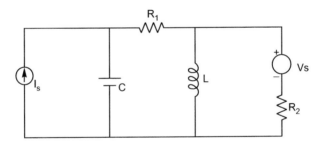

FIGURE 3.48

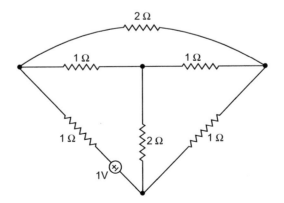

FIGURE 3.49

9. Draw the graph of the network shown in Figure 3.48. How many trees are possible for this graph? Draw the trees.

10. For the network shown in Figure 3.49, calculate the branch voltages and branch currents using nodal analysis.

11. For the given network, draw the graph and a tree. Select suitable tree-branch voltages and write the cut-set schedule. Write equations for the branch voltages in terms of tree branch voltages. All resistance are 1 ohms.

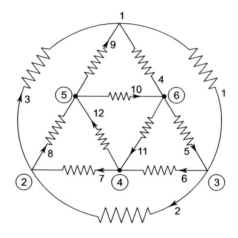

FIGURE 3.50

12. For the given resistive network, write a tie-set schedule and obtain equilibrium equations on the current basis. Solve these equations and hence calculate values of branch currents and branch voltages.

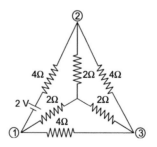

FIGURE 3.51

Chapter 4

THE LAPLACE TRANSFORM

4.1 INTRODUCTION

The Laplace Transform is a mathematical tool by which time domain is transformed into frequency domain or frequency domain to time domain.

In order to transform a given function of time $f(t)$ into its corresponding Laplace transform first multiply $f(t)$ by e^{-st}, where s is a complex number ($s = \sigma + J\omega$). Integrate this product with respect to time with a limit from zero to infinity. This integration results in a Laplace transform of $f(t)$ and this is denoted $F(s)$ or $\mathcal{L}f(t)$.

s is a complex frequency ($\sigma + J\omega$).

σ is known as a Neper frequency and ω is known as a real or natural frequency.

$$\mathcal{L}f(t) = F(s) = \int_0^\infty e^{-st} F(t) dt. \tag{4.1}$$

The term "Laplace transform of $f(t)$" is used for the letter $\mathcal{L}f(t)$.

Similarly,

$$\mathcal{L}^{-1}F(s) = f(t).$$

The term "$\mathcal{L}^{-1}$" is called the inverse Laplace transformation.

The time function $f(t)$ and its Laplace transform $F(s)$ are a transform pair.

4.2 LAPLACE TRANSFORM OF SOME COMMONLY USED FUNCTIONS

(1) The Laplace transform of e^{-at} is

$$\mathcal{L}e^{-at} = \int_0^\infty e^{-st} e^{-at} dt$$

$$= \int_0^\infty e^{-(s+a)t} dt$$

$$= -\left(\frac{e^{-(s+a)t}}{(s+a)}\right)\Bigg|_0^\infty = -\left[0 - \frac{1}{s+a}\right].$$

$$\mathcal{L}[e^{-at}] = \left(\frac{1}{s+a}\right) \tag{4.2}$$

The inverse Laplace transform of $\left(\frac{1}{s+a}\right)$ is e^{-at}

$$\mathcal{L}^{-1}\left(\frac{1}{s+a}\right) = e^{-at}. \tag{4.3}$$

(2) Similarly,

$$\mathcal{L}[e^{at}]. \tag{4.4}$$

Put $a = -1$ into Equation (4.2) and find

$$\mathcal{L}[e^{at}] = \frac{1}{s+a}.$$

(3) $f(t) = 1 = e^{0t}$.
Put $a = 0$ into Equation (4.2) and

$$\mathcal{L}[1] = \frac{1}{s+a} = \frac{1}{s}.$$

(4) In the function $f(t) = e^{-at}$ put $a = J\omega$.
Hence, $f(t) = e^{-J\omega t}$.
By Equation (4.2),

$$\mathcal{L}[e^{-J\omega t}] = \frac{1}{s+J\omega} = \frac{1}{s+J\omega} \times \frac{s-J\omega}{s-J\omega} = \frac{s}{s^2+\omega^2} - J\frac{\omega}{s^2+\omega^2}$$

and we know $e^{-J\omega t} = \cos \omega t - J \sin \omega t$.

$$\mathcal{L}[\cos \omega t - J \sin \omega t] = \frac{s}{s^2+\omega^2} - J\frac{\omega}{s^2+\omega^2}. \tag{4.5}$$

Separating into real and imaginary parts

$$\mathcal{L}[\cos \omega t] = \frac{s}{s^2+\omega^2} \Rightarrow \mathcal{L}^{-1}\left[\frac{s}{s^2+\omega^2}\right] = \cos \omega t.$$

$$\mathcal{L}[\sin \omega t] = \frac{\omega}{s^2+\omega^2} \Rightarrow \mathcal{L}^{-1}\left[\frac{\omega}{s^2+\omega^2}\right] = \sin \omega t.$$

(5) Now put $a = -\alpha + J\omega$ into Equation (4.2),

$$\mathcal{L}[e^{-(-\alpha+J\omega)t}] = \frac{1}{s + (-\alpha + J\omega)}$$

$$\mathcal{L}[e^{\alpha t}e^{-J\omega t}] = \frac{1}{(s-\alpha) + J\omega} \times \frac{s-\alpha-J\omega}{s-\alpha-J\omega}$$

$$\mathcal{L}[e^{\alpha t}[\cos\omega t - J\sin\omega t]] = \frac{s-\alpha}{(s-\alpha)^2 + \omega^2} - J\frac{\omega}{(s-\alpha)^2 + \omega^2}. \qquad (4.6)$$

Separating into real and imaginary parts

$$\mathcal{L}[e^{\alpha t}\cos\omega t] = \frac{s-\alpha}{(s-\alpha)^2 + \omega^2}$$

$$\mathcal{L}[e^{\alpha t}\sin\omega t] = \frac{\omega}{(s-\alpha)^2 + \omega^2}.$$

Similarly, if a is negative

$$\mathcal{L}[e^{-\alpha t}\cos\omega t] = \frac{s+\alpha}{(s+\alpha)^2 + \omega^2} \text{ and}$$

$$\mathcal{L}[e^{-\alpha t}\sin\omega t] = \frac{\omega}{(s+\alpha)^2 + \omega^2}.$$

(6) Put $a = 1$ into Equation (4.2),

$$\mathcal{L}[e^{-t}] = \frac{1}{s+1}.$$

Example 4.1. *Find the Laplace transform of the following functions:*

(a) $t^2 e^{at}$ (b) $e^{-n\alpha t}\sin\beta t$ (c) $e^{-\alpha t}\sinh t$.

Ans.

(a) $f(t) = t^2 e^{at}$

$$\mathcal{L}[f(t)] = \frac{\angle 2}{(s-a)^3} = \frac{2}{(s-a)^3}$$

(b) $f(t) = e^{-n\alpha t}\sin\beta t$

$$\mathcal{L}[f(t)] = \frac{\beta}{(s+n\alpha)^2 + \beta^2}$$

(c) $f(t) = e^{-\alpha t}\sinh t$

$$\mathcal{L}[f(t)] = \frac{1}{(s+\alpha)^2 - 1} = \frac{1}{s^2 + \alpha^2 + 2s\alpha - 1}$$

TABLE 4.1   Laplace Transform Pairs

S. No.	$f(t) = L^{-1}F(s)$	$F(s) = L[f(t)]$
1.	$\delta(t)$ unit impulse at $t = 0$	1
2.	$u(t)$ unit step at $t = 0$	$\dfrac{1}{s}$
3.	$u(t - T)$ shifted unit step	$\dfrac{1}{s}e^{-sT}$
4.	t unit ramp	$\dfrac{1}{s^2}$
5.	$r(t - 1) = (t - 1)u(t - 1)$ shifted ramp	$\dfrac{1}{s^2}e^{-s}$
6.	t^2 unit parabolic	$\dfrac{\angle 2}{s^3}$
7.	t^n	$\dfrac{\angle n}{s^{n+1}}$
8.	e^{-at}	$\dfrac{1}{s + a}$
9.	e^{at}	$\dfrac{1}{s - a}$
10.	te^{-at}	$\dfrac{1}{(s + a)^2}$
11.	te^{at}	$\dfrac{1}{(s - a)^2}$
12.	$t^n e^{at}$	$\dfrac{\angle n}{(s - a)^{n+1}}$
13.	$t^n e^{-at}$	$\dfrac{\angle n}{(s + a)^{n+1}}$
14.	$\sin \omega t$	$\dfrac{\omega}{s^2 + \omega^2}$
15.	$\cos \omega t$	$\dfrac{s}{s^2 + \omega^2}$
16.	$e^{-\alpha t} \sin \omega t$	$\dfrac{\omega}{(s + \alpha)^2 + \omega^2}$
17.	$e^{-\alpha t} \cos \omega t$	$\dfrac{s + \alpha}{(s + \alpha)^2 + \omega^2}$
18.	$e^{\alpha t} \sin \omega t$	$\dfrac{\omega}{(s - \alpha)^2 + \omega^2}$
19.	$e^{\alpha t} \cos \omega t$	$\dfrac{s - \alpha}{(s - \alpha)^2 + \omega^2}$
20.	$\sinh \alpha t \left[\dfrac{e^{\alpha t} - e^{-\alpha t}}{2} \right]$	$\dfrac{\alpha}{s^2 - \alpha^2}$
21.	$\cosh \alpha t \left[\dfrac{e^{\alpha t} + e^{-\alpha t}}{2} \right]$	$\dfrac{s}{s^2 - \alpha^2}$

TABLE 4.1   Continued

S. No.	$f(t) = L^{-1}F(s)$	$F(s) = L[f(t)]$
22.	$e^{\pm\beta t} \sinh \alpha t$	$\dfrac{\alpha}{(s \pm \beta^2) + \alpha^2}$
23.	$e^{\pm\beta t} \cosh \alpha t$	$\dfrac{\alpha}{(s \pm \beta^2) + \alpha^2}$
24.	$t \sin \omega t$	$\dfrac{2\omega s}{(s^2 + \omega^2)^2}$
25.	$t \cos \omega t$	$\dfrac{s^2 - \omega^s}{(s^2 + \omega^2)^2}$

4.3   BASIC LAPLACE TRANSFORM THEOREM

The basic theorems of the Laplace transform are given below:

1. The Laplace transform of a linear combination (linearity)

$$\mathcal{L}[\alpha f_1(t) \pm \beta f_2(t)] = \alpha F_1(s) \pm \beta F_2(s), \tag{4.7}$$

where $f_1(t)$ and $f_2(t)$ are functions of time and α, β are constant.

Example 4.2. *Find the Laplace transform of* $f(t) = 2t + 5e^t \cos t$.

Solution: In this case $f_1(t) = t$ and $f_2(t) = e^t \cos t$ and $\alpha = 2$, $\beta = 5$. Applying the linearity theorem,

$$\mathcal{L}[F(t)] = F(s) = \frac{2}{s} + \frac{5(s-1)}{(s-1)^2 + 1}.$$

2. The Laplace transform of differentiation with respect to 't'

(i) $\mathcal{L}\dfrac{d}{dt}f(t) = [sF(s) - F(0+)]$ \hfill (4.8) $[\mathcal{L}F(t) = F(s)]$

(ii) $\mathcal{L}\dfrac{d^2 f(t)}{dt^2} = [s^2 F(s) - sF(0+) - f^1(0+)]$ \hfill (4.9)

(iii) $\mathcal{L}\dfrac{d^3 f(t)}{dt^3} = s^3 F(s) - s^2 f(0+) - sf'(0+) - f''(0+),$ \hfill (4.10)

where $f(0+), f'(0+), f''(0+)$ are the values of $f(t), \frac{df(t)}{dt}, \frac{d^2 f(t)}{dt^2}$ at $t = (0+)$.

Example 4.3. $\mathcal{L}[F(t)] = (\frac{2as}{s^2+a^2})$. *Assume all initial conditions are zero then find* $\mathcal{L}[\frac{d^2f(t)}{dt^2}]$ *and* $\mathcal{L}[\frac{df(t)}{dt}]$.

Ans.

(i) $\mathcal{L}\left[\dfrac{df(t)}{dt}\right] = sF(s) - F(0) = s \times \left(\dfrac{2as}{s^2+a^2}\right) - 0 = \dfrac{2as}{s^2+a^2}$

(ii) $\mathcal{L}\left[\dfrac{d^2f(t)}{dt^2}\right] = [s^2F(s) - sF(0+) - f'(0+)]$

$$= s^2 \times \frac{2as}{(s^2+a^2)} = \frac{2as^3}{(s^2+a^2)}$$

Given all initial conditions are zero $[f(0+) = f'(0+) = 0]$.

3. The Laplace transform integration

(i) $\mathcal{L}\displaystyle\int f(t)dt = \left[\dfrac{F(s)}{s} + \dfrac{f^{-1}(0+)}{s}\right]$ \hfill (4.11)

(ii) $\mathcal{L}\displaystyle\iint f(t)df \cdot dt = \left[\dfrac{F(s)}{s^2} + \dfrac{f^{-1}(0+)}{s^2} + \dfrac{f^{-2}(0+)}{s}\right]$ \hfill (4.12)

(iii) $\mathcal{L} = \displaystyle\iiint f(t)dt \cdot df \cdot dt$

$$= \left[\frac{F(s)}{s^3} + \frac{f^{-1}(0+)}{s^3} + \frac{f^{-2}(0+)}{s^2} + \frac{f^{-3}(0+)}{s}\right] \hspace{1cm} (4.13)$$

Given

$$\mathcal{L}[f(t)] = \frac{2as}{s^2+a^2}, \text{ find } \mathcal{L}\left[\int f(t)dt\right] \text{ and } \mathcal{L}\left[\iint f(t)dt \cdot dt\right].$$

Example 4.4. *Assume all initial conditions are zero.*

(i) $\mathcal{L}\left[\displaystyle\int f(t)dt\right] = \dfrac{F(s)}{s} + \dfrac{F^{-1}(0+)}{s} = \dfrac{2a}{s^2+a^2}[f - 1(0+) = 0]$

(ii) $\mathcal{L}\left[\displaystyle\iint f(t)dt \cdot dt\right] = \dfrac{F(s)}{s^2} + \dfrac{F^{-1}(0+)}{s^2} + \dfrac{F^{-2}(0+)}{s}$

$$= \frac{2as}{s^2+a^2} \times \frac{1}{s^2} + 0 + 0 = \frac{2a}{s(s^2+a^2)}$$

4. If the Laplace transform of $f(t)$ is F(s), then

$$\mathcal{L}[e^{\mp at}f(t)] = \text{F}(s \pm a). \tag{4.14}$$

Example 4.5. $\mathcal{L}[f(t)] = \frac{2as}{s^2+a^2} = f(s)$. *Find the Laplace transform of*

$$\text{F}(t) = e^{-2t}f(t).$$

Ans.

$$\mathcal{L}[e^{-2t}f(t)] = f(s+2) = \frac{2a(s+2)}{(s+2)^2+a^2}$$

$$= \frac{2as+4a}{(s+2)^2+a^2}$$

5. If the Laplace transform of $f(t)$ is $f(s)$, then

$$\mathcal{L}t\text{F}(t) = -\frac{d}{ds}\text{F}(s)\left[\text{L}t^n f(t) = (-1)^n \frac{d^n}{ds^n}\text{F}(s)\right]. \tag{4.15}$$

Example 4.6. $\mathcal{L}[\text{F}(t)] = \frac{2as}{s^2+a^2} = \text{F}(s)$. *Find the Laplace transform of* $[t\text{F}(t)]$.

Ans.

$$\mathcal{L}[t\text{F}(t)] = -\frac{d}{ds}\text{F}(s) = -\frac{d}{ds}\left[\frac{2as}{s^2+a^2}\right]$$

$$= -\frac{d}{ds}[2as(s^2+a^2)^{-1}]$$

$$= -2a[(s^2+a^2)^{-1} + s \times -1 \times (s^2+a^2)^{-2} \times 2s]$$

$$= \frac{2a(s^2-a^2)}{(s^2+a^2)^2}$$

4.4 LAPLACE TRANSFORM OF SOME OTHER FUNCTIONS

The Laplace transform of periodic function.
   For periodic function,

$$f(t) = f(t+\text{T}) \text{ for } t > 0,$$

where T is time period.

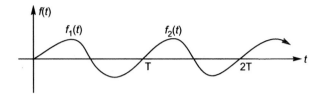

FIGURE 4.1

Let $f_1(t), f_2(t) \ldots$ be the functions describing the first cycle, second cycle $\ldots$ let $F_1(s), F_2(s) \ldots$ be their Laplace transforms, respectively, then

$$f(t) = f_1(t) + f_2(t) + f_3(t) + \ldots$$
$$f(t) = f_1(t)u(t) + f_1(t-T)u(t-T) + f_1(t-2T)u(t-2T) + \ldots \quad (4.16)$$

Taking the Laplace transform

$$f(s) = F_1(s) + F_1(s)e^{-Ts} + F_1(s)e^{-2Ts} + \ldots$$
$$F(s) = F_1(s)[1 + e^{-Ts} + e^{-2Ts} + \cdots]$$
$$F(s) = F_1(s)\frac{1}{1 - e^{-Ts}}F_1(s). \quad (4.17)$$

The equation shows that the Laplace transform of a periodic function with time period T is equal to $\frac{1}{1-e^{-Ts}}$ times the Laplace transform of the first cycle.

Example 4.7. *Find the Laplace transform of a train of pulses of width a, amplitude A, and periodic time T as shown in Figure 4.2.*

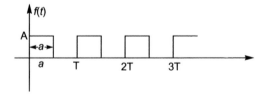

FIGURE 4.2

Solution: $F_1(t) = A[u(t) - u(t-a)]$
Taking the Laplace transform,

$$F_1(s) = A\left[\frac{1}{s} - \frac{e^{-as}}{s}\right].$$

For the periodic wave,

$$F(s) = \frac{F_1(s)}{1 - e^{-Ts}} = \frac{A}{s} \frac{1 - e^{-as}}{(1 - e^{-Ts})}.$$

4.5   INITIAL VALUE THEOREM

It is clear by the name that the value of the function at the beginning is known as the initial value of the function. So we put $t = 0$ as into Equation (4.18) and we know time and frequency are inverse to each other

$$f(0) = \lim_{t \to 0} f(t) = \lim_{s \to \infty} sF(s), \tag{4.18}$$

where $\mathcal{L}[F(t)] = F(s)$.

4.6   FINAL VALUE THEOREM

$$f(\infty) = \lim_{t \to \infty} f(t) = \lim_{s \to 0} sF(s) \tag{4.19}$$

The initial value theorem is applicable everywhere but the final value theorem is not applicable everywhere. If the denominator of $sF(s)$ has any root having a real part as zero or positive, then the final value theorem is not valid.

Example 4.8. *Find the initial and final value of* $F(s) = \frac{(s+a)}{s(s+b)}$.

Solution: Applying the initial value theorem,

$$f(0) = \lim_{t \to 0} f(t) = \lim_{s \to \infty} sF(s) = \lim_{s \to \infty} s \times \frac{s+a}{s(s+b)} = 1.$$

By the final value theorem,

$$f(\infty) = \lim_{t \to \infty} f(t) = \lim_{s \to 0} sF(s) = \lim_{s \to 0} \left(\frac{s+a}{s+b} \right) = \frac{a}{b}.$$

4.7   INVERSE LAPLACE TRANSFORMATION

This can be done with the help of the partial fraction technique and convolution integral.

4.7.1 Partial Fraction Method

Consider the network function

$$F(s) = \frac{a_m s^m + a_{m-1}s^{m-1} + \cdots + a_1 s + a_0}{b_n s^n + b_{n-1}s^{n-1} + \cdots + b_1 s + b_0} = \frac{N(s)}{D(s)}$$
$$= \frac{K(s - z_1)(s - z_2)\cdots(s - z_m)}{(s - P_1)(s - pa_2)\cdots(s - P_n)},$$

where N(s) is the numerator polynomial and D(s) is the denominator polynomial. In this case F(s) is a proper fraction if F(s) is not a proper function. We shall divide N(s) by D(s) until the degree of the remainder numerator is one less than the denominator.

$$F(s) = Q(s) + \frac{N_1(s)}{D(s)},$$

where Q(s) is the quotient and $N_1(s)/D(s)$ is a proper fraction and $N_1(s)$ is of degree $(n - 1)$.

Case 1: If the roots are simple and real,
we write,

$$F(s) = \frac{A}{s - P_1} + \frac{B}{s - P_2} + \frac{C}{s - P_3} + \cdots,$$

where A, B, C, ... are residues at $P_1, P_2, P_3, \ldots$ respectively

$$A = [(s - P_1)F(s)]s = P_1$$
$$B = [(s - P_2)F(s)]s = P_2$$
$$C = [(s - P_3)F(s)]s = P_3$$

Example 4.9. *Find the Laplace inverse of* $F(s) = \frac{s^2+3s+1}{(s+1)(s+2)(s+3)}$.

Ans.

$$F(s) = \frac{s^2 + 3s + 1}{(s + 1)(s + 2)(s + 3)} = \frac{A}{(s + 1)} + \frac{B}{(s + 2)} + \frac{C}{(s + 3)}$$

$$A = [(s + 1) \cdot F(s)]_{s=-1} = \left[\frac{s^2 + 3s + 1}{(s + 2)(s + 3)}\right]_{s=-1} = -\frac{1}{2}$$

$$B = [(s + 2)F(s)]_{s=-2} = \left[\frac{s^2 + 3s + 1}{(s + 1)(s + 3)}\right]_{s=-2}$$

$$= \frac{4 - 6 + 1}{-1 \times 1} = 1$$

$$C = [(s + 3)F(s)]_{s=-3} = \left[\frac{s^2 + 3s + 1}{(s + 1)(s + 2)} \right]_{s=-3}$$

$$= \frac{9 - 9 + 1}{-2 \times -1} = \frac{1}{2}$$

Now

$$F(s) = -\frac{1}{2(s + 1)} + \frac{1}{(s + 2)} + \frac{1}{2(s + 3)}.$$

Now taking the Laplace inverse

$$F(t) = -\frac{1}{2}e^{-t} + e^{-2t} + \frac{1}{2}e^{-3t}.$$

Case 2: If the roots are repeated

$$F(s) = \frac{N(s)}{(s - p)^3} = \frac{A}{(s - p)} + \frac{B}{(s - p)^2} + \frac{C}{(s - p)^3}$$

$$C = \frac{1}{\lfloor 0} [(s - p)^3 F(s)]_{s=p}$$

$$B = \frac{1}{\lfloor 1} \left[\frac{d}{ds} \{(s - p)^3 F(s)\} \right]_{s=p}$$

$$A = \frac{1}{\lfloor 2} \left[\frac{d^2}{ds^2} \{(s - p)^3 F(s)\} \right]_{s=p}.$$

Example 4.10. *Find the Laplace inverse of* $F(s) = \frac{2s+1}{(s+1)^3}$.

Ans.

$$F(s) = \frac{2s + 1}{(s + 1)^3} = \frac{A}{(s + 1)} + \frac{B}{(s + 1)^2} + \frac{C}{(s + 1)^3}$$

$$C = [(s + 1)^3 F(s)]_{s=-1} = [2s + 1]_{s=-1} = -1$$

$$B = \left[\frac{d}{ds}(s + 1)^3 F(s) \right]_{s=-1} = \left[\frac{d}{ds}(2s + 1) \right]_{s=-1} = 2$$

$$A = \frac{1}{\angle 2} \left[\frac{d^2}{ds^2} [(s + 1)^3 F(s)] \right] = \frac{1}{2} \left[\frac{d^2}{ds^2}(2s + 1) \right] = 0$$

$$F(s) = \frac{2}{(s + 1)^2} - \frac{1}{(s + 1)^3}.$$

Taking the Laplace inverse

$$F(t) = 2te^{-t} - \frac{1}{2}t^2e^{-t}.$$

NOTE ▶ Students can solve this problem and calculate A, B, and C using the comparison of order method.

4.7.2 Convolution Integral

If $f_1(t)$ and $f_2(t)$ are two functions of time which are zero for $t < 0$ and if $\mathcal{L}[f_1(t)] = F_1(s)$ and

$$\mathcal{L}[f_2(t)] = f_2(s),$$

the convolution of $f_1(t)$ and $f_2(t)$ is denoted by a special notation [star between $f_1(t)$ and $f_2(t)$] as

$$f_1(t)^*f_2(t) = \int_0^t f_1(\tau)f_2(t - \tau)d\tau$$

$$= \int_0^t f_1(t - \tau)f_2(\tau)d\tau$$

$$= f_2(t)^*f_1(t),$$

where τ is a dummy variable for t.

Hence, the convolution integral may be given as

$$\mathcal{L}[f_1(t)^*f_2(t)] = \mathcal{L}\left[\int_0^t f_1(\tau)f_2(t - \tau)d\tau\right]$$

$$= \mathcal{L}\left[\int_0^t f_1(t - \tau)f_2(\tau)d\tau\right]$$

$$= F_1(s) \cdot F_2(s) = F_2(s) \cdot F_1(s) = F(s)$$

or

$$f_1(t)^*f_2(t) = \mathcal{L}^{-1}[F_1(s) \cdot F_2(s)] = \mathcal{L}^{-1}[F(s)]$$

$$= \int_0^t f_1(\tau)f_2(t - \tau)d\tau.$$

Example 4.11. *Find the Laplace inverse using the convolution integral of*

$$F(s) = \frac{1}{(s + 1)(s + 3)}.$$

Ans.

$$F(s) = \left(\frac{1}{s+1}\right) \times \left(\frac{1}{s+3}\right)$$

$$F_1(s) = \frac{1}{(s+1)} \Rightarrow F_1(t) = e^{-t}$$

$$F_2(s) = \frac{1}{(s+3)} \Rightarrow F_2(t) = e^{-3t}$$

$$\mathcal{L}^{-1}[F(s)] = \int_0^t f_1(\tau)f_2(t-\tau)d\tau$$

$$= \int_0^t e^{-\tau}e^{-3(t-\tau)}d\tau$$

$$= \int_0^t e^{-3t}e^{2\tau}d\tau = e^{-3t}\left[\frac{e^{2\tau}}{2}\right]_0^t$$

$$= \frac{e^{-3t}}{2}[e^{2t}-1] = \frac{1}{2}[e^{-t}-e^{-3t}]$$

SOLVED PROBLEMS

Problem 4.1. *Find the inverse Laplace transformation of given functions:*

(a) $\mathcal{L}^{-1}\dfrac{3s}{(s^2+1)(s^2+4)}$ (b) $\mathcal{L}^{-1}\left(\dfrac{s+1}{s^2+2s}\right)$

(c) $\mathcal{L}^{-1}\dfrac{1}{(s+1)(s+2)^2}$ (d) $\mathcal{L}^{-1}\dfrac{s^2+2s+1}{(s+2)(s^2+4)}$

(e) $\mathcal{L}^{-1}\dfrac{s^2}{(s^2+1)^2}$.

Solution:

(a) $$F(s) = \frac{3s}{(s^2+1)(s^2+4)} = \frac{As+B}{s^2+1} + \frac{Cs+D}{s^2+4}$$

$$= \frac{(As+B)(s^2+4) + (Cs+D)(s^2+1)}{(s^2+1)(s^2+4)}$$

$$= \frac{As^3 + 4As + Bs^2 + 4B + Cs^3 + Cs + Ds^2 + D}{(s^2+1)(s^2+4)}$$

$$= \frac{s^3(A+C) + s^2(B+D) + s(4A+C) + 4B+D}{(s^2+1)(s^2+4)}$$

Now,
$$3s = s^3(A + C) + s^2(B + D) + s(4A + C) + 4B + D.$$

After comparing,

$$
\begin{aligned}
A + C = 0 \quad &\Rightarrow \quad A = -C \text{ or } C = -A \\
B + D = 0 \quad &\Rightarrow \quad B = -D \\
4A + C = 3 \quad &\Rightarrow \quad 4A - A = 3 \qquad\qquad \Rightarrow \qquad A = 1 \\
4B + D = 0 \quad &\Rightarrow \quad B = D = 0
\end{aligned}
$$

Now $A = 1, C = -1$

$$F(s) = \frac{s}{s^2 + 1} - \frac{s}{s^2 + 4}.$$

Taking the Laplace inverse,

$$F(t) = \cos t - \cos 2t.$$

(b) $$F(s) = \frac{(s + 1)}{s^2 + 2s} = \frac{(s + 1)}{s(s + 2)} = \frac{A}{s} + \frac{B}{s + 2}$$

$$A = [sF(s)]_{s=0} = \left[\frac{s + 1}{s + 2}\right]_{s=0} = \frac{1}{2}$$

$$B = [(s + 2)F(s)]_{s=-2} = \left[\frac{s + 1}{s}\right]_{s=-2} = \frac{-1}{-2} = \frac{1}{2}$$

$$F(s) = \frac{1}{2}\left[\frac{1}{s} + \frac{1}{(s + 2)}\right]$$

Taking the Laplace inverse,

$$F(t) = \frac{1}{2}[1 + e^{-2t}]$$

(c) $$F(s) = \frac{1}{(s + 1)(s + 2)^2} = \frac{A}{(s + 1)} + \frac{B}{(s + 2)} + \frac{C}{(s + 2)^2}$$

$$A = [(s + 1)F(s)]_{s=-1} = \left[\frac{1}{(s + 2)^2}\right]_{s=-1} = 1$$

$$C = [(s + 2)^2 F(s)]_{s=-2} = \left[\frac{1}{s + 1}\right]_{s=-2} = -1$$

$$B = \left[\frac{d}{ds}(s + 2)^2 F(s)\right]_{s=-2} = \left[\frac{d}{ds}\left\{\frac{1}{s + 1}\right\}\right]_{s=-2} = -1$$

$$F(s) = \frac{1}{(s + 1)} - \frac{1}{(s + 2)} - \frac{1}{(s + 2)^2}.$$

Taking the Laplace inverse,

$$F(t) = e^{-t} - e^{-2t} - te^{-2t}$$

(d)
$$F(s) = \frac{s^2 + 2s + 1}{(s+2)(s^2+4)} = \frac{A}{(s+2)} + \frac{Cs + D}{(s^2+4)}$$

$$= \frac{A(s^2+4) + (Cs+D)(s+2)}{(s+2)(s^2+4)}$$

$$s^2 + 2s + 1 = (A+C)s^2 + s(2C+D) + 4A + 2D.$$

After comparing,

$$A + C = 1$$
$$2C + D = 2$$
$$4A + 2D = 1.$$

After solving,

$$A = \frac{1}{8}, \quad C = \frac{7}{8}, \quad D = \frac{1}{4}$$

$$F(s) = \frac{1}{8(s+2)} + \frac{\frac{7}{8}s + \frac{1}{4}}{(s^2+4)} = \frac{1}{8(s+2)} + \frac{1}{8}\left(\frac{7s+2}{s^2+4}\right)$$

$$F(s) = \frac{1}{8(s+2)} + \frac{7}{8}\left(\frac{s}{s^2+4}\right) + \frac{2}{8(s^2+4)}.$$

Taking the Laplace transform,

$$F(t) = \frac{1}{8}e^{-2t} + \frac{7}{8}\cos 2t + \frac{1}{8}\sin 2t$$

(e)
$$F(s) = \frac{s^2}{(s^2+1)^2} = \frac{s^2+1-1}{(s^2+1)^2} = \frac{1}{(s^2+1)} - \frac{1}{(s^2+1)^2}.$$

Taking the Laplace inverse,

$$F(t) = \sin t - t \sin t.$$

Problem 4.2. *Find the Laplace transform of the following functions:*

(a) $F(t) = (t^2 + 2t + 1)u(t-1)$
(b) $F(t) = \sin \omega(t+1)$
(c) $F(t) = t \sin \omega(t+1).$

Solution:

(a) $f(t) = (t^2 + 2t + 1)u(t - 1)$

$f(t) = (t + 1)^2 u(t - 1)$

$= (t - 1 + 2)^2 u(t - 1)$

$= [(t - 1)^2 + 4 + 4(t - 1)]u(t - 1)$

$= (t - 1)^2 u(t - 1) + 4u(t - 1) + 4(t - 1)u(t - 1)$

Now taking the Laplace transform,

$$F(s) = \frac{|2}{s^3}e^{-s} + \frac{4}{s}e^{-s} + \frac{4}{s^2}e^{-s}$$

$$F(s) = \frac{e^{-s}}{s}\left(\frac{2}{s^2} + 4 + \frac{4}{s}\right)$$

(b) $f(t) = \sin \omega(t + 1) = \sin(\omega t + \omega)$

$= \sin\omega t \cos \omega + \cos \omega t \sin \omega.$

Now, taking the Laplace transform,

$$f(s) = \left(\frac{\omega}{s^2 + \omega^2}\right)\cos \omega + \left(\frac{s}{s^2 + \omega^2}\right)\sin \omega$$

$$= \frac{1}{s^2 + \omega^2}(\omega \cos \omega + s \sin \omega).$$

(c) $F(t) = t \sin \omega(t + 1)$

Applying the formula

$$tf(t) = -\frac{d}{ds}F(s)$$

here $f(t) = \sin \omega(t + 1)$.
and

$$F(s) = \left(\frac{\omega}{s^2 + \omega^2}\right)\cos \omega + \left(\frac{s}{s^2 + \omega^2}\right)\sin \omega$$

$$\mathcal{L}[t \sin \omega(t + 1)] = -\frac{d}{ds}\left[\left(\frac{\omega}{s^2 + \omega^2}\right)\cos \omega + \left(\frac{s}{s^2 + \omega^2}\right)\sin \omega\right]$$

$$= -\left[\omega \cos \omega \times -(s^2 + \omega^2)^{-2} \times 2s\right.$$

$$\left. + \frac{1}{(s^2 + \omega^2)^2}\{s^2 + \omega^2 - 2s^2\}\sin \omega\right]$$

$$= -\frac{1}{(s^2 + \omega^2)^2}[-2\omega s \cos \omega + (\omega^2 - s^2)\sin \omega].$$

Problem 4.3. *Solve the following equations using the Laplace transformation method.*

(a) $\dfrac{d^2 i}{dt^2} - i = 25 + e^{2t}$ *all initial conditions are zero.*

(b) $2\dfrac{di}{dt} + i = \delta(t)$ $i(0) = 5\text{A}.$

Solution: (a)

$$\frac{d^2 i}{dt^2} - i = 25 + e^{2t}$$

Taking the Laplace transform,

$$s^2 \mathrm{F}(s) - \mathrm{I}(s) = \frac{25}{s} + \frac{1}{s-2} \text{ given all initial conditions are zero.}$$

$$\mathrm{I}(s) = \frac{25(s-2) + s}{s(s-2)(s+1)(s-1)}$$

$$\mathrm{I}(s) = \frac{25s - 50 + s}{s(s-1)(s-2)(s+1)} = \frac{26s - 50}{s(s-1)(s-2)(s+1)}$$

$$\mathrm{I}(s) = \frac{\mathrm{A}}{s} + \frac{\mathrm{B}}{(s-1)} + \frac{\mathrm{C}}{(s-2)} + \frac{\mathrm{D}}{s+1}$$

$$\mathrm{A} = \frac{26s - 50}{(s-1)(s+1)(s-2)}\bigg|_{s=0} = \frac{-50}{-1 \times 1 \times -2} = -25$$

$$\mathrm{B} = \frac{26s - 50}{s(s-2)(s+1)}\bigg|_{s=1} = \frac{-24}{1 \times -1 \times 2} = 12$$

$$\mathrm{C} = \frac{26s - 50}{s(s-1)(s+1)}\bigg|_{s=2} = \frac{2}{2 \times 1 \times 3} = \frac{1}{3}$$

$$\mathrm{D} = \frac{26s - 50}{s(s-1)(s-2)}\bigg|_{s=-1} = \frac{-76}{-1 \times -2 \times -3} = \frac{38}{3}$$

$$\mathrm{I}(s) = -\frac{25}{s} + \frac{12}{s-1} + \frac{1}{3(s-2)} + \frac{38}{3(s+1)}.$$

Taking the Laplace inverse,

$$\mathrm{I}(t) = -25 + 12e^t + \frac{1}{3}e^{2t} + \frac{38}{3}e^{-t}.$$

(b) $\qquad 2\dfrac{di}{dt} + i = \delta(t), i(0) = 5\text{A}.$ $\qquad\qquad [\mathcal{L}\delta(t) = 1]$

Taking the Laplace transform,

$$2[sI(s) - I(0)] + I(s) = 1$$
$$2sI(s) - 2 \times 5 + I(s) = 1$$

$$I(s) = \frac{11}{(2s + 1)} = \frac{11}{2}\left(\frac{1}{s + \frac{1}{2}}\right).$$

Taking the Laplace inverse

$$I(t) = \frac{11}{2}e^{-\frac{1}{2}t}.$$

Problem 4.4. *At* $t = 0$, *a switch is closed, connecting a voltage source* $v = V \sin \omega t$ *to a series RL circuit. By the method of Laplace transformation, show that the current is given by the equation*

$$i = \frac{V}{Z}\sin(\omega t - \phi) + \frac{\omega LV}{Z^2}e^{-\frac{Rt}{L}},$$

where

$$Z = \sqrt{R^2 + (\omega L)^2} \text{ and } \phi = \tan -1\left(\frac{\omega L}{R}\right).$$

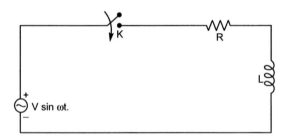

FIGURE 4.3

Solution: Applying KVL,

$$V \sin \omega t = iR + L\frac{di}{dt}.$$

Taking the Laplace transform [assume all initial conditions are zero]

$$\frac{V\omega}{s^2 + \omega^2} = I(s)R + LsI(s)$$

$$I(s) = \frac{V\omega}{L\left(s + \frac{R}{L}\right)(s^2 + \omega^2)}.$$

Now for

$$\frac{1}{\left(s + \frac{R}{L}\right)(s^2 + \omega^2)} = \frac{A}{\left(s + \frac{R}{L}\right)} + \frac{Cs + D}{(s^2 + \omega^2)}$$

$$= \frac{A(s^2 + \omega^2) + (Cs + D)\left(s + \frac{R}{L}\right)}{\left(s + \frac{R}{L}\right)(s^2 + \omega^2)}$$

$$1 = s^2(A + C) + s\left(\frac{R}{L}C + D\right) + A\omega^2 + \frac{DR}{L}.$$

After comparison

$$A + C = 0 \Rightarrow A = -C$$

$$\frac{R}{L}C + D = 0 \Rightarrow D = \frac{-R}{L}C$$

$$D = \frac{R}{L}A \qquad\qquad\qquad\qquad \text{(A)}$$

$$A\omega^2 + \frac{DR}{L} = 1. \qquad\qquad\qquad \text{(B)}$$

By Equations (A) and (B),

$$A = \frac{L^2}{R^2 + L^2\omega^2}, \quad D = \frac{R}{L}\left(\frac{L^2}{R^2 + L^2\omega^2}\right)$$

$$C = \frac{-L^2}{R^2 + L^2\omega^2}.$$

Now,

$$I(s) = \frac{V\omega}{L}\left[\frac{L^2}{R^2 + L^2\omega^2}\left(\frac{1}{s + \frac{R}{L}}\right) - \frac{L^2}{(R^2 + L^2\omega^2)}\left(\frac{s - \frac{R}{L}}{s^2 + \omega^2}\right)\right].$$

Taking the Laplace inverse

$$I(t) = \frac{V\omega L}{R^2 + L^2\omega^2}\left[e^{-\frac{R}{L}t} - \cos\omega t + \frac{R}{L\omega}\sin\omega t\right]$$

$$I(t) = \frac{V\omega L}{Z^2}e^{-\frac{R}{L}t} + \left(\frac{V\omega L}{R^2 + L^2\omega^2}\right)\left[-\cos\omega t + \frac{R}{L\omega}\sin\omega t\right]$$

$$I(t) = \frac{V\omega L}{Z^2}e^{-\frac{R}{L}t} + \frac{V}{Z}\sin(\omega t - \phi),$$

where

$$Z = \sqrt{R^2 + L^2\omega^2}$$

$$\phi = \tan^{-1}\left(\frac{\omega L}{R}\right). \quad \textbf{Proved.}$$

Problem 4.5. *Find the initial and final values of the following function:*

$$F(s) = \frac{4e^{-2s}(s + 2)}{s}.$$

Solution: The initial value $f(0) = \lim_{s\to\infty} sf(s) = \lim_{s\to\infty} 4e^{-2s}(s + 2)$

$$f(0) = 0.$$

The final value is

$$f(\infty) = \lim_{s\to 0} sF(s) = \lim_{s\to 0} 4e^{-2s}(s + 2) = 8.$$

Problem 4.6. *Use partial fractional expansion to determine the inverse Laplace transform of*

$$F(s) = \frac{5s + 40}{s(s^2 + 12s + 27)}.$$

Solution: The given function has to be split up by partial fractions as follows:

$$F(s) = 5\left\{\frac{s + 8}{s(s + 3)(s + 9)}\right\} = 5\left[\frac{A}{s} + \frac{B}{s + 3} + \frac{C}{s + 9}\right]$$

$$A = \frac{(s + 8)}{(s + 3)(s + 9)}\bigg|_{s=0} = \frac{8}{27}$$

$$B = \frac{s + 8}{s(s + 9)}\bigg|_{s=-3} = -\frac{5}{18}$$

$$C = \frac{s + 8}{s(s + 3)}\bigg|_{s=-9} = -\frac{1}{54}$$

$$F(s) = 5\left[\frac{8}{27s} - \frac{5}{18(s + 3)} - \frac{1}{54(s + 9)}\right].$$

Taking the Laplace inverse transform, we get

$$f(t) = 5\left[\frac{8}{27} - \frac{5}{18}e^{-3t} - \frac{1}{54}e^{-9t}\right].$$

Problem 4.7. *Obtain an expression for current i(t) from the differential equation* $\frac{d^2 i(t)}{dt^2} + 10\frac{di}{dt} + 25i(t) = 0$ *with the initial conditions* $i(0+) = 2$; $\frac{di(0+)}{dt} = 0$.

Solution: Taking the Laplace transform of the given differential equation, we get

$$[s^2 I(s) - sI(0+) - \frac{di(0+)}{dt}] + 10[sI(s) - I(0+)] + 25I(s) = 0$$

$$I(s)[s^2 + 10s + 25] = 2s + 20$$

$$I(s) = \frac{2(s+10)}{(s+5)^2} = \frac{2(s+5)}{(s+5)^2} + \frac{10}{(s+5)^2}.$$

Taking the Laplace inverse,

$$I(t) = 2e^{-5t} + 10te^{-5t}.$$

Problem 4.8. *(a) Find the initial and final values of the following functions:*

(i) $\dfrac{s-1}{(s+1)(s+2)}$;   *(ii)* $\dfrac{1}{s^2 + 4s + 5}$.

(b) Find the inverse Laplace transform of the functions stated above.

Solution: (a)

(i) Initial value $f(0) = \lim_{s \to \infty} sF(s)$

$$= \lim_{s \to \infty} \frac{s(s-1)}{(s+1)(s+2)} = \lim_{s \to \infty} \frac{1 - \frac{1}{s}}{\left(1 + \frac{1}{s}\right)\left(1 + \frac{2}{s}\right)}$$

$$= \frac{1}{1 \times 1} = 1.$$

The final value $F(\infty) = \lim_{s \to 0} sF(s) = 0$.

(ii) The initial value $= \lim_{s \to \infty} sF(s) = \lim_{s \to \infty} \frac{s}{s^2 + 4s + 5} = \lim_{s \to \infty} \frac{1}{s + 4 + \frac{5}{s}}$

$$F(0) = 0.$$

The final value $F(\infty) = \lim_{s \to 0} sF(s) = \lim_{s \to 0} \frac{s}{(s^2 + 4s + 5)} = 0.$

(b) (i) $F(s) = \dfrac{s-1}{(s+1)(s+2)} = \dfrac{A}{(s+1)} + \dfrac{B}{(s+2)}$

$A = \left.\dfrac{s-1}{(s+2)}\right|_{s=-1} = -2$

$B = \left.\dfrac{s-1}{(s+1)}\right|_{s=-2} = 3$

$F(s) = -\dfrac{2}{s+1} + \dfrac{3}{s+2}.$

Taking the inverse transform,

$$F(t) = -2e^{-t} + 3e^{-2t}$$

(ii) $F(s) = \dfrac{1}{s^2 + 4s + 5} = \dfrac{1}{(s+2)^2 + 1}.$

Taking the Laplace inverse

$$F(t) = e^{-2t} \sin t.$$

Problem 4.9. *Solve the following differential equation using the Laplace transform*

$$\frac{d^2 i}{dt^2} + 4\frac{di}{dt} + 5i = 5u(t) \text{ given that } i(0) = 1, \frac{di}{dt}(0) = 2.$$

Solution: Taking the Laplace transform, we get

$$\left[s^2 I(s) - sI(0) - \frac{di}{dt}(0)\right] + 4[sI(s) - I(0)] + 5I(s) = \frac{5}{s}.$$

Putting the initial values,

$$I(s)[s^2 + 4s + 5] = \frac{5}{s} + s + 2 + 4$$

$$I(s) = \frac{5 + 6s + s^2}{s(s^2 + 4s + 5)} = \frac{(s^2 + 6s + 5)}{s[(s+2)^2 + 1]} = \frac{A}{s} + \frac{Bs + C}{(s^2 + 4s + 5)}$$

$$= \frac{A(s^2 + 4s + 5) + Bs^2 + Cs}{s(s^2 + 4s + 5)} = \frac{s^2(A + B) + s(4A + C) + 5A}{s(s^2 + 4s + 5)}.$$

Comparing coefficients of s^2, s, and constant

$$A + B = 1$$
$$4A + C = 6$$
$$5A = 5 \quad \Rightarrow \quad A = 1$$
$$B = 0$$
$$C = 2$$

$$I(s) = \frac{1}{s} + \frac{2}{(s^2 + 4s + 5)} = \frac{1}{s} + \frac{2}{[(s + 2)^2 + 1]}.$$

Taking the Laplace inverse

$$I(t) = 1 + 2e^{-2t} \sin t.$$

Problem 4.10. *State and prove the initial and final value theorems. Find the initial value and final value of the following function:*

$$F(s) = \frac{s + 2}{s(s + 3)(s + 4)}.$$

Solution: According to the initial value theorem,

$$f(0) = \lim_{s \to \infty} sF(s) = \lim_{s \to \infty} \frac{s + 2}{(s + 3)(s + 4)} = 0.$$

According to the final value theorem,

$$F(\infty) = \lim_{s \to 0} sF(s) = \lim_{s \to 0} \frac{s + 2}{(s + 3)(s + 4)} = \frac{2}{3 \times 4} = \frac{1}{6}.$$

Problem 4.11. *For the given Laplace transform* $F(s) = \frac{17s^3 + 7s^2 + s + 6}{s^5 + 3s^4 + 5s^3 + 4s^2 + 2s}$. *Find the initial of the final value of the function.*

Solution:

$$Initial\ value = f(0) \lim_{s \to \infty} sF(s) = \lim_{s \to \infty} \left(\frac{17s^4 + 7s^3 + s^2 + 6s}{s^5 + 3s^4 + 5s^3 + 4s^2 + 2s} \right)$$

$$= \lim_{s \to \infty} \left(\frac{17 + \frac{7}{s} + \frac{1}{s^2} + \frac{6}{s^3}}{s + 3 + \frac{5}{s} + \frac{4}{s^2} + \frac{2}{s^3}} \right).$$

$$\boxed{f(0) = 0} \quad \text{[dividing by } s^4 \text{ in the numerator and denominator]}$$

$$\text{Final value} = f(\infty) = \lim_{s \to 0} sF(s) = \lim_{s \to 0} \left(\frac{17s^4 + 7s^3 + s^2 + 6s}{s^5 + 3s^4 + 5s^3 + 4s^2 + 2s} \right).$$

$$= \frac{1}{0} = \infty$$

$$\boxed{F(\infty) = \infty}$$

Problem 4.12. *Given* $F(s) = \frac{s+2}{s(s+1)}$, *the initial and final values of* $f(t)$ *will be, respectively*

(a) 1, 2   (b) 2, 1   (c) 1, 1   (d) 2, 2.

Solution: Initial value $f(0) = \lim_{s \to \infty} sF(s) = \lim_{s \to \infty} \frac{s+2}{s+1} = 1.$

Final value $f(\infty) = \lim_{s \to 0} sF(s) = \lim_{s \to 0} \frac{s+2}{s+1} = 2.$

Problem 4.13. *The Laplace transform of the function* $i(t)$ *is*

$$I(s) = \frac{10s + 4}{s(s+1)(s^2 + 4s + 5)}. \text{ Its final value will be}$$

(a) $\frac{4}{5}$ (b) $\frac{5}{4}$ (c) 4   (d) 5.

Solution: Final value $f(\infty) = \lim_{s \to 0} sF(s) = \lim_{s \to 0} \frac{10s+4}{(s+1)(s^2+4s+5)} = \frac{4}{5}.$

Problem 4.14. *Find the inverse Laplace transform of the following functions:*

(a) $F(s) = \dfrac{1}{s^2 + 4s + 8}$ (b) $F(s) = \dfrac{5}{s(s^2 + 4s + 5)}$

(c) $F(s) = \dfrac{s^2 + 2s + 3}{s^3 + 6s^2 + 12s + 8}.$

Solution: (a) $F(s) = \dfrac{1}{s^2 + 4s + 8} = \dfrac{1}{(s+2)^2 + (2)^2} = \dfrac{1}{2} \cdot \dfrac{2}{(s+2)^2 + (2)^2}.$

Taking the Laplace inverse,

$$F(t) = \frac{1}{2} e^{-2t} \sin 2t$$

(b) $$F(s) = \frac{5}{s(s^2 + 4s + 5)} = \frac{A}{s} + \frac{Bs + C}{s^2 + 4s + 5}.$$

$$5 = A(s^2 + 4s + 5) + s(Bs + C)$$

$$5 = s^2(A + B) + s(4A + C) + 5A$$

After comparing,

$$5A = 5 \Rightarrow A = 1, A + B = 0$$
$$B = -A = -1$$
$$4A + C = 0$$
$$C = -4A = -4$$

$$F(s) = \frac{1}{s} - \frac{s+4}{(s+2)^2 + (1)^2} = \frac{1}{s} - \frac{s+2+2}{(s+2)^2 + (1)^2}$$

$$= \frac{1}{s} - \frac{s+2}{(s+2)^2 + (1)^2} - \frac{2}{(s+2)^2 + (1)^2}.$$

Taking the Laplace inverse,

$$F(t) = (1 - e^{-2t} \cos t - 2e^{-2t} \sin t)u(t).$$

(c) $$F(s) = \frac{s^2 + 2s + 3}{(s+2)^3} = \frac{A}{(s+2)} + \frac{B}{(s+2)^2} + \frac{C}{(s+2)^3}.$$

After solving $A = 1$, $B = -2$, $C = 3$

$$F(s) = \frac{1}{(s+2)} - \frac{2}{(s+2)^2} + \frac{3}{(s+2)^3}.$$

Taking the inverse Laplace transform,

$$F(t) = e^{-2t} - 2te^{-2t} + \frac{3}{2}t^2 e^{-2t}.$$

Problem 4.15. *Find the solution of the differential equation given below using the Laplace transform*

$$2\frac{dx}{dt} + 8x = 10 \text{ given } x(0+) = 2.$$

Solution: Taking the Laplace transform,

$$2[sX(s) - X(0)] + 8X(s) = \frac{10}{s}$$

$$2[sX(s) - 2] + 8X(s) = \frac{10}{s}$$

$$X(s) = \frac{2s+5}{s(s+4)} = \frac{A}{s} + \frac{B}{s+4}.$$

After solving A = 1.25, B = 0.75,

$$X(s) = \frac{1.25}{s} + \frac{0.75}{s+4}.$$

Taking the Laplace inverse,

$$X(t) = (1.25 + 0.75e^{-4t})u(t).$$

Problem 4.16. *The Laplace transformed equation the charging current of a capacitor arranged in series with a resistance is given by,*

$$I(s) = \frac{Cs}{RCs+1} \cdot E(s).$$

If E = 100 V, R = 2 MΩ, C = 1μF, calculate the initial and final value of the charging current.

Solution:

$$E = 100\,\text{V}, E(s) = \frac{100}{s}.$$

Now by the given equation,

$$I(s) = \frac{Cs}{RCs+1} \cdot E(s).$$

Substituting the given values

$$I(s) = \frac{10^{-6}s}{2 \times 10^6 \times 1 \times 10^{-6}s + 1} \times \frac{100}{s} = \frac{10^{-4}}{2s+1}.$$

Now the initial value

$$I(0) = \lim_{s \to \infty} sI(s) = \lim_{s \to \infty} \frac{s \times 10^{-4}}{2s+1} = \lim_{s \to \infty} \frac{10^{-4}}{2 + \frac{1}{s}} = 0.5 \times 10^{-4}$$

$$I(0) = 50\mu\text{A}.$$

The final value

$$I(\infty) = \lim_{s \to 0} sI(s) = \lim_{s \to 0} \frac{s \times 10^{-4}}{2s+1} = 0 \text{ Amp.}$$

Problem 4.17. *Find the initial and final values of the following function, using the initial value and final value theorem, respectively.*

$$F(s) = \frac{(s-1)}{(s+1)(s+2)}$$

Solution: Initial value

$$f(0) = \lim_{s \to \infty} sF(s) = \lim_{s \to \infty} \frac{s(s-1)}{(s+1)(s+2)} = 1.$$

Final value

$$f(\infty) = \lim_{s \to \infty} sF(s) = \lim_{s \to 0} \frac{s(s-1)}{(s+1)(s+2)} = 0.$$

Problem 4.18. *Show that $\mathcal{L}[tf(t)] = -\frac{d}{ds}F(s)$. Using this result, find the Laplace transform of $t\sin a\ t^2 e^{-at}$ and $t\sinh \beta t$, where a, β are constants.*

Solution: For $t\sin at$

$$f(t) = \sin at \Rightarrow F(s) = \frac{a}{s^2 + a^2}$$

$$\mathcal{L}[t\sin at] = -\frac{d}{ds} \times \frac{a}{(s^2 + a^2)} = \frac{2as}{(s^2 + a^2)^2},$$

for $t^2 e^{-at}$

$$f(t) = e^{-at} \Rightarrow F(s) = \frac{1}{(s+a)}$$

$$\mathcal{L}[t^2 e^{-at}] = (-1)^2 \frac{d^2}{ds^2}\left[\frac{1}{(s+a)}\right]$$

$$= \frac{2}{(s+a)^3},$$

for $t\sinh \beta t$

$$f(t) = \sinh \beta t = \frac{\beta}{s^2 - \beta^2}$$

$$\mathcal{L}[t\sinh \beta t] = (-1)\frac{d}{ds}\left(\frac{\beta}{s^2 - \beta^2}\right)$$

$$= \frac{2\beta s}{(s^2 - \beta^2)^2}.$$

Problem 4.19. *Find $i(t)$ for the given equation,*

$$\frac{di}{dt} + 2\int i\,dt = 5. \text{ Assume all initial conditions are zero.}$$

Solution: Taking the Laplace transform

$$sI(s) + 2\frac{I(s)}{s} = \frac{5}{s}$$

$$I(s)[s^2 + 2] = 5$$

$$I(s) = \frac{5}{\sqrt{2} + 2}$$

$$I(s) = \frac{5}{\sqrt{2}} \left(\frac{\sqrt{2}}{s^2 + 2} \right).$$

Taking the Laplace inverse,

$$I(t) = \frac{5}{\sqrt{2}} \sin \sqrt{2} t.$$

Problem 4.20. *Given the function* $F(s) = \frac{5s+3}{s(s+1)}$, *find the initial value* $f(0+)$, *final value* $f(\infty)$, *and the corresponding time function* $f(t)$.

Solution: Initial value

$$f(0+) = \lim_{s \to \infty} [sF(s)]$$

$$= \lim_{s \to \infty} \left[\frac{5s + 3}{s + 1} \right] = \lim_{s \to \infty} \left[\frac{5 + \frac{3}{s}}{1 + \frac{1}{s}} \right] = 5,$$

final value

$$f(\infty) = \lim_{s \to 0} [sF(s)] = \lim_{s \to 0} \left[\frac{5s + 3}{s + 1} \right],$$

$$= 3$$

and

$$F(s) = \frac{5s + 3}{s(s + 1)} = \frac{3}{s} + \frac{2}{s + 1}.$$

Taking the Laplace inverse

$$F(t) = (3 + 2e^{-t})u(t)$$

$$f(0+) = \lim_{t \to 0} f(t) = \lim_{t \to 0} (3 + 2e^{-t}) = 5$$

$$f(\infty) = \lim_{t \to 0} f(t) = \lim_{t \to \infty} (3 + 2e^{-t}) = 3 + 0 = 3.$$

Problem 4.21. *Without finding the inverse Laplace transform of* $F(s)$, *determine* $f(0+)$ *and* $f(\infty)$ *for each of the following functions:*

(i) $\dfrac{4e^{-2s}(s + 50)}{s}$ (ii) $\dfrac{s^2 + 6}{s^2 + 7}$

Solution:

(*i*) $f(0+) = \lim\limits_{s\to\infty} sF(s) = \lim\limits_{s\to\infty} 4e^{-2s}(s+50) = 0$

$f(\infty) = \lim\limits_{s\to 0} sF(s) = \lim\limits_{s\to 0} 4e^{-2s}(s+50) = 4 \times 50 = 200$

(*ii*) $f(0) = \lim\limits_{s\to\infty} sF(s) = \lim\limits_{s\to\infty} \dfrac{s^3 + 6s}{s^2 + 7} = \lim\limits_{s\to\infty} \dfrac{1 + \frac{6}{s^2}}{\frac{1}{s} + \frac{7}{s^3}}$

$= \infty$

$f(\infty) = \lim\limits_{s\to 0} \dfrac{s(s^2 + 6)}{(s^2 + 7)} = 0$

Problem 4.22. *Without finding the inverse Laplace transform of $f(s)$, determine $f(0+)$ and $f(\infty)$ for the following function:*

$$F(s) = \frac{5s^3 - 1600}{s(s^3 + 18s^2 + 90s + 800)}.$$

Solution:

$F(0+) = \lim\limits_{s\to\infty} sF(s) = \lim\limits_{s\to\infty}\left[\dfrac{5s^3 - 1600}{s^3 + 18s^2 + 90s + 800}\right]$

$= \lim\limits_{s\to\infty} \dfrac{5 - \frac{1600}{s^3}}{1 + \frac{18}{s} + \frac{90}{s^2} + \frac{800}{s^3}} = 5$

$F(\infty) = \lim\limits_{s\to 0} sF(s) = \lim\limits_{s\to 0}\left[\dfrac{5s^3 - 1600}{s^3 + 18s^2 + 90s + 800}\right] = \dfrac{-1600}{800} = -2$

Problem 4.23. *Find the initial and final values of the following functions:*
(a) $f(t) = 2t + \sin^2 t$, *(b)* $F(s) = \frac{2s+5}{(s+1)^2}$.

Solution:

(*a*) $F(t) = 2t + \sin^2 t$

Initial value $= f(0) = \lim\limits_{t\to 0} f(t) = 0.$

Final value $= f(\infty) = \lim\limits_{t\to\infty} f(t) = \infty$

(*b*) $F(s) = \dfrac{2s + 5}{(s + 1)^2}.$

Initial value $= f(0) = \lim\limits_{s\to\infty} sF(s) = \lim\limits_{s\to\infty} \dfrac{2 + \frac{5}{s}}{(1 + \frac{1}{s})^2} = 2.$

Final value $= f(\infty) = \lim\limits_{s\to 0} sF(s) = 0.$

Problem 4.24. *Find the convolution between*

$$f_1(t) = e^{-2t}u(t) \text{ and } f_2(t) = tu(t).$$

Solution: $F_1(s) = \frac{1}{s+2}$ and $F_2(s) = \frac{1}{s^2}$

$$F(s) = F_1(s).F_2(s) = \frac{1}{s^2(s+2)}$$

$$F(s) = \frac{A}{s^2} + \frac{B}{s} + \frac{C}{s+2}$$

$$A = \left.\frac{1}{s+2}\right|_{s=0} = 0.5$$

$$B = \left.\frac{d}{ds}\left[\frac{1}{s+2}\right]\right|_{s=0} = -0.25$$

$$C = \left.\frac{1}{s^2}\right|_{s=-2} = 0.25$$

$$F(s) = \frac{0.5}{s^2} - \frac{0.25}{s} + \frac{0.25}{s+2}.$$

Taking the Laplace inverse,

$$F(t) = (-0.25 + 0.5t + 0.25e^{-2t})u(t).$$

Problem 4.25. *Find the inverse Laplace transfrom of the given function* $F(s) = \frac{1}{s^2(s+2)}$, *using convolution integral.*

Solution: $F(s) = F_1(s).F_2(s) = \frac{1}{s^2(s+2)}$.
   Assume

$$F_1(s) = \frac{1}{s^2} \Rightarrow f_1(t) = tu(t)$$

$$F_2(s) = \frac{1}{s+2} \Rightarrow f_2(t) = e^{-2t}u(t).$$

We know that,

$$f(t) = f^{-1}[F(s)] = f_1(t) * f_2(t) = \int_0^t f_1(t-\tau)f_2 t \, d\tau$$

$$= \int_0^t (t-\tau)e^{-2t} d\tau$$

$$= \int_0^t te^{-2\tau} d\tau - \int_0^t \tau e^{-2\tau} d\tau \, t\frac{e^{-2\tau}}{-2}\Big|_0^t - \left[\tau.\frac{e^{-2\tau}}{-2}\right]_0^t - \int_0^t 1 * \frac{e^{-2\tau}}{-2} d\tau$$

$$= \frac{t}{2}(1 - e^{-2t}) + \frac{t}{2}e^{-2t} - \frac{1}{2}\frac{e^{-2\tau}}{-2}\bigg|_0^t$$

$$F(t) = \frac{t}{2}(1 - e^{-2t}) + \frac{t}{2}e^{-2t} + \frac{1}{4}(e^{-2t} - 1).$$

Problem 4.26. *Find the convolution between $f_1(t) = u(t)$ and $f_2(t) = e^{-t}u(t)$.*

Solution: We know that $f(t) = f_1(t) * f_2(t) = \int_0^t f_1(\tau)f_2(t - \tau)\,d\tau = \int_0^t f_2(\tau)f_1(t - \tau)\,d\tau$

$$= \int_0^t e^{-\tau} \cdot 1\,d\tau = \frac{e^{-\tau}}{-1}\bigg|_0^t$$

$$= -1[e^{-t} - 1] = (1 - e^{-t})u(t).$$

Problem 4.27. *Find the inverse Laplace of $F_1(s)$, $F_2(s)$ by convolution integral for the following:*

(i) $F_1(s) = \dfrac{1}{s + a}, F_2(s) = \dfrac{1}{(s + b)(s + c)}.$

(ii) $F_1(s) = \dfrac{s}{s + 1}, F_2(s) = \dfrac{1}{s^2+1}.$

Solution: (i) $F_1(t) = e^{-at}, F_2(t) = \frac{1}{c-b}(e^{-bt} - e^{-ct})$

$$\mathcal{L}^{-1}F(s) = f_1(t) * f_2(t) = f(t)$$

$$= \int_0^t f_1(t - \tau) \cdot f_2(t)\,d\tau$$

$$= \int_0^t e^{-a(t-\tau)} \cdot \left(\frac{e^{-b\tau} - e^{-c\tau}}{c - b}\right)d\tau$$

$$= \frac{e^{-at}}{c - b}\left[\int_0^t e^{(a-b)\tau}d\tau - \int_0^t e^{(a-c)\tau}d\tau\right]$$

$$= \frac{e^{-at}}{c - b}\left[\frac{1}{(a - b)}e^{(a-b)\tau}\bigg|_0^t - \frac{1}{a - c}e^{(a-c)\tau}\bigg|_0^t\right]$$

$$= \frac{e^{-at}}{c - b}\left[\frac{1}{(a - b)}\{e^{(a-b)t} - 1\} - \frac{1}{a - c}\{e^{(a-c)t} - 1\}\right]$$

$$= \frac{e^{-bt}}{(a - b)(c - b)} + \frac{e^{-ct}}{(a - c)(b - c)} + \frac{1}{(b - a)(c - a)}$$

(ii) $F_1(s) = \frac{s+1-1}{s+1} \Rightarrow F_1(t) = \delta(t) - e^{-t}$

$$F_2(s) = \frac{1}{s^2+1} \Rightarrow F_2(t) = \sin t$$

$$\mathcal{L}^{-1}F(s) = F(t) = \int_0^t [\delta(t) - e^{-\tau}] \cdot \sin(t-\tau)\,d\tau$$

$$= \sin(t-\tau)|_{\tau=0} - \int_0^t e^{-\tau}\sin(t-\tau)\,d\tau$$

$$= \sin t - \int_0^t e^{-\tau}\left\{\frac{e^{j(t-\tau)} - e^{-j(t-\tau)}}{2j}\right\}d\tau$$

$$= \sin t - \frac{1}{2j}\int_0^t e^{jt}e^{-(1+j)\tau}\,d\tau + \frac{1}{2j}\int_0^t e^{-jt}\cdot e^{-(1-j)\tau}\,d\tau$$

$$= \sin t - \frac{e^{jt}}{2j}\left\{\frac{e^{-(1+j)\tau}}{-(1+j)}\right\}\Big|_0^t - \frac{e^{-jt}}{2j}\left\{\frac{e^{-(1-j)\tau}}{1-j}\right\}\Big|_0^t$$

$$= \sin t - \frac{1}{2j}\left[\frac{e^{-t}-e^{-jt}}{-(1+j)} - \frac{e^{-t}-e^{-jt}}{-(1-j)}\right]$$

$$= \sin t - \frac{1}{2j}\left[\frac{(1-j)(e^{-t}-e^{jt}) - (1+j)(e^{-t}-e^{-jt})}{-(1+j)(1-j)}\right]$$

$$= \sin t - \frac{1}{2j}\left[\frac{e^{-t}-e^{jt}-je^{-t}+je^{jt}-e^{-t}+e^{-jt}-je^{-t}+je^{-jt}}{-2}\right]$$

$$= \sin t + \frac{1}{2}\left[-\frac{(e^{jt}-e^{-jt})}{2j} - \frac{2je^{-t}}{2j} + \frac{e^{jt}+e^{-jt}}{2}\right]$$

$$= \sin t + \frac{1}{2}[-\sin t - e^{-t} + \cos t]$$

$$= \frac{1}{2}[\sin t + \cos t - e^{-t}].$$

Problem 4.28. *Determine the inverse Laplace transform of the following function using convolution integral:*

$$F(s) = F_1(s)F_2(s) = \frac{s+1}{s(s^2+4)}$$

Solution:

$$F(s) = F_1(s) \cdot F_2(s) = \frac{(s+1)}{s(s^2+4)}$$

$$F_1(s) = \frac{1}{s} \text{ and } F_2(s) = \frac{s+1}{s^2+4}$$

$$f_1(t) = u(t) \text{ and } f_2(t) = \cos 2t + \frac{1}{2}\sin 2t$$

$$f(t) = \int_0^t f_1(t-\tau)f_2(\tau)\,d\tau = \int_0^t u(t-\tau)\left(\cos 2\tau + \frac{1}{2}\sin 2\tau\right)d\tau$$

$$= \int_0^t \cos 2\tau\,d\tau + \frac{1}{2}\int_0^t \sin 2\tau\,d\tau = \left.\frac{\sin 2\tau}{2}\right|_0^t + \frac{1}{2}\left[\frac{-\cos 2\tau}{2}\right]_0^t$$

$$f(t) = \frac{1}{2}\sin 2t - \frac{1}{4}(\cos 2t - 1) = \frac{1}{4} + \frac{1}{2}\sin 2t - \frac{1}{4}\cos 2t$$

$$f(t) = \frac{1}{4}(1 + 2\sin 2t - \cos 2t).$$

QUESTIONS FOR DISCUSSION

1. Define initial value theorem.
2. Define final value theorem.
3. Find the Laplace transform of $\sinh \beta t$.
4. Is the initial value theorem applicable everywhere?
5. Is the final value theorem applicable everywhere?
6. What is the Laplace inverse of 1?
7. What is the initial value of $\frac{s}{s^2+1}$?
8. What is the final value of $\frac{2s}{s+1}$?
9. What is the Laplace transform of $\sin 2t$?
10. What is the Laplace transform of $\cos at$?
11. What is the Laplace transform of e^{at} ?
12. What do you mean by 's' in the Laplace transform?
13. What is the difference between the Laplace tranform and the Fourier transform?

OBJECTIVE QUESTIONS

1. The Laplace transformation of $f(t)$ is F(s).

   Given $F(s) = \dfrac{\omega}{s^2 + \omega^2}$, the final value of $f(t)$ is

   (a) Infinity   (b) Zero   (c) One   (d) None of the above.

2. The Laplace transform of $(t^2 - 2t)u(t - 1)$ is

   (a) $\dfrac{2}{s^3}e^{-s} - \dfrac{2}{s^2}e^{-s}$ (b) $\dfrac{2}{s^3}e^{-2s} - \dfrac{2}{s^2}e^{-s}$

   (c) $\dfrac{2}{s^3}e^{-s} - \dfrac{1}{s}e^{-s}$ (d) None of the above

3. A rectangular current pulse of duration t and magnitude I has the Laplace transform

   (a) $\dfrac{I}{s}$ (b) $\dfrac{I}{s}e^{-sT}$ (c) $\left(\dfrac{I}{s}\right)e^{sT}$ (d) $\left(\dfrac{I}{s}\right)(1 - e^{-sT})$.

4. With symbols having the usual meanings, the Laplace transform of $u(t - a)$ is

   (a) $\dfrac{1}{s}$ (b) $\dfrac{1}{(s-a)}$ (c) $\dfrac{e^{-as}}{s}$ (d) $\dfrac{e^{as}}{s}$.

5. If $F(s) = \dfrac{1}{s}\dfrac{(s+1)}{s+k}$ and a $t \to \infty$ is $\dfrac{1}{2}$, the value of k is

   (a) $\dfrac{1}{2}$ (b) 1   (c) 2   (d) ∞.

6. If $\mathcal{L}f(t) = F(s)$, then what is the value of $\mathcal{L}t^2f(t)$

   (a) $\dfrac{d^2F(s)}{ds^2}$ (b) $(-1)2\dfrac{dF(s)}{ds}$

   (c) $(-1)\dfrac{dF(s)}{ds}$ (d) $\dfrac{(-1)^2}{\lfloor 2}\dfrac{d^2F(s)}{ds^2}$.

7. $\mathcal{L}^{-1}[\dfrac{e^{-2s}}{s^3}]$ is

   (a) $\dfrac{1}{2} \times t^2u(t - 2)$ (b) $\dfrac{(t-2)^2}{2}u(t-2)$

   (c) $t^2u(t-2)$ (d) $(t-2)^2u(t-2)$.

8. The Laplace transform of $e^{-2t}\sinh 2t$ is

   (a) $\dfrac{2}{s(s+4)}$ (b) $\dfrac{2}{s^2-4}$

   (c) $\dfrac{2}{(s-2)^2-4}$ (d) $\dfrac{2}{(s-2)^2+4}$.

9. $\mathcal{L}[\delta(t)]$ is –
   where $\delta(t)$ is the impulse function

   (a) 1   (b) e^{-s} (c) e^s (d) $\dfrac{1}{s}$.

10. The Laplace transform of $\dfrac{e^{-at}-e^{-bt}}{(b-a)}$ is

   (a) $\dfrac{1}{(s+a)(s-b)}$ (b) $\dfrac{1}{(s-a)(s+b)}$

   (c) $\dfrac{1}{(s-a)(s-b)}$ (d) $\dfrac{1}{(s+a)(s+b)}$.

UNSOLVED PROBLEMS

1. Find the Laplace transform of the following time functions:
    (a) $e^{-at} \sinh t$ (b) $e^{2a} \cos \omega t$ (c) $t^3 e^{5t}$ (d) $5(1 - e^{-2t})$
    (e) $t^2 + 2 \sin 4t$ (f) $te^{5t} \cos t$.

2. If $\mathcal{L}F(t) = F(s)$, show that $\mathcal{L}[e^{-at}f(t)] = F(s + a)$.

3. If $\mathcal{L}F(t) = F(s)$, show that

$$\mathcal{L}[t^n f(t)] = (-1)^n \frac{d^n F(s)}{ds^n}.$$

4. Determine the inverse Laplace transform of the following functions:

    (a) $\dfrac{s}{(s+1)(s+2)}$ (b) $\dfrac{4}{s(s^2 + 6s + 9)}$ (c) $\dfrac{3s}{(s^2 + 4)(s + 5)}$

    (d) $\dfrac{2}{s^2 + 7s + 12}$ (e) $\dfrac{s+5}{s^2 + 2s + 5}$ (f) $\dfrac{5s}{s^2 + 3s + 2}$

    (g) $\dfrac{1}{s(s^2 - b^2)}$ (h) $\dfrac{2(s+1)}{(s+2)^2(s+3)}$ (i) $\dfrac{4 - 2s}{s^3 + 4s}$

    (j) $\dfrac{s^3 + 3s^2 + 3s + 1}{s^2 + 2s + 2}$.

5. Find the initial and final value of following functions:

    (a) $F(s) = \dfrac{s^3 + 3s^2 + 3s + 1}{s^2 + 2s + 2}$ (b) $F(s) = \dfrac{s^2 + 5}{s^3 + 2s + 1}$

    (c) $F(s) = \dfrac{s^3 + 2s^2 + 3}{s^4 + 4s + 5}$ (d) $F(s) = \dfrac{2s + 5}{(s^2 + 1)(s + 1)}$

    (e) $F(s) = \dfrac{3s^3 + 2}{(s^2 + 2)(s + 1)}$ (f) $F(s) = \dfrac{2as + b}{(s + a)(s + b)}$.

6. State the initial and final value theorems and find the initial value of
    I, where $I(s) = 6.67\left[\frac{s+250}{s(s+166.7)}\right]$.

7. Find the inverse Laplace transform of the given functions using
    convolution integral.

    (a) $\dfrac{1}{s^2}$ (b) $\dfrac{1}{s^2(s^2 + 4)}$ (c) $\dfrac{1}{s(s^2 + 4)}$ (d) $\dfrac{s}{(s + 1)(s + 2)}$.

8. Find the solution of the given differential equations:

    (a) $2\dfrac{d^2V}{dt^2} + 5\dfrac{dV}{dt} + 3V = 10$ (b) $\dfrac{d^2V}{dt^2} + 3\dfrac{dV}{dt} + 2V = 5$.

    Assume all initial conditions are zero.

5 BASIC SIGNALS AND WAVEFORM SYNTHESIS

Chapter **5**

5.1 INTRODUCTION

A signal is a physical quantity, or quality, which conveys information. Electrical engineers normally consider a signal to be an electric current or voltage, and these currents and voltages are functions of time. The step, ramp, impulse, exponential, and sinusoidal functions, etc., are the basic signals. These signals may be combined by addition or subtraction to build a variety of general waveforms used in practice.

Signals play an important role in life. They are used in communication, speech processing, medical science, etc.

5.2 STEP FUNCTION

Suppose that a constant voltage or a current is suddenly applied on a circuit at time $t = 0$. The switching process can be described mathematically by defining a function called the step function. The step function is denoted $U(t)$ and defined by the equations

$$f(t) = \begin{cases} 0; & t < 0 \\ A; & t > 0. \end{cases} \tag{5.1}$$

A graph of the step function is shown in the figure.

For a unit step function

$$A = 1.$$

Now

$$u(t) = \begin{cases} 0; & t < 0 \\ 1; & t > 0 \end{cases} \tag{5.2}$$

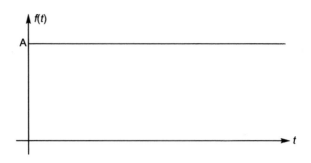

FIGURE 5.1

and the graph of a unit step function is

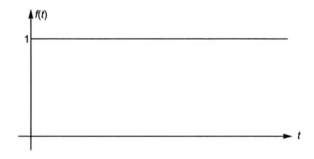

FIGURE 5.2

The Laplace transform of a step function is

$$f(s) = \int_0^\infty e^{-st} f(t)dt = \int_0^\infty e^{-st} \cdot A\,dt = \left.\frac{Ae^{-st}}{-s}\right|_0^\infty.$$

$$f(s) = \frac{A}{s}$$

Now the Laplace transform of a unit step function is

$$A = 1; f(s) = \frac{1}{s}. \qquad\qquad \text{[for a unit step function } A = 1]$$

5.2.1 Shifted or delayed unit step function

A shifted or delayed unit step function is shown in Figure 5.3(a).

$$f(t) = \begin{cases} 0; & t < a \\ 1; & t \geq a \end{cases} \qquad\qquad (5.3)$$

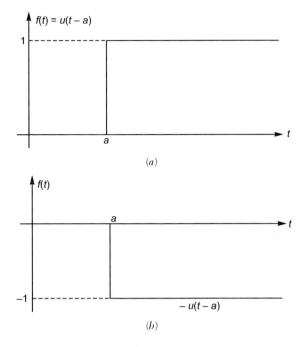

FIGURE 5.3

The Laplace transform of a shifted unit step function is

$$\mathcal{L}[u(t-a) = \frac{e^{-as}}{s} \tag{5.4}$$

$$f(t) = \begin{cases} 0 & t < -a \\ 1 & t > -a \end{cases}. \tag{5.5}$$

and in reversed direction.

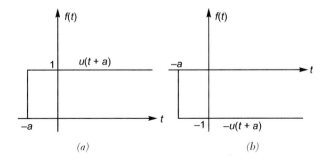

FIGURE 5.4

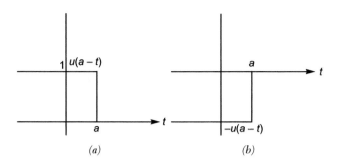

(a)   *(b)*

FIGURE 5.5

Example 5.1. *Show how a unit pulse of width 'a' can be obtained from two unit step functions.*

Solution: It is clear by the Figure 5.2 and 5.3(*a*).

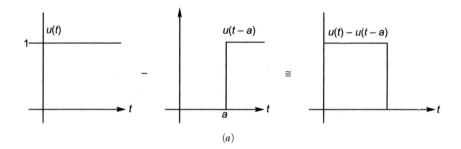

(a)

FIGURE 5.6

This method is known as the summation and subtraction method. We can also obtain a pulse of width '*a*' by the multiplication method.

Multiplication of Figure 5.2 and 5.5(*a*).

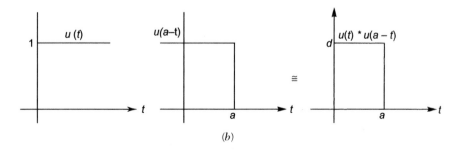

(b)

FIGURE 5.6

Example 5.2. *Sketch the following functions* $u(t-2)$, $u(3-t)$, $u(t-2) - u(t-4)$, *and* $u(5+t)$.

5.2.2 Multiplication of a Function by a Step Function

When any function of time is multiplied by the unit step function a new function of time is produced. Consider the function described by the relationship

$$f_2(t) = f_1(t)u(t).$$

Suppose

$$f_1(t) = t$$

and the unit step function

$$f_2(t) = tu(t).$$

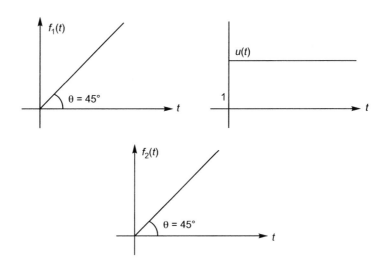

FIGURE 5.7

It will be clear with another example.
Suppose

$$f_2(t) = f_1(t)u(t-2).$$

Since $u(t-2)$ is zero for $t < 2$ the product of $u(t-2)$ and t is zero for $t < 2$ when $t > 2u(t-2)$ is equal to 1.

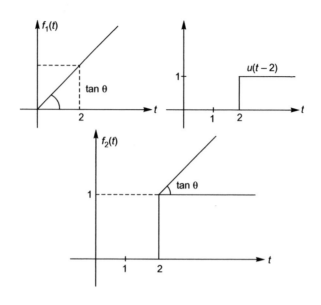

FIGURE 5.8

If $f_1(t) = K$ a constant
k may be any constant 1, 2, 3, 4, and 5, etc.

Similarly, K exists for all (positive and negative time) but a unit step function exists after $t = 0$. Now the multiplication is clear by Figure 5.9 for $t < 0, u(t) = 0$ so the multiplication is zero and for $t > 0\, f_2(t)$ is K times $u(t)$.

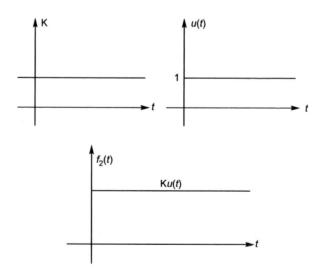

FIGURE 5.9

5.2.3   Addition and Subtraction of Step Functions

By addition and subtraction, we can build a variety of more general waveforms used in practice. The sum of two or more step functions is the algebraic sum of the individual functions.

Example 5.3. *With the help of mathematical expressions represent the forcing functions with their characteristic curves starting at $t = 0$ and lasting for 2 seconds.*

Solution: The given function shown in the figure.

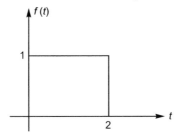

This is a gate pulse of width 2, hence $f(t) = u(t) - u(t - 2)$.

Example 5.4. *Write the mathematical equation for a gate pulse starting at $t = 2$ and lasting for 4 seconds.*

Ans. $f(t) = u(t - 2) - u(t - 4)$

5.3   RAMP FUNCTION

The basic ramp function is a straight line beginning at the origin and increasing or decreasing linearly with time as shown in Figure 5.10.

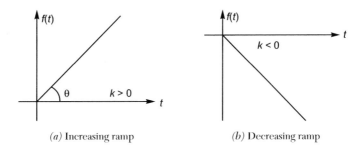

(*a*) Increasing ramp   (*b*) Decreasing ramp

FIGURE 5.10

Let $f(t)$ be the ramp function. The ramp function is mathematically expressed as follows:

$$f(t) = 0 \begin{cases} 0 & \text{for } t < 0 \\ kt & \text{for } t \geq 0 \end{cases}. \qquad (5.6)$$

The constant k represents the slope of the function

$$\tan \theta = k.$$

If k is positive, the slope is upward (increasing ramp).
If k is negative, the slope is downward (decreasing ramp).
If $k = 1 = \tan \theta$ or $\theta = 45°$ the ramp is called a unit ramp function, and the unit ramp function is denoted $r(t)$. Thus,

$$r(t) = \begin{cases} 0 & \text{for } t < 0 \\ t & \text{for } t \geq 0 \end{cases} \qquad (5.7)$$

and $r(t) = tu(t)$. It is clear by Figure 5.7 that multiplication of t and $u(t)$ is the same as $r(t)$.

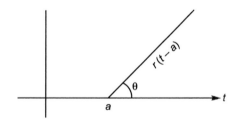

FIGURE 5.11

5.3.1 Shifted Ramp Function

A shifted or delayed ramp function is shown in Figure 5.11.

$$r(t - a) = \begin{cases} 0 & \text{for } t < a \\ kt & \text{for } t \geq a \end{cases} \qquad (5.8)$$

for the shifted unit ramp function $k = 1$ and $\tan \theta = k = 1$,

$$\theta = 45°.$$

The shifted ramp function in terms of the shifted unit step function is

$$r(t - a) = (t - a)u(t - a). \qquad (5.9)$$

Example 5.5. *Draw a delayed unit ramp function starting at 2 seconds.*

Solution: The waveform is shown in Figure 5.12(a).

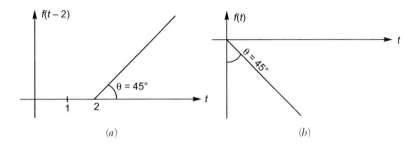

(a)   (b)

FIGURE 5.12

Example 5.6. *Find the mathematical expressions for the given waveform [Figure 5.12(b)]*

Ans. $f(t) = -t$

Example 5.7. *Find the Laplace transform of the ramp and shifted ramp function.*

Ans. $f_1(s) = \frac{1}{s^2}$ and $f_2(s) = \frac{e^{-as}}{s^2}$

5.3.2 Addition of a Two Ramp Functions with Different Slopes

The following points are to be noted while adding two ramp functions.

1. The slope of the resulting function in a given interval is equal to the algebraic sum of all the ramp functions that have already been switched on.
2. The starting point in a given interval is the value reached by the resulting function at the end of the preceding interval.

Let us consider some ramp functions with different slopes.

Example 5.8. *Find the value of $f(t) = f_1(t) + f_2(t)$ and draw its waveform.*

Solution:

$$f_1(t) = m_1 t u(t) [\text{Figure } 5.13(a)]$$
$$f_2(t) = m_2(t-a)u(t-a)[\text{Figure } 5.13(b)]$$
$$f(t) = m_1 t u(t) + m_2(t-a)u(t-a)$$

and it is shown in Figure 5.14.

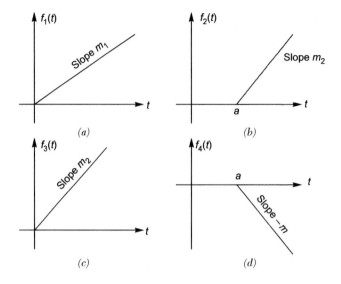

FIGURE 5.13 $[m_2 > m_1]$

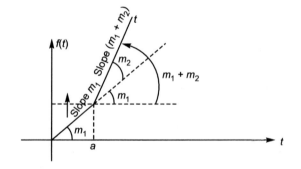

FIGURE 5.14

Example 5.9. *Find the value of* $f(t) = f_3(t) + f_4(t)$ *and draw the waveforms.*

Ans. $f(t) = m_2 t u(t) - m_3(t - a)u(t - a)$

5.4 PARABOLIC FUNCTION

The parabolic function is defined as follows:

$$f(t) = \begin{cases} 0 & t < 0 \\ k\frac{t^2}{2} & t > 0 \end{cases}. \tag{5.10}$$

If $k = 1$, this function is known as a unit parabolic function. This function is shown in Figure 5.15.

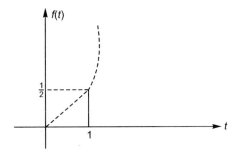

FIGURE 5.15

It is zero for the negative value of t and a parabola for the positive value of t.

The Laplace transform is

$$\mathcal{L}\left[\frac{t^2}{2}\right] = \frac{1}{s^3}.$$

5.5   IMPULSE FUNCTION OR DIRAC DELTA FUNCTION

The impulse function or delta function was first used by physicist Dirac in 1930. Therefore, it is also called the Dirac Delta function. The impulse or the delta function is denoted $\delta(t)a$ and is shown in Figure 5.16.

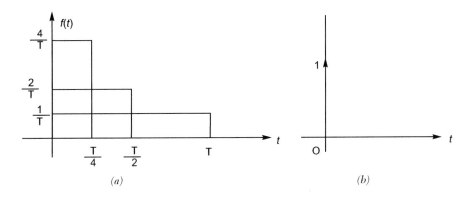

FIGURE 5.16

Mathematically, the expression for an impulse function is

$$f(t) = \frac{1}{T}[u(t) - u(t - T)]. \tag{5.11}$$

This pulse has a width T and a magnitude $\frac{1}{T}$ such that the area of the pulse is unity,

$$T \times \frac{1}{T} = 1.$$

The pulse has an amplitude that is inversely proportional to its duration. In another case time duration is halved and double the amplitude, therefore area is unity.

It is seen that the areas under all the three waveforms remain constant at unity.

When the duration (width) of the pulse approaches to zero the amplitude of the pulse approaches infinity but the area of the pulse still remains unity. The pulse for which the duration tends to be zero and the amplitude tends to be infinity is called the impulse function or the delta function.

Since it is impossible to show an infinite amplitude (magnitude) and zero duration as shown in Figure 5.16(b) the unit impulse occuring at $t = 0$ is represented graphically by a vertical arrow as shown in Figure 5.16(b).

Mathematical expression for the unit impulse function is

$$\delta(t) = \begin{cases} 0 & \text{at} \quad t \neq 0 \\ \infty & \text{at} \quad t = 0 \end{cases}. \tag{5.12}$$

Since the area under the impulse is unity,

$$\int_{-\infty}^{\infty} \delta(t)\, dt = 1.$$

5.5.1 Shifted Impulse Function

A unit impulse function occurs at $t = t_0$ is $\delta(t - t_0)$. It is defined by the relations

$$\delta(t - t_0) = \begin{cases} 0 & \text{at} \quad t \neq t_0 \\ \infty & \text{at} \quad t = t_0 \end{cases}. \tag{5.13}$$

A shifted or delayed impulse function is shown in Figure 5.17(a).

If the area of the impulse function is k times unity, it is known as a general impulse function. It is shown in Figure 5.17(b).

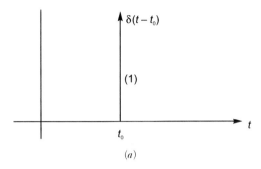

(a)

FIGURE 5.17

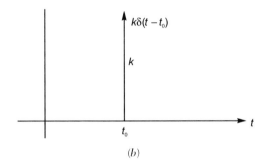

(b)

FIGURE 5.17

In this case,

$$\int_{-\infty}^{\infty} \delta(t - t_0)dt = k.$$

Example 5.10. *Find the Laplace transform of the impulse function.*

Ans. 1

5.6   EXPONENTIAL FUNCTION

A very commonly used exponential function is given by

$$f(t) = Ae^{\pm at}.$$

The positive $(+)$ sign shows for the exponential rise function and the negative $(-)$ sign shows for exponential decay, both exponential functions (rise and decay) are shown in Figure 5.18.

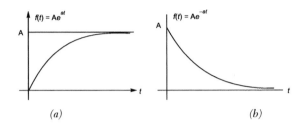

FIGURE 5.18

Mathematical expression for the exponential function is

$$f(t) = \begin{cases} 0 & \text{at} \quad t < 0 \\ Ae^{\pm at} & \text{at} \quad t \geq 0 \end{cases}. \tag{5.14}$$

If $A = 1$, this function is known as a unit exponential function.

Example 5.11. *Find the Laplace transform of the general exponential function (decaying).*

Ans. $\frac{A}{s+a}$

5.7   SINUSOIDAL SIGNAL

The sinusoidal signal $f(t)$ is shown in Figure 5.19 and is defined as

$$f(t) = \begin{cases} 0 & \text{at} \quad t < 0 \\ V_m \sin \omega t & \text{at} \quad t \geq 0 \end{cases},$$

where V_m is maximum amplitude and ω is angular frequency in rad/sec. $(\omega = \frac{2\pi}{T})$.

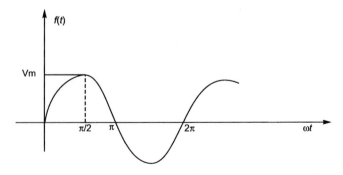

FIGURE 5.19

5.8   RELATION BETWEEN PARABOLIC, RAMP, UNIT STEP, AND IMPULSE FUNCTIONS

We know for the unit function and $t > 0$

$$u(t) = 1, \tag{5.15}$$

$$r(t) = t, \tag{5.16}$$

$$P(t) = \frac{t^2}{2}, \tag{5.17}$$

the derivative of a ramp signal = step signal, and
the derivative of a parabolic signal = a ramp signal,
or we can say,

$$\frac{d}{dt}P(t) = t = r(t) \tag{5.18}$$

$$\frac{d}{dt}r(t) = 1 = u(t) \tag{5.19}$$

and

$$\frac{d}{dt}u(t) = \delta(t), \tag{5.20}$$

the derivative of the unit step signal = the impulse signal.

TABLE 5.1

Signal	Laplace Transform	Remark
Impulse signal $\delta(t)$	1	Differentiate
Unit step signal $u(t)$	$\frac{1}{s}$	
Ramp signal $r(t)$	$\frac{1}{s^2}$	
Parabolic signal $P(t)$	$\frac{1}{s^3}$	Integrate

5.9   SYNTHESIS OF GENERAL WAVEFORMS

The basic waveforms (step, ramp, and parabolic) can be combined to obtain a desired waveform. Waveform synthesis is a method to determine the component waveforms of a desired waveform.

5.9.1 Useful Formula for Waveform Synthesis

The basic formula used in this book is very easy and convenient. First break any waveform into small intervals of time (especially where the magnitude or slope of the waveform is changing).

$$f(t) = M_1 G_{0,1}(t) + M_2 G_{1,2}(t) + M_3 G_{2,3}(t) + \cdots \qquad (5.21)$$

where, M_1 is the magnitude between 0 to 1 ($G_{0,1}(t)$ see $G_{0,1}(t)$ is a gate pulse starting at $t = 0$ and lasting at $t = 1$)

Similarly, for M_2 and $G_{1,2}(t)$ and so on. Let us consider an example shown in Figure 5.20.

$$M_1 = 1 \quad \text{and} \quad G_{0,1}(t) = u(t) - u(t - 1).$$
$$M_2 = 2 \quad \text{and} \quad G_{1,2}(t) = u(t - 1) - u(t - 2).$$

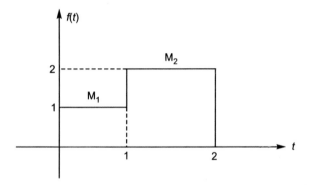

FIGURE 5.20

Hence,

$$f(t) = 1[u(t) - u(t - 1)] + 2[u(t - 1) - u(t - 2)].$$

Now for a triangular pulse where the magntiude (M) is changing with respect to time for a ramp function or straight we use the straight line formula,

$$y - y_1 = \frac{y_2 - y_1}{x_2 - x_1}(x - x_1). \qquad (5.22)$$

In this case $y = M$ and $x = t$.

x_1 and y_1 are the initial values of the slope and x_2 and y_2 are the final values of the slope. In this case, for Figure 5.21,

$$x_1 = 0, \quad y_1 = 0$$
$$x_2 = 1, \quad y_2 = 2.$$

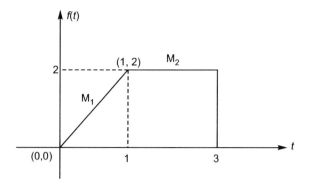

FIGURE 5.21

Put this value into Equation 5.22,

$$M_1 - 0 = \frac{2-0}{1-0}(t-0).$$
$$M_1 = 2t \text{ and } G_{0,1}(t) = u(t) - u(t-1).$$

The gate pulse will remain the same, only magnitude (M_1) is changing with respect to time,

$$M_2 = 2 \text{ and } G_{1,3}(t) = u(t-1) - u(t-3).$$

Now put these values in the basic formula (equation 5.21),

$$f(t) = 2t[u(t) - u(t-1)] + 2[u(t-1) - u(t-3)].$$

Example 5.12. *Synthesize the waveform as shown in Figure 5.22 in terms of unit and ramp function.*

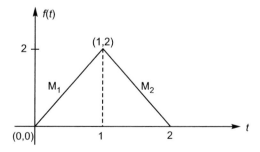

FIGURE 5.22

Solution: $f(t) = M_1 G_{0,1}(t) + M_2 G_{1,2}(t)$, where M_1,

$$M_1 - 0 = \frac{2-0}{1-0}(t-0) \text{ by Equation 5.22,}$$

$$M_1 = 2t \text{ and } G_{0,1}(t) = u(t) - u(t-1),$$

$$M_2 - 2 = \frac{0-2}{2-1}(t-1),$$

$$M_2 = 2 - 2t + 2 = 4 - 2t \text{ and } G_{1,2}(t) = u(t-1) - u(t-2).$$

Now

$$f(t) = 2t[u(t) - u(t-1)] + (4-2t)[u(t-1) - u(t-2)].$$

We can further solve

$$f(t) = 2tu(t) - 2tu(t-1) + (4-2t) + u(t-1) - (4-2t)(ut-2)$$
$$f(t) = 2r(t) - u(t-1)[2t - 4 + 2t] - (4-2t)u(t-2)[r(t) = tu(t)]$$
$$f(t) = 2r(t) - 4(t-1)u(t-1) - (4-2t)u(t-2)$$
$$= 2r(t) - 4r(t-1) - (4-2t)u(t-2)$$
$$[r(t-1) = (t-1)u(t-1)].$$

Example 5.13. *Synthesize the waveform as shown in Figure 5.23.*

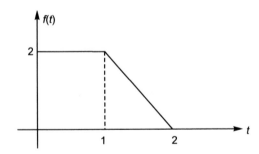

FIGURE 5.23

Ans. $f(t) = 2[u(t) - u(t-1)] + (4-2t)[u(t-1) - u(t-2)]$

SOLVED PROBLEMS

Problem 5.1. *Synthesize the waveform as shown in Figure 5.24 using standard signal.*

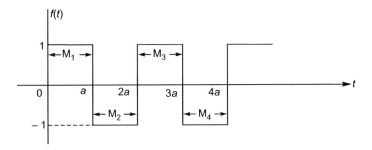

FIGURE 5.24

Solution:

$$f(t) = M_1 G_{0a}(t) + M_2 G_{a,2a}(t) + M_3 G_{2a,3a}(t) + M_4 G_{3a,4a}(t)$$
$$f(t) = 1[u(t) - u(t-a)] - 1[u(t-a) - u(t-2a)]$$
$$+ 1[u(t-2a) - u(t-3a)] - 1[u(t-3a) - u(t-4a)] + \cdots$$
$$f(t) = u(t) - 2u(t-a) + 2U(t-2a) - 2U(t-3a) + \cdots$$

Problem 5.2. *Synthesize the waveform as shown in Figure 5.25 using step functions.*

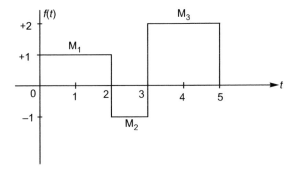

FIGURE 5.25

Solution:

$$M_1 = +1$$
$$M_2 = -1$$
$$M_3 = +2$$

$$f(t) = M_1 G_{0,2}(t) + M_2 G_{2,3}(t) + M_3 G_{3,5}(t)$$
$$= 1[u(t) - u(t-2)] + (-1) - u(t-2) - u(t-3)]$$
$$+ 2[u(t-3) - u(t-5)]$$
$$= u(t) - 2u(t-2) + 3u(t-3) - 2u(t-5).$$

Problem 5.3. *Synthesize the waveform as shown in Figure 5.26.*

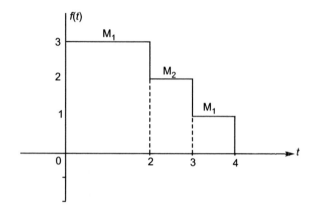

FIGURE 5.26

Solution:

$$f(t) = M_1 G_{0,2}(t) + M_2 G_{2,3}(t) + M_3 G_{3,4}(t)$$
$$f(t) = 3[u(t) - u(t-2)] + 2[u(t-2) - u(t-3)] + 1[u(t-3) - u(t-4)]$$
$$= 3u(t) - u(t-2) - u(t-3) - 4(t-4).$$

Problem 5.4. *The accompaning Figure 5.27 shows a waveform made up of straight line segments. For this waveform, write an equation for the v(t) in terms of steps, ramps, and other related functions as needed.*

Solution: Use the straight line formula, $M - y_1 = \frac{y_2 - y_1}{x_2 - x_1}(t - x_1)$.

$$f(t) = M_1 G_{0,1}(t) + M_2 G_{1,2}(t) + M_3 G_{2,3}(t) + M_4 G_{3,4}(t)$$

for M_1: $M_1 - 0 = \dfrac{2-0}{1-0}(t-0) \Rightarrow M_1 = 2t$

$M_2 = 2$

for M_3: $M_3 - 2 = \dfrac{-2-2}{-3-2}(t-2)$

$M_3 = 2 + 8 - 4t = (-4t + 10)$

for M$_4$: $\mathrm{M}_4 - (-2) = \dfrac{0 - (-2)}{4 - 3}(t - 3)$

$$\mathrm{M}_4 = -2 + 2t - 6 = 2t - 8$$
$$f(t) = 2t[u(t) - u(t - 1)] + 2[u(t - 1) - u(t - 2)]$$
$$+ (-4t + 10)[u(t - 2) - u(t - 3)]$$
$$+ (2t - 8)[u(t - 3) - u(t - 4)].$$

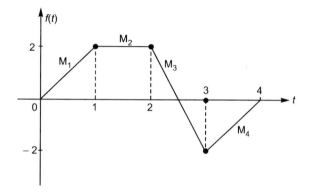

FIGURE 5.27

Problem 5.5. *Synthesize the waveform as shown in Figure 5.28.*

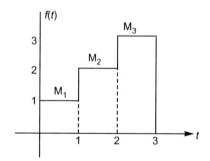

FIGURE 5.28

Solution:

$$f(t) = \mathrm{M}_1 \mathrm{G}_{0,1}(t) + \mathrm{M}_2 \mathrm{G}_{1,2}(t) + \mathrm{M}_3 \mathrm{G}_{2,3}(t)$$
$$= 1[u(t) - u(t - 1)] + 2[u(t - 1) - u(t - 2)] + 3[u(t - 2) - u(t - 3)]$$
$$f(t) = u(t) + u(t - 1) + u(t - 2) - 3u(t - 3).$$

Problem 5.6. *Synthesize the waveform as shown in Figure 5.29.*

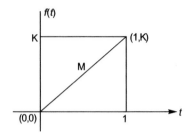

FIGURE 5.29

Solution:

$$f(t) = MG_{0,1}(t),$$

where M, using the straight line equation, is

$$M - y_1 = \frac{y_2 - y_1}{x_2 - x_1}(t - x_1)$$

$$M - 0 = \frac{K - 0}{1 - 0}(t - 0)$$

$$M = Kt$$

and

$$G_{0,1}(t) = [u(t) - u(t - 1)].$$

Therefore,

$$f(t) = Kt[u(t) - u(t - 1)].$$

Problem 5.7. *Synthesize the waveform as shown in Figure 5.30.*

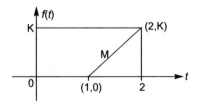

FIGURE 5.30

Solution:

$$f(t) = MG_{1,2}(t)$$

$$M - 0 = \frac{K - 0}{2 - 1}(t - 1)$$

$$M = K(t - 1)$$

$$f(t) = K(t - 1)[u(t - 1) - u(t - 2)]$$

Problem 5.8. *Synthesize the following waveforms as shown in Figure 5.31.*

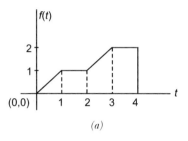 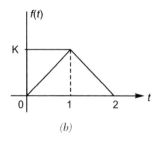

(a)   *(b)*

FIGURE 5.31

Solution: $(a) f(t) = M_1G_{0,1}(t) + M_2G_{1,2}(t) + M_3G_{2,3}(t) + M_4G_{3,4}(t)$
for M_1:

$$M_1 - 0 = \frac{1 - 0}{1 - 0}(t - 0)$$

$$M_1 = t$$

$$M_2 = 1.$$

for M_3:

$$M_3 - 1 = \frac{2 - 1}{3 - 2}(t - 2)$$

$$M_3 = (t - 1)$$

$$M_4 = 2$$

$$f(t) = t[u(t) - u(t - 1)] + 1[u(t - 1) - u(t - 2)]$$
$$+ (t - 1)[u(t - 2) - u(t - 3)] + 2[u(t - 3) - u(t - 4)].$$

$(b) f(t) = M_1G_{0,1}(t) + M_2G_{1,2}(t)$
for M_1:

$$M_1 - 0 = \frac{K - 0}{1 - 0}(t - 0)$$

$$M_1 = Kt$$

for M_2:

$$M_2 - K = \frac{0 - K}{2 - 1}(t - 1)$$

$$M_2 = 2K - Kt = -K(t - 2)$$

$$f(t) = K(t)[u(t) - u(t - 1)] - K(t - 2)[u(t - 1) - u(t - 2)]$$

Problem 5.9. *Synthesize the following waveform as shown in Figure 5.32.*

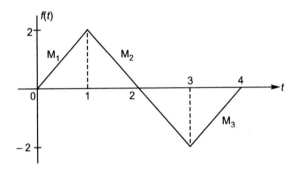

FIGURE 5.32

Solution:

$$f(t) = M_1 G_{0,1}(t) + M_2 G_{1,3}(t) + M_3 G_{3,4}(t) \tag{1}$$

Using the straight line formula,

$$M_1 - y_1 = \frac{y_2 - y_1}{x_2 - x_1}(t - x_1) \qquad \text{[This is from 0 to 1 slope]}$$

$$M_1 - 0 = \frac{2 - 0}{1 - 0}(t - 0).$$

$$M_1 = 2t$$

Similarly,

$$M_2 = 4 - 2t \qquad \text{[In this case } x_1 = 1 \text{ and } y_1 = 2, x_2 = 3 \text{ and } y_2 = -2]$$

and

$$M_3 = 2t - 8$$

$$G_{0,1}(t) = u(t) - u(t - 1)$$

$$G_{2,3}(t) = u(t - 1) - u(t - 3)$$

$$G_{3,4}(t) = u(t - 3) - u(t - 4).$$

Now put all values into equation (1),

$$f(t) = 2t[u(t) - u(t - 1)] + (4 - 2t)[u(t - 1)$$
$$- u(t - 3) + (2t - 8)[u(t - 3) - u(t - 4)].$$

Problem 5.10. *Synthesize the given waveform.*

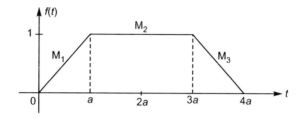

FIGURE 5.33

Solution:

$$f(t) = M_1 G_{0,a}(t) + M_2 G_{a,3a}(t) + M_3 G_{3a,4a}(t)$$

Similarly, by the last example

$$M_1 - 0 = \frac{1 - 0}{a - 0}(t - 0) \Rightarrow M_1 = \frac{t}{a}$$

$$M_2 = 1$$

$$M_3 - 1 = \frac{0 - 1}{4a - 3a}(t - 3a)$$

$$M_3 = \left(\frac{4a - t}{a}\right)$$

$$f(t) = \frac{t}{a}[u(t) - u(t - a)] + 1[u(t - a) - u(t - 3a)]$$
$$+ \left(\frac{4a - t}{a}\right) \times [u(t - 3a) - u(t - 4a)]$$

Problem 5.11. *Synthesize the waveform as shown in Figure 5.34.*

Solution:

$$f(t) = M_1 G_{0,1}(t) + M_2 G_{1,2}(t) + M_3 G_{2,3}(t) + \cdots$$

$$M_1 = \frac{t}{T}$$

$$M_2 = \frac{1}{T}(t - T) \text{ and } M_3 = \frac{1}{T}(t - 2T).$$

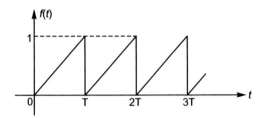

FIGURE 5.34

Now,

$$f(t) = \frac{t}{T}[u(t) - u(t - T)] + \frac{1}{T}[u(t - T) - u(t - 2T)]$$
$$+ \frac{1}{T}(t - 2t)[u(t - 2t) - u(t - 3t)] + \cdots$$

Problem 5.12. *Synthesize the waveform as shown in Figure 5.35.*

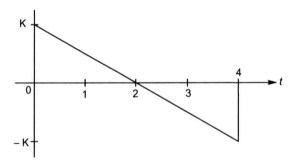

FIGURE 5.35

Solution:

$$f(t) = M_1 G_{0,4}(t)$$
$$M_1 - K = \frac{-K - K}{4 - 0}(t - 0)$$
$$M_1 = K - \frac{K}{2}t = \frac{K}{2}(2 - t)$$
$$f(t) = \frac{K}{2}(2 - t)[u(t) - u(t - 4)]$$

NOTE Students can also solve in two steps (0 to 2 then 2 to 4),

$$f(t) = M_1 G_{0,2}(t) + M_2 G_{2,4}(t).$$

Problem 5.13. *Synthesize the waveform as shown in Figure 5.36.*

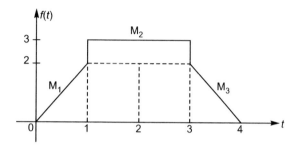

FIGURE 5.36

Solution:

$$f(t) = M_1 G_{0,1}(t) + M_2 G_{1,3}(t) + M_3 G_{3,4}(t).$$

Now,

$$M_1 - 0 = \frac{2 - 0}{1 - 0}(t - 0)$$

$$M_1 = 2t$$

$$M_2 = 3$$

$$M_3 - 2 = \frac{0 - 2}{4 - 3}(t - 3)$$

$$M_3 = 2 - 2(t - 3) = 2 - 2t + 6 = 8 - 2t$$

$$f(t) = 2t[u(t) - u(t - 1)] + 3[u(t - 1) - u(t - 3)]$$
$$\quad + (8 - 2t)[u(t - 3) - u(t - 4)]$$

$$f(t) = 2tu(t) - 2tu(t - 1) + 3u(t - 1) - 3u(t - 3)$$
$$\quad + (8 - 2t)u(t - 3) - (8 - 2t)u(t - 4)$$

$$f(t) = 2tu(t) + u(t - 1)[-2t + 3] + u(t - 3)[-3 + 8 - 2t]$$
$$\quad - (8 - 2t)u(t - 4)$$

$$f(t) = 2tu(t) + (3 - 2t)u(t - 1) + (5 - 2t)u(t - 3) - (8 - 2t)u(t - 4)$$

Problem 5.14. *Synthesize the given waveform as shown in Figure 5.37.*

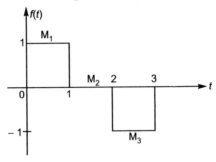

FIGURE 5.37

Solution:

$$f(t) = M_1 G_{0,1}(t) + M_2 G_{1,2}(t) + M_3 G_{2,3}(t).$$

Now

$$M_1 = 1, M_2 = 0, M_3 = -1$$
$$f(t) = 1[(u(t) - u(t-1)] - 1[u(t-2) - u(t-3)]$$
$$f(t) = u(t) - u(t-1) - u(t-2) + u(t-3)$$

Problem 5.15. *Synthesize the waveform as shown in Figure 5.38.*

Solution: By Figure 5.38,

$$f_1(t) = 10 \sin \pi t.$$

Now the shifted sinusoidal waveform

$$f_2(t) = 10 \sin \pi (t-1).$$

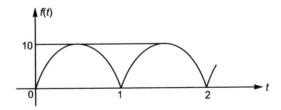

FIGURE 5.38

Adding Figure 5.38(*a*) and 5.38(*b*) we get

$$f_3(t) = f_1(t) + f_2(t),$$

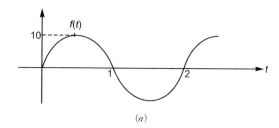

(a)

FIGURE 5.38

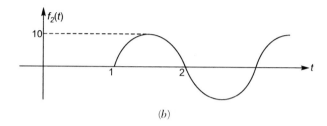

(b)

FIGURE 5.38

where

$$f_3(t) = 10 \sin \pi t + 10 \sin \pi (t - 1),$$

or

$$f_3(t) = 10[\sin \pi t + \sin \pi (t - 1)].$$

Similarly, for

$$f_4(t) = 10[\sin \pi (t - 1) + \sin \pi (t - 2)].$$

Now,

$$f(t) = f_3(t) + f_4(t) + \cdots$$
$$f(t) = 10[\sin \pi t + 2 \sin \pi (t - 1) + 2 \sin \pi (t - 2) + \cdots].$$

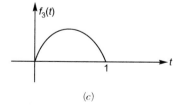

(c)

FIGURE 5.38

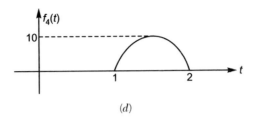

(d)

FIGURE 5.38

Problem 5.16. *Synthesize the following waveform as shown in Figure 5.39.*

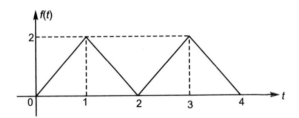

FIGURE 5.39

Solution:

$$f(t) = M_1 G_{0,1}(t) + M_2 G_{1,2}(t) + M_3 G_{2,3}(t) + M_4 G_{3,4}(t)$$

$$M_1 - 0 = \frac{2 - 0}{1 - 0}(t - 0) \Rightarrow M_1 = 2t$$

$$M_2 - 2 = \frac{0 - 2}{2 - 1}(t - 1) \Rightarrow M_2 = 4 - 2t$$

$$M_3 - 0 = \frac{2 - 0}{3 - 2}(t - 2) \Rightarrow M_3 = (2t - 4)$$

$$M_4 - 2 = \frac{0 - 2}{4 - 3}(t - 3) \Rightarrow M_4 = 8 - 2t$$

$$f(t) = 2t[u(t) - u(t - 1)] + (4 - 2t)[u(t - 1) - u(t - 2)]$$
$$+ (2t - 4)[u(t - 2) - u(t - 3)] + (8 - 2t)[u(t - 3) - u(t - 4)].$$

NOTE ➤ Students can solve further. It is necessary and convenient if we find the Laplace transform of any waveform.

Problem 5.17. *Find the Laplace transform of the time function in Figure 5.40.*

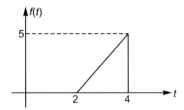

FIGURE 5.40

Solution:

$$f(t) = MG_{2,4}(t)$$

$$M - 0 = \frac{5 - 0}{4 - 2}(t - 2) = \frac{5}{52}(t - 2)$$

$$f(t) = \frac{5}{2}(t - 2)[u(t - 2) - u(t - 4)]$$

$$= \frac{5}{2}(t - 2)u(t - 2) - \frac{5}{2}(t - 2)u(t - 4)$$

$$= \frac{5}{2}(t - 2)u(t - 2) - \frac{5}{2}(t - 4 + 2)u(t - 4)$$

$$= \frac{5}{2}(t - 2)u(t - 2) - \frac{5}{2}(t - 4)u(t - 4) - 5u(t - 4)$$

$$f(t) = \frac{5}{2}r(t - 2) - \frac{5}{2}r(t - 4) - 5u(t - 4).$$

Taking the Laplace transform,

$$f(s) = \frac{5}{2}\frac{e^{-2s}}{s^2} - \frac{5}{2}\frac{e^{-4s}}{s^2} - \frac{5e^{-4s}}{s}$$

$$F(s) = \frac{5e^{-2s}}{s^2}[1 - e^{-2s} + 2se^{-2s}].$$

Problem 5.18. *Determine Laplace transform of the following waveform shown in Figure 5.41.*

Solution:

$$f(t) = M_1G_{0,3}(t) + M_2G_{3,5}(t)$$

$$M_1 = 4$$

$$M_2 - 4 = \frac{0-4}{5-3}(t-3) \Rightarrow M_2 = 4 - 2(t-3)$$

$$M_2 = 10 - 2t$$

$$f(t) = 4[u(t) - u(t-3)] + (-2t + 10)[u(t-3) - u(t-5)]$$

$$f(t) = 4u(t) - 2(t-3)u(t-3) + 2(t-5)u(t-5)$$

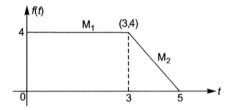

FIGURE 5.41

Taking the Laplace transform

$$f(s) = \frac{4}{s} - \frac{2e^{-3s}}{s^2} + \frac{2e^{-5s}}{s}.$$

Problem 5.19. *Synthesize the following waveform and find the Laplace transform.*

Solution:

$$f(t) = M_1 G_{0,1}(t) + M_2 G_{1,2}(t) + M_3 G_{2,3}(t)$$

$$M_1 = 2, M_2 = 1, M_3 = 2$$

$$f(t) = 2[u(t) - u(t-1)] + 1[u(t-1) - u(t-2)] + 2[u(t-2) - u(t-3)]$$

$$f(t) = 2u(t) + u(t-1)[-2+1] + u(t-2)[-1+2] - 2u(t-3)$$

$$f(t) = 2u(t) - u(t-1) + u(t-2) - 2u(t-3)$$

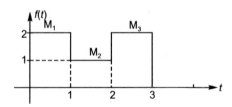

FIGURE 5.42

Taking the Laplace transform,

$$f(s) = \frac{2}{s} - \frac{1}{s}e^{-s} - \frac{2}{s}e^{-3s}.$$

$$= \frac{1}{s}[2 - e^{-s} + e^{-2s} - 2e^{-3s}].$$

Problem 5.20. *With the help of mathematical expressions, represent the following forcing functions with their characteristic curves.*

   (*i*)   *A pulse starting at t = 0 and lasting for 2 seconds.*
   (*ii*) *A Unit ramp and a delayed unit ramp.*
   (*iii*) *An impulse function occuring at t = a.*
   (*iv*) *Gate functions.*

Solution:
(i)

$$f(t) = \begin{cases} 0 & t < 0 \\ A & 0 \le t \le 2 \\ 0 & \text{at } t > 2. \end{cases}$$

$$f(t) = A[u(t) - u(t - 2)] \text{ A is amplitude of pulse.}$$

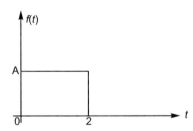

FIGURE 5.43

(ii) For the ramp function,

$$f(t) = \begin{cases} 0; & t < 0 \\ Kt; & t \ge 0 \end{cases}.$$

for the unit ramp function, K = 1

$$f(t) = \begin{cases} 0; & t < 0 \\ t; & t \ge 0 \end{cases},$$

and the shifted unit ramp function,

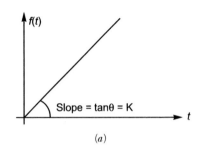

(a)

FIGURE 5.44

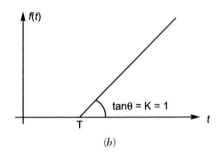

(b)

FIGURE 5.44

(iii)

$$\delta(t - a) = \begin{cases} 0; & t \neq a \\ 1; & t = a \end{cases}.$$

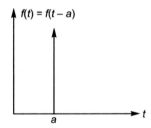

FIGURE 5.45

(iv) Gate functions,

$$f(t) = u(t) - u(t - a)$$
$$f(t) = u(t - a) - u(t - b).$$

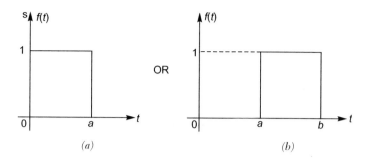

FIGURE 5.46

Problem 5.21. *Two ramp functions are given by $f_1(t) = mt\,u(t)$ $f_2(t) = m'(t-a)u(t-a)$, where m and m' are two slopes $(+ve)$ and $m > m'$. Draw the final waveform adding these two functions.*

Solution:

$$f_1(t) = mtu(t) \quad f_2(t) = m'(t-a)u(t-a)$$

Given $m > m'$,

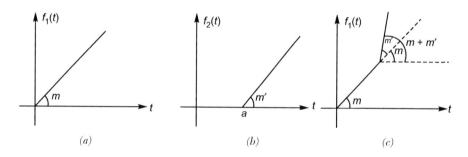

FIGURE 5.47

now,

$$f(t) = f_1(t) + f_2(t).$$

QUESTIONS FOR DISCUSSION

1. What is a unit step function?
2. What is a ramp function?

3. What is the relation between a unit step function and a ramp function?
4. What is a parabolic function?
5. What is an impulse function?
6. What is the relation between a unit step function and an impulse function?
7. What is the relation between ramp and parabolic function?
8. What is a shifted unit step function?
9. What is a shifted unit ramp function?
10. Make a pulse of width 'a' by using two unit step functions.
11. What is $f(t)$ in Figure 5.48?

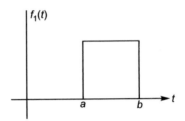

FIGURE 5.48

12. What is an exponential function?
13. What is $f(t)$ in Figure 5.49?

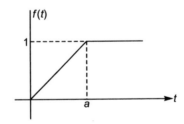

FIGURE 5.49

OBJECTIVE QUESTIONS

1. The impulse response of an initially replaced linear system is $e^{-2t}u(t)$. To produce a response of $te^{-2t}u(t)$, the input must be equal to

   (a) $2e^{-t}u(t)$ (b) $\dfrac{1}{2}e^{-2t}u(t)$ (c) $e^{-2t}u(t)$ (d) $e^{-t}u(t)$.

2. The unit impulse response of a system is given as $c(t) = -4e^{-t} + 6e^{-2t}$, the step response of the same system for $t \geq 0$ is equal to
    (a) $-3e^{-2t} + 4e^{-t} + 1$ (b) $-3e^{-2t} + 4e^{-t} - 1$
    (c) $-3e^{-2t} - 4e^{-t} - 1$ (d) $3e^{-2t} + 4e^{-2t} - 1$.

3. In the circuit shown in the figure, it is desired to have a constant direct current $i(t)$ through the ideal inductor L, so the nature of the voltage source $v(t)$ must be
    (a) constant voltage   (b) linearly increasing voltage
    (c) an ideal impulse   (d) exponentially increasing voltage.

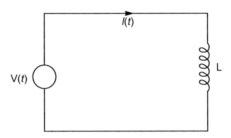

FIGURE 5.50

4. A rectangular voltage pulse of magnitude V and duration T is applied to a series combination of resistance R and capacitance C. The maximum voltage developed across the capacitor is
    (a) $V(1 - e^{-T/RC})$ (b) VT/RC   (c) V   (d) $Ve^{-T/RC}$.

5. A voltage $V(t) = 12t^2$ is applied across a 1 H inductor for $t \geq 0$, with the initial current through it being zero. The current through the inductor for $t \geq 0$ is given by
    (a) $12t$ (b) $24t$ (c) $12t^3$ (d) $4t^3$.

6. The voltage across R after $t = 0$ and $t = 1.0$ sec. will be
    (a) 100 V, 632 V   (b) 0 V, 63.2 V
    (c) 100 V, 36.8 V   (d) 0 V, 36.8 V.

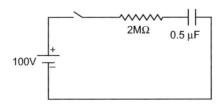

FIGURE 5.51

7. If the step response of an initially relaxed circuit is known, then the ramp response can be obtained by
    (a) integrating the step response
    (b) differentiating the step response
    (c) integrating the step response twice
    (d) differentiating the step response twice.

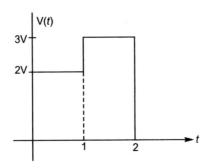

FIGURE 5.52

8. The Laplace transform of V(t) shown in Figure 5.52 is
    (a) $\frac{V}{s}e^{-s} - \frac{3V}{s}e^{-2s}$ (b) $\frac{2V}{s} - \frac{3V}{s}e^{-2s}$
    (c) $\frac{2V}{s} + \frac{V}{s}e^{-s}$ (d) $\frac{2V}{s} + \frac{V}{s}e^{-s} - \frac{3V}{s}e^{-2s}$.
9. The equation for V(t) as shown in Figure 5.53 is
    (a) $u(t-1) + u(t-2) + u(t-3)$
    (b) $u(t-1) + 2u(t-2) + 3u(t-3)$
    (c) $u(t) + u(t-1) + u(t-2) + u(t-4)$
    (d) $u(t-1) + u(t-2) + u(t-3) - 3u(t-4)$.

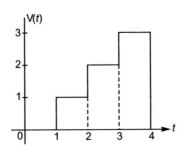

FIGURE 5.53

10. The mathematical expression for the given waveform in Figure 5.54 is:
$(a) f(t) = 2[u(t) - u(t - 1)]$ $(b) f(t) = 1[u(t) - u(t - 2)]$
$(c) f(t) = [ut - u(t - 1)]$ $(d) f(t) = 2[u(t - 1) - u(t - 2)]$.

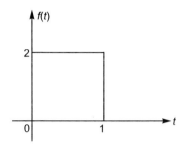

FIGURE 5.54

11. The unit step function is the first derivative of the
(a) ramp function (b) impulse function
(c) gate function (d) parabolic function.

UNSOLVED PROBLEMS

1. Define unit step, unit impulse, unit ramp, and shifted unit step functions and indicate their inter-relationship.

2. Find the current $i(t)$ in a series RLC circuit comprising R $= 3\Omega, L = 1$ H, and $C = 0.1\,\mu F$ when each of the following driving force voltages is applied:
(i) ramp voltage $12r(t - 2)$ (ii) step voltage $2u(t - 3)$.

3. Synthesize the waveform shown in Figure 5.55.

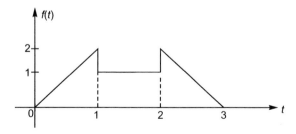

FIGURE 5.55

4. Show how a unit pulse of width '*a*' can be obtained from two unit step functions.

5. Define impulse function and represent it in mathematical and graphical forms. Show that it is a derivative of the unit step function. How is it realized practically?

6. Show that the derivative of a parabolic function is a ramp function and the derivative of a ramp function is a step function.

7. Write the mathematical expressions for the waveforms shown in Figure 5.56.

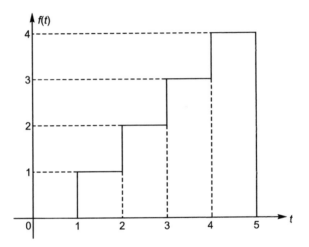

FIGURE 5.56

8. Synthesize the waveforms shown in Figure 5.57.

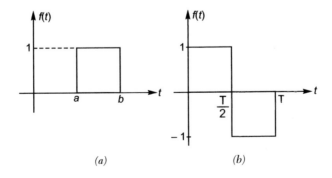

(a)   *(b)*

FIGURE 5.57

9. Synthesize the waveform shown in Figure 5.58.

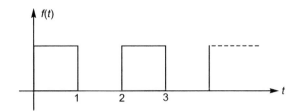

FIGURE 5.58

10. Synthesize the following waveforms:

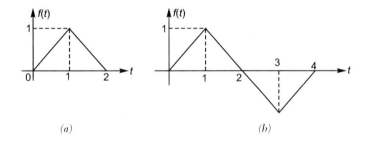

FIGURE 5.59

11. A sinusoidal current with a period of $2s$ and an amplitude of $100\,mA$ begins at $t = 3s$. Write an equation for the current.

12. The waveform shown in Figure 5.60 occurs only once. Write an equation for $v(t)$ in terms of step and related functions as needed.

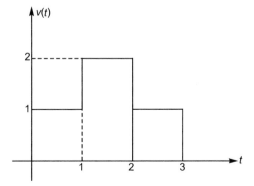

FIGURE 5.60

13. Look at the waveform shown in Figure 5.61. Write an equation for this waveform $v(t)$.

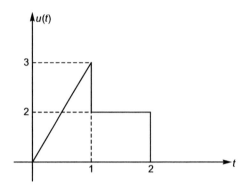

FIGURE 5.61

14. Write an equation for the waveform shown in Figure 5.62 as a linear combination of step functions.

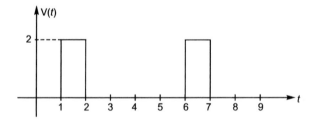

FIGURE 5.62

15. Synthesize the waveform.

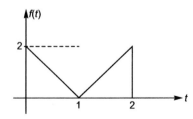

FIGURE 5.63

16. Write the mathematical expressions for the given waveform in terms of step function and ramp function.

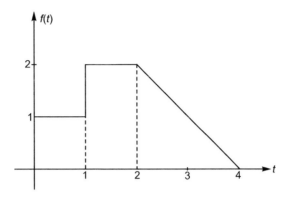

FIGURE 5.64

17. Synthesize the waveform in Figure 5.65.

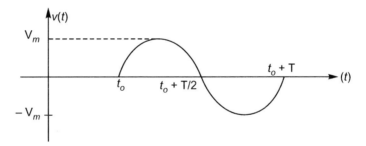

FIGURE 5.65

18. Synthesize the following waveform in Figure 5.66.

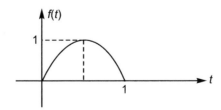

FIGURE 5.66

19. Write the mathematical expressions of the waveform shown in Figure 5.67.

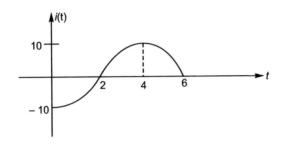

FIGURE 5.67

CIRCUIT ANALYSIS BY THE CLASSICAL METHOD

Chapter **6**

6.1 INTRODUCTION

Suppose a system is present. Whether it is working or not does not matter. If we want to change the present condition of the system; we will apply some force or signal to the system or remove it from the system. Now the system is, unbalanced and trying to create a stable or steady state condition from the start to before achieving the steady-state period is known as the transition period. During this period the elements, current, and voltages change from their former values to new ones. This period is known as transient. After the transition period, the circuit is said to be in a steady state.

Now, the linear differential equation, that is, the drive from the circuit, will have two components in its solution:

1. The complementary function (corresponds to the transient).
2. The particular integral (corresponds to the steady state).

Basically, the classical method is a mathematical approach to solve a circuit.

The circuit changes (ON or OFF) are assumed to occur at time $t = 0$ and are represented by a switch as shown in Figure 6.1.

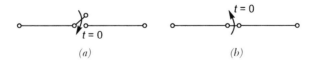

(a)　　　　　(b)

FIGURE 6.1

The arrowhead shows the change in direction. In Figure 6.1(a) initially the circuit was open, at $t = 0$ the circuit closed. Similarly, in Figure 6.1(b)

initially fine circuit was closed, at $t = 0$ the circuit opened.

$t = 0-$, the instant prior to $t = 0$

$t = 0+$, the instant immediately after switching

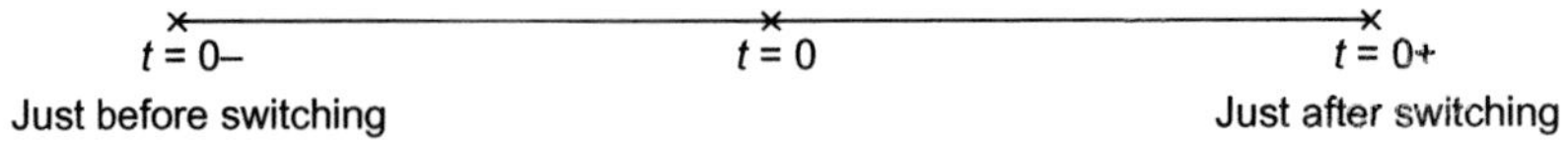

| $t = 0-$ | $t = 0$ | $t = 0+$ |
| Just before switching | | Just after switching |

FIGURE 6.2

Suppose, $i_L(0 -)$ is the value of the current flowing through the inductor before switching.

Example 6.1. *Find the differential equations for the given electrical system.*

Solution: Applying KVL,

$$V = V_R + V_C.$$

Put the value of V_R and V_C using Table 6.1.

$$V = Ri(t) + \frac{1}{C} \int_{-\infty}^{t} i(t)dt \tag{6.1}$$

[Value of current is the same]

TABLE 6.1   V-I Relationship for the Parameters

Parameter	V-I Relationship
Resistor (R)	$v_R(t) = Ri_R(t) \Rightarrow i_R(t) = \frac{v_R(t)}{R}$
Inductor (L)	$v_L(t) = L\frac{di_L}{dt} \Rightarrow i_L(t) = \int_{-\infty}^{t} v_L(t)dt$
Capacitor (C)	$v_C(t) = \frac{1}{C} \int_{-\infty}^{t} i_C(t)dt \Rightarrow i_C(t) = C\frac{dv_C(t)}{dt}$

Now differentiating the Equation (6.1),

$$\left[\frac{dV}{dt} = 0 \right]$$

$$0 = R\frac{di(t)}{dt} + \frac{1}{C}i(t)\frac{di(t)}{dt} + \frac{1}{RC}i(t) = 0. \tag{6.1a}$$

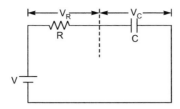

FIGURE 6.3

Example 6.2. *Find the differential equation for the given electrical system.*

Solution: Applying KVL,

$$V_m \sin \omega t = Ri(t) + L\frac{di(t)}{dt}. \tag{6.2}$$

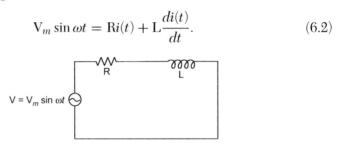

FIGURE 6.4

6.2   LINEAR DIFFERENTIAL EQUATIONS

To study the transients condition in electric circuits, it is necessary to be familiar with the mathematical concept of linear differential equations and solution techniques.

The order of the linear differential equation represents the highest derivative involved and is equal to the number of energy-storing elements. It is clear by Equations (6.1) and (6.2), the differential equations that are formed for transient conditions should be linear, ordinary differential equations with constant coefficients.

In this chapter we will discuss only first and second order differential equations.

TYPE I (First-order Homogeneous Differential Equations)

$$\frac{dy(t)}{dt} + Py(t) = 0, \quad \text{where P is a constant} \tag{6.3}$$

auxiliary equation of Equation (6.3).

Equation (6.3) can be expressed by

$$\left(\frac{d}{dt} + P\right)y(t) = 0y(t) \neq 0 \left[\frac{d}{dt} = m\right]$$

$$m + P = 0 \Rightarrow m = -p.$$

Hence, the solution is, $y(t) = Ke^{-mt} = Ke^{-Pt}$, where K is a constant.

TYPE II (First-Order Non-Homogeneous Differential Equations)

$$\frac{dy(t)}{dt} + Py(t) = Q \tag{6.4}$$

The auxiliary equation of Equation (6.4) is

$$m + P = 0$$

$$m = -P.$$

The complementary function (C.F.) $= K_1 e^{-mt} = K_1 e^{-Pt}$.
Now the particular integral

$$\text{(P.I.)} = \frac{Q}{F(D)} = \frac{Q}{D + P}. \qquad \left[\text{In this case, } \frac{d}{dt} = D\right]$$

The P.I. will vary according to the value of function Q.
The final solution

$$y(t) = \text{C.F.} + \text{P.I.} \tag{6.5}$$

TYPE III (Second is order Differential Equation)

$$A\frac{d^2y(t)}{dt^2} + B\frac{dy(t)}{dt} + Cy(t) = 0 \tag{6.6}$$

The auxiliary equation of Equation (6.6) is

$$Am^2 + Bm + c = 0$$

$$m^2 + \frac{B}{A}m + \frac{C}{A} = 0$$

$$m = \frac{-\frac{B}{A} \pm \sqrt{(\frac{B}{A})^2 - \frac{4C}{A}}}{2} = -\frac{B}{2A} \pm \frac{1}{2A}\sqrt{B^2 - 4AC}$$

$$m_1 = -\frac{B}{2A} + \frac{1}{2A}\sqrt{B^2 - 4AC}, \quad m_2 = -\frac{B}{2A} - \frac{1}{2A}\sqrt{B^2 - 4AC}.$$

The roots of the auxiliary equation (A.E.) are distinct so the solution of the equation will be

$$y(t) = K_1 e^{m_1 t} + K_2 e^{m_2 t}. \tag{6.7}$$

Put the values of m_1 and m_2. This is the final solution. There are no P.I. terms because the right-hand function [Q] is zero.

6.3   DETERMINATION OF THE COMPLEMENTARY FUNCTION (C.F.)

Case I: *If all the roots of the auxiliary equation are distinct* $(m_1, m_2, m_3, \ldots, m_n)$, the complementary function in this case is:

$$y = K_1 e^{m_1 t} + K_2 e^{m_2 t} + \cdots + K_n e^{m_n t}. \tag{6.8}$$

Suppose the auxiliary equation has one root only, $m_1 = -P$,

$$\text{C.F.} = K_1 e^{-Pt}.$$

Case II: *Auxiliary equation having equal roots*
   If two roots are equal, say $m_1 = m_2$, then the complementary function becomes

$$y = (K_1 + K_2 t)e^{mt}. \qquad [m_1 = m_2 = m]$$

Example 6.3. *Find the solution of equation* $\frac{d^2 i}{dt^2} + 2\frac{di}{dt} + 4 = 0$.

Solution: The A.E. is $m^2 + 2m + 4 = 0$

$$(m + 2)^2 = 0$$
$$m = -2, -2 (\text{roots are equal})$$
$$\text{C.F.} = (K_1 + K_2 t)e^{mt} [m = -2]$$
$$= (K_1 + K_2 t)e^{-2t}$$
$$\text{P.I.} = 0. \qquad\qquad [Q \text{ term is zero}]$$

The final solution is $i(t) = (K_1 + K_2 t)e^{-2t}$.

Case III: *Auxiliary equation having complex roots*
   Let the two roots of the auxiliary equation be complex, say $m_1 = \alpha + J\beta$ and $m_2 = \alpha - J\beta$ (where $J = \sqrt{-1}$), and the solution corresponding to these

two roots will be

$$y = K_1 e^{(\alpha + J\beta)t} + K_2 e^{(\alpha - J\beta)t}.$$

After solving,

$$y = (K_1 + K_2) e^{\alpha t} \cos \beta t + J(K_1 - K_2) e^{\alpha t} \sin \beta t$$

or

$$y = e^{\alpha t} [A_1 \cos \beta t + A_2 \sin \beta t] \tag{6.9}$$
$$A_1 = K_1 + K_2, A_2 = J(K_1 - K_2).$$

This is the final solution.

Example 6.4. *Solve:* $\frac{d^2i}{dt^2} - 7\frac{di}{dt} + 12i = 0.$

Solution: The auxiliary equation is

$$m^2 - 7m + 12 = 0$$
$$(m - 3)(m - 4) = 0$$
$$m_1 = 3, m_2 = 4.$$

Hence, the solution is

$$i(t) = K_1 e^{m_1 t} + K_2 e^{m_2 t} = K_1 e^{3t} + K_2 e^{4t}.$$

NOTE ▶ In this case the final solution and complementary function are the same because Q is zero, hence, the P.I. is zero.

We know, the final solution = C.F. + P.I.

Example 6.5. *Solve for i(t).*

$$\frac{d^3 i(t)}{dt^3} - 3\frac{di}{dt} + 2i = 0$$

Solution: The auxiliary equation is $m^3 - 3m + 2 = 0$

$$(m - 1)(m - 1)(m + 2) = 0$$
$$m_1 = 1, m_2 = 1, m_3 = -2.$$

Two roots, m_1 and m_2, are the same.
Hence, the C.F. for the final solution is $i(t) = (K_1 + K_2 t)e^t + K_3 e^{-2t}.$

Example 6.6. *Find the solution for $i(t)$.*

$$\frac{d^3i}{dt^3} - 8i = 0$$

Solution: The auxiliary equation is $m^3 - 8 = 0$

$$(m - 2)(m^2 + 2m + 4) = 0$$

$$m_1 = 2, m_{2,3} = -1 \pm J\sqrt{3}. \qquad [\alpha = -1, \beta = \sqrt{3}]$$

Two roots are complex conjugates.

Hence, the solution is $i(t) = K_1 e^{2t} + e^{-t}[A_1 \cos \sqrt{3}t + A_2 \sin \sqrt{3}t]$.

6.4 DETERMINATION OF THE PARTICULAR INTEGRAL (P.I.)

Suppose a differential equation,

$$A\frac{d^2i}{dt^2} + B\frac{di}{dt} + C_i = Q, \qquad (6.10)$$

Where A, B, and C are constant.

NOTE ▶ P.I. exists, if the value of Q is non-zero.

Case I: *To find the P.I. when Q is of the form e^{at}, where a is a constant.*

$$\text{P.I.} = \frac{1}{F(D)}e^{at} = \frac{1}{F(a)}e^{at}[F(a) \neq 0] \qquad (6.11)$$

Example 6.7. *Find the solution of the given equation,*

$$\frac{d^2i}{dt^2} - 3\frac{di}{dt} + 2i = 5e^{5t}.$$

Solution: In this case Q is non-zero [$Q = 5e^{5t}$], hence, the complete solution = C.F. + P.I.

For C.F. the auxiliary equation is

$$m^2 - 3m + 2 = 0$$

$$m = 1, 2,$$

hence,

$$\text{C.F.} = K_1 e^t + K_2 e^{2t}$$

For the P.I.

$$\text{P.I.} = \frac{1}{F(D)} \times 5e^{5t} = \frac{1}{(D^2 - 3D + 2)} \times 5e^{5t} \qquad [D = a = 5]$$

$$\text{P.I.} = \left(\frac{5}{25 - 3 \times 5 + 2}\right)e^{5t} = \frac{5}{12}e^{5t}.$$

Hence, the complete solution is $i(t) = K_1 e^t + K_2 e^{2t} + \frac{5}{12}e^{5t}$.

Case II: *To find the P.I. when Q is of the form of sin at or cos at*

$$\text{P.I.} = \frac{1}{F(D^2)} \sin at = \frac{1}{F(-a^2)} \sin at[F(-a^2) \neq 0]. \qquad (6.12)$$

Similarly,

$$\text{P.I.} = \frac{1}{F(D^2)} \cos at = \frac{1}{F(-a^2)} \cos at[F - (a^2) \neq 0]. \qquad (6.13)$$

Example 6.8. *Solve* $\frac{d^2i}{dt^2} - \frac{di}{dt} - 2i = \sin 2t$

Solution: For the C.F. the auxiliary equation is $m^2 - m - 2 = 0$

$$m = -1, 2$$

$$\text{C.F.} = K_1 e^{-t} + K_2 e^{2t}.$$

For the P.I.,

$$\text{P.I.} = \frac{1}{D^2 - D - 2} \sin 2t = \frac{1}{-4 - D - 2} \sin 2t \qquad [D = a = 2]$$

$$= -\frac{1}{(D + 6)} \sin 2t = -\frac{(D - 6)}{(D^2 - 36)} \sin 2t$$

$$= -\frac{(D - 6)}{-4 - 36} \sin 2t[D = a = 2] \left[\frac{1}{F(D^2)} = \frac{1}{F(-a^2)}\right]$$

$$= \frac{1}{40}[D(\sin 2t) - 6 \sin 2t] \qquad [D \text{ means differentiating}]$$

$$= \frac{1}{40}[2 \cos 2t - 6 \sin 2t].$$

Hence, the complete solution,

$$i(t) = K_1 e^{-t} + K_2 e^{2t} + \frac{1}{40}[2 \cos 2t - 6 \sin 2t].$$

Case III: *To find the P.I. when Q is of the form of* $(t)^m$, *where m is a positive integer,*

$$\text{P.I.} = \left(\frac{1}{D-a}\right)t^m = -\frac{1}{a\{1-(\frac{D}{a})\}}t^m = -\frac{1}{a}\left(1-\frac{D}{a}\right)^{-1}t^m$$

$$= -\frac{1}{a}\left(1+\frac{D}{a}+\frac{D^2}{a^2}+\cdots\right)t^m. \tag{6.14}$$

Expanding by the binomial theorem.

Example 6.9. *Solve* $\frac{d^2i}{dt^2} - 4i = t^2$.

Solution: For the C.F. the A.E. is $m^2 - 4 = 0$; $m = \pm 2$.

$$\text{C.F.} = K_1 e^{2t} + K_2 e^{-2t}.$$

For the

$$\text{P.I.} = \left(\frac{1}{D^2-4}\right)t^2 = -\frac{1}{4\left[1-\frac{1}{4}D^2\right]}t^2 = -\frac{1}{4}\left[1-\frac{1}{4}D^2\right]^{-1}t^2$$

$$= -\frac{1}{4}\left[1+\frac{1}{4}D^2+\cdots\right]t^2.$$

Expand by the binomial theorem up to the terms containing D^2 because all the other terms vanish

$$= -\frac{1}{4}\left(t^2 + \frac{1}{4}D^2 \cdot t^2\right) \qquad [D^2 \text{ means two times differentiating}]$$

$$= -\frac{1}{4}\left[t^2 + \frac{1}{4} \times 2\right] = -\frac{1}{4}\left[t^2 + \frac{1}{2}\right].$$

Hence, the complete solution is $i(t) = \text{C.F.} + \text{P.I.}$

$$i(t) = K_1 e^{2t} + K_2 e^{-2t} - \frac{1}{4}\left[t^2 + \frac{1}{2}\right],$$

where K_1 and K_2 are constants.

6.5 INITIAL CONDITIONS IN CIRCUITS

We need initial conditions $(t = 0)$ to evaluate the arbitrary constants $(K_1, K_2,$ etc.) in the complete solution of differential equations. The number of initial

TABLE 6.2   Initial and Final Behavior of R, L, C

Element with Initial Conditions	Equivalent Circuit at $t = 0+$	Equivalent Circuit at $t = \infty$
R	R	R
L	OC	SC
(capacitor)	SC	O/C

conditions required is equal to the number of energy-storing elements in the circuit.

Procedure to evaluate the initial conditions. There are two positions of the switch: just before switching and just after switching. Transients start just after switching and move to stability. The initial position (just before switching) gives the initial value.

NOTE ▶ Inductance is a property of material, which opposes any change in current. If in any circuit, the value of the current is changing, the inductor will try to flow the same current in the circuit due to the energy stored in the inductor. It means initial and final values for the inductor is current. Similarly, the initial and final value for the capacitor is voltage [$V_C(0)$ and $V_C(\infty)$].

Example 6.10. *Find the initial value of current in the circuit shown in Figure 6.5.*

Solution: Initially the circuit is open. At $t = 0$ the switch is closed.

Initially in the circuit $i(0) = 0$. There was no initial current because the circuit was open.

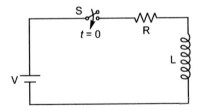

FIGURE 6.5

Example 6.11. *Find the initial value of current in the circuit shown in the figure.*

Solution: Initially the circuit is closed at $t = 0$ and the switch moves from position 1 to 2.

$$i(0) = \frac{V}{R_1} \text{ [It is assumed the first position was for a very long time}$$

therefore the inductor behaves as a short circuit].

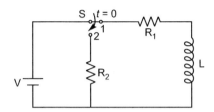

FIGURE 6.6

6.6 TRANSIENT RESPONSE OF SERIES RL CIRCUIT HAVING D.C. EXCITATION

Initially the circuit is open which means $i(0) = 0$ and it is closed at $t = 0$.

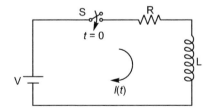

FIGURE 6.7

Now applying KVL,

$$V = i(t)R + L\frac{di(t)}{dt}$$

$$\frac{di(t)}{dt} + \frac{R}{L}i(t) = \frac{V}{L},$$

for the general solution, C.F. + P.I.

$$\text{C.F.} = K_1 e^{-\frac{R}{L}t}$$

$$\text{P.I.} = \frac{\frac{V}{L}e^{0t}}{D + \frac{R}{L}} \qquad\qquad [D = a = 0]$$

$$= \frac{V}{L} \times \frac{L}{R} = \frac{V}{R}$$

$$i(t) = \frac{V}{R} + K_1 e^{-\frac{R}{L}t} \quad \text{and } i(0) = 0.$$

$$0 = \frac{V}{R} + K_1 \Rightarrow K_1 = \frac{-V}{R}$$

$$i(t) = \frac{V}{R} - \frac{V}{R}e^{-\frac{R}{L}t} = \frac{V}{R}\left[1 - e^{-\frac{R}{L}t}\right] \qquad (6.15)$$

$$V_R(t) = Ri(t) = V\left[1 - e^{-\frac{R}{L}t}\right] \qquad (6.16)$$

$$V_L(t) = L\frac{di(t)}{dt} = L\frac{V}{R}\left[0 - \left(-\frac{R}{L}\right)e^{-\frac{R}{L}t}\right]$$

$$V_L(t) = Ve^{-\frac{R}{L}t}. \qquad (6.17)$$

It is clear by Equation (6.15),

$$i(0) = \frac{V}{R}[1 - e^{-0}] = \frac{V}{R}[1 - 1] = 0$$

$$V_L(0) = V, V_R(0) = 0$$

at

$$t = \infty, i(\infty) = \frac{V}{R} \text{ and } V_L(\infty) = 0, V_R(\infty) = V$$

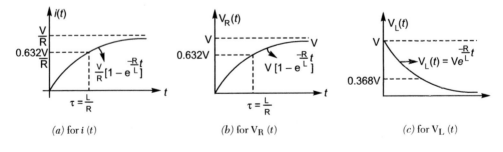

(a) for $i(t)$ (b) for $V_R(t)$ (c) for $V_L(t)$

FIGURE 6.8

at
$$t = \tau = \frac{L}{R}$$
$$i(\tau) = \frac{V}{R}[1 - e^{-1}] = 0.632\frac{V}{R}$$

and
$$V_L(t) = Ve^{-1} = 0.368V, V_R(t) = 0.632V.$$

NOTE ▶ $\tau = \frac{L}{R}$ is known as the time constant of the circuit and is defined as the interval after which current or voltage changes 63.2% of the final value.

Example 6.12. *Find i(t) for the given RL circuit.*

Solution: Initially, the switch was at position 1 and at $t = 0$ the switch moved to position 2.

$$i(0) = \frac{V}{R} \text{ [It is assumed, the first position was for a very long time,}$$

therefore, the inductor behaves as a short circuit.]

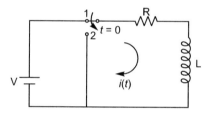

FIGURE 6.9

Now applying KVL, for position 2,

$$i(t)R + L\frac{di(t)}{dt} = 0 \Rightarrow \frac{di(t)}{dt} + \frac{R}{L}i(t) = 0$$
$$i(t) = \text{C.F.} + \text{P.I.}$$

For the C.F., the A.E. is $m + \frac{R}{L} = 0$

$$m = -\frac{R}{L}$$
$$\text{C.F.} = K_1 e^{-\frac{R}{L}t}$$

P.I. = 0 because Q is zero.

$$i(t) = K_1 e^{-\frac{R}{L}t} \quad \text{[Exponential decay]}$$

$$i(0) = \frac{V}{R} = K_1 e^{-\frac{R}{L} \times 0}$$

$$K_1 = \frac{V}{R}$$

$$\boxed{i(t) = \frac{V}{R} e^{-\frac{R}{L} t}.}$$

(6.18)

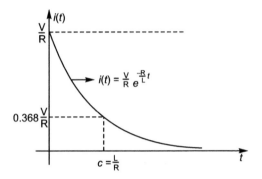

FIGURE 6.10

6.7 TRANSIENT RESPONSE OF SERIES RC CIRCUIT HAVING D.C. EXCITATION

Applying KVL,

$$V = Ri(t) + \frac{1}{C} \int_{-\infty}^{t} i(t) dt$$

or

$$V = Ri(t) + \frac{1}{C} \int_{0}^{t} i(t) dt + V_C(0).$$

Put $V_C(0) = 0$ and differentiating, we get

$$R\frac{di(t)}{dt} + \frac{1}{C}i(t) = 0 \Rightarrow \frac{di(t)}{dt} + \frac{1}{RC}i(t) = 0.$$

(6.19)

The general solution of this differential equation is C.F. + P.I.

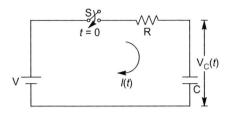

FIGURE 6.11

For the C.F. the A.E. is $m + \frac{1}{RC} = 0 \Rightarrow m = -\frac{1}{RC}$.

$$\text{C.F.} = K_1 e^{-\frac{t}{RC}}.$$
$$\text{P.I.} = 0 \text{ because Q [right-hand term] is zero.}$$
$$i(t) = K_1 e^{-\frac{t}{RC}}$$

at $t = 0+, i(0+) = \frac{V}{R}$ (since the capacitor behaves as a short circuit at switching)

$$v_C(t) = \frac{1}{C} \int_0^t i(t)dt = \frac{1}{C} \int_0^t K_1 e^{-\frac{t}{RC}} dt \quad \left[i(0+) = \frac{V}{R} = K_1 e^{-0} \right]$$

$$= \frac{K_1}{C} \times \left. \frac{e^{-\frac{t}{RC}}}{-\frac{1}{RC}} \right|_0^t \left[K_1 = \frac{V}{R} \right]$$

$$v_C(t) = -RK_1 \left[e^{-\frac{t}{RC}} - 1 \right]$$

$$v_C(t) = RK_1 \left[1 - e^{-\frac{t}{RC}} \right] = R \times \frac{V}{R} \left[1 - e^{-\frac{t}{RC}} \right]$$

$$v_C(t) = V \left[1 - e^{-\frac{t}{RC}} \right]$$

$$v_R(t) = i(t)R = \frac{V}{R} e^{-\frac{t}{RC}} \times R = V e^{-\frac{t}{RC}}$$

at $t = 0, i(t) = \frac{V}{R}$ and $V_R(t)2 = V; V_C(t) = 0$

at $t = \infty, i(t) = 0$ and $V_R(t) = 0; V_C(t) = V$

at $t = R_C$, [time constant $= \tau$]; $i(t) = \frac{V}{R} e^{-1} = 0.368 \frac{V}{R}$

$$V_C(t) = V(1 - e^{-1}) = 0.632V, V_R(t) = 0.368V.$$

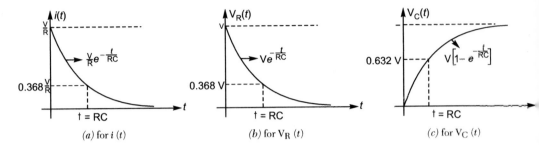

(a) for $i(t)$ (b) for $V_R(t)$ (c) for $V_C(t)$

FIGURE 6.12

Example 6.13. *Find the $i(t)$ for the circuit shown in Figure 6.13.*

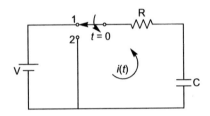

FIGURE 6.13

Solution: Initially the switch is at position 1. It moves from position 1 to 2 at $t = 0$. The capacitor keeps the steady state voltage $V_C(0) = V$, and the direction of $i(t)$ is negative because the capacitor is discharging in this case.

Applying KVL, for position 2,

$$\frac{1}{C} \int_0^t i(t)dt + V_C(0) + i(t)R = 0$$

$$\frac{1}{C} \int_0^t i(t)dt + Ri(t) = -V[V_C(0) = V].$$

Differentiating, we get

$$R\frac{di(t)}{dt} + \frac{1}{C}i(t) = 0 \Rightarrow \frac{di(t)}{dt} + \frac{1}{RC}i(t) = 0.$$

Now the solution is $i(t) = \text{C.F.} + \text{P.I.}$

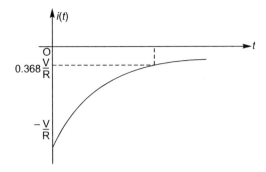

FIGURE 6.14

For the C.F., the A.E. is $m + \frac{1}{RC} = 0 \Rightarrow m = -\frac{1}{RC}$

$$\text{C.F.} = K_1 e^{-\frac{t}{RC}}$$

and

$$\text{P.I.} = 0. \qquad\qquad [Q = 0]$$

Therefore, the final solution is $i(t) = K_1 e^{-\frac{t}{RC}}$

$$i(0+) = -\frac{V}{R} = K_1$$

$$K_1 = -\frac{V}{R}$$

$$\boxed{i(t) = -\frac{V}{R} e^{-\frac{t}{RC}}}. \qquad\qquad (6.20)$$

6.8   TRANSIENT RESPONSE OF SERIES RLC CIRCUIT HAVING D.C. EXCITATION

Initial values are assumed to be zero $V_C(0) = 0, i_L(0) = 0$.

Applying KVL,

$$V = i(t)R + L\frac{di(t)}{dt} + \frac{1}{C}\int_0^t i(t)dt + V_C(0). \qquad\qquad (6.21)$$

Differentiating, we get

$$L\frac{d^2 i(t)}{dt^2} + R\frac{di(t)}{dt} + \frac{1}{C}i(t) = 0$$

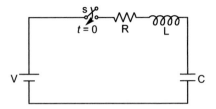

FIGURE 6.15

or

$$\frac{d^2i(t)}{dt^2} + \frac{R}{L}\frac{di(t)}{dt} + \frac{1}{LC}i(t) = 0.$$

Now the solution is $i(t) = $ C.F. $+$ P.I.
For the C.F., the A.E. is $m^2 + \frac{R}{L}m + \frac{1}{LC} = 0$

$$m_{1,2} = \frac{-\frac{R}{L} \pm \sqrt{\left(\frac{R}{L}\right)^2 - \frac{4}{LC}}}{2}.$$

Now the roots are distinct

$$i(t) = K_1 e^{m_1 t} + K_2 e^{m_2 t} \quad [\text{P.I.} = 0]$$

$$m_1 = \frac{-\frac{R}{L} + \sqrt{\left(\frac{R}{L}\right)^2 - \frac{4}{LC}}}{2} \text{ and } m^2 = \frac{-\frac{R}{L} - \sqrt{\left(\frac{R}{L}\right)^2 - \frac{4}{LC}}}{2}$$

$$V_R(t) = Ri(t) = R[K_1 e^{m_1 t} + K_2 e^{m_2 t}] \qquad (6.22)$$

$$V_L(t) = L\frac{di(t)}{dt} = L[K_1 m_1 e^{m_1 t} + K_2 m_2 e^{m_2 t}]$$

$$V_C(t) = \frac{1}{C}\int_0^t i(t)dt = \frac{1}{C}\left[\frac{K_1}{m_1}e^{m_1 t} + \frac{K_2}{m_2}e^{m_2 t}\right]_0^t$$

$$V_C(t) = \frac{1}{C}\left[\frac{K_1}{m_1}e^{m_1 t} + \frac{K_2}{m_2}e^{m_2 t} - \frac{K_1}{m_1} - \frac{K_2}{m_2}\right]. \qquad (6.23)$$

Example 6.14. *Find $i(t)$ for the given circuit, where*

$$R = 2\Omega, L = 1H \text{ and } C = 1F.$$

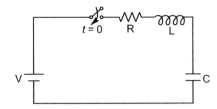

FIGURE 6.16

Solution: Applying KVL,

$$V = 2i(t) + 1\frac{di(t)}{dt} + \frac{1}{1}\int_0^t i(t)dt + V_C(0)$$

$[V_C(0) = 0]$.

Differentiating, we get

$$\frac{d^2i(t)}{dt^2} + 2\frac{di(t)}{dt} + i(t) = 0.$$

Now the solution is $i(t) = $ C.F. + P.I.   [P.I. = 0].
For the C.F. the A.E. is $m^2 + 2m + 1 = 0 \Rightarrow (m+1)^2 = 0$

$$m_1 = -1, m_2 = -1.$$

The roots are equal, hence, the solution is

$$i(t) = (K_1 + K_2t)e^{-t}$$
$$i(0) = 0 = K_1 \Rightarrow K_1 = 0$$
$$i(t) = K_2te^{-t}$$

and

$$V_C(t) = \int_0^t i(t)dt = K_2\int_0^t te^{-t}dt$$

$$= K_2\left[t\frac{e^{-t}}{-1} - \int \frac{e^{-t}}{-1}dt\right]_0^t$$

$$= K_2\left[-te^{-t} + \frac{e^{-t}}{-1}\right]_0^t$$

$$V_C(t) = K_2[-te^{-t} - e^{-t} + 1]$$
$$V_C(0) = K_2 \times 0 = 0$$
$$K_2 \neq 0.$$

SOLVED PROBLEMS

Problem 6.1. *In the given circuit shown in Figure 6.17, the switch s is changed from position '1' to '2' at $t = 0$. Find $i(t)$ and $\frac{di(t)}{dt}$ at $t = 0+$.*

Solution: Initially the switch is at position 1 for a very long time, therefore, the inductor behaves as a short circuit.

$$i(0+) = \frac{100}{1} = 100\,\text{A}$$

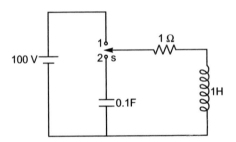

FIGURE 6.17

At position 2, there is no battery in this circuit but the inductor tries to flow the same current in the circuit.

Applying KVL,

$$i(t) \times 1 + i\frac{di(t)}{dt} + \frac{1}{0.1}\int_0^t i(t)dt + V_C(0) = 0 [V_C(0) = 0]. \qquad (6.24)$$

Since initially the capacitor is uncharged and at switching it behaves as a short circuit, the voltage drop across the capacitor will be zero.

Now Equation (6.25) becomes

$$\frac{di(t)}{dt} = -i(t) \Rightarrow \frac{di(0+)}{dt} = -i(0+) = -100\,\text{A/sec}.$$

and differentiating Equation (6.25), we get

$$\frac{d^2i(t)}{dt^2} + \frac{di(t)}{dt} + 10i(t) = 0$$

$$\frac{d^2i(0+)}{dt^2} = -\frac{di(0+)}{dt} - 10i(0+)$$

$$= 100 - 1000 = -900\,\text{A}^2/\text{sec}^2.$$

Problem 6.2. *In the circuit shown in Figure 6.18, the switch is opened at* $t = 0$, *a steady state current having previously been attained. Find* $i(t)$.

Solution: Initially the switch is short, therefore, no current is flowing through resistor R_2,

$$i(0+) = \frac{V}{R_1}.$$

Since the inductor behaves as a short circuit at $t = \infty$, when the switch is opened at $t = 0$, applying KVL,

$$V = i(t)(R_1 + R_2) + L\frac{di(t)}{dt}$$

or

$$\frac{di(t)}{dt} + \left(\frac{R_1 + R_2}{L}\right)i(t) = \frac{V}{L}.$$

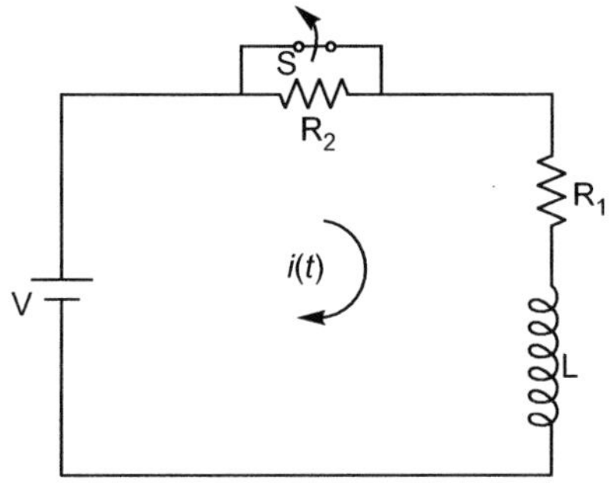

FIGURE 6.18

Now $i(t) = $ C.F. + P.I.
For the C.F., the A.E. is $m + \left(\frac{R_1+R_2}{L}\right) = 0 \Rightarrow m = -\left(\frac{R_1+R_2}{L}\right)$

$$\text{C.F.} = K_1 e^{-\left(\frac{R_1+R_2}{L}\right)t}.$$

For the P.I.,

$$\text{P.I.} = \frac{\frac{V}{L}e^{0t}}{D + \left(\frac{R_1+R_2}{L}\right)} \qquad [D = a = 0]$$

$$\text{P.I.} = \frac{V}{L} \times \frac{L}{R_1 + R_2} = \frac{V}{R_1 + R_2}.$$

Now the final solution is

$$i(t) = K_1 e^{-\left(\frac{R_1+R_2}{L}\right)t} + \frac{V}{R_1 + R_2}. \qquad (6.25)$$

By the initial condition and Equation (6.26),

$$i(0+) = \frac{V}{R_1} = K_1 + \frac{V}{R_1 + R_2}$$

$$K_1 = V\left[\frac{R_1 + R_2 - R_1}{R_1(R_1 + R_2)}\right] = \frac{VR_2}{R_1(R_1 + R_2)}.$$

Put the value of K_1 into Equation (6.26),

$$i(t) = \frac{VR_2}{R_1(R_1 + R_2)}e^{-\left(\frac{R_1+R_2}{L}\right)t} + \frac{V}{R_1 + R_2}$$

$$i(t) = \frac{V}{(R_1 + R_2)}\left[\frac{R_2}{R_1}e^{-\left(\frac{R_1+R_2}{L}\right)t} + 1\right].$$

Problem 6.3. *Find the expression for $i(t)$ in the circuit shown in Figure 6.19. Initial conditions are $i(t) = 5$ and $\frac{di(t)}{dt} = -10$ A/sec. at $t = 0$.*

Solution: Applying KVL,

$$0.05\frac{di(t)}{dt} + \frac{5}{4}i(t) + \frac{1}{0.2}\int_0^t i(t)dt + V_C(0) = 0.$$

Differentiating, we get

$$0.05\frac{d^2i(t)}{dt^2} + \frac{5}{4}\frac{di(t)}{dt} + 5i(t) = 0$$

or

$$\frac{d^2i(t)}{dt^2} + 25\frac{di(t)}{dt} + 100i(t) = 0.$$

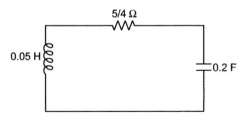

FIGURE 6.19

The A.E. is $m^2 + 25m + 100 = 0$

$$m = \frac{-25 \pm \sqrt{625 - 400}}{2} = \frac{-25 \pm \sqrt{225}}{2}$$

$$m = \frac{-25 \pm 15}{2}$$

$$m_1 = -5, m_2 = -20.$$

Hence,

$$i(t) = K_1 e^{-5t} + K_2 e^{-20t} \tag{6.26}$$

$$i(0) = 5 = K_1 + K_2 \tag{6.27}$$

$$\frac{di(t)}{dt} = -5K_1 e^{-5t} - 20K_2 e^{-20t}$$

at $t = 0$

$$\frac{di(0)}{dt} = -10 = -5K_1 - 20K_2$$

or

$$K_1 + 4K_2 = 2. \tag{6.28}$$

By Equations (6.28) and (6.29),

$$-3K_2 = 3 \Rightarrow K_2 = -1$$

$$K_1 = 6.$$

Putting the values of K_1 and K_2 into Equation (6.27)

$$\boxed{i(t) = 6e^{-5t} - e^{-20t}.}$$

Problem 6.4. *The switch in Figure 6.20 has been in position 1 for a long time. It is moved to 2 at $t = 0$. Obtain the expression for $i(t)$, for $t > 0$.*

Solution: For position 1,

$$i(0-) = i(0) = i(0+) = \frac{50}{40} = 1.25\text{A}$$

when the switch moves to position 2.

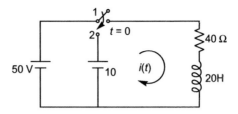

FIGURE 6.20

Applying KVL,

$$10 = 40i(t) + 20\frac{di(t)}{dt}$$

$$\frac{di(t)}{dt} + 2i(t) = 0.5.$$

For C.F., the A.E. is $m + 2 = 0$

$$m = -2$$
$$\text{C.F} = K_1 e^{-2t},$$

for the P.I.,

$$\text{P.I.} = \frac{0.5e^{0t}}{D + 2}[D = a = 0]$$
$$\text{P.I.} = 0.25.$$

Therefore, the complete solution is

$$i(t) = K_1 e^{-2t} + 0.25 \tag{6.29}$$
$$i(0) = K_1 + 0.25 = 1.25 \text{ (by initial condition)}$$
$$K_1 = 1.$$

Put K_1 into Equation (6.30),

$$\boxed{i(t) = e^{-2t} + 0.25.}$$

Problem 6.5. *For the network shown in Figure 6.21, find i_1, i_2, and V_1 at*
(a) $t = 0-$ (b) $t = 0+$ (c) $t = \infty$ (d) $t = 50\,m\,sec.$

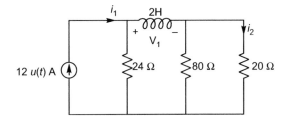

FIGURE 6.21

Solution: By the source transformation method the current source is converted into a voltage source. By the definition of step signal

$$u(t) = \begin{cases} 1 & t > 0 \\ 0 & t < 0 \end{cases}.$$

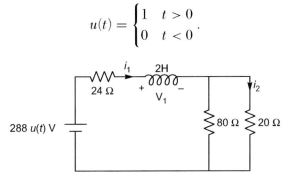

FIGURE 6.22

(a) At $t = 0-$, $u(t) = 0$ means there is no energy source,

$$i_1(0-) = 0, i_2(0-) = 0, V_1(0-) = 0.$$

(b) At $t = 0+$: Applying KVL in the outer loop,

$$288u(t) = 24i_1(t) + 2\frac{di(t)}{dt} + 20i_2(t). \tag{6.30}$$

By using the current division rule,

$$i_2 = \frac{80}{80 + 20} \times i_1 \Rightarrow i_2 = 0.8i_1. \tag{6.31}$$

By Equations (6.31) and (6.32),

$$288u(t) = 24i_1(t) + 2\frac{di_1(t)}{dt} + 16i_1(t)$$

$$288u(t) = 40i_1(t) + 2\frac{di_1(t)}{dt}$$

or

$$\frac{di_1(t)}{dt} + 20i_1(t) = 144u(t).$$

Hence, the complete solution $= i_1(t) = $ C.F. + P.I.
For C.F., the A.E. is $m + 20 = 0$,

$$m_1 = -20$$
$$\text{C.F.} = K_1 e^{-20t}(u(t) = 1 \text{ for } t > 0].$$

For the P.I., P.I. $= \frac{144e^{0t}}{D+20}[D = a = 0]$

$$\text{P.I.} = 7.2.$$

Therefore, $i_1(t) = K_1 e^{-20t} + 7.2$.
At $t = 0+$, the inductor behaves as an open circuit,

$$i_1(0+) = 0 = K_1 + 7.2 \Rightarrow K_1 = -7.2$$
$$i_1(t) = 7.2(1 - e^{-20t}). \tag{6.32}$$

By Equations (6.32) and (6.33),

$$i_2 = 0.8 \times 7.2(1 - e^{-20t}) = 5.76(1 - e^{-20t}) \tag{6.33}$$

$i_1(0+) = 0$, therefore, $i_2(0+) = 0$ and $V_1(0+) = 288 \text{ V}$.
(c) At $t = \infty$ from Equations (6.33) and (6.34)

$$i_1(\infty) = 7.2, i_2(\infty) = 5.76$$

$V_1(\infty) = 2\frac{di_i(\infty)}{dt} = 2 \times 0 = 0$ [inductor behaves SC at $t = \infty$].
(d) At $t = 50$ m sec.,

$$i_1(5 \text{ m sec.}) = 7.2(1 - e^{-20 \times 50 \times 10^{-3}}) = 4.55 \text{ A}$$
$$i_2(5 \text{ m sec.}) = 5.76(1 - e^{-1}) = 3.64 \text{ A}$$
$$V_1 = 2\left.\frac{di_1}{dt}\right|_{\text{at } t = 50 \text{ m/sec}} = 2 \times 7.2 \times 20e^{-20 \times 50 \times 10^{-3}}$$
$$= 105.95 \text{ V}.$$

Problem 6.6. *In the given circuit of Figure 6.23 the switch is closed at $t = 0$ (assume initial current in the inductor is zero). Find the values of $i(t)$, $\frac{di(t)}{dt}$ and $\frac{d^2i(t)}{dt^2}$ at $t = 0+$.*

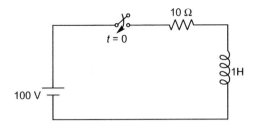

FIGURE 6.23

Solution: Given $i(0-) = i(0) = i(0+) = 0$, applying KVL,

$$100 = 10i(t) + 1\frac{di(t)}{dt}.$$

At $t = 0+$, the inductor behaves as an open circuit, hence, $i(0+) = 0$

$$100 = 10i(0+) + 1\frac{di(0+)}{dt}$$

$$100 = 100 \times 0 + 1\frac{di(0+)}{dt} \Rightarrow \frac{di(0+)}{dt} = 100\,\text{A/sec}.$$

Differentiating Equation (6.32), we get

$$\frac{d^2i(t)}{dt^2} + 10\frac{di(t)}{dt} = 0,$$

at $t = 0+$

$$\frac{d^2i(0+)}{dt^2} = -10\frac{di(0+)}{dt} = -10 \times 100 = -1000\,\text{A}^2/\text{sec}^2.$$

Problem 6.7. *In the circuit shown in Figure 6.24 the capacitor C has an initial voltage* $V_C = 10$ *volts and at the same instant, the current through inductor L is zero, and the switch K is closed at time $t = 0$. Find the expression for the voltage $v(t)$ across the inductor L using differential equation formulation.*

Solution: Applying KVL,

$$\frac{1}{c}\int_0^t i(t)dt \pm V_C(0) + V(t) = 0$$

$+V_C(0)$ means capacitor is charging.
$-V_C(0)$ means capacitor is discharging.

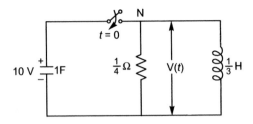

FIGURE 6.24

In this case the capacitor is discharging [Given $V_C(0) = 10\,\text{V}$]

$$10 = \frac{1}{1}\int_0^t i(t)dt + V(t).$$

On differentiating

$$0 = i(t) + \frac{dV(t)}{dt}. \qquad (6.34)$$

Applying KCL at node N,

$$i(t) = \frac{v(t)}{\frac{1}{4}} + \frac{1}{\frac{1}{3}}\int_0^t v(t)dt$$

$$i(t) = 4.v(t) + 3\int_0^t v(t)dt. \qquad (6.35)$$

From Equations (6.35) and (6.36), we find

$$0 = 4v(t) + 3\int_0^t v(t)dt + \frac{dv(t)}{dt}.$$

Once again, differentiating,

$$\frac{d^2v(t)}{dt} + 4\frac{dv(t)}{dt} + 3v(t) = 0.$$

Hence, the final solution is $v(t) = \text{C.F.} + \text{P.I.} \,[\text{P.I.} = 0]$.
For the C.F. the A.E. is $m^2 + 4m + 3 = 0$

$$m_1 = -1, m_2 = -3.$$

Therefore, the solution is

$$v(t) = K_1 e^{-t} + K_2 e^{-3t} \tag{6.36}$$

at

$$t = 0 + v(0+) = 10 = K_1 + K_2, \tag{6.37}$$

$i_L(0+) = 0$ (since the inductor behaves as an open circuit at $t = 0$)

$$i_L(t) = 3 \int v(t)dt = 3 \int [K_1 e^{-t} + K_2 e^{-3t}]dt$$

$$i_L(t) = 3 \left[\frac{K_1 e^{-t}}{-1} + \frac{K_2 e^{-3t}}{-3} \right]$$

$$i_L(t) = 3 \left[-K_1 e^{-t} - \frac{K_2}{3} e^{-3t} \right]$$

$$i_L(0+) = 0 = 3 \left[-K_1 - \frac{K_2}{3} \right]$$

$$K_1 = -\frac{K_2}{3} \tag{6.38}$$

By Equations (6.37) and (6.38), $K_1 = -5$, $K_2 = 15$.
Put the value of K_1 and K_2 into Equation (6.37),

$$\boxed{v(t) = -5e^{-t} + 15e^{-3t} \text{ V}}$$

Problem 6.8. *Find the differential equation relating $v(t)$ and $i(t)$ in the network shown in Figure 6.25.*

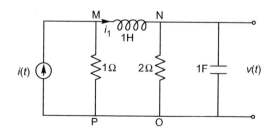

FIGURE 6.25

Solution: Applying KCL at node N,

$$i_1(t) = \frac{v(t)}{2} + L\frac{dv(t)}{dt}. \tag{6.39}$$

Applying KVL in the center loop MNOP,

$$L\frac{di_1(t)}{dt} + v(t) - \{i(t) - i_1(t)\} \times 1 = 0$$

$$i_1(t) - i(t) + \frac{di_1(t)}{dt} + v(t) = 0. \tag{6.40}$$

From Equations (6.40) and (6.41),

$$\frac{v(t)}{2} + \frac{dv(t)}{dt} - i(t) + \frac{d}{dt}\left[\frac{v(t)}{2} + \frac{dv(t)}{dt}\right] + v(t) = 0$$

$$\boxed{3\frac{v(t)}{2} + \frac{3}{2}\frac{dv(t)}{dt} + \frac{d^2v(t)}{dt^2} - i(t) = 0.}$$

Problem 6.9. *For the circuit shown in Figure 6.26 find $V_C(t)$ at $t = 0-$, $t = 0+$, $t = \infty$, and $t = 0.08\,sec$.*

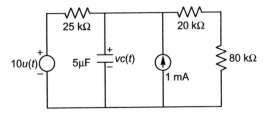

FIGURE 6.26

Solution: 1 mA current source is connected with $(20\,K\Omega + 80\,K\Omega)$ resistors in parallel. Apply the source transformation formula.

By the definition of unit step signal

$$u(t) = \begin{cases} 1 & t > 0 \\ 0 & t < 0 \end{cases}.$$

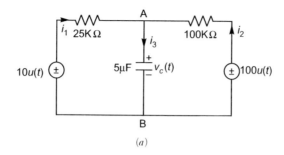

(a)

FIGURE 6.26

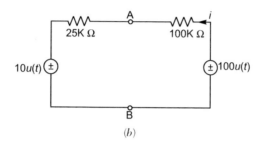

(b)

FIGURE 6.26

(a) at $t = 0-$ the applied potential is zero, hence,

$$V_C(0-) = 0\,\text{V}.$$

(b) at $t = 0+$: Just after switching, the capacitor behaves as a short circuit, therefore, $V_C(0+) = 0\,\text{V}.$

(c) at $t = \infty$: The capacitor behaves as an open circuit

$$-100 + 125 \times 10^3 i + 10 = 0$$

$$i = \frac{90}{125}\,\text{mA}$$

$$V_{AB} = V_C(\infty) = 100 - 100 \times 10^3 \times \frac{90}{125} \times 10^{-3}$$

$$= 28\,\text{V}.$$

(d) at $t = 0.08$ sec: Applying KVL at node A,

$$i_1 + i_2 = i_3 \tag{6.41}$$

$$i_1 = \frac{10 - V_C}{25 \times 10^3} \quad [\text{Potential at A} = V_C \text{ with respect to B}]$$

$$i_2 = \frac{100 - V_C}{100 \times 10^3} \quad \text{and} \quad i_3 = C\frac{dv}{dt} = 5 \times 10^{-6}\frac{dv_C}{dt}.$$

Put the values i_1, i_2, and i_3 into Equation (6.42),

$$\frac{10 - V_C}{25 \times 10^3} + \frac{100 - V_C}{100 \times 10^3} = 5 \times 10^{-6} \frac{dv_C}{dt}$$

or

$$2(140 - 5V_C) = \frac{dv_C}{dt}$$

$$\frac{dv_C}{dt} = 280 \times 10V_C$$

or

$$\frac{dv_C}{dt} + 10V_C = 280.$$

The general solution is $V_C(t) = \text{C.F.} + \text{P.I.}$
For the C.F., the A.E. is $m + 10 = 0 \Rightarrow m = -10$.

$$\text{C.F.} = K_1 e^{-10t}$$

For the P.I., the P.I. $= \frac{280e^{0t}}{D+10}[D = a = 0]$

$$\text{P.I.} = 28.$$

Hence, the final solution is $V_C(t) = K_1 e^{-10t} + 28$

$$V_C(0) = 0 = K_1 + 28$$
$$K_1 = -28$$

$$\boxed{V_C(t) = 28(1 - e^{-10t})}$$

$$V_C(\infty) = 28(1 - e^{-\infty}) = 28(1 - 0) = 28\,\text{V}$$
$$V_C(0.08) = 28(1 - e^{-0.8}) = 28(1 - 0.449)$$
$$V_C(0.08\,\text{sec}) = 15.4\,\text{V}.$$

OBJECTIVE QUESTIONS

1. The solution of the given equation is $\left[\frac{d^2 i}{dt^2} - 7\frac{di}{dt} + 12i = 0\right]$,
   (a) $i(t) = K_1 e^{3t} + K_2 e^{-4t}$ (b) $i(t) = K_1 e^{3t} + K_2 e^{4t}$
   (c) $i(t) = K_1 e^{-3t} + K_2 e^{-4t}$ (d) $i(t) = K_1 e^{-3t} + K_2 e^{4t}$.

2. The particular integral (P.I.) of the given equation $\left[\frac{d^2i}{dt^2} + 3\frac{di}{dt} + 2i = 5e^{5t}\right]$ is

   (a) $\frac{5}{42}e^{5t}$ (b) $\frac{5}{12}e^{5t}$ (c) $\frac{1}{42}e^{5t}$ (d) $\frac{1}{12}e^{5t}$.

3. The value of $i(t)$ for the given circuit is (Figure 6.27)

   (a) $\frac{V}{R}(e^{-\frac{R}{L}t} - 1)$ (b) $\frac{V}{R}(1 - e^{-\frac{R}{L}t})$ (c) $\frac{V}{R}e^{-\frac{R}{L}t}$ (d) None of these.

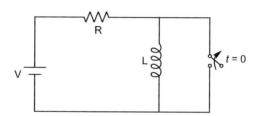

FIGURE 6.27

4. In the circuit shown in Figure 6.28 assume the switch was initially at position L. The time constant for the given network is

   (a) $\frac{L}{R_1}$ (b) $\frac{L}{R_2}$ (c) $\frac{L}{R_1+R_2}$ (d) $\frac{R_1+R_2}{L}$.

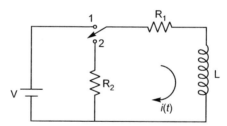

FIGURE 6.28

5. The general solution of the differential equation is equation $\frac{d^2i}{dt^2} + 3\frac{di}{dt} + 2i = 0$

   (a) $i(t) = K_1e^{-t} + K_2e^{-2t}$ (b) $i(t) = K_1e^{t} + K_2e^{2t}$
   (c) $i(t) = K_1e^{t} + K_2e^{-2t}$ (d) $i(t) = K_1e^{-t} - K_2e^{-2t}$.

6. The general solution of the differential equation $\left[\frac{d^2i}{dt^2} + 3\frac{di}{dt} + 2i = 5t\right]$ is,

   (a) $i = K_1e^{t} + K_2e^{-2t}$ (b) $i = K_1e^{-t} + K_2e^{-2t}$
   (c) $i = K_1e^{-t} + K_2e^{-2t} + \frac{15}{4}$ (d) $i = K_1e^{-t} + K_2e^{-2t} - \frac{15}{4}$.

7. The Complementary function (C.F.) of the given differential equation $\left[\frac{di}{dt} + 5i = 2\cos t\right]$ is
   (a) $K_1 e^{-5t}$ (b) $K_1 e^{-2t}$ (c) $\frac{5}{12}\sin 2t$ (d) $K_2 \sin 2t$.

8. The final solution for $i(t)$ is (Figure 6.29) and
   (a) $1 - e^{-t}$ (b) $1 + e^{-t}$ (c) $e^{-t} - 1$ (d) $1 - e^t$.

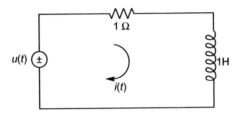

FIGURE 6.29

9. The expression for $i(t)$ in the given circuit is assumed to be $V_C(0) = 10\,\text{V}$ (Figure 6.30) and
   (a) $10\sin t$ (b) $-10\sin t$ (c) $10e^{-t}$ (d) $10\cos t$.

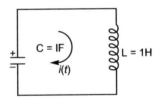

FIGURE 6.30

UNSOLVED PROBLEMS

1. Find the value of $i(t)$ in the circuit in Figure 6.31.
2. In the series RL circuit in Figure 6.32, an exponential voltage $v = 50e^{-100t}\,(\text{V})$ is applied by closing the switch at $t = 0$. Find the resulting current.
3. In Figure 6.33, find the loop currents, or $i_1(t)$ and $i_2(t)$.
4. In Figure 6.34, find $V_C(s)$ for $t \geq 0$.
5. In Figure 6.35 the switch is thrown from position 1 to 2 at $t = 0$. Just before the switch is thrown the initial conditions are $i_L(0^-) = 2\,\text{amp}$ and $V_C(0^-) = 2\,\text{volt}$. Find the current $i(t)$ after the switching action.

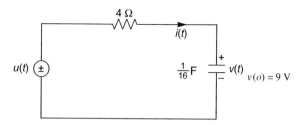

FIGURE 6.31

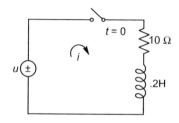

FIGURE 6.32

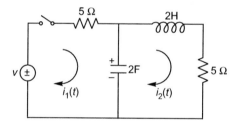

FIGURE 6.33

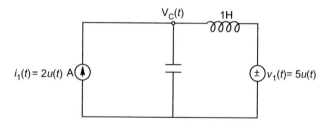

FIGURE 6.34

6. In the network of the figure, the switch 'K' is moved from position 1 to position 2 at $t = 0$. Find $i(t)$.

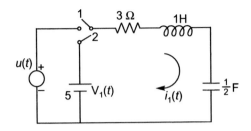

FIGURE 6.35

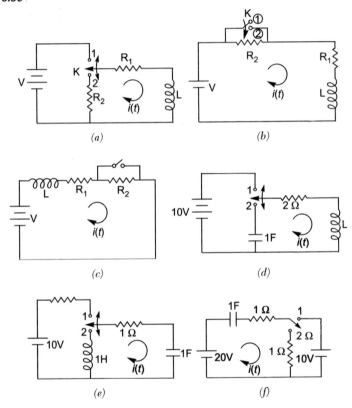

FIGURE 6.36

7. Find the final expression for $i(t)$ as shown in Figure 6.37. Assume all initial conditions are zero.

8. Find the final expression for $i(t)$ as shown in Figure 6.38. Assume all initial conditions are zero.

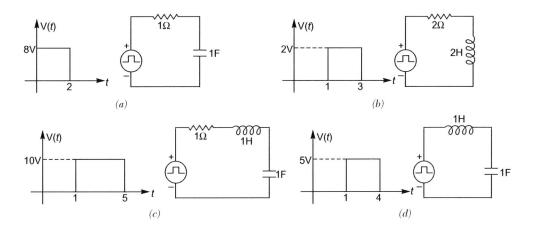

FIGURE 6.37

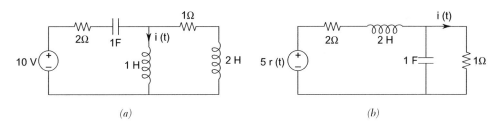

FIGURE 6.38

9. Find the final expression for $v(t)$ as shown in Figure 6.39. Assume all initial conditions are zero.

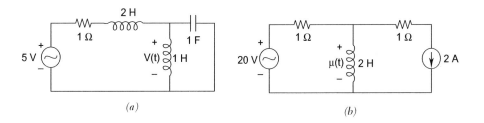

FIGURE 6.39

Chapter 7

APPLICATIONS OF THE LAPLACE TRANSFORM

7.1 INTRODUCTION

The Laplace transform is a very important mathematical tool. By using the Laplace transform, any electrical circuit can be solved and calculations are very easy for transient and steady state conditions. The following steps involve the analysis of a linear system (electrical or mechanical, etc.). In this chapter we will consider only the electrical system.

1. Apply KVL and make a differential or integro-differential equation form.
2. Take the Laplace transform of the system differential or integro-differential equation together with the input excitation to obtain an algebraic equation in the s-domain.
3. Now take the Laplace inverse transform to get the solution in the time domain.

7.2 SOLUTION OF LINEAR DIFFERENTIAL EQUATIONS

In the last chapter we discussed

$$\mathcal{L}\frac{dI(t)}{dt} = s.I.(s) - I(0) \text{ and}$$

$$\mathcal{L}\left[\int i(t)dt\right] = \frac{I(s)}{s} - \frac{I^{-1}(0)}{s}$$

and the Laplace transform of second order or higher. These formulas are very important to solve a linear differential equation.

Now consider a second differential equation

$$\frac{d^2i(t)}{dt^2} + a\frac{di(t)}{dt} + bi(t) = V(t). \tag{7.1}$$

We know,

$$\mathcal{L}\frac{d^2i(t)}{dt^2} = s^2I(s) - sI(0) - I'(0) \tag{7.2}$$

and

$$\mathcal{L}\frac{di(t)}{dt} = sI(s) - I(0). \tag{7.3}$$

Taking the Laplace transform on both sides of Equation (7.1),

$$s^2I(s) - sI(0) - I'(0) + a[sI(s) - I(0)] + bI(s) = V(s),$$

where $I(0) = $ value of $i(t)$ at $t = 0$

$$I'(0) = \text{ Value of } \frac{di(t)}{dt} \text{ at } t = 0$$

$$I(s) = \frac{s(I_0 + a) + I'(0) + V(s)}{s^2 + as + b}. \tag{7.4}$$

Now taking the Laplace inverse,

$$I(t) = \mathcal{L}^{-1}\left[\frac{s(a + I_0) + I'(0) + V(s)}{s^2 + as + b}\right]. \tag{7.5}$$

The result, as in Equation (7.5), not only depends on the characteristic roots but also on the excitation function and initial conditions. The Laplace inverse of Equation (7.5) gives the time response, if we can modify the right-hand side of Equation (7.5) to a known form of the available Laplace inverse transform and convert the response in the time domain.

Example 7.1. *Solve the differential equation*

$$\frac{d^2i(t)}{dt^2} + 3\frac{di(t)}{dt} + 2i(t) = 0$$

and

$$i(0) = 2, \quad i'(0) = 0.$$

Solution: Taking the Laplace transform, $s^2I(s) - sI(0) - I'(0) + 3[sI(s) - I(0)] + 2I(s) = 0$

$$I(s)[s^2 + 3s + 2] = 2s$$

$$I(s) = \frac{2s}{s^2 + 3s + 2} = \frac{2s}{(s + 1)(s + 2)}.$$

Now taking the partial fraction,

$$I(s) = \frac{2s}{(s+1)(s+2)} = \frac{A}{(s+1)} + \frac{B}{(s+2)}$$

$$A = [(s+1)I(s)]_{s=-1} = \left[\frac{2s}{s+2}\right]_{s=-1} = -2$$

$$B = [(s+2)I(s)]_{s=-2} = \left[\frac{2s}{(s+1)}\right]_{s=-2} = 4$$

$$I(s) = -\frac{-2}{(s+1)} + \frac{4}{(s+2)}.$$

Taking the Laplace inverse

$$I(t) = -2e^{-t} + 4e^{-2t}.$$

7.3 TRANSFORMED CIRCUIT COMPONENTS REPRESENTATION

7.3.1 Active Components

The sources $v(t)$ and $i(t)$ may be represented by their transformations, $V(s)$, and $I(s)$, respectively, as shown in Figure 7.1.

$$\mathcal{L}v(t) = V(s) \quad \text{and} \quad \mathcal{L}I(t) = I(s) \quad \text{or} \quad I(t) \Leftrightarrow I(s)$$

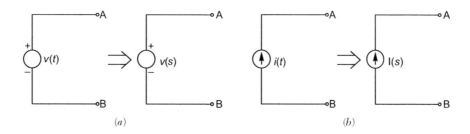

FIGURE 7.1   Representation of (*a*) voltage source; (*b*) current source.

7.3.2 Passive Components

The passive components are (R, L, and C.)
(*a*) *Resistance Parameter:*
The *v-i* relationship for a resistor in the *t*-domain is

$$v_R(t) = Ri_R(t) \quad \text{[Ohm's law]}. \tag{7.6}$$

Taking the Laplace transform of Equation (7.6),

$$v_R(s) = Ri_R(s). \tag{7.7}$$

By Equations (7.6) and (7.7), we observe that the representation of a resistor in the t-domain and the s-domain are the same as shown in Figure 7.2.

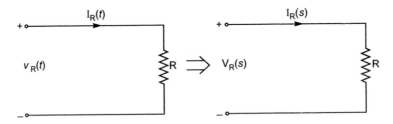

FIGURE 7.2   Representation of a resistor in the t- and s-domain.

(b) *Inductance Parameter:*
The v-i relationship for an inductor is

$$v_L(t) = L\frac{di_L(t)}{dt} \quad \text{[Faraday's law].} \tag{7.8}$$

Taking the Laplace transform of Equation (7.8),

$$v_L(s) = L[sI_L(s) - I_L(0)]$$
$$v_L(s) = LsI_L(s) - LI_L(0) \tag{7.9}$$

or

$$I_L(s) = \frac{v_L(s)}{Ls} + \frac{I_L(0)}{s}. \tag{7.10}$$

From the above equations, we get the transformed circuit representation for the inductor as shown in Figure 7.3.

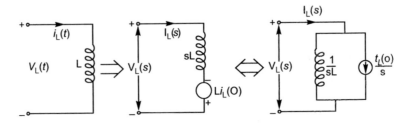

FIGURE 7.3   Representation of an inductor in the t- and s-domain.

(c) *Capacitance Parameter:* For a capacitor

$$q = CV = it,$$

the *v-i* relationship is

$$i_C(t) = C\frac{dV_C(t)}{dt}.$$

Taking the Laplace transform

$$i_C(s) = C[sV_C(s) - V_C(0)]$$
$$= C_s V_C(s) - CV_C(0) \qquad (7.11)$$

or

$$V_C(s) = \frac{I_C(s)}{Cs} + \frac{V_C(0)}{s}. \qquad (7.12)$$

From the above equations, we get the transformed circuit representation for the capacitor as shown in Figure 7.4.

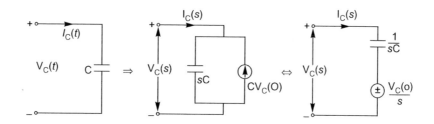

FIGURE 7.4   Representation of a capacitor in the *t*- and *s*-domain.

Example 7.2. *Consider the RL circuit with $R = 4\,\Omega$ and $L = 1\,H$ excited by a 20 V D.C. source as shown in Figure 7.5(a). Assume the initial value of the current in the inductor is 2 A. Using the Laplace transform, determine the current $i(t)$. Also draw the s-domain representation of the circuit.*

Solution: Applying KVL,

$$20 = 4i(t) + L\frac{di(t)}{dt}. \qquad [\text{L} = 1\,\text{H}]$$

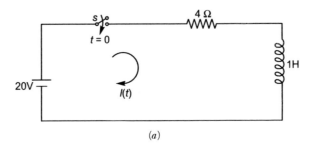

(a)

FIGURE 7.5

Taking the Laplace transform,

$$\frac{20}{s} = 4I(s) + sI(s) - I(0) \qquad\qquad [I(0) = 2\,\text{A}]\ (7.13)$$

$$I(s) \times (s + 4) = \frac{20}{s} + 2$$

$$I(s) = \frac{20 + 2s}{s(s + 4)} = \frac{A}{s} + \frac{B}{(s + 4)}.$$

By using the partial fraction,

$$A = [sI(s)]_{s=0} = \left[\frac{20 + 2s}{s + 4}\right]_{s=0} = 5$$

$$B = [(s + 4)I(s)]_{s=-4} = \left[\frac{20 + 2s}{s}\right]_{s=-4} = -3$$

$$I(s) = \frac{5}{s} - \frac{3}{(s + 4)}.$$

Taking the Laplace inverse,

$$i(t) = (5 - 3e^{-4t})u(t)\text{A}.$$

By Equation (7.13), the s-domain representation is shown in Figure 7.5 (b). Apply the KVL in Figure 7.5(b) and we get Equation (7.13) as

$$\frac{20}{s} = 4I(s) + sI(s) - 2.$$

Example 7.3. *In the network shown in Figure 7.6, (a) find $v_2(t)$ using the Laplace transform technique.*

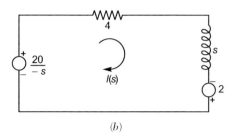

(b)

FIGURE 7.5

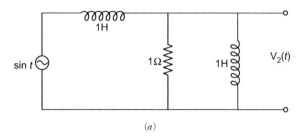

(a)

FIGURE 7.6

Solution: First draw the transformed circuit diagram as shown in Figure 7.6(b).

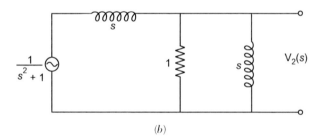

(b)

FIGURE 7.6

Now 1 and s are in parallel

$$Z_{eq} = \frac{s}{s+1}$$

as shown in Figure 7.6(c).

By using the voltage division formula,

$$V_2(s) = \left[\frac{\left(\frac{s}{s+1}\right)}{\left(\frac{s}{s+1}\right) + s}\right]\left(\frac{1}{s^2 + 1}\right) = \frac{1}{(s^2 + 1)(s + 2)}.$$

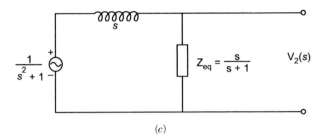

(c)

FIGURE 7.6

By using the partial fraction method,

$$V_2(s) = \frac{As + B}{s^2 + 1} + \frac{C}{s + 2}.$$

After solving

$$A = -\frac{1}{5}, \quad B = \frac{2}{5}, \quad C = \frac{1}{5}.$$

Now,

$$V_2(s) = \frac{-\frac{1}{5}s + \frac{2}{5}}{s^2 + 1} + \frac{\frac{1}{5}}{s + 2}$$

or

$$V_2(s) = -\frac{1}{5}\left(\frac{s}{s^2 + 1}\right) + \frac{2}{5}\left(\frac{1}{s^2 + 1}\right) + \frac{1}{5}\left(\frac{1}{s + 2}\right).$$

Taking the Laplace inverse transform

$$V_2(t) = -\frac{1}{5}\cos t + \frac{2}{5}\sin t + \frac{1}{5}e^{-2t}.$$

Example 7.4. *Consider the* RC *circuit with* R = 4 Ω *and* C = 1 F *excited by a* 20 VD.C. *source as shown in Figure 7.7(a). Assume the initial value of voltage in the capacitor is* 5 V. *Using the Laplace transform determine the current* i(t). *Also draw the s-domain representation of the circuit.*

Solution: Applying KVL,

$$20 = 4i(t) + \frac{1}{1}\int_0^t i\,dt \pm V_C(0).$$

$+ \rightarrow$ for capacitor charging
$- \rightarrow$ for capacitor discharging

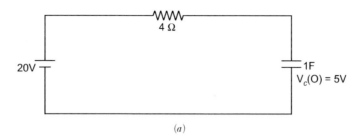

FIGURE 7.7

In this case the capacitor is further charged because the capacitor's initial voltage is less than the source voltage,

$$20 = 4i(t) + \int_0^t i\,dt + 5.$$

Taking the Laplace transform,

$$\frac{20}{s} = 4I(s) + \frac{I(s)}{s} + \frac{5}{s} \tag{7.14}$$

$$\frac{15}{s} = I(s)\frac{[4s + 1]}{s}$$

$$I(s) = \frac{15}{4\left(s + \frac{1}{4}\right)}.$$

Taking the Laplace inverse transform,

$$I(t) = \frac{15}{4}e^{-\frac{1}{4}t}.$$

By Equation (7.14) the s-domain representation is shown in Figure 7.7(b).

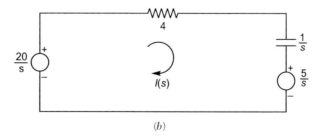

(b)

FIGURE 7.7

SOLVED PROBLEMS

Problem 7.1. *Find the current $i(t)$ in a series RLC circuit comprising $R = 5\,\Omega, L = 1\,H$, and $C = \frac{1}{4}F$ when the impulse voltage $3\delta(t-1)$ is applied.*

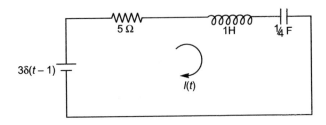

FIGURE 7.8

Solution: Applying KVL,

$$3\delta(t-1) = 5i(t) + 1\frac{di(t)}{dt} + 4\int_0^t i(t)dt.$$

Taking the Laplace transform,

$$3e^{-s} = \left(5 + s + \frac{4}{s}\right)I(s).$$

[Assume all initial conditions are zero.]

$$I(s) = \frac{3se^{-s}}{s^2 + 5s + 4} = e^{-s}\left[\frac{3s}{(s+1)(s+4)}\right]$$

$$= e^{-s}\left[\frac{4}{(s+4)} - \frac{1}{(s+1)}\right]$$

Taking the inverse Laplace transform, we have $I(t) = 4e^{-4(t-1)} - e^{-(t-1)}$.

Problem 7.2. *Find the current $i(t)$ for $t > 0$ for the circuit shown in Figure 7.9. Assume that the circuit has reached a steady state at $t = 0$.*

Solution: From the source transformation formula the current source is converted into the voltage source as shown in Figure 7.9(a).

At a steady state, the inductor behaves as a short circuit and the capacitor as an open circuit. So $i_L(0) = 0$ and $v_C(0) = 40 - 4 = 36\,V$ when the switch is connected.

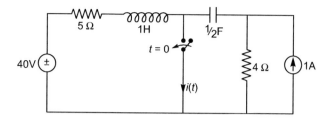

FIGURE 7.9

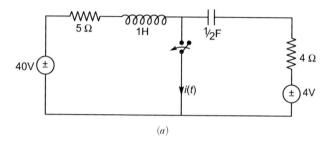

(a)

FIGURE 7.9

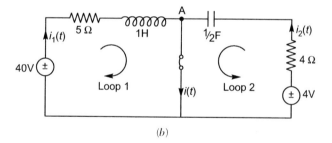

(b)

FIGURE 7.9

By KCL at node A,

$$i(t) = i_1(t) + i_2(t).$$

By KVL at loop 1,

$$40 = 5i_1(t) + 1 \cdot \frac{di_1(t)}{dt}.$$

Taking the Laplace transform,

$$\frac{40}{s} = (5 + s)I_1 s \implies I_1(s) = \frac{40}{s(s + 5)}.$$

Taking the Laplace inverse,

$$I_1(t) = 8(1 - e^{-5t}).$$

Applying KVL at loop 2,

$$4 = 4i_2(t) + 2\int i_2(t)dt - 36$$

$$40 = 4i_2(t) + 2\int i_2(t)dt.$$

Taking the Laplace transform,

$$\frac{40}{s} = 4i_2(s) + 2\left[\frac{I_2(s)}{s}\right]$$

$$I_2(s) = \frac{40}{4s + 2} = \frac{10}{s + 0.5}.$$

By the Laplace inverse,

$$i_2(t) = 10e^{-0.5t}$$

$$i(t) = i_1(t) + i_2(t) = 8(1 - e^{-5t}) + 10e^{-0.5t}.$$

Problem 7.3. *Obtain an expression for i(t) for the RLC circuit shown and obtain its value 2 ms after the switch is closed. Assume all initial conditions to be zero.*

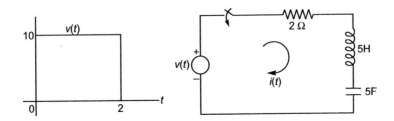

FIGURE 7.10

Solution: Given $v(t) = 10[u(t) - u(t - 2)]$,
    By KVL,

$$v(t) = 2i(t) + 5\frac{di(t)}{dt} + \frac{1}{5}\int i(t)dt$$

$$10[u(t) - u(t - 2)] = 2i(t) + 5\frac{di(t)}{dt} + 0.2\int i(t)dt.$$

Taking the Laplace transform,

$$10\left[\frac{1}{s} - \frac{e^{-2s}}{s}\right] = 2I(s) + 5sI(s) + 0.2\frac{I(s)}{s}$$

$$\frac{10}{s}(1 - e^{-2s}) = I(s)\left[2 + 5s + \frac{0.2}{s}\right]$$

$$\frac{10}{s}(1 - e^{-2s}) = I(s)\left[\frac{2s + 5s^2 + 0.2}{s}\right]$$

$$I(s) = \frac{10(1 - e^{-2s})}{5s^2 + 2s + 0.1} = \frac{2(1 - e^{-2s})}{s^2 + 0.4s + 0.04}$$

$$= \frac{2(1 - e^{-2s})}{(s + 0.2)^2}$$

$$I(s) = \frac{2}{(s + 0.2)^2} - \frac{2e^{-2s}}{(s + 0.2)^2}.$$

Taking the Laplace inverse,

$$I(t) = 2te^{-0.2t} - 2(t - 2)e^{-0.2(t-2)}.$$

Problem 7.4. *A voltage pulse of magnitude 8 volts and duration 2 seconds extending from t = 2 sec. to t = 4 sec. is applied to a series* RL *circuit as shown in Figure 7.11. Obtain the expression for the current i(t).*

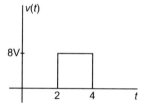
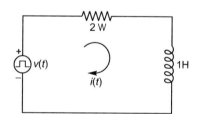

FIGURE 7.11

Solution: $v(t) = 8[u(t - 2) - u(t - 4)]$

By KVL,

$$v(t) = 2i(t) + 1\frac{di(t)}{dt},$$

$$8[u(t - 2) - u(t - 4)] = 2i(t) + \frac{di(t)}{dt}$$

$$8\left[\frac{e^{-2s}}{s} - \frac{e^{-4s}}{s}\right] = 2I(s) + sI(s) - 0$$

$$\frac{8}{s}(e^{-2s} - e^{-4s}) = I(s)[s + 2]$$

$$I(s) = \frac{8}{s(s + 2)}(e^{-2s} - e^{-4s}).$$

Now,

$$\frac{8}{s(s + 2)} = \frac{A}{s} + \frac{B}{s + 2}$$

$$A = s \times \frac{8}{\frac{I}{s}(s + 2)}\Big|_{s=0}$$

$$A = \frac{8}{2} = 4$$

$$B = \frac{8}{s}\Big|_{s=-2}$$

$$B = \frac{8}{-2} = -4$$

$$I(s) = (e^{-2s} - e^{-4s})\left[\frac{4}{s} - \frac{4}{s + 2}\right]$$

$$I(s) = \frac{4}{s}(e^{-2s} - e^{-4s}) - \frac{4}{s + 2}(e^{-2s} + e^{-4s})$$

$$I(s) = \frac{4}{s}e^{-2s} - \frac{4}{s}e^{-4s} - \frac{4}{s + 2}e^{-2s} + \frac{4}{s + 2}e^{-4s}.$$

Taking the Laplace inverse,

$$I(t) = 4u(t - 2) - 4u(t - 4) - 4e^{-2(t-2)}u(t - 2) + 4e^{-2(t-4)}u(t - 4).$$

Problem 7.5. *Find the impulse response of the network shown in Figure 7.12.*

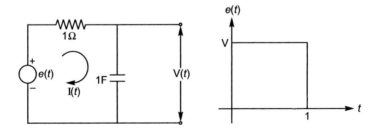

FIGURE 7.12

Solution: $$i(t) = V[u(t) - u(t-1)]$$

By KVL,

$$e(t) = 1 \cdot i(t) + \frac{1}{1}\int i(t)dt$$

$$V[u(t) - u(t-1)] = i(t) + \int i(t)dt.$$

Taking the Laplace transform,

$$V\left[\frac{1}{s} - \frac{e^{-s}}{s}\right] = I(s) + \frac{I(s)}{s}$$

$$\frac{V}{s}(1 - e^{-s}) = I(s)\left[1 + \frac{1}{s}\right]$$

$$\frac{V}{s}(1 - e^{-s}) = I(s)\left[\frac{s+1}{s}\right]$$

$$I(s) = \frac{V(1 - e^{-s})}{(s+1)}$$

$$I(s) = \frac{V}{s+1} - \frac{Ve^{-s}}{s+1}$$

Taking the Laplace inverse,

$$I(t) = Ve^{-t}u(t) - Ve^{-(t-1)}u(t-1).$$

Problem 7.6. *In the network shown in Figure 7.13, the switch K is in position 1 long enough to establish a steady state condition. At time $t = 0$, the switch K is moved from position 1 to 2. Find the expression for the current $i(t)$.*

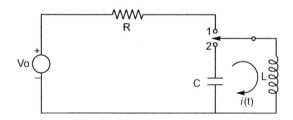

FIGURE 7.13

Solution:

$$i(0) = \frac{V_0}{R}$$

Now by KVL,

$$\frac{1}{C}\int i(t)dt + L\frac{di}{dt} = 0.$$

Taking the Laplace transform,

$$\frac{1}{C}\left[\frac{I(s)}{s} - 0\right] + L[sI(s) - I(0)] = 0$$

$$\frac{I(s)}{Cs} + LsI(s) - L\frac{V_0}{R} = 0$$

$$I(s)\left[Ls + \frac{1}{Cs}\right] = \frac{LV_0}{R}$$

$$I(s) = \frac{LV_0}{R}\frac{1}{\left(Ls + \frac{1}{Cs}\right)} = \frac{V_0}{R}\left(\frac{s}{s^2 + \frac{1}{LC}}\right).$$

Taking the Laplace inverse,

$$I(t) = \frac{V_0}{R}\cos\sqrt{\frac{1}{LC}}t.$$

Problem 7.7. *Find $i(t)$ for $t > 0$ in the circuit shown in Figure 7.14. The switch is opened at $t = 0$.*

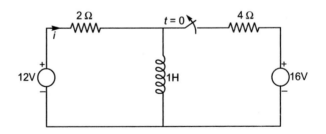

FIGURE 7.14

Solution: The steady state current in the inductor

$$i(0) = \frac{12}{2} + \frac{16}{4} = 6 + 4 = 10\,\text{A}$$

at $t = 0$ the switch is opened.

Applying KVL in the left mesh,

$$12 = 2i(t) + 1\frac{di(t)}{dt}.$$

Taking the Laplace transform,

$$\frac{12}{s} = 2I(s) + sI(s) - i(0)$$

$$\frac{12}{s} + 10 = I(s)(s+2)$$

$$I(s) = \frac{12 + 10s}{s(s+2)} = \frac{A}{s} + \frac{B}{s+2}$$

$$A = [sI(s)]_{s=0} = \left[\frac{12 + 10s}{s+2}\right]_{s=0} = 6$$

$$B = [(s+2)I(s)]_{s=-2} = \left[\frac{12 + 10s}{s}\right]_{s=-2} = 4.$$

Now,

$$I(s) = \frac{6}{s} + \frac{4}{(s+2)}.$$

Taking the Laplace inverse,

$$I(t) = 6u(t) + 4e^{-2t}u(t) = [6 + 4e^{-2t}]u(t).$$

Problem 7.8. *In the circuit shown in Figure 7.15 capacitor C has an initial voltage* $V_C = 10\,V$ *and at the same instant, the current through inductor* L *is zero. The switch* K *is closed at time* $t = 0$. *Find the expression for the voltage* $V(t)$ *across the inductor* L.

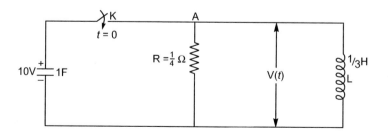

FIGURE 7.15

Solution: Applying KCL at node A,

$$I_C = I_R + I_L \qquad \left[\begin{array}{l} I_C = C\dfrac{dv}{dt} \\ \text{and } v = 10 - V(t) \end{array} \right]$$

$$\frac{d}{dt}[10 - V(t)] = 4V(t) + \frac{1}{\frac{1}{3}} \int v(t)dt$$

$$-\frac{dV(t)}{dt} = 4V(t) + 3 \int V(t)dt. \qquad (A)$$

Rearrange equation (A),

$$\frac{dV(t)}{dt} + 4V(t) + 3 \int V(t)dt = 0.$$

Now taking the Laplace transform,

$$sV(s) - V(0) + 4V(s) + 3\frac{V(s)}{s} = 0$$

$$V(s) = \frac{10s}{(s^2 + 4s + 3)} = \frac{10s}{(s+1)(s+3)} = \frac{A}{s+1} + \frac{B}{(s+3)}$$

$$A = -5, B = 15.$$

$$V(s) = -\frac{5}{(s+1)} + \frac{15}{(s+3)}$$

Taking the Laplace inverse,

$$V(t) = (-5e^{-t} + 15e^{-3t})u(t).$$

Problem 7.9. *A step voltage $3u(t-3)$ is applied to a series RLC circuit comprised of resistor $R = 5\ \Omega$, inductor $L = 1H$, and capacitor $C = \frac{1}{4}F$. Find the expression for current $i(t)$ in the circuit.*

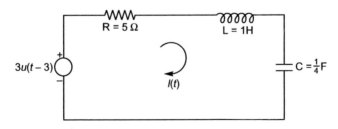

FIGURE 7.16

Solution: Applying KVL,

$$3u(t - 3) = 5i(t) + 1\frac{di(t)}{dt} + 4\int_0^t i(t)dt.$$

Taking the Laplace transform,

$$\frac{3}{s}e^{-s} = \left[5 + s + \frac{5}{s}\right]I(s) - I(0)[I(0) = 0]$$

$$I(s) = \frac{3e^{-s}}{(s^2 + 5s + 4)} = \frac{3e^{-s}}{(s + 1)(s + 4)}.$$

Using partial fraction expansion, we have

$$I(s) = e^{-3s}\left[\frac{1}{(s + 1)} - \frac{1}{(s + 4)}\right]$$

$$= [e^{-(t-3)} - e^{-4(t-3)}]u(t - 3).$$

Problem 7.10. *Initially the switch is connected with position 1. After a long time the switch moves to position 2. Find i(t).*

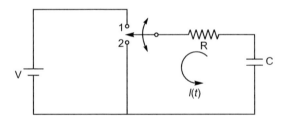

FIGURE 7.17

Solution: $v_c(0) = V$

Now by KVL,

$$i(t)R + \frac{1}{C}\int i(t)dt - V = 0$$

$$V = i(t)R + \frac{1}{C}\int i(t)dt.$$

Taking the Laplace transform,

$$\frac{V}{s} = I(s)R + \frac{1}{C}\frac{I(s)}{s}$$

$$\frac{V}{s} = I(s)\left(R + \frac{1}{Cs}\right) = I(s)\left[\frac{RCs + 1}{Cs}\right]$$

$$I(s) = \frac{VC}{RCs + 1} = \frac{VC}{RC(s + \frac{1}{RC})}$$

$$I(s) = \frac{V}{R}\left(\frac{1}{s + \frac{1}{RC}}\right).$$

Taking the Laplace inverse,

$$I(t) = \frac{V}{R}e^{-\frac{t}{RC}}.$$

Problem 7.11. *Find i (t). Initially the switch is connected after a long time. At t = 0 the switch is opened.*

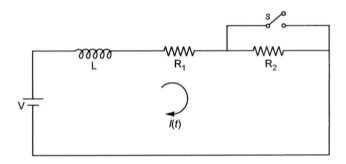

FIGURE 7.18

Solution: Initially the switch is connected. The initial value for the inductor is

$$i_L(0) = \frac{V}{R_1}.$$

at $t = 0$ the switch is opened.

Now by KVL,

$$V = i(t)R_1 + i(t)R_2 + L\frac{di(t)}{dt}$$

$$V = (R_1 + R_2)i(t) + L\frac{di(t)}{dt}.$$

Taking the Laplace transform,

$$\frac{V}{s} = (R_1 + R_2)I(s) + L[sI(s) - i(0)]$$

$$\frac{V}{s} = (R_1 + R_2)I(s) + L\left[sI(s) - \frac{V}{R_1}\right]$$

$$\frac{V}{s} + \frac{LV}{R_1} = I(s)[R_1 + R_2 + Ls]$$

$$I(s) = \frac{V}{s(R_1 + R_2 + Ls)} + \frac{LV}{R_1(R_1 + R_2 + Ls)}$$

$$I(s) = \frac{V}{Ls(s + \frac{R_1+R_2}{L})} + \frac{V}{R_1(s + \frac{R_1+R_2}{L})}.$$

Taking the Laplace inverse,

$$I(t) = \frac{V}{R_1 + R_2} + \frac{VR_2}{R_1(R_1 + R_2)}e^{-(\frac{R_1+R_2}{L})t}.$$

Problem 7.12. *The switch is opened after a long time. Find i(t) as shown in Figure 7.19.*

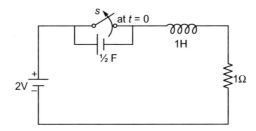

FIGURE 7.19

Solution: When the switch s is closed,

$$i(0) = \frac{2}{1} = 2A.$$

When the switch is opened at $t = 0$, applying KVL,

$$2 = 2\int_0^t i(t)dt + \frac{di}{dt} + i(t).$$

Taking the Laplace transform,

$$\frac{2}{s} = \frac{2I(s)}{s} + sI(s) - I(0) + I(s)$$

$$\frac{2}{s} + 2 = I(s)\left[\frac{2}{s} + s + 1\right]$$

$$I(s) = \frac{2(1 + s)}{(2 + s^2 + s)} = \frac{2(s + 1)}{(s^2 + s + 2)}$$

$$I(s) = \frac{2(s+1)}{(s+\frac{1}{2})^2 + (\frac{\sqrt{7}}{2})^2} = \frac{2(s+\frac{1}{2}+\frac{1}{2})}{(s+\frac{1}{2})^2 + (\frac{\sqrt{7}}{2})^2}$$

$$I(s) = \frac{2(s+\frac{1}{2})}{(s+\frac{1}{2})^2 + (\frac{\sqrt{7}}{2})^2} + \left(\frac{2}{\sqrt{7}}\right)\frac{(\frac{\sqrt{7}}{2})}{(s+\frac{1}{2})^2 + (\frac{\sqrt{7}}{2})^2}.$$

Taking the Laplace inverse,

$$I(t) = 2e^{-\frac{1}{2}t}\cos\frac{\sqrt{7}}{2}t + \frac{2}{\sqrt{7}}\sin\frac{\sqrt{7}}{2}t.$$

Problem 7.13. *In the network of Figure 7.20 the switch s is opened at t = 0. After the network has attained a steady state with the switch closed, find i(t) and $v_L(t)$.*

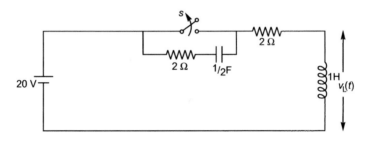

FIGURE 7.20

Solution: When the switch s is closed,

$$i(0) = \frac{20}{2} = 10\text{A}.$$

When the switch s is opened at $t = 0$, applying KVL,

$$20 = 2i(t) + \frac{1}{C}\int idt + 2i(t) + 1 \times \frac{di(t)}{dt}.$$

Taking the Laplace transform,

$$\frac{20}{s} = 2I(s) + 2\frac{I(s)}{s} + 2I(s) + sI(s) - I(0)$$

$$\frac{20}{s} + 10 = I(s)\left[4 + \frac{2}{s} + s\right]$$

$$I(s) = \frac{20 + 10s}{s^2 + 4s + 2} = \frac{10(s+2)}{(s+2)^2 - 2}.$$

Taking the Laplace inverse transform,

$$I(t) = 10e^{-2t} \cosh \sqrt{2}t.$$

Problem 7.14. *In the network shown, C is initially charged to V_0 and the switch K is closed at $t = 0$. Solve for the current $i(t)$, using the Laplace transformation method.*

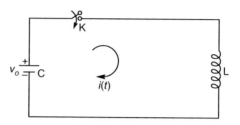

FIGURE 7.21

Solution: Applying KVL,

$$\frac{1}{C} \int i\,dt + L\frac{di}{dt} = 0.$$

Taking the Laplace transform,

$$\frac{1}{C}\left[\frac{I(s)}{s} - \frac{q(0)}{s}\right] + L[sI(s) - I(0)] = 0$$

$$\frac{q(0)}{C} = V(0) \quad \text{and} \quad I(0) = 0$$

$$\frac{1}{Cs}I(s) - \frac{V_0}{s} + LsI(s) = 0$$

$$I(s)\left[Ls + \frac{1}{Cs}\right] = \frac{V_0}{s}$$

$$I(s) = \frac{V_0 C}{LCs^2 + 1} = \frac{V_0}{L\left(s^2 + \frac{1}{LC}\right)}$$

$$I(s) = \frac{V_0}{L} \times \frac{\sqrt{LC}}{1}\left(\frac{\frac{1}{\sqrt{LC}}}{s^2 + \frac{1}{LC}}\right).$$

Now taking the Laplace inverse,

$$I(t) = V_0\sqrt{\frac{C}{L}} \sin\left(\frac{t}{\sqrt{LC}}\right).$$

Problem 7.15. *At $t = 0$, a pulse of width 'a' is applied to the RL network. Determine an expression for the current $i(t)$.*

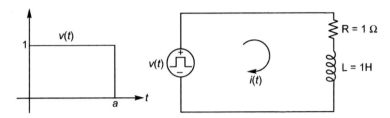

FIGURE 7.22

Solution: Applying KVL,

$$v(t) = i(t)\text{R} + \text{L}\frac{di(t)}{dt}$$

$$[u(t) - u(t - 9)] = i(t)\text{R} + \frac{\text{L}di(t)}{dt}.$$

Taking the Laplace transform,

$$\left(\frac{1}{s} - \frac{e^{-as}}{s}\right) = \text{R}I(s) + \text{L}[sI(s) - I(0)]$$

$$i(0) = 0$$

$$\frac{1 - e^{-as}}{s} = I(s)[\text{R} + \text{L}s].$$

Put

$$\text{R} = 1\,\Omega, \quad \text{L} = 1\,\text{H}$$

$$I(s) = \frac{1 - e^{-as}}{s(\text{R} + \text{L}s)} = \frac{1 - e^{-as}}{s\text{L}(s + \frac{\text{R}}{\text{L}})} = \frac{1 - e^{-as}}{s(s + 1)}.$$

After partial fraction,

$$I(s) = \frac{1}{s} - \frac{1}{s + 1} - \frac{e^{-as}}{s} + \frac{e^{-as}}{s + 1}.$$

Taking the Laplace inverse,

$$I(t) = u(t) - e^{-t}u(t) - u(t - a) + e^{-(t-a)}u(t - a)$$

$$= (1 - e^{-t})u(t) - [1 - e^{-(t-a)}]u(t - a).$$

Problem 7.16. *The switch in Figure 7.23 has been in position 1 for a long time. It is moved to 2 at $t = 0$. Obtain the expression for $i(t)$.*

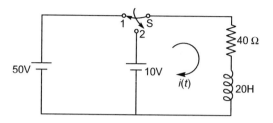

FIGURE 7.23

Solution: When the switch is in position 1,

$$i(0^-) = \frac{50}{40} = 1.25 \text{ A.}$$

When the switch is connected with position 2, applying KVL,

$$10 = 40i(t) + 20\frac{di(t)}{dt}.$$

Taking the Laplace transform,

$$\frac{10}{s} = 40I(s) + 20[sI(s) - i(0)]$$

$$I(s) = \frac{10 + 25s}{20s(s + 2)} = \frac{A}{s} + \frac{B}{s + 2}$$

$$A = \frac{10 + 25s}{20(s + 2)}\bigg|_{s=0} = 0.25$$

$$B = \frac{10 + 25s}{20s}\bigg|_{s=-2} = \frac{10 - 50}{-40} = 1$$

$$I(s) = \frac{0.25}{s} + \frac{1}{s + 2} \quad \Rightarrow \quad I(t) = 0.25\, u(t) + e^{-2t}u(t).$$

Problem 7.17. *The switch in the circuit has been closed for a very long time. It moved to position 2 at $t = 0$. Find the voltage across the capacitor for $t > 0$.*

Solution: When the switch s is in position 1 for a very long time, the capacitor charges up to 6 V,

$$v_C(0) = 6 \text{ V.}$$

When the switch s is in position 2,

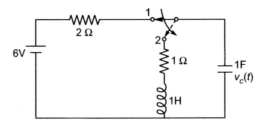

FIGURE 7.24

now applying KVL,

$$\frac{1}{C}\int i(t)dt + i(t) + 1.\frac{di(t)}{dt} = 0.$$

Taking the Laplace transform,

$$\frac{1}{C}\left[\frac{I(s)}{s} - \frac{q(0)}{s}\right] + I(s) + sI(s) - I(0) = 0.$$

Now,

$$\frac{q(0)}{C} = V(0) = 6 \text{ V and } I(0) = 0$$

$$\frac{I(s)}{Cs} + I(s) + sI(s) = \frac{6}{s}$$

$$I(s) = \frac{6}{s^2 + s + 1} = \frac{6}{s^2 + 2 \times \frac{1}{2} \times s + \frac{1}{4} + \frac{3}{4}}$$

$$I(s) = \frac{6}{\left(s + \frac{1}{2}\right)^2 + \left(\frac{\sqrt{3}}{2}\right)^2} = \frac{6}{\left(\frac{\sqrt{3}}{2}\right)}\left[\frac{\left(\frac{\sqrt{3}}{2}\right)}{\left(s + \frac{1}{2}\right)^2 + \left(\frac{\sqrt{3}}{2}\right)^2}\right].$$

Taking the Laplace inverse,

$$I(t) = 4\sqrt{3}e^{-\frac{1}{2}t} \sin \frac{\sqrt{3}}{2}t.$$

Problem 7.18. *In the circuit shown in Figure 7.25 eg(t) = 2.5 t volts. What are the values of i(t) and $V_L(t)$ at t = 4 seconds?*

Solution: Applying KVL,

$$eg(t) = 2i(t) + 4\frac{di(t)}{dt}$$

$$2.5t = 2i(t) + 4\frac{di(t)}{dt}.$$

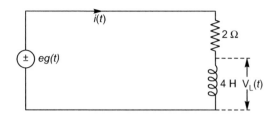

FIGURE 7.25

Taking the Laplace transform,

$$\frac{2.5}{s^2} = 2I(s) + 4sI(s) \qquad\qquad [i(0) = 0]$$

$$I(s) = \frac{2.5}{4.s^2(s + 0.5)}$$

$$= \frac{2.5}{4}\left[\frac{A}{s} + \frac{B}{s^2} + \frac{C}{s + 0.5}\right]$$

$$\frac{1}{s^2(s + 0.5)} = \frac{A}{s} + \frac{B}{s^2} + \frac{C}{(s + 0.5)}.$$

By the partial fraction, we get

$$A = -3, \quad B = 2, \quad C = 4$$

$$I(s) = -\frac{7.5}{4s} + \frac{1.25}{s^2} + \frac{2.5}{s + 0.5}.$$

Taking the Laplace inverse,

$$I(t) = -\frac{7.5}{4} + 1.25t + 2.5e^{-0.5t}$$

$$I(4) = 3.46 \text{ Amp.}$$

and

$$v_{\mathrm{L}}(t) = 4\frac{di\,t}{dt} \quad \text{and} \quad \text{put } t = 4 \text{ sec.}$$

Problem 7.19. *The switch s in the circuit of Figure 7.26 has been open for a long time before being closed at t = 0. Find i(t) for t ≥ 0.*

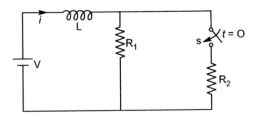

FIGURE 7.26

Solution: When the switch is open

$$i_L(0) = \frac{V}{R_1}.$$

When the switch is closed

$$V = L\frac{di}{dt} + iR_{eq} \quad \left(R_{eq} = \frac{R_1 \cdot R_2}{R_1 + R_2} \right).$$

Taking the Laplace transform,

$$\frac{V}{s} = L[sI(s) - I(0)] + I(s)R_{eq}$$

$$\frac{V}{s} + \frac{LV}{R_1} = I(s)[Ls + R_{eq}]$$

$$I(s) = \frac{V}{(Ls + R_{eq})s} + \frac{LV}{R_1(Ls + R_{eq})}$$

$$I(s) = \frac{V}{Ls\left(s + \frac{R_{eq}}{L}\right)} + \frac{V}{R_1\left(s + \frac{R_{eq}}{L}\right)}$$

$$I(s) = \frac{V}{R_{eq}}\left(\frac{1}{s} - \frac{1}{\left(s + \frac{R_{eq}}{L}\right)} \right) + \frac{V}{R_1\left(s + \frac{R_{eq}}{L}\right)}.$$

Taking the Laplace inverse,

$$I(t) = \frac{V}{R_{eq}}\left[1 - e^{-\frac{L}{R_{eq}}t} \right] + \frac{V}{R_1}e^{-\frac{L}{R_{eq}}t}.$$

Problem 7.20. *In the network shown in Figure 7.27, the switch s is open and closed at t = 0. Determine i(t).*

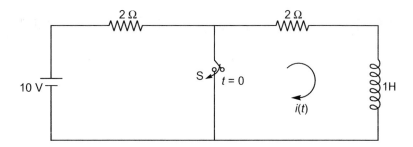

FIGURE 7.27

Solution: $i(0) = \frac{10}{4} = 2.5$ A.

By KVL,

$$2i + \frac{di}{dt} = 0.$$

Taking the Laplace transform,

$$2\mathrm{I}(s) + s\mathrm{I}(s) - \mathrm{I}_0 = 0$$
$$\mathrm{I}(s)(s + 2) = 2.5$$
$$\mathrm{I}(s) = \frac{2.5}{(s + 2)}.$$

Taking the Laplace inverse,

$$\mathrm{I}(t) = 2.5e^{-2t}.$$

Problem 7.21. *In the network shown in Figure 7.28 the circuit was initially in the steady-state condition with the switch s closed. At the instant when the switch is opened determine the value of current $i(t)$.*

Solution: Initially the switch S is closed for a very long time.

The inductor behaves as a short circuit,

$$i_1 = \frac{2}{1}\left[\mathrm{R}_{eq} = \frac{2 \times 2}{2 + 2} = 1\Omega\right]$$

and

$$i = \frac{i_1}{2} = 1 \text{ Amp,}$$

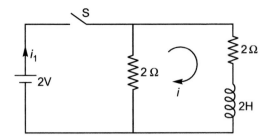

FIGURE 7.28

hence,

$$i(0) = 1 \text{ Amp.}$$

When the switch is open, applying KVL,

$$2i + 2\frac{di}{dt} + 2i = 0$$

$$\frac{di}{dt} + 2i = 0.$$

Taking the Laplace transform,

$$sI(s) - i(0) + 2I(s) = 0$$

$$I(s)(s + 2) = 1$$

$$I(s) = \frac{1}{s + 2}.$$

Taking the Laplace inverse transform,

$$I(t) = e^{-2t} \text{ Amp.}$$

Problem 7.22. *Using Laplace transformation, solve the following differential equation:*

$$\frac{d^2i}{dt^2} + 4\frac{di}{dt} + 8i = 8U(t),$$

given that $\mu(0^+) = 3$ *and* $\frac{di}{dt}(0^+) = -4.$

Solution: $\dfrac{d^2i(t)}{dt^2} + 4\dfrac{di}{dt} + 8i = 8U(t)$

Taking the Laplace transform,

$$s^2I_s - sI(0^+) - I'(0^+) + 4[sI(s) - I(0^+)] + 8I(s) = \frac{8}{s}$$

$$(s^2 + 4s + 8)I(s) = \frac{8}{s} + sI(0^+) + I'(0^+) + 4I(0^+).$$

Putting the values $i(0^+) = 3$ and $\frac{di}{dt}(0^+) = -4$,

$$(s^2 + 4s + 8)I(s) = \frac{8}{s} + 3s + 1 \times -4 + 4 \times 3$$

$$(s^2 + 4s + 8)I(s) = \frac{8}{s} + 3s + 8$$

$$I(s) = \frac{3s^2 + 8s + 8}{s(s^2 + 4s + 8)}.$$

Using partial fraction expansion, we get

$$I(s) = \frac{1}{s} + \frac{2(s + 2)}{(s + 2)^2 + (2)^2}.$$

By Laplace inverse transform,

$$I(t) = [1 + 2e^{-2t}\cos 2t]U(t).$$

QUESTIONS FOR DISCUSSION

1. What do you mean by transient condition?
2. What do you mean by steady state condition?
3. In how much time will a capacitor be charged fully?
4. What is the value of energy stored in a capacitor?
5. What is the value of energy stored in an inductor?
6. A capacitor is initially charged with a battery of 5 V for a very long time. Now it is connected with (*a*) 4 V and (*b*) 10 V battery in which case the capacitor is discharging. Why?
7. After a very long time, a capacitor will behave as a short circuit or open circuit. Why?
8. After a very long time, an inductor behaves as a short circuit or open circuit. Why?

9. Initially, which element behaves as an open circuit, (*a*) Resistor (*b*) Inductor, or (*c*) Capacitor?

10. Initially, which element behaves as a short circuit, (*a*) Resistor (*b*) Inductor, or (*c*) Capacitor?

OBJECTIVE QUESTIONS

1. The switch *s* in Figure 7.29 is closed at $t = 0$. If $V_2(0) = 10$ V and $Vg(0) = 0$ V, respectively, the voltage across the capacitors at a steady state will be
   (*a*) $V_2(\infty) = Vg(\infty) = 0$ (*b*) $V_2(\infty) = 2V, Vg(\infty) = 8$ V
   (*c*) $V_2(\infty) = Vg(\infty) = 8$ V   (*d*) $V_2(\infty) = Vg(\infty) = 2$ V.

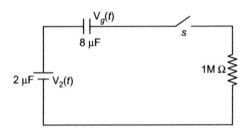

FIGURE 7.29

2. The time constant of the network shown in Figure 7.30 is
   (*a*) 2RC   (*b*) 3RC   (*c*) $\dfrac{RC}{2}$ (*d*) $\dfrac{2RC}{3}$.

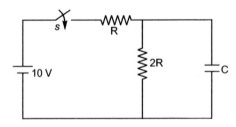

FIGURE 7.30

3. In the given circuit, the switch *s* is closed at $t = 0$ with $i_L(0) = 0$ and $V_C(0) = 0$. The steady state V_C equal.
   (*a*) 200 V   (*b*) 100 V   (*c*) Zero   (*d*) -100 V.

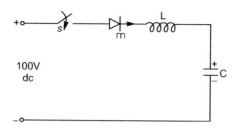

FIGURE 7.31

4. In the series RC circuit shown in Figure 7.32 the voltage across C starts increasing. The D.C. source is switched on the rate of increase of voltage C at the instant. Just after the switched is closed (*i.e.*, at $t = 0^+$), will be

    (*a*) Zero   (*b*) Infinity   (*c*) RC   (*d*) $\dfrac{1}{RC}$.

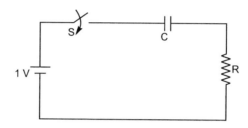

FIGURE 7.32

5. The transient response of the initially relaxed network shown in the Figure 7.33 is

    (*a*) $i = \dfrac{V}{R} e^{-t/RC}$

    (*b*) $i = \dfrac{V}{R} e^{t/RC}$

    (*c*) $i = \dfrac{V}{R}(1 - e^{-t/RC})$

    (*d*) $i = \dfrac{V}{R}(1 + e^{-t/RC})$.

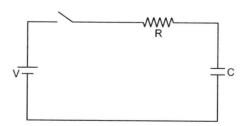

FIGURE 7.33

6. In a circuit having 90 ohms resistance in series with 90 ohms capacitive reactance, driven by a sine wave, the angle of phase difference between the applied voltage and the current is
    (a) $-90°$ (b) $-45°$ (c) $90°$ (d) $0°$.

7. An RC circuit has a capacitor $C = 2\,\mu F$ in series with a resistance $R = 1M\Omega$. The time of 6 secs. will be equal to
    (a) one time constant   (b) two time constants
    (c) three time constants   (d) none of these.

8. The Laplace transform of a delayed unit impulse function $\delta(t - 2)$ is
    (a) 1   (b) 0   (c) e^{-2s} (d) s.

9. The unit impulse response of a system is given as $C(t) = -4e^{-t} + 6e^{-2t}$. The stop response of the same system for $t \geq 0$ is equal to
    (a) $-3e^{-t} = 4e^{-t} + 1$ (b) $-3e^{-2t} + 4e^{-t} - 1$
    (c) $-3e^{-2t} - 4e^{-t} + 1$ (d) $3e^{-2t} + 4e^{-t} - 1$.

10. If the step response of an initially relaxed circuit is known then the ramp response can be obtained by
    (a) integrating the step response
    (b) differentiating the step response
    (c) integrating the step response twice
    (d) differentiating the step response twice.

11. In the network shown in Figure 7.34, the circuit was initially in the steady-state condition with the switch K closed. At the instant when the switch is opened, the rate of decay of current through the inductor will be
    (a) zero   (b) 0.5 A/sec.   (c) 1 A/sec.   (d) 2 A/sec.

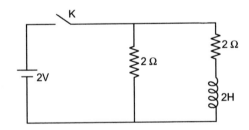

FIGURE 7.34

12. In the circuit shown in Figure 7.35, the switch S has been open for a long time. It is closed at $t = 0$. For $t > 0$ the current flowing through the inductor will be given by
    (a) $1.2 + 0.8e^{-2t}$ (b) $0.8 + 1.2e^{-2t}$
    (c) $1.2 - 0.8e^{-2t}$ (d) $0.8 - 1.2e^{-2t}$.

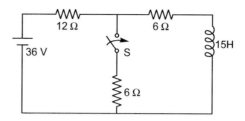

FIGURE 7.35

UNSOLVED PROBLEMS

1. Give examples of applications of Laplace transform in system analysis.
2. In the network shown in Figure 7.36 find $v_2(t)$ using the Laplace transform technique.

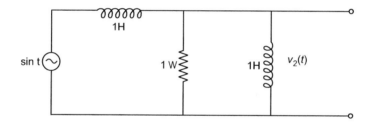

FIGURE 7.36

3. A triangular wave is applied as input to a series RL circuit with R = 2 Ω, L = 2 H, as shown in Figure 7.37. Calculate the current $i(t)$ through the circuit.

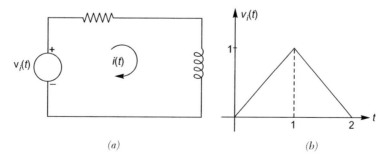

(a)   *(b)*

FIGURE 7.37

4. A voltage pulse, of unit magnitude and width T, is applied to the RC circuit as shown in Figure 7.38. Determine the current $i(t)$.

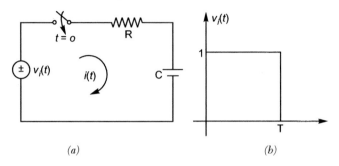

(a)   (b)

FIGURE 7.38

5. Find $i_1(t)$ and $i_2(t)$ by the mesh analysis method. Assume the initial values for the inductor and capacitor are zero. Also draw the s-domain representation of the circuit.

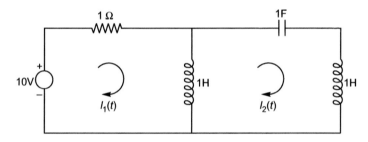

FIGURE 7.39

6. A voltage pulse, as shown in Figure 7.40(a), is applied on the RL circuit as shown in Figure 7.40(b). Find $i(t)$.

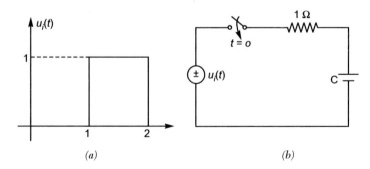

(a)   (b)

FIGURE 7.40

7. Draw the s-domain circuit for the network shown in Figure 7.41 and obtain the expressions for i_1 and i_2.

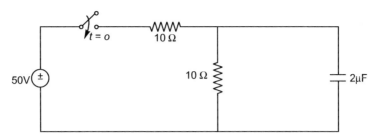

FIGURE 7.41

8. Transform the circuit shown in Figure 7.42 into the s-domain. $v_C(t)$ of $v(t)$ is:
    (a) $4\delta(t)$ (b) $4\delta(t-2)$.

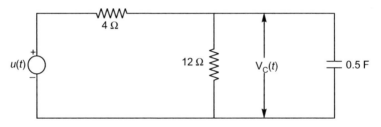

FIGURE 7.42

9. A coil has a resistance of 50 ohm and an inductance of 0.2 H. Find the expression for the current if the coil is excited by a voltage $150 \sin (500\, t)$.
10. Initially the switch is connected with position 1. After a long time the switch moves to position 2. Find the expression for $i(t)$.

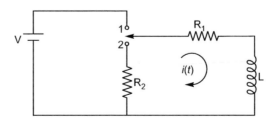

FIGURE 7.43

11. Initially the switch is open. At $t = 0$ the switch is closed. Find the expression for $i(t)$.

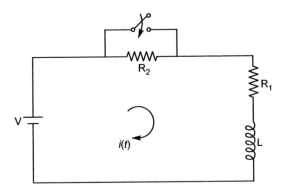

FIGURE 7.44

12. Solve the above problem. Assume initially the switch is closed and open at $t = 0$. Find the expression for $i(t)$.
13. Find the Laplace transform of the waveforms shown in Figure 7.45(a) and (b).

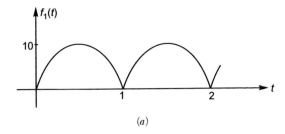

(a)

FIGURE 7.45

14. Obtain the solution of the differential equation given below using the Laplace transform,

$$\frac{d^2x}{dt^2} + 2\frac{dx}{dt} + 2x = 0,$$

given

$$x(0+) = 0 \text{ and } x'(0+) = 1.$$

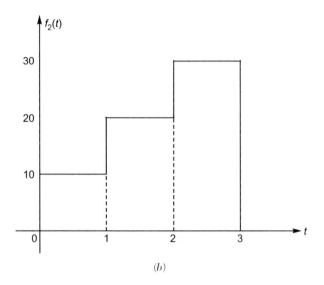

(b)

FIGURE 7.45

15. Initially the switch is connected with position 1 for a very long at $t = 0$. The switch moves to position 2. Find the expression for $i(t)$.

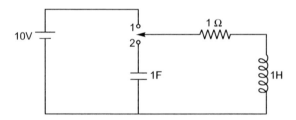

FIGURE 7.46

16. Initially the switch is connected with position 1 for a very long time at $t = 0$. The switch moves to position 2. Find the expression for $i(t)$.

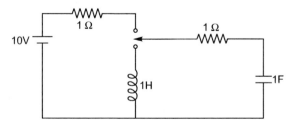

FIGURE 7.47

17. Using Laplace transform, solve the differential equation

$$\frac{d^2 i(t)}{dt^2} + \frac{di(t)}{dt} = t^2 + 2t \quad \text{given that} \quad i(0) = 4 \text{ and } \frac{di(0)}{dt} = -2.$$

18. Determine the inductor current for $t > 0$ in the network shown in Figure 7.48. The switch s has been open for a long time and is then closed at $t = 0$.

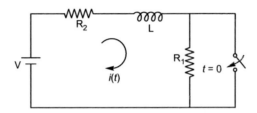

FIGURE 7.48

19. The switch S in the circuit of Figure 7.49 has been in position 1 for a long time compared to RC [time constant]. At $t = 0$, it is moved to position 2. Determine $i(t)$ for $t > 0$.

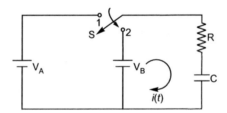

FIGURE 7.49

20. For the circuit shown in Figure 7.50 determine the current $i(t)$ for $t > 0$. The switch s has been closed for a long time and is opened at $t = 0$.

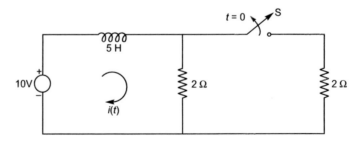

FIGURE 7.50

21. Determine the current through the inductor for $t \geq 0$ of a parallel RL circuit shown in Figure 7.51. The switch s has been in position 1 for a long time and is then moved to position 2 at $t = 0$.

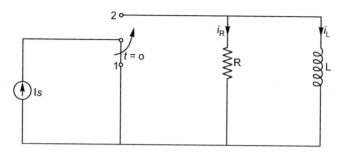

FIGURE 7.51

NETWORK THEOREMS

Chapter **8**

8.1 INTRODUCTION TO NETWORK THEOREMS

Anyone who has studied geometry should be familiar with the concept of a *theorem*: a relatively simple rule used to solve a problem, derived from a more intensive analysis using fundamental rules of mathematics. At least hypothetically, any problem in mathematics can be solved just by using the simple rules of arithmetic (in fact, this is how modern digital computers carry out the most complex mathematical calculations: by repeating many cycles of addition and subtraction!), but human beings aren't as consistent or as fast as a digital computer. We need "shortcut" methods in order to avoid procedural errors.

In electric network analysis, the fundamental rules are Ohm's Law and Kirchhoff's Laws. While these humble laws may be applied to analyze just about any circuit configuration (even if we have to resort to complex algebra to handle multiple unknowns), there are some "shortcut" methods of analysis to make the mathematics easier for the average human.

As with any theorem of geometry or algebra, these network theorems are derived from fundamental rules. In this chapter, we are not going to delve into the formal proofs of any of these theorems. If you doubt their validity, you can always empirically test them by setting up example circuits and calculating values using the "old" (simultaneous equation) methods versus the "new" theorems, to see if the answers coincide. They always should!

8.2 MILLMAN'S THEOREM

In Millman's Theorem, the circuit is re-drawn as a parallel network of branches, each branch containing a resistor or series battery/resistor combination. Millman's Theorem is applicable only to those circuits which can be re-drawn accordingly.

By considering the supply voltage within each branch and the resistance within each branch, Millman's Theorem will tell us the voltage across all branches. Please note that we've labeled the battery in the rightmost branch

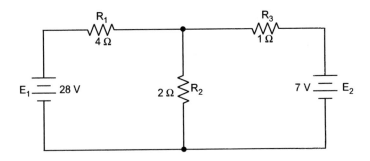

FIGURE 8.1

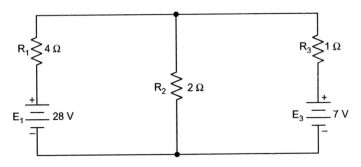

FIGURE 8.2

as "E_3" to clearly denote it as being in the third branch, even though there is no "E_2" in the circuit! [$E_2 = 0\,V$]

Millman's Theorem is nothing more than a long equation, applied to any circuit drawn as a set of parallel-connected branches, each branch with its own voltage source and series resistance:

Millman's Theorem Equation is

$$E_{eq} = \frac{\frac{E_1}{R_1} + \frac{E_2}{R_2} + \frac{E_3}{R_3} + \cdots}{\frac{1}{R_1} + \frac{1}{R_2} + \frac{1}{R_3} + \cdots}$$

$$\frac{1}{R_{eq}} = \frac{1}{R_1} + \frac{1}{R_2} + \frac{1}{R_3} + \cdots .$$

For the given circuit,

$$\frac{\frac{E_1}{R_1} + \frac{E_2}{R_2} + \frac{E_3}{R_3}}{\frac{1}{R_1} + \frac{1}{R_2} + \frac{1}{R_3}} = \text{Voltage across all branches.}$$

Substituting the actual voltage and resistance figures from our example circuit for the variable terms of this equation, we get the following expression:

$$\frac{\frac{28\,\text{V}}{4\,\Omega} + \frac{0\,\text{V}}{2\,\Omega} + \frac{7\,\text{V}}{1\,\Omega}}{\frac{1}{4\,\Omega} + \frac{1}{2\,\Omega} + \frac{1}{1\,\Omega}} = 8\,\text{V}.$$

The final answer of 8 volts is the voltage seen across all parallel branches as shown in Figure 8.3.

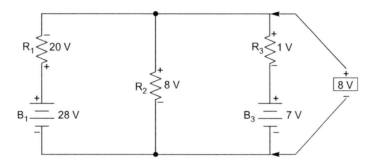

FIGURE 8.3

The polarity of all voltages in Millman's Theorem are referenced to the same point. In the example circuit above, we used the bottom wire of the parallel circuit as our reference point, and so the voltages within each branch (28 for the R_1 branch, 0 for the R_2 branch, and 7 for the R_3 branch) were inserted into the equation as positive numbers. Likewise, when the answer came out to 8 volts (positive), this meant that the top wire of the circuit was positive with respect to the bottom wire (the original point of reference). If both batteries had been connected backwards (negative ends up and positive ends down), the voltage for branch 1 would have been entered into the equation as −28 volts, the voltage for branch 3 as −7 volts, and the resulting answer of −8 volts would have told us that the top wire was negative with respect to the bottom wire (our initial point of reference).

To solve for resistor voltage groups the Millman voltage (across the parallel network) must be compared against the voltage source within each branch, using the principle of voltages adding in series to determine the magnitude and polarity of voltage across each resistor:

$$E_{R1} = 8\,\text{V} - 28\,\text{V} = -20\,\text{V} \text{ (negative on top)}$$
$$E_{R2} = 8\,\text{V} - 0\,\text{V} = 8\,\text{V} \text{ (positive on top)}$$
$$E_{R3} = 8\,\text{V} - 7\,\text{V} = 1\,\text{V} \text{ (positive on top)}$$

To solve for branch currents, each resistor voltage drop can be divided by its respective resistance $(I = E/R)$:

$$I_{R1} = \frac{20\,V}{4\,\Omega} = 5\,A$$

$$I_{R2} = \frac{8\,V}{2\,\Omega} = 4\,A$$

$$I_{R3} = \frac{1\,V}{1\,\Omega} = 1\,A$$

The direction of current through each resistor is determined by the polarity across each resistor, *not* by the polarity across each battery, as current can be forced backwards through a battery, as is the case with E_3 in the example circuit. This is important to keep in mind, since Millman's Theorem doesn't provide as direct an indication of "wrong" current direction as does the Branch Current or Mesh Current methods. You must pay close attention to the polarities of resistor voltage drops as given by Kirchhoff's Voltage Law, determining the direction of currents from that.

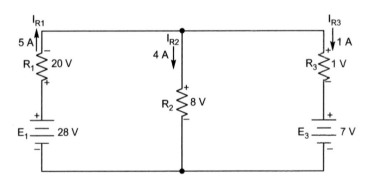

FIGURE 8.4

Millman's Theorem is very convenient for determining the voltage across a set of parallel branches, where there are enough voltage sources present to preclude a solution via regular series-parallel reduction methods. It also is easy in the sense that it doesn't require the use of simultaneous equations. However, it is limited in that it only applies to circuits which can be re-drawn to fit this form. It cannot be used, for example, to solve an unbalanced bridge circuit.

 • Millman's Theorem treats circuits as a parallel set of series-component branches and such types of networks easily convert into simple networks.

- All voltages entered and solved for in Millman's Theorem are polarity-referenced at the same point in the circuit.

8.2.1 Millman Theorem for an A.C. Circuit

This theorem can be used either for A.C. or D.C. and it is also useful for voltage sources and current sources connected in series and parallel, respectively.

Millman Theorem for voltage sources: Consider a circuit containing n voltage sources $E_1, E_2, \ldots E_n$ and their internal impedances are $Z_1, Z_2, \ldots Z_n$, respectively, as shown in Figure 8.5.

$$E = \frac{E_1 \times \frac{1}{Z_1} + E_2 \times \frac{1}{Z_2} + \cdots + E_n \times \frac{1}{Z_n}}{\frac{1}{Z_1} + \frac{1}{Z_2} + \cdots + \frac{1}{Z_n}} \tag{8.1}$$

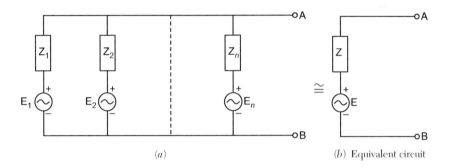

(a)   (b) Equivalent circuit

FIGURE 8.5 **Millman Theorem for voltage sources.**

and

$$\frac{1}{Z} = \frac{1}{Z_1} + \frac{1}{Z_2} + \cdots + \frac{1}{Z_n}. \tag{8.2}$$

By using sources transformation, we can convert the voltage source into the current source, as shown in Figure 8.6(a).

$$I_1 = \frac{E_1}{Z_1}, I_2 = \frac{E_2}{Z_2}.$$

Similarly,

$$I_n = \frac{E_n}{Z_n}.$$

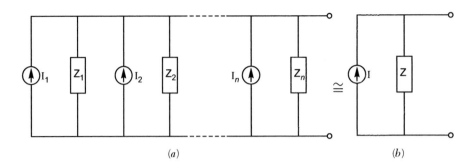

FIGURE 8.6   Millman Theorem for current sources.

8.2.2   Steps for Millman's Theorem

Step 1. First convert all current sources into voltage sources as shown in the figure by using source transformation. If there is no source in any branch, assume a zero potential source in series with the resistance or impedance.

Step 2. Calculate the equivalent potential (E) of the circuit by using the formula in Equation 8.1

Step 3. Now calculate the equivalent impedance (Z) or resistance by using the formula in Equation 8.2.

Step 4. Now, draw the equivalent circuit diagram as shown in Figure 8.5(b) with the load.

Step 5. Load the current $I_L = \frac{E}{Z+Z_L}$.

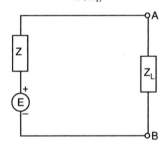

Example 8.1. *Determine I as shown in Figure 8.7(b) using Millman's Theorem.*

Solution:

$$E = \frac{10 \times \frac{1}{2} + 5 \times \frac{1}{1} + 0 \times \frac{1}{2}}{\frac{1}{2} + \frac{1}{1} + \frac{1}{2}}.$$

In this case ($E_3 = 0$, no voltage source is connected),

$$E = 5\,\mathrm{V}.$$

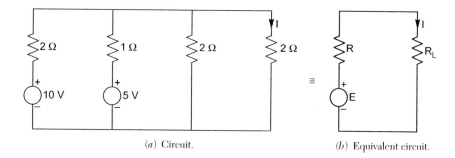

(a) Circuit.   (b) Equivalent circuit.

FIGURE 8.7   Problem for Millman Theorem.

Now the equivalent resistance $\frac{1}{R} = \frac{1}{2} + \frac{1}{1} + \frac{1}{2}$

$$R = \frac{1}{2} = 0.5\,\Omega.$$

The equivalent circuit diagram is shown in Figure 8.7(b),

$$E = 5\,V, R = 0.5\,\Omega, R_L = 2\,\Omega$$

$$I = \frac{E}{R + R_L} = \frac{5}{0.5 + 2} = 2\ \text{Amp}.$$

Example 8.2. *Determine I as shown in Figure 8.8 using Millman's Theorem.*

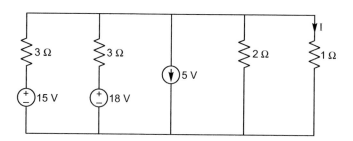

FIGURE 8.8

Solution: First convert the current source (5A) into the voltage source using source transformation as shown in Figure 8.8(a).

$$E = \frac{15 \times \frac{1}{3} + 18 \times \frac{1}{3} - 10 \times \frac{1}{2}}{\frac{1}{3} + \frac{1}{3} + \frac{1}{2}}$$

$$E = \frac{36}{7}\ V$$

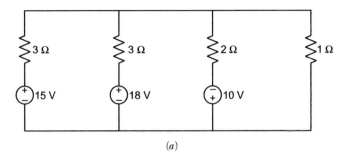

(a)

FIGURE 8.8

NOTE ► E₃ is negative because the polarity of E₃ is opposite. Students can remember that generally we take the upper sign. In the case of E₃ the upper sign is negative.

$$\frac{1}{R} = \frac{1}{3} + \frac{1}{3} + \frac{1}{2} = \frac{7}{6}$$
$$R = \frac{6}{7}\,\Omega$$

The equivalent circuit is shown in Figure 8.8(b).

$$I = \frac{E}{R + R_L} = \frac{\frac{36}{7}}{\frac{6}{7} + 1} = \frac{36}{13}\ \text{Amp.}$$

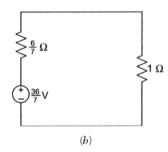

(b)

FIGURE 8.8

8.3 SUPERPOSITION THEOREM

A theorem like Millman's certainly works well, but it is not quite obvious *why* it works so well. Superposition, on the other hand, is obvious.

The strategy used in the Superposition Theorem is to eliminate all but one source of power within a network at a time, using series/parallel analysis to determine voltage drops (and/or currents) within the modified network for each power source separately and then voltage drops or currents have been determined for each power source working separately. The values are all "superimposed" (algebraically) to find the actual voltage drops/currents with all sources active. Let's look at our example circuit again and apply Superposition Theorem to it.

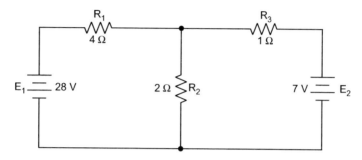

FIGURE 8.9

Since we have two sources of power in this circuit, we will have to calculate two sets of values for voltage drops and/or currents. One for the circuit with only the 28 volt battery in effect and one for the circuit with only the 7 volt battery in effect:

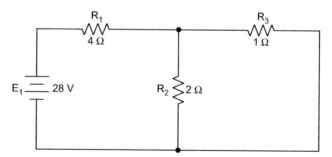

FIGURE 8.10

When re-drawing the circuit for series/parallel analysis with one source, all other voltage sources are replaced by wires (shorts), and all current sources with open circuits (breaks). Since we only have voltage sources (batteries) in our example circuit, we will replace every inactive source during analysis with a wire.

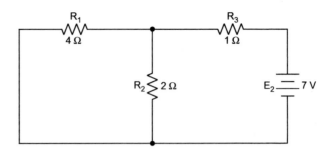

FIGURE 8.11

Analyzing the circuit with only the 28 volt battery, we obtain the following values for voltage and current:

	R_1	R_2	R_3	$R_2 // R_3$	$R_1 + R_2 // R_3$ Total	
E	24	4	4	4	28	Volts
I	6	2	4	6	6	Amps
R	4	2	1	0.667	$4.667 \left(\dfrac{14}{3}\right)$	Ohms

$$I = \frac{E_1}{R_{eq}} = \frac{28}{14} \times 3 = 6A$$

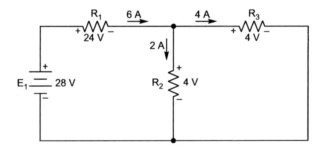

FIGURE 8.12

By the table, it is clear

$$I_{R2} = 2 \text{ Amp}$$
$$I_{R3} = 4 \text{ A}$$
$$E_{R3} = 4 \, \Omega.$$

Analyzing the circuit with only the 7 volt battery, we obtain another set of values for voltage and current:

	R_1	R_2	R_3	$R_1//R_2$	$R_3 + R_1//R_2$ Total	
E	4	4	3	4	7	Volts
I	1	2	3	3	3	Amps
R	4	2	1	1.333	2.333	Ohms

Similarly,

$$I = \frac{E}{R_{eq}} = \frac{7}{7/3} = 3\,A.$$

It is clear by the table,

$$I_{R1} = 1A \text{ and } E_{R1} = 4\,V.$$

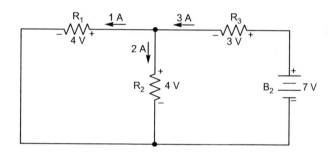

FIGURE 8.13

When superimposing these values of voltage and current, we have to be very careful to consider polarity (voltage drop) and direction of current as the values have to be added *algebraically*.

With 28 V battery	*With 7 V battery*	*With both batteries*
24 V $-+\!\!\bigwedge\!\!-$ E_{R1}	4 V $--\!\!\bigwedge\!\!+$ E_{R1}	20 V $E_{R1}\ +\!\!\bigwedge\!\!-$ $24V - 4\,V = 20\,V$
$E_{R2} \gtrless 4\,V$	$E_{R2} \gtrless 4\,V$	$E_{R2} \gtrless 8\,V$ $4V + 4V = 8V$
4 V $-+\!\!\bigwedge\!\!-$ E_{R3}	3 V $--\!\!\bigwedge\!\!+$ E_{R3}	1 V $E_{R3}\ +\!\!\bigwedge\!\!-$ $4V - 3\,V = 1V$

Applying these superimposed voltage figures to the circuit, the end result looks something like this:

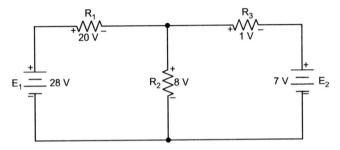

FIGURE 8.14

Currents add up algebraically as well, and can either be superimposed as done with the resistor voltage drops, or simply calculated from the final voltage drops and respective resistances $(I = E/R)$. Either way, the answers will be the same. Here we will show the superposition method applied to the current:

With 28 V battery	*With 7 V battery*	*With both batteries*
$\rightarrow$ 6 A I_{R1}	$\leftarrow$ 1 A I_{R1}	I_{R1} $\rightarrow$ 5 A $6A - 1A = 5A$
I_{R2} $\downarrow$ 2 A	I_{R2} $\downarrow$ 2 A	I_{R2} $\downarrow$ 4 A $2A + 2A = 4A$
$\rightarrow$ 4 A I_{R3}	$\leftarrow$ 3 A I_{R3}	I_{R3} $\rightarrow$ 1 A $4A - 3A = 1A$

Once again applying these superimposed figures to our circuit:

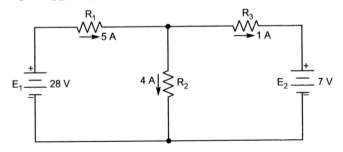

FIGURE 8.15

Quite simple and elegant, don't you think? It must be noted, though, that the Superposition Theorem works only for circuits that are reducible to series/parallel combinations for each of the power sources at a time (thus, this theorem is useless for analyzing an unbalanced bridge circuit), and it only works where the underlying equations are linear (no mathematical powers or roots). The requisite of linearity means that the Superposition Theorem is only applicable for determining voltage and current, *not power!!!* Power dissipations, being nonlinear functions, do not algebraically add to an accurate total when only one source is considered at a time. The need for linearity also means this theorem cannot be applied in circuits where the resistance of a components changes with voltage or current. Hence, networks containing components like lamps (incandescent or gas-discharge) or varistors could not be analyzed.

Another prerequisite for the Superposition Theorem is that all components must be "bilateral," meaning that they behave the same with electrons flowing either direction through them.

The Superposition Theorem finds use in the study of alternating (A.C. circuits and semiconductors) amplifier circuits, where sometimes A.C. is often mixed (superimposed) with D.C. Because A.C. voltage and current equations (Ohm's law) are linear just like D.C., we can use superposition to analyze the circuit with just the D.C. power source, then just the A.C. power source, combining the results to tell what will happen with both A.C. and D.C. sources in effect. For now, though, superposition will suffice as a break from having to do simultaneous equations to analyze a circuit.

NOTE
- The Superposition Theorem states that a circuit can be analyzed with only one source of power at a time. The corresponding component voltages and currents are algebraically added to find out what they'll do with all power sources in effect.
- To negate all but one power source for analysis, replace any source of voltage (batteries) with a wire; replace any current source with an open (break), or we can say set any voltage source to zero (means short circuit) and set any current source to zero (means open circuit).

8.3.1 Steps for the Superposition Theorem.

The Superposition Theorem is applicable for both A.C. and D.C. and the steps are the same in both cases.

Step 1. Select one source (voltage or current) at a time and the other sources are set to zero. The voltage source set at zero means short circuit and the current source set at zero means it is open circuit (no current flowing). Calculate the value of current and direction in the desired branch.

Step 2. Apply Step 1 for other sources (select another source and set to zero). Calculate the value of current.

Step 3. Now we have calculated the value of the current and the direction of each source. Add all the currents algebraically. This is the final current, which is flowing in the desired branch, when all the sources are acting or are in the 'ON' stage.

Example 8.3. *Determine the value of current in the 2 Ω resistor as shown in the figure using the Superposition Theorem.*

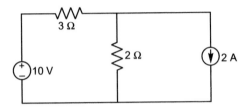

FIGURE 8.16

Solution: Case I: A 10 V battery is operating and a 2 A current source is set at zero (open) as shown in Figure 8.16(*a*).

$$I_1 = \frac{10}{5} = 2 \text{ Amp.}$$

Case II: 2 A current source is operating and now a 10 V battery is set to zero (short) as shown in Figure 8.16(*b*).

By using the current division formula,

$$I_2 = \left(\frac{R_1}{R_1 + R_2} \right) I = \left(\frac{3}{3+2} \right) \times 2.$$

$$I_2 = 1.2A$$

Now the resultant current in the 2 Ω resistor = $I_1 - I_2$

$$= 2 - 1.2 = 0.8 \text{ A.}$$

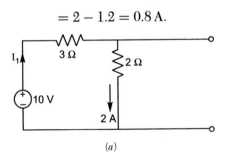

(*a*)

FIGURE 8.16

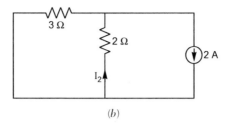

(b)

FIGURE 8.16

NOTE ➤ It is clear I_1 and I_2 are in opposite directions.

Example 8.4. *Determine the value of current in the 2 Ω resistor using the Superposition Theorem as shown in Figure 8.17.*

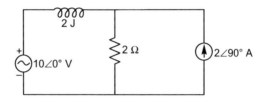

FIGURE 8.17

Solution: Case I: A $10\angle 0°$ V battery is operating and $2\angle 90°$ is set to zero as shown in Figure 8.17(*a*).

$$I_1 = \frac{10}{2J + 2} = \frac{5}{\sqrt{2}}\angle - 45°$$

$$I_1 = 3.53\angle - 45° \text{ Amp.}$$

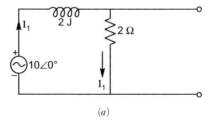

(a)

FIGURE 8.17

Case II: A $2\angle 90°$ current source is operating and $10\angle 0°$ V is set to zero (short) as shown in Figure 8.17(*b*).

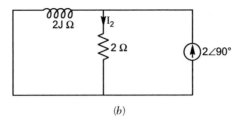

(b)

FIGURE 8.17

By the current division formula,

$$I_2 = \frac{2J}{2 + 2J} \times 2J = -\frac{2}{1 + J}. \qquad\qquad [2\angle 90° = 2J]$$

The resultant current in the $2\,\Omega$ resistor $= I_1 + I_2$ (because the direction of the current is the same)

$$= \frac{5}{1 + 5} - \frac{2}{1 + J} = \frac{3}{1 + J} = \frac{3}{\sqrt{2}}\angle 45° = 2.12\angle 45°\,\text{A}.$$

Example 8.5. *Determine I using the Superposition Theorem as shown in Figure 8.18.*

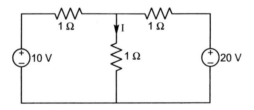

FIGURE 8.18

Solution: Case I: A 10 V battery is operating and a 20 V battery is set to zero (short) as shown in Figure 8.18(*a*).

The current flowing due to the 10 V battery $= \frac{10}{1.5} = \frac{20}{3}\text{A}.$

Now I_1 by the current division formula,

$$I_1 = \frac{1}{2} \times \frac{20}{30} = \frac{10}{3}\text{A}.$$

Case II: A 20 V battery is operating and a 10 V battery is set to zero (short) as shown in Figure 8.18(*b*). The current flowing by the 20 V battery $= \frac{40}{3}\text{A}.$

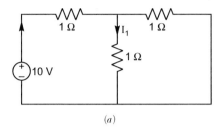

(a)

FIGURE 8.18

Now

$$I = I_1 + I_2 = \frac{10}{3} + \frac{40}{3}$$

$$I = \frac{50}{3} \text{ Amp.}$$

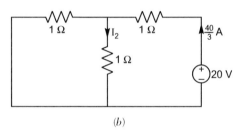

(b)

FIGURE 8.18

8.4 THEVENIN'S THEOREM

Thevenin's Theorem states that it is possible to simplify any linear circuit, no matter how complex, to an equivalent circuit with just a single voltage source and series resistance connected to a load. The qualification of "linear" is identical to that found in the Superposition Theorem, where all the underlying equations must be linear (no exponents or roots). If we're dealing with passive components (such as resistors and later inductors and capacitors), this is true. However, there are some components (especially certain gas discharge and semiconductor components) which are non-linear that is their opposition to current changes with voltage and/or current. As such, we would call circuits containing these types of components, non-linear circuits.

Thevenin's Theorem is especially useful in analyzing power systems and other circuits where one particular resistor in the circuit (called the "load" resistor) is subject to change and recalculation of the circuit is necessary with each trial value of load resistance, to determine voltage across it and current through it. Lets take another look at our example circuit:

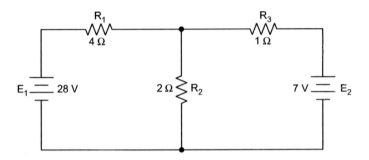

FIGURE 8.19

Let's suppose that we decide to designate R_2 as the "load" resistor in this circuit. We already have four methods of analysis (Branch Current, Mesh Current, Millman's Theorem, and Superposition Theorem) to use in determining voltage across R_2 and current through R_2, but each of these methods are time-consuming. Imagine repeating any of these methods over and over again to find what would happen if the load resistance changed (changing load resistance is *very* common in power systems, as multiple loads get switched on and off as needed; the total resistance of their parallel connections changes depending on how many are connected at a time). This could potentially involve a *lot* of work!

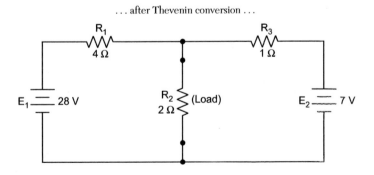

FIGURE 8.20

Thevenin's Theorem makes this easy by temporarily removing the load resistance from the original circuit and reducing what's left to an equivalent circuit composed of a single voltage source and series resistance. The load resistance can then be re-connected to this "Thevenin equivalent circuit" and calculations carried out as if the whole network were nothing but a simple series circuit.

8.4.1 Thevenin Equivalent Circuit

The "Thevenin Equivalent Circuit" is the electrical equivalent of E_1, R_1, R_3, and E_2 as seen from the two points where our load resistor (R_2) connects.

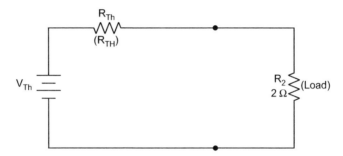

FIGURE 8.21

The Thevenin equivalent circuit, if correctly derived, will behave exactly the same as the original circuit formed by E_1, R_1, R_3, and E_2. In other words, the load resistor (R_2) voltage and current should be exactly the same for the same value of load resistance in the two circuits. The load resistor R_2 cannot "tell the difference" between the original network of E_1, R_1, R_3, and E_2, and the Thevenin equivalent circuit of E_{Th} and R_{Th}, provided that the values for E_{Th} and R_{Th} have been calculated correctly. The advantage of performing the "Thevenin conversion" to the simpler circuit, of course, is that it makes load voltage and load current so much easier to solve than in the original network. Calculating the equivalent Thevenin source voltage and series resistance is actually quite easy. First, the chosen load resistor is removed from the original circuit, replaced with a break (open circuit), and this is step one.

Step two: the voltage between the two points where the load resistor used to be attached is determined. Use whatever analysis methods are at your disposal to do this. In this case, the original circuit with the load resistor removed is nothing more than a simple series circuit with opposing batteries, and so we can determine the voltage across the open load terminals by applying the

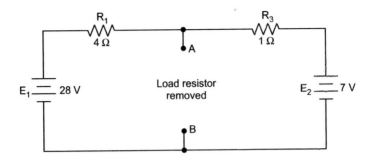

FIGURE 8.22

rules of series circuits, Ohm's Law, and Kirchhoff's Voltage Law:

	R_1	R_3	Total	
E	16.8	4.2	21	Volts
I	4.2	4.2	4.2	Amps
R	4	1	5	Ohms

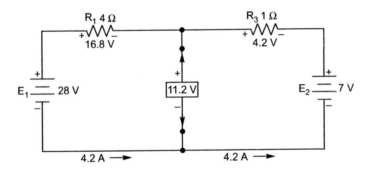

FIGURE 8.23

The voltage between the two load connection points can be figured from the one of the battery's voltage and one of the resistor's voltage drops, and comes out to 11.2 volts. This is our "Thevenin voltage" (V_{Th}) in the equivalent circuit:

Thevenin Equivalent Circuit

To find the Thevenin series resistance for our equivalent circuit, we need to take the original circuit (with the load resistor still removed), remove the power sources (in the same style as we did with the Superposition Theorem: voltage sources replaced with wires (short circuit) and current sources

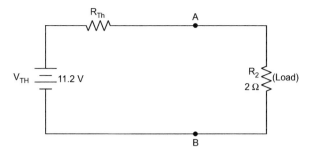

FIGURE 8.24

replaced with breaks, (open circuit)), and figure the resistance from one load terminal to the other:

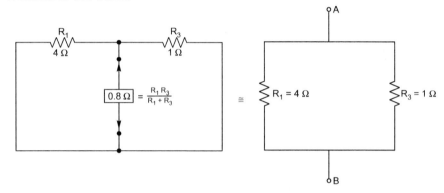

FIGURE 8.25

With the removal of the two batteries, the total resistance measured at this location is equal to R_1 and R_3 in parallel: 0.8 Ω. This is our "Thevenin resistance" (R_{Th}) for the equivalent circuit:

Thevenin Equivalent Circuit

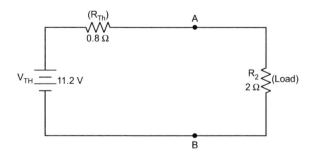

FIGURE 8.26

With the load resistor (2 Ω) attached between the connection points, we can determine voltage across it and current through it as though the whole network were nothing more than a simple series circuit:

	R_{TH}	R_{Load}	Total	
V	3.2	8	11.2	Volts
I	4	4	4	Amps
R	0.8	2	2.8	Ohms

Notice that the voltage and current figures for R_2 (8 volts, 4 amps) are identical to those found using other methods of analysis. Also notice that the voltage and current figures for the Thevenin series resistance and the Thevenin source (*total*) do not apply to any component in the original, complex circuit. Thevenin's Theorem is only useful for determining what happens to a *single* resistor in a network: the load.

The advantage, of course, is that you can quickly determine what would happen to that single resistor if it were of a value other than 2 Ω without having to go through a lot of analysis again. Just plug in that other value for the load resistor into the Thevenin equivalent circuit and a little bit of series circuit calculation will give you the result.

- Thevenin's Theorem is a way to reduce a network to an equivalent circuit composed of a single voltage source, series resistance, and series load.
- Steps to follow for Thevenin's Theorem:
  1. Find the Thevenin source voltage by removing the load resistor from the original circuit and calculating voltage across the open connection points where the load resistor used to be (V_{TH}).
  2. Find the Thevenin resistance by removing all power sources in the original circuit (voltage sources shorted and current sources open) and calculating total resistance between the open connection points (R_{TH}).
  3. Draw the Thevenin equivalent circuit, with the Thevenin voltage source in series with the Thevenin resistance. The load resistor re-attaches between the two open points of the equivalent circuit.
  4. Analyze voltage and current for the load resistor following the rules for series circuits.

8.5 NORTON'S THEOREM

Norton's Theorem states that it is possible to simplify any linear circuit, no matter how complex, to an equivalent circuit with just a single current source and parallel resistance connected to a load. Just as with Thevenin's Theorem, the qualification of "linear" is identical to that found in the Super-position Theorem: all underlying equations must be linear (no exponents or roots).

Contrasting our original example circuit against the Norton equivalent: it looks something like this:

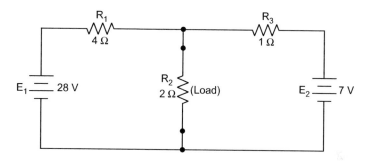

FIGURE 8.27

after Norton conversion.

8.5.1 Norton Equivalent Circuit

Remember that a *current source* is a component whose job is to provide a constant amount of current, outputting as much or as little voltage necessary to maintain that constant current.

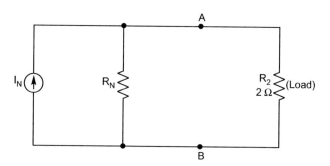

FIGURE 8.28

As with Thevenin's Theorem, everything in the original circuit except the load resistance has been reduced to an equivalent circuit that is sim-pler to analyze. Also smilar to Thevenin's Theorem are the steps used in Norton's Theorem to calculate the Norton source current (I_{Norton}) and Norton resistance (R_{Norton}).

As before, the first step is to identify the load resistance and remove it from the original circuit:

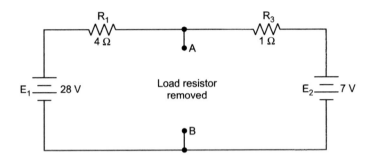

FIGURE 8.29

Then, to find the Norton current (for the current source in the Norton equivalent circuit), place a direct wire (short) connection between the load points, and determine the resultant current. Note that this step is exactly the opposite of the respective step in Thevenin's Theorem, where we replaced the load resistor with a break (open circuit):

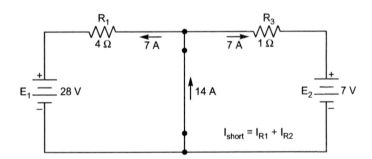

FIGURE 8.30

With zero voltage dropped between the load resistor connection points, the current through R_1 is strictly a function of E_1's voltage and R_1's resistance: 7 amps ($I = E/R$). Likewise, the current through R_3 is now strictly a function of E_2's voltage and R_3's resistance: 7 amps ($I = E/R$). The total current through the short between the load connection points is the sum of these two currents: 7 amps +7 amps = 14 amps. This figure of 14 amps becomes the Norton source current (I_{Norton}) in our equivalent circuit:

Norton Equivalent Circuit

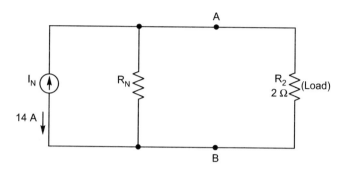

FIGURE 8.31

Remember, the arrow notation for a current source points in the direction *opposite* that of electron flow. For better or for worse, this is standard electronic symbol notation.

To calculate the Norton resistance (R_{Norton}), we do the exact same thing as we did for calculating Thevenin resistance ($R_{Thevenin}$): take the original circuit (with the load resistor still removed), remove the power sources (in the same style as we did with the Superposition Theorem: voltage sources are replaced with wires and current sources are replaced with breaks), and figure total resistance from one load connection point to the other:

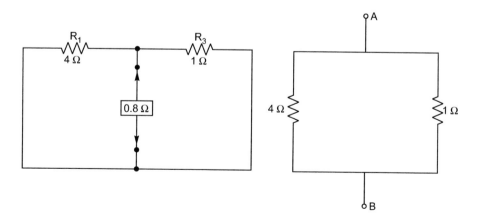

FIGURE 8.32

Now our Norton equivalent circuit looks like this:

Norton Equivalent Circuit

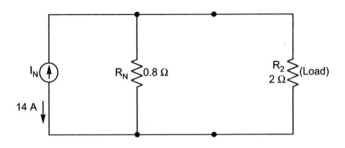

FIGURE 8.33

If we re-connect our original load resistance of $2\,\Omega$, we can analyze the Norton circuit as a simple parallel arrangement:

	R_{Norton}	R_{Load}	Total	
E	8	8	8	Volts
I	10	4	14	Amps
R	0.8	2	571.43×10^{-3}	Ohms

As with the Thevenin eqivalent circuit, the only useful information from this analysis is the voltage and current values for R_2; the rest of the information is irrelevant to the original circuit. However, the same advantages seen with Thevenin's Theorem apply to Norton's as well: if we wish to analyze load resistor voltage and current over several different values of load resistance, we can use the Norton equivalent circuit again and again, applying nothing more complex than simple parallel circuit analysis to determine what's happening with each trial load.

- Norton's Theorem is a way to reduce a network to an equivalent circuit composed of a single current source, parallel resistance, and parallel load.
- Steps to follow for Norton's Theorem:
  1. Find the Norton source current by removing the load resistor from the original circuit and calculating current through a short (wire) jumping across the open connection points where the load resistor used to be.
  2. Find the Norton resistance by removing all power sources in the original circuit (voltage sources shorted and current sources open)

and calculating total resistance between the open connection points.

3. Draw the Norton equivalent circuit, with the Norton current source in parallel with the Norton resistance. The load resistor re-attaches between the two open points of the equivalent circuit.

4. Analyze voltage and current for the load resistor following the rules for parallel circuits.

8.5.2 Steps for the Norton and Thevenin Theorems

Steps for the Thevenin and Norton theorems are the same.

Step 1. Open the terminal AB, where we need current or voltage and calculate V_{AB} using any method (KVL, mesh, etc.)

$$\boxed{V_{AB} = V_{TH}}$$

Step 2. Short the terminal AB and calculate the short circuit current $(I_{sc} = I_N)$.

Step 3. Now calculate $R_{TH} = \frac{V_{TH}}{I_{sc}} = R_N$.

NOTE ▶ Equivalent resistance across AB is $R_{AB} = R_{TH}$. (Students can calculate R_{TH} by this method also.)

Step 4. Draw the Thevenin or Norton equivalent circuit as required and calculate the load current.

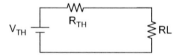 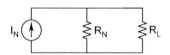

Example 8.6. *Determine I_L using the Thevenin and Norton theorems as shown in Figure 8.34.*

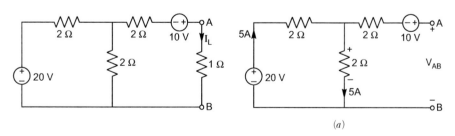

(a)

FIGURE 8.34

Solution: First open the terminal AB as shown in Figure 8.34(a) and calculate V_{AB}.

NOTE ▶ No current is flowing through the 2 Ω resistor (near the 10 V battery). Applying KVL,

$$-V_{AB} + 10 + 2 \times 0 + 2 \times 5 = 0$$
$$V_{AB} = 20\,\text{V} = V_{TH}.$$

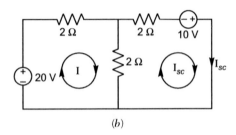

(b)

FIGURE 8.34

Now short the terminal AB.
Applying mesh analysis,

$$-20 + 2I + 2(I - I_{sc}) = 0$$
$$4I - 2I_{sc} = 20$$
$$2I - I_{sc} = 10 \tag{1}$$
$$2(I_{sc} - I) + 2I_{sc} - 10 = 0$$
$$2I_{sc} - I = 5. \tag{2}$$

By Equations (1) and (2), we get $I_{sc} = \frac{20}{3}\text{A} = I_N$.

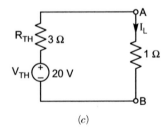

(c)

FIGURE 8.34

Now

$$R_{TH} = R_N = \frac{V_{TH}}{I_{sc}} = \frac{20 \times 3}{20} = 3\,\Omega.$$

The Thevenin equivalent circuit is shown in Figure 8.34(c),

$$I_L = \frac{V_{TH}}{R_{TH} + R_L} = \frac{20}{3+1} = 5\,A.$$

The Norton equivalent circuit is shown in Figure 8.34(d). By the current division formula,

$$I_L = \left(\frac{R_N}{R_N + R_L}\right) \times I_N = \frac{3}{4} \times \frac{20}{3} = 5A.$$

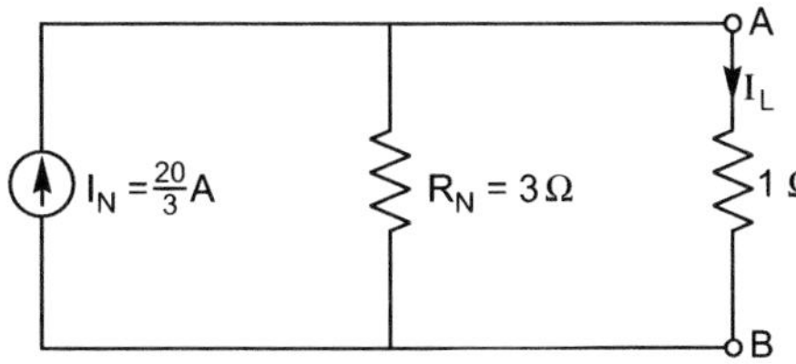

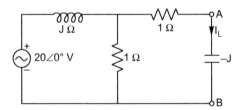

(d)

FIGURE 8.34

Example 8.7. *Determine I_L using the Thevenin and Norton theorems as shown in Figure 8.35.*

FIGURE 8.35

Solution: Open the terminal AB and calculate V_{AB} as shown in Figure 8.35(a).

$$I = \frac{20}{1 + J}$$

$$V_{AB} = 1 \times \frac{20}{1 + J} = \frac{20}{1 + J} V = V_{TH}$$

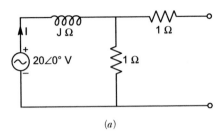

(a)

FIGURE 8.35

Short the terminal AB as shown in Figure 8.35(b). By using mesh analysis,

$$-20\angle 0°V + JI + (I - I_{sc}) = 0$$
$$I(1 + J) - I_{sc} = 20 \tag{1}$$
$$1(I_{sc} - I) + I_{sc} = 0$$
$$I = 2I_{sc}. \tag{2}$$

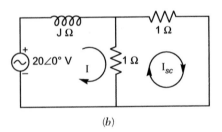

(b)

FIGURE 8.35

By Equations (1) and (2),

$$I_{sc}[2 + 2J - 1] = 20$$
$$I_{sc} = \frac{20}{1 + 2J} = I_N.$$

Now

$$Z_{TH} = \frac{V_{TH}}{I_{sc}} = \frac{20}{1 + J} \times \frac{1 + 2J}{20}$$

$$= \frac{1 + 2J}{1 + J} = \left(\frac{3 + J}{2}\right) = Z_N.$$

NOTE ▶ In the case of the AC circuit we use impedance (Z) in place of resistance (R).

The Thevenin equivalent circuit is shown in Figure 8.35(c),

$$I_L = \frac{V_{TH}}{Z_{TH} + Z_L}$$

$$= \frac{20}{1+J} \times \frac{1}{\frac{3+J}{2} + (-J)} = \frac{20}{2+J} A = 8.94\angle -26.56° \text{Amp.}$$

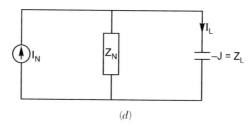

(c)

FIGURE 8.35

The Norton equivalent circuit is shown in Figure 8.35(d),

$$I_L = \left(\frac{Z_L}{Z_N + Z_L}\right) I_N$$

$$= \frac{\left(\frac{3+J}{2}\right)}{\left(\frac{3+J}{2}\right) + (-J)} \times \frac{20}{1+2J}$$

$$I_L = 8.94\angle -26.56° \text{Amp.}$$

(d)

FIGURE 8.35

8.5.3 Thevenin-Norton Equivalencies

Since Thevenin's and Norton's Theorems are two equally valid methods of reducing a complex network down to something simpler to analyze, there must be some way to convert a Thevenin equivalent circuit to a Norton equivalent circuit, and visa-versa (just what you were dying to know, right?). Well, the procedure is very simple.

You may have noticed that the procedure for calculating Thevenin resistance is identical to the procedure for calculating Norton resistance: remove all power sources and determine resistance between the open load connection points. As such, Thevenin and Norton resistances for the same original network must be equal. Using the example circuits from the last two sections, we can see that the two resistances are indeed equal:

Thevenin Equivalent Circuit

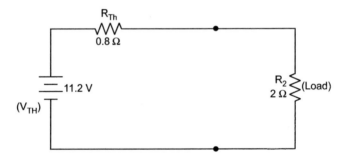

FIGURE 8.36

Norton Equivalent Circuit

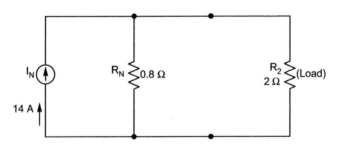

FIGURE 8.37

$$R_{TH} = R_N$$

Considering the fact that both Thevenin and Norton equivalent circuits are intended to behave the same as the original network in supplying voltage and current to the load resistor (as seen from the perspective of the load connection points), these two equivalent circuits, having been derived from the same original network, should behave identically.

This means that both Thevenin and Norton equivalent circuits should produce the same voltage across the load terminals with no load resistor attached. With the Thevenin equivalent, the open-circuited voltage would be equal to the Thevenin source voltage (no circuit current present to drop voltage across the series resistor), which is 11.2 volts in this case. With the Norton equivalent circuit, all 14 amps from the Norton current source would have to flow through the 0.8 Ω Norton resistance, producing the exact same voltage, 11.2 volts ($E = IR$). Thus, we can say that the Thevenin voltage is equal to the Norton current times the Norton resistance.

$$\boxed{V_{TH} = I_N R_N} \quad \text{or} \quad \boxed{V_{TH} = I_N R_{TH}}$$

So, if we wanted to convert a Norton equivalent circuit to a Thevenin equivalent circuit, we could use the same resistance and calculate the Thevenin voltage with Ohm's law.

Conversely, both Thevenin and Norton equivalent circuits should generate the same amount of current through a short circuit across the load terminals. With the Norton equivalent, the short-circuit current would be exactly equal to the Norton source current, which is 14 amps in this case. With the Thevenin equivalent, all 11.2 volts would be applied across the 0.8 Ω Thevenin resistance, producing the exact same current through the short, 14 amps ($I = E/R$). Thus, we can say that the norton current is equal to the Thevenin voltage divided by the Thevenin resistance:

$$I_N = \frac{V_{TH}}{R_{Th}}.$$

This equivalence between Thevenin and Norton circuits can be a useful tool in itself, as we shall see in the next section.

NOTE
- Thevenin and Norton resistances are equal.
- Thevenin voltage is equal to Norton current times Norton resistance $[V_{TH} = I_N \times R_N]$.
- Norton current is equal to Thevenin voltage divided by Thevenin resistance.

8.6   MILLMAN'S THEOREM REVISITED

You may have wondered where we got that strange equation for the determination of "Millman Voltage" across parallel branches of a circuit where each branch contains a series resistance and voltage source.

Millman's Theorem Equation

$$\frac{\frac{E_1}{R_1} + \frac{E_2}{R_2} + \frac{E_3}{R_3}}{\frac{1}{R_1} + \frac{1}{R_2} + \frac{1}{R_3}} = \text{Voltage across all branches}$$

Parts of this equation seem familiar to equations we've seen before. For instance, the denominator of the large fraction looks conspicuously like the denominator of our parallel resistance equation. And, of course, the E/R terms in the numerator of the large fraction should give figures for current, Ohm's Law being what it is $(I = E/R)$.

Now that we've covered Thevenin and Norton source equivalencies, we have the tools necessary to understand Millman's equation. What Millman's equation is actually doing is treating each branch (with its series voltage source and resistance) as a Thevenin equivalent circuit and then converting each one into equivalent Norton circuits.

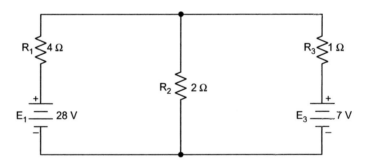

FIGURE 8.38

Thus, in the circuit above, battery E_1 and resistor R_1 are seen as a Thevenin source to be converted into a Norton source of 7 amps (28 volts/4 Ω) in parallel with a 4 Ω resistor. The rightmost branch will be converted into a 7 amp current source (7 volts/1 Ω) and a 1 Ω resistor in parallel. The center branch, containing no voltage source at all, will be converted into a Norton source to 0 amps in parallel with a 2 Ω resistor:

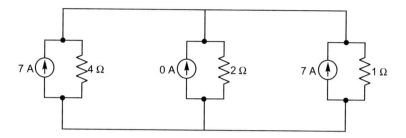

FIGURE 8.39

Since current sources directly add their respective currents in parallel, the total circuit current will be $7 + 0 + 7$, or 14 amps. This addition of Norton source currents is what's being represented in the numerator of the Millman equation:

Millman's Theorem Equation

$$I_{total} = \frac{E_1}{R_1} + \frac{E_2}{R_2} + \frac{E_3}{R_3} \rightarrow \frac{\frac{E_1}{R_1} + \frac{E_2}{R_2} + \frac{E_3}{R_3}}{\frac{1}{R_1} + \frac{1}{R_2} + \frac{1}{R_3}}$$

All the Norton resistances are in parallel with each other as well in the equivalent circuit, so they diminish to create a total resistance. This diminishing of source resistances is what's being represented in the denominator of the Millman's equation:

Millman's Theorem Equation

$$R_{total} = \frac{1}{\frac{1}{R_1} + \frac{1}{R_2} + \frac{1}{R_3}} \rightarrow \frac{\frac{E_1}{R_1} + \frac{E_2}{R_2} + \frac{E_3}{R_3}}{\frac{1}{R_1} + \frac{1}{R_2} + \frac{1}{R_3}}$$

In this case, the resistance total will be equal to 571.43 milliohms (571.43 mΩ). We can re-draw our equivalent circuit now as one with a single Norton current source and Norton resistance:

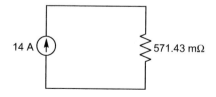

FIGURE 8.40

Ohm's Law can tell us the voltage across these two components now $(E = I/R)$:

$$E_{total} = (14A)(571.43 m\Omega)$$

$$E_{total} = 8\,V$$

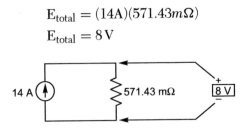

FIGURE 8.41

Let's summarize what we know about the circuit thus far. We know that the total current in this circuit is given by the sum of all the branch voltages divided by their respective currents. We also know that the total resistance is found by taking the reciprocal of all the branch resistance reciprocals. Furthermore, we should be well aware of the fact that total voltage across all the branches can be found by multiplying total current by total resistance ($E = I/R$). All we need to do is put together the two equations we had earlier for total circuit current and total resistance, multiplying them to find total voltage:

Ohm's Law:

$$I \times R = E$$

$$(\text{total current}) \times (\text{total resistance}) = (\text{total voltage})$$

$$\frac{E_1}{R_1} + \frac{E_2}{R_2} + \frac{E_3}{R_3} \times \frac{1}{\frac{1}{R_1} + \frac{1}{R_2} + \frac{1}{R_3}} = (\text{total voltage})$$

or

$$\frac{\frac{E_1}{R_1} + \frac{E_2}{R_2} + \frac{E_3}{R_3}}{\frac{1}{R_1} + \frac{1}{R_2} + \frac{1}{R_3}} = (\text{total voltage})$$

The Millman's equation is nothing more than a Thevenin-to-Norton conversion matched together with the parallel resistance formula to find total voltage across all the branches of the circuit.

8.7   MAXIMUM POWER TRANSFER THEOREM

The Maximum Power Transfer Theorem is not so much a means of analysis as it is an aid to system design. Simply stated, the maximum amount of power will be dissipated by a load resistance when that load resistance is equal to the Thevenin/Norton resistance of the network supplying the power. If the load resistance is lower or higher than the Thevenin/Norton resistance of the source network, its dissipated power will be less than maximum.

This is essentially what is aimed for in stereo system design, where speaker "impedance" is matched to amplifier "impedance" for maximum sound power output. Impedance, the overall opposition to A.C. and D.C. current, is very similar to resistance, and must be equal between source and load for the greatest amount of power to be transferred to the load. A load impedance that is too high will result in low power output. A load impedance that is too low will not only result in low power output, but possibly overheating of the amplifier due to the power dissipated in its internal (Thevenin or Norton) impedance.

Taking our Thevenin equivalent example circuit, the Maximum Power Transfer Theorem tells us that the load resistance resulting in the greatest power dissipation is equal in value to the Thevenin resistance (in this case, 0.8Ω):

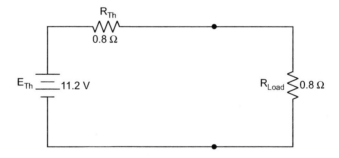

FIGURE 8.42

With this value of load resistance, the dissipated power will be 39.2 watts:

	R_{Th}	R_{Load}	Total	
E	5.6	5.6	11.2	Volts
I	7	7	7	Amps
R	0.8	0.8	1.6	Ohms
P	39.2	39.2	78.4	Watts

If we were to try a lower value for the load resistance ($0.5\,\Omega$ instead of $0.8\,\Omega$, for example), our power dissipated by the load resistance would decrease:

	R_{Th}	R_{Load}	Total	
E	6.892	4.308	11.2	Volts
I	8.615	8.615	8.615	Amps
R	0.8	0.5	1.3	Ohms
P	59.38	37.11	96.49	Watts

Power dissipation increased for both the Thevenin resistance and the total circuit, but it decreased for the load resistor. Likewise, if we increase the load resistance (1.1 Ω instead of 0.8 Ω, for example), power dissipation will also be less than it was at 0.8 Ω exactly:

	R_{Th}	R_{Load}	Total	
E	4.716	6.484	11.2	Volts
I	5.895	5.895	5.895	Amps
R	0.8	1.1	1.9	Ohms
P	27.80	38.22	66.02	Watts

If you were designing a circuit for maximum power dissipation at the load resistance, this theorem would be very useful. Having reduced a network down to a Thevenin voltage and resistance (or Norton current and resistance), you simply set the load resistance equal to that Thevenin or Norton equivalent (or vise-versa) to ensure maximum power dissipation at the load. Practical applications of this might include stereo amplifier design (seeking to maximize power delivered to speakers) or electric vehicle design (seeking to maximize power delivered to drive motor).

 • The *Maximum Power Transfer Theorem* states that the maximum amount of power will be dissipated by a load resistance if it is equal to the Thevenin or Norton resistance of the network supplying power.

8.7.1 Maximum Power Transfer Theorem for A.C. Circuits

Maximum power at output is obtained when the load impedance is equal to the complex conjugate of the internal impedance of the circuit.

$$I = \frac{V_{TH}}{Z_{TH}} [Z_{TH} = R_{TH} + JX_{TH}]$$

$$[Z_L = R_L + JX_L]$$

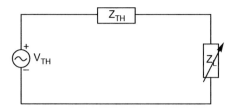

FIGURE 8.43   Circuit illustrating maximum power transfer theorem.

Power dissipated in the load is

$$P_L = I^2 R_L.$$

So,

$$P_L = \frac{V_{TH}^2 \times R_L}{(R_{TH} + R_L)^2 + (X_{TH} + X_L)^2}.$$

For maximum power, $\frac{dP_L}{dX_L}$, must be zero

$$\frac{dP_L}{dX_L} = \frac{0 - V_{TH}^2 R_L . 2 (X_{TH} + X_L)}{\left[(R_{TH} + R_L)^2 + (X_{TH} + X_L)^2 \right]^2} = 0.$$

$$X_{TH} + X_L = 0$$
$$X_L = -X_{TH}$$

The reactance of the load impedance is of opposite sign to the reactance of internal impedance of the circuit (X_{TH}).

Putting

$$X_L = -X_{TH} \text{ into eqn.} \tag{8.1}$$

$$P_L = \frac{V_{TH}^2 R_L}{(R_{TH} + R_L)^2}, \tag{8.2}$$

again for maximum power, $\frac{dP_L}{dR_L} = 0$

$$\frac{dP_L}{dR_L} = \frac{(R_{TH} + R_L)^2 \cdot V_{TH}^2 - V_{TH}^2 R_L \cdot 2 (R_{TH} + R_L)}{\left[(R_{TH} + R_L)^2 \right]^2} = 0.$$

$$R_{TH} + R_L = 2R_L$$
$$R_L = R_{TH}$$

Therefore, maximum power will be transferred from source to load, if $R_L = R_{TH}$ and $X_L = -X_{TH}$ (*i.e.*, for maximum power transfer, load impedance should be the complex conjugate of the internal impedance (Z_{TH}) of the circuit, *i.e.*, $Z_L = Z_{TH}^*$).

The maximum power transferred will be

$$P_{\max} = \frac{V_{TH}^2}{4R_L} = \frac{V_{TH}^2}{4R_{TH}}.$$

So, the overall efficiency of a circuit supplying maximum power is 50%.

Example 8.8. *Determine Z_L for maximum power transfer and P_{max} as shown in Figure 8.44.*

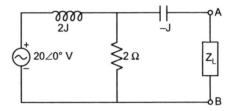

FIGURE 8.44

Solution: First open the terminal AB and calculate $Z_{AB} = Z_{TH}$ as shown in Figure 8.44(a).

$$Z_{AB} = \frac{2 \times 2J}{2(1 + J)} - J$$

$$Z_{AB} = \frac{2J - J(1 + J)}{(1 + J)}$$

$$Z_{AB} = Z_{TH} = 1\,\Omega$$

$$Z_L = Z_{TH} = 1\,\Omega$$

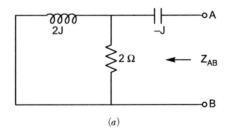

(a)

FIGURE 8.44

NOTE In this case $X_{TH} = 0$.

$$V_{AB} = \frac{20}{2(1 + J)} \times 2 = V_{TH}$$

$$V_{TH} = \frac{20}{1 + J} = 10\sqrt{2}\angle - 45°V$$

$$P_{max} = \frac{V_{TH}^2}{4R_{TH}} = \frac{200}{4 \times 1} = 50W$$

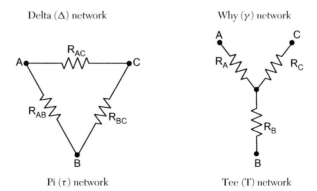

(b)

FIGURE 8.44

8.8 Δ-Y AND Y-Δ CONVERSIONS

In many circuit applications, we encounter components connected together in one of two ways to form a three-terminal network: the "Delta," or Δ (also known as the "Pi," or π) configuration, and the "Y" (also known as the "T") configuration.

It is possible to calculate the proper values of resistors necessary to form one kind of network (Δ or Y) that behaves identically to the other kind, as

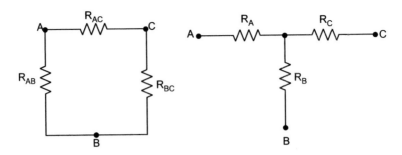

FIGURE 8.45

analyzed from the terminal connections alone. That is, if we had two separate resistor networks, one Δ and one Y, each with its resistors hidden from view, with nothing but the three terminals (A, B, and C) exposed for testing, the resistors could be sized for the two networks so that there would be no way to electrically determine one network apart from the other. In other words, equivalent Δ and Y networks behave identically.

There are several equations used to convert one network to the other:

To convert a Delta (Δ) to a Wye (Y)

$$R_A = \frac{R_{AB}R_{AC}}{R_{AB} + R_{AC} + R_{BC}}$$

$$R_B = \frac{R_{AB}R_{BC}}{R_{AB} + R_{AC} + R_{BC}}$$

$$R_C = \frac{R_{Ac}R_{BC}}{R_{AB} + R_{AC} + R_{BC}}$$

To convert a Wye (Y) to a Delta (Δ)

$$R_{AB} = \frac{R_A R_B + R_A R_C + R_B R_C}{R_C}$$

$$R_{BC} = \frac{R_A R_B + R_A R_C + R_B R_C}{R_A}$$

$$R_{AC} = \frac{R_A R_B + R_A R_C + R_B R_C}{R_B}$$

Δ and Y networks are seen frequently in 3-phase A.C. power systems, but even then they're usually balanced networks (all resistors equal in value) and conversion from one to the other need not involve such complex calculations. When would the average technician ever need to use these equations?

A prime application for Δ-Y conversion is in the solution of unbalanced bridge circuits, such as in Figure 8.46.

The Solution of this circuit with Branch Current or Mesh Current analysis is fairly involved, and neither the Millman nor Superposition theorems are of any help, since there's only one source of power. We could use Thevenin's or Norton's theorem, treating R_3 as our load.

If we were to treat resistors R_1, R_2, and R_3 as being connected in a Δ configuration (R_{ab}, R_{ac}, and R_{bc}, respectively) and generate an equivalent Y

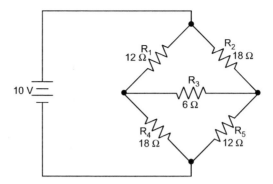

FIGURE 8.46

network to replace them, we could turn this bridge circuit into a (simpler) series/parallel combination circuit:

Selecting the Delta (Δ) network to convert:

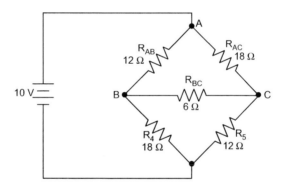

FIGURE 8.47

After the Δ-Y conversion ... Δ converted to a Y

If we perform our calculations correctly, the voltages between points A, B, and C will be the same in the converted circuit as in the original circuit, and we can transfer those values back to the original bridge configuration.

$$R_A = \frac{(12\,\Omega)(18\,\Omega)}{(12\,\Omega) + (18\,\Omega) + (6\,\Omega)} = \frac{216}{36} = 6\,\Omega$$

$$R_B = \frac{(12\,\Omega)(6\,\Omega)}{(12\,\Omega) + (18\,\Omega) + (6\,\Omega)} = \frac{72}{36} = 2\,\Omega$$

$$R_C = \frac{(18\,\Omega)(6\,\Omega)}{(12\,\Omega) + (18\,\Omega) + (6\,\Omega)} = \frac{108}{36} = 3\,\Omega$$

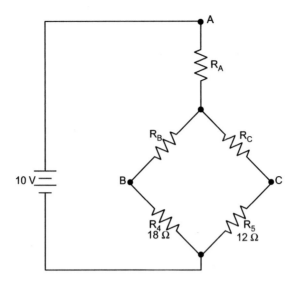

FIGURE 8.48

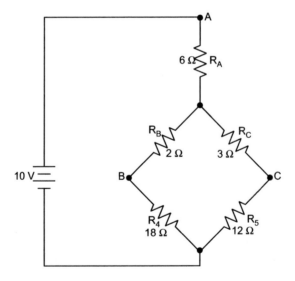

FIGURE 8.49

Resistors R_4 and R_5, of course, remain the same at $18\,\Omega$ and $12\,\Omega$, respectively.

Analyzing the circuit now as a series/parallel combination, we arrive at the following figures:

	R_A	R_B	R_C	R_4	R_5	
E	4.118	588.24 m	1.176	5.294	4.706	Volts
I	68.27 m	294.12 m	392.16 m	294.12 m	392.16	Amps
R	6	2	3	18	12	Ohms

$$R_B + R_4$$
$$//$$

	$R_B + R_4$	$R_C + R_5$	$R_C + R_5$	Total	
E	5.882	5.882	5.882	0	Volts
I	294.12 m	392.16 m	686.27 m	686.27 m	Amps
R	20	15	8.571	14.571	Ohms

We must use the voltage drops figures from the table above to determine the voltages between points A, B, and C, seeing how they add up (or subtract, as is the case with voltage between points B and C):

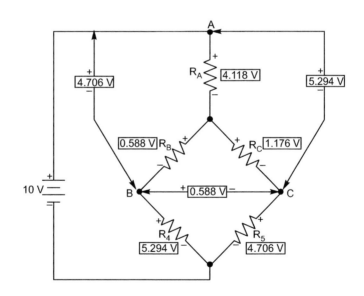

FIGURE 8.50

$$E_{A-B} = 4.706\,V$$
$$E_{A-C} = 5.294\,V$$
$$E_{B-C} = 588.24\,mV$$

Now that we know these voltages, we can transfer them to the same points A, B, and C in the original bridge circuit:

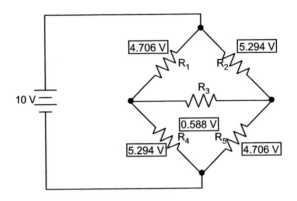

FIGURE 8.51

Voltage drops across R$_4$ and R$_5$, of course, are exactly the same as they were in the converted circuit.

At this point, we could take these voltages and determine resistor currents through the repeated use of Ohm's Law (I = E/R):

$$I_{R1} = \frac{4.706\,V}{12\,\Omega} = 392.16\,mA$$

$$I_{R2} = \frac{5.294\,V}{18\,\Omega} = 294.12\,mA$$

$$I_{R3} = \frac{588.24\,mV}{6\,\Omega} = 98.04\,mA$$

$$I_{R4} = \frac{5.294\,V}{18\,\Omega} = 294.12\,mA$$

$$I_{R5} = \frac{4.706\,V}{12\,\Omega} = 392.16\,mA$$

A quick simulation with SPICE will serve to verify our work:

The voltage figures, as read from left to right, represent voltage drops across the five respective resistors, R$_1$ through R$_5$ we could have shown currents as well, but since that would have required the insertion of "dummy" voltage sources in the SPICE netlist, and since we're primarily interested in validating the Δ-Y conversion equations and not Ohm's Law, this will suffice.

NOTE

- "Delta" (Δ) networks are also known as "Pi" (π) networks.
- "Y" networks are also known as "T" networks.

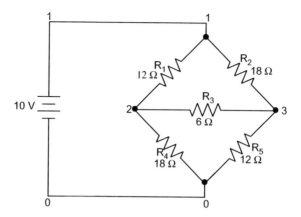

FIGURE 8.52

- Δ and Y networks can be converted to their equivalent counterparts with the proper resistance equations. By "equivalent," we mean that the two networks will be electrically identical as measured from the three terminals (A, B, and C).
- A bridge circuit can be simplified to a series/parallel circuit by converting half of it from a Δ to a Y network. After voltage drops between the original three connection points (A, B, and C) have been solved for, those voltage can be transferred back to the original bridge circuit, across those same equivalent points.

8.9 COMPENSATION THEOREM

The Compensation Theorem states that in a linear network, if an impedance Z carrying a current I is changed to $(Z + \Delta Z)$, then the change in current (ΔI) in any branch of the network can be found by replacing the voltage source by its internal impedance and placing a compensation voltage source of magnitude $V_C = -I \cdot \Delta Z$ in series with impedance.

By the Figure 8.53(a),

$$I = \frac{V}{Z_{TH} + Z}.$$ (8.3)

By the Figure 8.53(b),

$$I' = \frac{V}{Z_{TH} + (Z + \Delta Z)}.$$ (8.4)

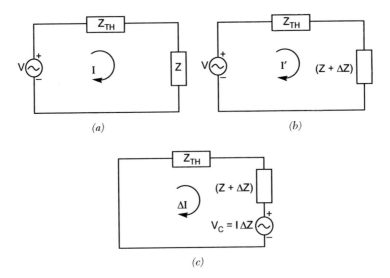

FIGURE 8.53

From the equations, the change in the current

$$\Delta I = I' - I = \frac{V}{Z_{TH} + (Z + \Delta Z)} - \frac{V}{Z_{TH} + Z}$$

$$\Delta I = \frac{-V \Delta Z}{[Z_{TH} + (Z + \Delta Z)][Z_{TH} + Z]}$$

$$= \frac{-I \Delta Z}{[Z_{TH} + (Z + \Delta Z)]} \qquad \left[I = \frac{V}{Z_{TH} + Z} \right]$$

$$\Delta I = \frac{V_C}{[Z_{TH} + (Z + \Delta Z)]}.$$

The Compensation Theorem is useful in determining the incremental changes in voltages and currents in the branches of a circuit due to a change in impedance in one branch. The Compensation Theorem is based on the principle of superposition. Therefore, the Compensation Theorem is not applicable to the non-linear circuits.

Example 8.9. *In the circuit of Figure 8.54, the resistance* R *is changed from 5 to 10 Ω. Verify the Compensation Theorem.*

Solution: By applying mesh analysis for mesh 1,

$$-20 + 10I_1 + 10(I_1 - I_2) = 0$$

$$2I_1 - I_2 = 2. \tag{1}$$

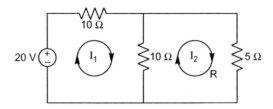

FIGURE 8.54

For mesh 2,

$$10(I_2 - I_1) + 5I_2 = 0$$

$$2I_2 - 2I_1 + I_2 = 0$$

$$2I_1 = 3I_2$$

$$I_1 = 1.5I_2. \tag{2}$$

By Equations (1) and (2) $I_1 = 1.5A, I_2 = 1A$.
Now the resistance R is changed to $10\,\Omega$ as shown in Figure 8.54(a).

$$I'_1 = \frac{20}{15} = \frac{4}{3}A$$

$$I'_2 = \frac{2}{3}A$$

$$\Delta I = I'_2 - I_2 = \frac{2}{3} - 1$$

$$= \frac{1}{3}A[\Delta Z = Z_2 - Z_1]$$

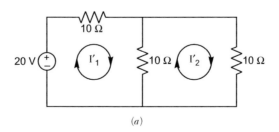

(a)

FIGURE 8.54

Using the Compensation Theorem, $V_C = I_2 \Delta Z = 1 \times 5 = 5\,V$.

$$I = \frac{5}{15} = \frac{1}{3} = -\Delta I[I = -\Delta I]$$

$$\Delta I = -\frac{1}{3}$$

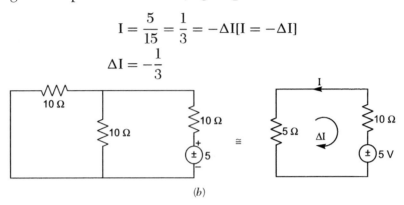

(b)

FIGURE 8.54

Hence, the Compensation Theorem is verified.

8.9.1 Steps for the Compensation Theorem.

Step 1. First calculate the value of current (I) in the desired branch where the resistance or impedance is changing.

Step 2. Once again, calculate the value of the current (I') in the desired branch for the changed resistance.

Step 3. Now calculate $\Delta I = I' - I$.

Step 4. Calculate the compensation voltage $V_C = I \Delta Z (\Delta Z = Z_2 - Z_1)$.

Step 5. Draw the compensation circuit, which has V_C and other impedances or resistances and calculate ΔI. It should be the same as we calculated in step 3.

Example 8.10. *In the circuit of Figure 8.55 the resistance R is changed from* $10\,\Omega$ *to* $5\,\Omega$. *Verify the Compensation Theorem.*

Solution: By applying mesh analysis,
    for mesh 1

$$-20 + 10I_1 + 10(I_1 - I_2) = 0$$
$$10(I_2 - I_1) + 10I_2 = 0$$
$$I_1 = 2I_2 = \frac{4}{3}A$$
$$I_2 = \frac{2}{3}A.$$

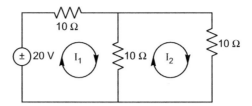

FIGURE 8.55

Now the resistance R changes to $5\,\Omega$ as shown in Figure 8.55(a). Once again applying mesh analysis, for mesh 1,

$$-20 + 10I_1' + 10(I_1' - I_2') = 0$$
$$2I_1' - I_2' = 2. \tag{1}$$

For mesh 2,

$$10(I_2' - I_1') + 5I_2' = 0$$
$$I_1' = 1.5I_2'. \tag{2}$$

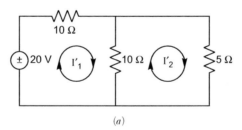

(a)

FIGURE 8.55

By Equations (1) and (2),

$$I_1' = 1.5A, I_2' = 1A$$

$$\Delta I = I_2' - I_2 = 1 - \frac{2}{3} = \frac{1}{3}A$$

Using the Compensation Theorem $V_C = I_2 \Delta Z [\Delta Z = 5 - 10]$

$$V_C = \frac{2}{3} \times -5 = -\frac{10}{3}V$$

$$I = -\Delta I = \frac{V_C}{10} = -\frac{10}{3 \times 10} = -\frac{1}{3}$$

$$\Delta I = \frac{1}{3}.$$

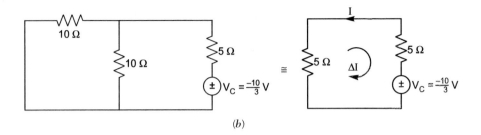

(b)

FIGURE 8.55

Hence, the Compensation Theorem is verified.

Example 8.11. *In the circuit of Figure 8.56, the resistance R is changed from 5 to 10 Ω. Verify the Compensation Theorem.*

Solution: By the current division formula,

$$I = \frac{5}{10} \times 10$$
$$I = 5A.$$

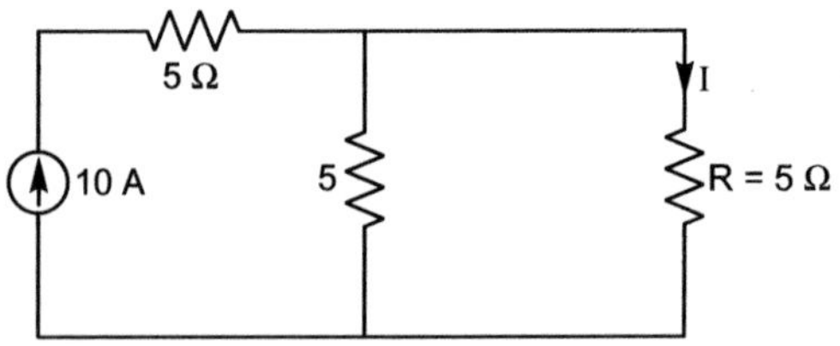

FIGURE 8.56

Now the resistance changes to 10 Ω as shown in Figure 8.56(a).

$$I' = \frac{5}{15} \times 10$$
$$I' = \frac{10}{3} A$$

Now ΔI (change in current) $= I' - I$

$$= \frac{10}{3} - 5 = -\frac{5}{3} A$$

and

$$V_C = I\Delta Z \text{ or } I\Delta R[\Delta R = R_2 - R_1]$$
$$= 5 \times (10 - 5) = 25 V.$$

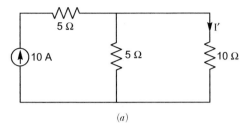

(a)

FIGURE 8.56

The current source is set to zero (open circuit).
Now

$$I_1 = \frac{V_C}{10 + 5} = \frac{25}{15} = \frac{5}{3}$$

and

$$I_1 = -\Delta I \text{ or } \Delta I = -I_1 = -\frac{5}{3}A.$$

(b)

FIGURE 8.56

Thus, we have verified the Compensation Theorem.

8.10 TELLEGEN'S THEOREM

This theorem, proposed by B.D. Tellegen, is known as the Tellegen Theorem.

Statement: Tellegen's Theorem states that the sum of the products of instantaneous branch voltages and branch currents in a network is zero.

$$\sum_{k=1}^{b} v_k i_k = 0,$$

where $v_1, v_2, \ldots v_b$ are instantaneous branch voltages and $i_1, i_2, \ldots i_b$ are instantaneous values of branch currents. We know that the product $(v_k \cdot i_k)$

is the instantaneous power. Tellegen's Theorem tells us that summation of instantaneous powers for all the branches of a network is zero.

We know the active elements give (deliver) the power and passive elements take (receive) the power. In simple terms Tellegen's Theorem states that the given power (by the battery) is equal to the taken power (by the passive elements).

NOTE Sometimes the battery is taking power. In this case, the battery is charging (current is entering into the battery positive terminal).

This theorem is applicable to all networks consisting of lumped elements which may be linear or non-linear, passive or active, time-invariant or time-varying.

Consider the network shown in Figure 8.57.

$$\sum_{k=1}^{6} v_k \cdot i_k = -v_1 i_1 + v_2 i_2 + v_3 i_3 + v_4 i_4 + v_5 i_5 + v_6 i_6 \qquad (8.5)$$

NOTE i_1 is taking as $-i_1$ because current is entering in the negative end of the active element (battery). By the KCL,

$$\text{at node A} - i_1 + i_2 + i_5 = 0$$
$$\text{at node B} - i_2 + i_3 + i_6 = 0$$
$$\text{at node C}, -i_3 - i_5 + i_4 = 0$$

and

$$v_1 = v_A$$
$$v_2 = v_A - v_B$$
$$v_3 = v_B - v_C$$
$$v_4 = v_C - v_D$$
$$v_5 = v_A - v_C$$
$$v_6 = v_B - v_D.$$

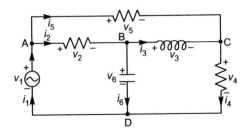

FIGURE 8.57

Now putting the values of $v_1, v_2, v_3, v_4, v_5,$ and v_6 into Equation (8.5),

$$\sum v_k \cdot i_k = -v_A i_1 + (v_A - v_B)i_2 + (v_B - v_C)i_3 + (v_C - v_D)i_4$$
$$+ (v_A - v_C)i_5 + (v_B - v_D)i_6$$
$$= v_A(-i_1 + i_2 + i_5) + v_B(-i_2 + i_3 + i_6) + v_C(-i_3 + i_4 - i_5)$$
$$= v_A \times 0 + v_B \times 0 + v_C \times 0$$
$$= 0.$$

Therefore,

$$\sum v_k \cdot i_k = 0.$$

Thus, Tellegen's Theorem is proved.

8.10.1 Steps for Tellegen's Theorem

Step 1. First calculate the value of current in each element using any method (KVL, Mesh, Nodal, etc.).

Step 2. Now calculate the value of voltage across each element.

Step 3. Now apply $\sum_1^b v_k \cdot i_k$ and verify the theorem. We can also say, given power = taken power.

NOTE ▶ Students can remember from Tellegen's Theorem, that we tally the given and taken powers.

Example 8.12. *Verify Tellegen's Theorem in the network shown in Figure 8.58.*

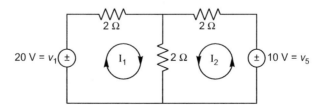

FIGURE 8.58

Solution: By using mesh analysis, for mesh 1,

$$-20 + 2I_1 + 2(I_1 + I_2) = 0$$
$$2I_1 + I_2 = 10. \tag{1}$$

For mesh 2,

$$-10 + 2I_2 + 2(I_1 + I_2) = 0$$
$$I_1 + 2I_2 = 5. \tag{2}$$

By Equations (1) and (2),

$$I_1 = 5A I_2 = 0A,$$

$I_2 = 0A$. It is clear that the 10 V battery is not delivering the power or receiving the power.

The value of the current in each branch and the voltage drop in each branch is shown in Figure 8.59.

$$\sum_1^5 v_k \cdot i_k = 0$$

$$= v_1 i_1 + v_2 i_2 + v_3 i_3 + v_4 i_4 + v_5 i_5$$
$$= -20 \times 5 + 10 \times 5 + 10 \times 5 + 0 \times 0 + 10 \times 0$$
$$= -100 + 100$$
$$= 0$$

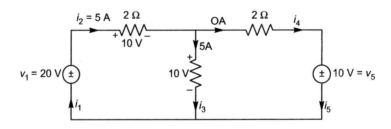

FIGURE 8.59

or in a simple way, we can solve as,

$$\text{Given power} = V_1 I_1 + v_5 i_5$$
$$= 20 \times 5 + 10 \times 0$$
$$= 100\,\text{W}.$$

Other elements (passive elements) take the power,

$$\text{taken power} = i_2^2 R_2 + i_3^2 R_3 + i_4^2 R_4$$
$$= (5)^2 \times 2 + (5)^2 \times 2 + (0)^2 \times 2$$
$$= 100\,\text{W}.$$

Now the given power = taken power
Hence, Tellegen's Theorem is verified.

Example 8.13. *Verify Tellegen's Theorem in the network shown in Figure 8.60.*

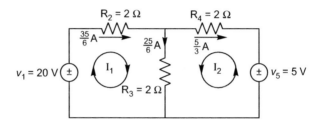

FIGURE 8.60

Solution: First calculate the value of current in each element.
By using mesh analysis,
for mesh 1,

$$20 = 2I_1 + 2(I_1 + I_2)$$

or

$$2I_1 + I_2 = 10. \tag{1}$$

For mesh 2,

$$I_1 + 2I_2 = 2.5. \tag{2}$$

By Equations (1) and (2),

$$I_1 = \frac{35}{6}\,\text{A}, \quad I_2 = -\frac{5}{3}\,\text{A}. \quad \text{[Negative sign shows the direction is opposite.]}$$

NOTE ► The source v_5 is not delivering the power. It is taking power because the current is entering in positive (charging condition).

Now, the given power

$$= V_1 \cdot I_1$$

$$= 20 \times \frac{35}{6} = \frac{350}{3} \text{W.}$$

The taken power

$$= I_2^2 R_2 + I_3^2 R_3 + I_4^2 R_4 + v_5 I_5$$

$$= \left(\frac{35}{6}\right)^2 \times 2 + \left(\frac{25}{6}\right)^2 \times 2 + \left(\frac{5}{3}\right)^2 + 5 \times \left(\frac{5}{3}\right)$$

$$= \frac{1225}{18} + \frac{625}{18} + \frac{50}{9} + \frac{25}{3}$$

$$= \frac{2100}{18} = \frac{350}{3} \text{W.}$$

Hence, Tellegen's Theorem is verified.

Example 8.14. *Determine the value of* R_2 *using Tellegen's Theorem as shown in Figure 8.61.*

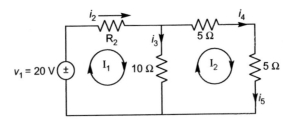

FIGURE 8.61

Solution: By using mesh analysis,

$$20 = I_1 R_2 + 10(I_1 - I_2) \tag{1}$$

$$5I_2 + 5I_2 + 10(I_2 - I_1) = 0$$

$$I_1 = 2I_2. \tag{2}$$

Now by Equations (1) and (2),

$$I_2 = \frac{10}{5 + R_2} \text{ and } I_1 = \frac{20}{5 + R_2}$$

Now the given power

$$= V_1 I_1 = \frac{20 \times 20}{5 + R_2}.$$ (3)

The taken power

$$= i_2^2 R_2 + i_3^2 R_3 + i_4^2 R_4 + i_5^2 R_5 \begin{bmatrix} i_2 = I_1 \\ i_3 = I_1 - I_2 \\ i_4 = i_5 = I_2 \end{bmatrix}$$

$$= \left(\frac{20}{5 + R_2}\right)^2 R_2 + \left(\frac{10}{5 + R_2}\right)^2 \times 10 + \left(\frac{10}{5 + R_2}\right)^2 \times 5 \times 2$$

$$= \left(\frac{20}{5 + R_2}\right)^2 R_2 + 20 \left(\frac{10}{5 + R_2}\right)^2.$$ (4)

By Equations (3) and (4) we know, given power = taken power,

$$\frac{400}{5 + R_2} = \left(\frac{400}{5 + R_2}\right)^2 R_2 + \frac{2000}{(5 + R_2)^2}$$

$$400(5 + R_2) = 160000 \, R_2 + 2000$$

$$R_2 = 0.$$

8.11   RECIPROCITY THEOREM

In a linear bilateral nework having one independent source and no dependent sources, an important relation exists between a source voltage in one branch and current in some other branch.

Statement: If a source of emf V, located at one point in a network composed of linear bilateral circuit elements produces a current I at a selected point in the network, the same source of emf V acting at the second point will produce the same current at the first point. Consider a reciprocal network N, with only two branches as shown in Figure 8.62.

Suppose a source of emf V_1 is inserted in branch 1 and that it produces current I_2 at branch 2.

Now, if the point of application of exciting the V_2 source is moved to branch 2 and the response I_1 is measured at 1,

according to the Reciprocity Theorem,

$$\frac{V_1}{I_2} = \frac{V_2}{I_1}.$$

If the source is the same $V_1 = V_2$.

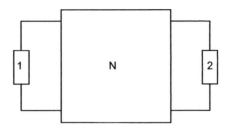

FIGURE 8.62

Then $I_1 = I_2$.

The Reciprocity Theorem is applicable to linear, time-invariant networks. It is not applicable to a nework consisting of any dependent source even if it is linear.

Example 8.15. *Verify the Reciprocity Theorem for the network shown in Figure 8.63.*

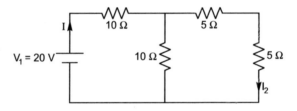

FIGURE 8.63

Solution: Case I: First calculate the value of I_2

$$I = \frac{20}{10 + 5} = \frac{4}{3}A \qquad V_1 = 20\,V$$

$$I_2 = \left(\frac{10}{10 + 10}\right) \times I \qquad \text{[by the current division formula]}$$

$$I_2 = \frac{I}{2} = \frac{2}{3}A.$$

Case II: Now the source shifted to point 2, as shown in Figure 8.63(*a*). Now calculate the value of current at point 1.

$$I = \frac{20}{15} = \frac{4}{3}A$$

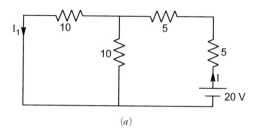

FIGURE 8.63

and

$$I_1 = \left(\frac{10}{10 + 10}\right) \times \frac{4}{3}$$

$$I_1 = \frac{2}{3}.$$

$I_1 = I_2$, hence, the Reciprocity Theorem is verified.

8.11.1   Steps for the Reciprocity Theorem

Step 1. First calculate the value of the current at point $2(I_2)$, when the source is at point 1.

Step 2. Now shift the source at point 2 and calculate the value of current at $1(I_1)$.

Step 3. Verify the theorem ($I_1 = I_2$).

NOTE ➤ If the current source is given at point 1, then we calculate the voltage at point $2(V_2)$ and shift the current source at point 2 and calculate the voltage at point 1 and $V_1 = V_2$ because the Reciprocity Theorem indicates that voltage (V) and current (I) are mutually interchangeable.

Example 8.16. *Verify the Reciprocity Theorem for the circuit shown in Figure 8.64.*

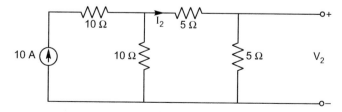

FIGURE 8.64

Solution: Case I: By the current division formula,

$$I_2 = \frac{10}{20} \times 10 = 5\,\text{A}$$
$$V_2 = 5 \times 5 = 25\,\text{V}.$$

Case II: Now the current source is shifted at point 2. Now calculate the value of current I_1

$$I_1 = \frac{5}{15+5} \times 10 = 2.5\,\text{A}$$
$$V_1 = 10 \times 2.5 = 25\,\text{V}.$$

$V_1 = V_2$, hence, the Reciprocity Theorem is verified.

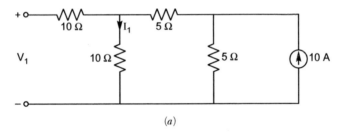

(a)

FIGURE 8.64

Example 8.17. *Verify the Reciprocity Theorem for the circuit shown in Figure 8.64(b).*

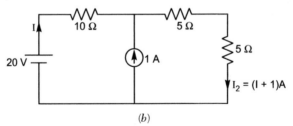

(b)

FIGURE 8.64

Solution: Case I: Applying KVL,

$$-20 + 10I + 10(I+1) = 0$$
$$I = 0.5\,\text{A}$$

and

$$I_2 = 1 + I = 1.5\,\text{A}.$$

Case II: Now the 20 V battery is shifted at point 2 as shown in Figure 8.64(c).

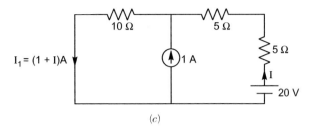

(c)

FIGURE 8.64

Now, calculate I_1 using KVL,

$$-20 + 10I + 10(1 + I) = 0$$
$$I = 0.5 \, \text{A}$$

and

$$I_1 = 1 + I = 1.5 \, \text{A}.$$

Hence, the Reciprocity Theorem is verified.

SOLVED PROBLEMS

Problem 8.1. *The network N in Figures 8.65(a) and (b) is passive and contains only linear resistors; the port currents in Figure 8.65(a) are as marked. Using these values and the principles of superposition and reciprocity, find I_x in Figure 8.65(b).*

Solution:
By the Superposition Theorem source V_1 is working alone by Figure 8.65(a). $I_x = 4$ A due to the 5 V source. Therefore, the response due to V_1 in Figure 8.65(b) is 8 A (I_1).

Now source V_2 is working alone. By using the Reciprocity Theorem the response and source can be interchanged, then the circuit will be as shown below.

From Figure 8.65(a) for 5 V the response is 1 A. Similarly, for 10 V it is 2 A

$$I_2 = 2 \, \text{A}.$$

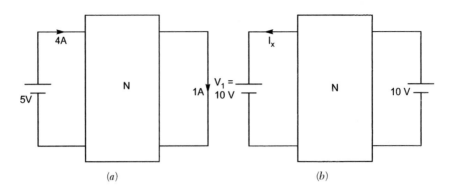

FIGURE 8.65

Now,

$$I_x = I_2 - I_1$$
$$= 2 - 8 = -6 \text{ Amp.}$$

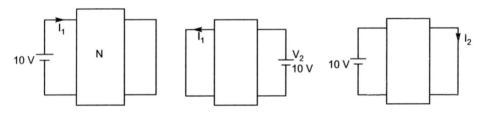

Problem 8.2. *Obtain the Thevenin equivalent circuit for the circuit shown in Figure 8.66.*

Solution: Applying nodal analysis at node C,

$$I_1 = 0.2\,V_x \text{ and } -V_x + I_1 + 4 = 0$$
$$V_x = 5\,V = V_{TH}.$$

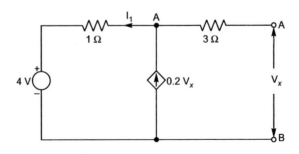

FIGURE 8.66

For R_{TH}: short the terminal AB.

Hence, $V_x = 0$, therefore the current by the dependent source is zero.

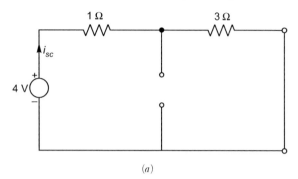

(a)

FIGURE 8.67

Now by KVL,

$$i_{SC} = \frac{4}{4} = 1 \text{ Amp.}$$

$$R_{TH} = \frac{V_{TH}}{i_{SC}} = \frac{5}{1} = 5\,\Omega.$$

(b)

FIGURE 8.67   Thevenin equivalent circuit.

Problem 8.3. *Determine the current in the capacitor branch by the Superposition Theorem in the circuit.*

Solution: Case I: When the voltage source is working and the current source is set to zero (open circuit),

$$I' = \frac{4\angle 0°}{3 + 4J + 3 - 4J}$$

$$I' = \frac{2}{3}\angle 0° \text{A}$$

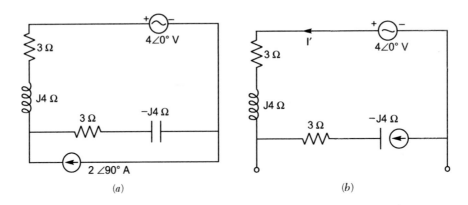

FIGURE 8.68

Case II: When the current source is working and the voltage source is set to zero (short circuit), by applying the current division formula,

$$I'' = \frac{3 + 4J}{3 + 4J + 3 - 4J} 2\angle 90°$$

$$I'' = \left(-\frac{4}{3} + J \right) A.$$

Total current in the capacitor branch,

$$I = I' + I''$$

$$I = \frac{2}{3} - \frac{4}{3} + J = -\frac{2}{3} + J = 1.2\angle 123.7°A.$$

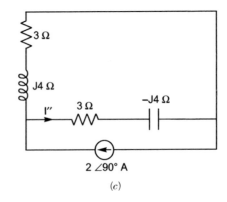

(c)

FIGURE 8.68

Problem 8.4. *In the circuit shown in Figure 8.69, two voltage sources act on the load impedance connected to terminal AB. If the load is variable in both reactance and resistance, what load* Z_L *will receive maximum power applying the Millman's Theorem? Also calculate maximum power.*

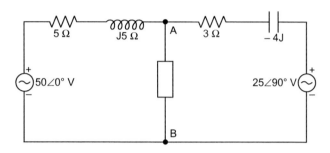

FIGURE 8.69

Solution: By the Millman Theorem,

$$E = \frac{E_1 \times \frac{1}{Z_1} + E_2 \times \frac{1}{Z_2}}{\frac{1}{Z_1} + \frac{1}{Z_2}}$$

where

$$E_1 = 50\angle 0°, Z_1 = (5 + 5J)\,\Omega$$

$$Y_1 = \frac{1}{Z_1} = \frac{1}{5 + 5J} = (0.1 - 0.1J)\,\mho$$

$$E_2 = 25\angle 90° = 25J\,V$$

$$Z_2 = (3 - 4J)\,\Omega.$$

$$Y_2 = \frac{1}{Z_2} = \frac{1}{3 - 4J} = (0.12 + J0.16)\,\mho$$

$$E = \frac{50(0.1 - J0.1) + 25J(0.12 + J0.16)}{(0.1 - J0.1) + (0.12 + J0.16)}$$

$$= \frac{5 - 5J + 3J - 4}{0.22 + J0.06} = \frac{2.236\angle -63.4°}{0.228\angle 15.25°}$$

$$= 9.807\angle -78.65°\,V$$

$$Z = \frac{1}{Y} = \frac{1}{Y_1 + Y_2} = \frac{1}{0.22 + J0.06}$$

$$= 4.385\angle -15.25° = (4.23 - J1.15)\,\Omega.$$

Therefore, for maximum power,

$$Z_L = Z^* = (4.23 + J1.15)\,\Omega.$$

Maximum power,

$$P_{max} = \frac{E^2}{4RL} = \frac{(9.807)^2}{4 \times 4.23}$$

$$P_{max} = 5.68\,W.$$

Problem 8.5. *In the network shown in Figure 8.70, find the value of* R_L *to which the maximum power can be delivered; hence, find the voltage across* R_L.

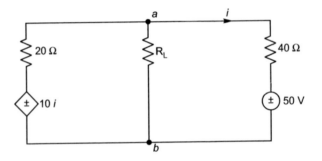

FIGURE 8.70

Solution: The condition for maximum power transfer,

$$Z_L = Z^* \text{ or for resistive circuit } R_L = R_{TH}.$$

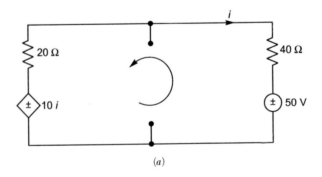

(a)

FIGURE 8.70

For V_{TH}, open the terminal $a - b$, and applying KVL,

$$-10i + 20i + 40i + 50$$
$$50i = -50 \Rightarrow i = -1 \text{ Amp.}$$
$$V_{TH} = 10i - 20i$$
$$V_{TH} = -10i = 10\,\text{V.}$$

For I_{SC}, short the terminal $a - b$,

$$i = -\frac{50}{40} = -1.25\,\text{A}$$
$$20I' = 10i \text{ by } KVL$$
$$I' = \frac{1}{2}i = 0.625\,\text{A}$$

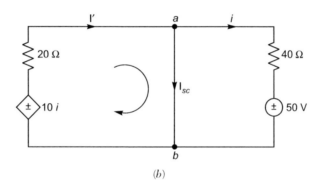

(b)

FIGURE 8.70

Total current

$$I_{SC} = i + i'$$
$$= +1.25 - 0.625$$
$$I_{SC} = 0.625\,\text{A.}$$

For R_{TH},

$$R_{TH} = \frac{V_{TH}}{I_{SC}} = \frac{10000}{0.625} = 16\,\Omega.$$

Problem 8.6. *Verify the Reciprocity Theorem for the network shown in Figure 8.71.*

Solution: Case I: Equivalent impedance

$$Z_{eq} = (2 - J)||(J) + 1$$

$$\frac{(2-J)J}{2-J+J} + 1 = \frac{3+2J}{2}$$

$$I_T = \frac{5}{Z_{eq}} = \frac{5.2}{3+2J}$$

$$= \frac{10}{3+J2}.$$

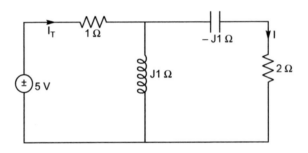

FIGURE 8.71

Now by the current division formula,

$$I = \left(\frac{J.1}{J.1 + 2 - JI}\right)\left(\frac{10}{3+J2}\right)$$

$$= \frac{5J}{3+J2}.$$

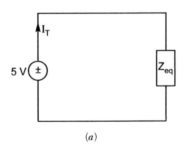

(a)

FIGURE 8.71

Case II:

$$Z'_{eq} = (1\|J1) - J1 + 2 = \frac{1 \times J}{1 + J} - J + 2 = 2.5 - J0.5$$

$$I'_T = \frac{5}{Z_{eq'}} = \frac{5}{(2.5 - J0.5)}$$

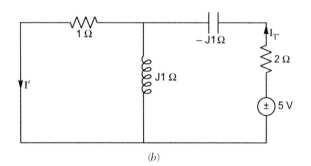

(b)

FIGURE 8.71

By the current division formula,

$$I' = \frac{J \cdot 1}{J \cdot 1 + 1} \times \frac{5}{2.5 - J0.5} = \frac{J5}{3 + 2J}.$$

Since $I = I'$, hence, the Reciprocity Theorem is verified.

Problem 8.7. *Find the Norton's equivalent of the network shown in Figure 8.72.*

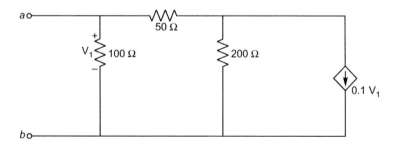

FIGURE 8.72

Solution: The circuit has only one dependent source, hence, $I_{SC} = I_N = 0\,A$. For R_N, a voltage source of V voltage is applied.

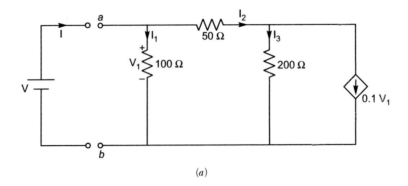

(a)

FIGURE 8.72

Applying nodal analysis at A and B,

$$I = I_1 + I_2 \Rightarrow I = \frac{V_1}{100} + \frac{V_A - V_B}{50}$$

$$I_2 = I_3 + 0.1\,V_1$$

$$\frac{V_A - V_B}{50} = \frac{V_B}{200} + 0.1\,V_1$$

and

$$V = V_1 = V_A$$

$$I = \frac{V_1}{100} + \frac{V_1 - V_B}{50} \qquad (1)$$

$$\frac{V_1 - V_B}{50} = \frac{V_B}{200} + 0.1\,V_1. \qquad (2)$$

After solving Equations (1) and (2),

$$V_B = -\frac{16}{5}\,V_1. \qquad (3)$$

Put the value of V_B into Equation (1),

$$I = \frac{V_1}{100} + \frac{V_1}{50} - \frac{V_B}{50} = \frac{V_1}{100} + \frac{V_1}{50} + \frac{16}{250}V_1$$

$$\frac{V_1}{I} = \frac{500}{47} = \frac{V}{I} = R_N$$

$$R_N = 10.63\ \Omega,$$

the Norton equivalent circuit.

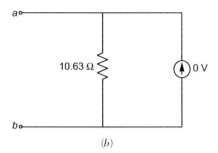

(b)

FIGURE 8.72

Problem 8.8. *Calculate the load current I in the circuit of Figure 8.73 by Millman's Theorem.*

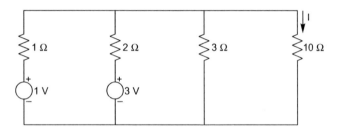

FIGURE 8.73

Solution: Assume the circuit is

$$E = \frac{E_1 \times \frac{1}{R_1} + E_2 \times \frac{1}{R_2} + E_3 \times \frac{1}{R_3}}{\frac{1}{R_1} + \frac{1}{R_2} + \frac{1}{R_3}}$$

$$E = \frac{1 \times 1 + 3 \times \frac{1}{2} + 0 \times \frac{1}{3}}{1 + \frac{1}{2} + \frac{1}{3}} = \frac{15}{11}$$

$$\frac{1}{R} = G = \frac{1}{R_1} + \frac{1}{R_2} + \frac{1}{R_3} = 1 + \frac{1}{2} + \frac{1}{3} = \frac{11}{6}$$

$$R = \frac{1}{G} = \frac{6}{11}.$$

Therefore, Millman's equivalent circuit is shown in the figure.

$$I = \frac{V}{R + R_L} = \frac{\frac{5}{11}}{\frac{6}{11} + 10}$$

$$I = \frac{15}{116} A.$$

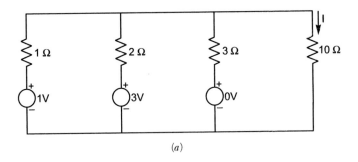

(a)

FIGURE 8.73

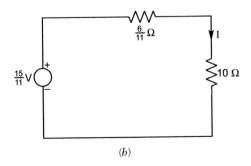

(b)

FIGURE 8.73

Problem 8.9. *Replace the network to the left of terminals AB in the figure by the Thevenin equivalent circuit. Find the current and power delivered to the load* Z_L *if* $Z_L = (4 + J4)\ \Omega$.

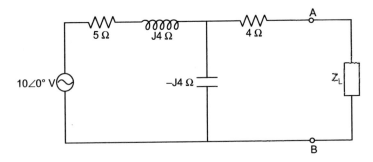

FIGURE 8.74

Solution: For V_{TH}, open the terminal AB and find $V_{AB} = V_{TH}$,

$$V_{TH} = iZ = \frac{10\angle 0}{5 + J4 - J4} \times -J4 = 8\angle -90° \text{ V}.$$

For Z_{TH}, short the terminal AB and find $I_{Sc} = I_N$

$$Z_{eq} = (4|| - J4) + 5 + 4J = -\frac{J16}{4 - 4J} + 5 + 4J$$

$$= -\frac{J16 + (5 + 4J)(4 - 4J)}{4 - 4J} = \frac{-16J + 20 - 20J + 16J - 16J^2}{4 - 4J}$$

$$= \frac{36 - 20J}{4 - 4J} = \frac{9 - 5J}{1 - J} \times \frac{1 + J}{1 + J}$$

$$Z_{eq} = \frac{9 + 9J - 5J + 5}{2} = \frac{14 + 4J}{2} = 7 + 2J$$

$$I = \frac{10}{Z_{eq}} = \frac{10}{7 + 2J}.$$

$$I_{SC} = \frac{-4J}{4 - 4J} \times I = \frac{-4J}{4 - 4J} \times \frac{10}{7 + 2J} = \frac{10}{5 + 9J}$$

$$Z_{TH} = \frac{V_{TH}}{I_{SC}} = \frac{-8J(5 + 9J)}{10} = 7.2 - 4J$$

$$I_L = \frac{-8J}{7.2 - 4J + 4 + 4J}$$

$$I_L = 0.71$$

$$P_L = I_L^2 R_L = (0.71)^2 \times 4 = 2\,\text{W}.$$

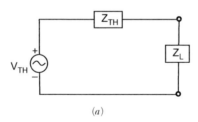

(a)

FIGURE 8.74

Problem 8.10. *Verify Tellegen's Theorem in the network shown in Figure 8.75.*

Solution: Applying mesh analysis for mesh (1) and (2),

$$33 = I_1 \times 1 + 3(I_1 + I_2)$$

$$4I_1 + 3I_2 = 33 \qquad\qquad (1)$$

$$22 = 2I_2 + 3(I_1 + I_2)$$
$$3I_1 + 5I_2 = 22. \qquad (2)$$

By Equations (1) and (2), $I_2 = -1\,\text{A}, I_1 = 9\,\text{A}.$

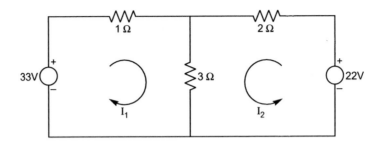

FIGURE 8.75

The I_2 direction is the opposite source, 22 V, and is not delivering power. It is taking (charging) by Tellegen's Theorem.

Power delivering $= VI = 33 \times 9 = 297.$

Power taking by $1\,\Omega, 2\,\Omega, 3\,\Omega$, and 22 V battery,

$$= (91^2 \times 1 + (-1)^2 \times 2 + (81^2 \times 3 + 22 \times 1) = 297.$$

Power delivering = Power taking verified.

Problem 8.11. *Find the Thevenin equivalent about AB for the circuit shown in Figure 8.76.*

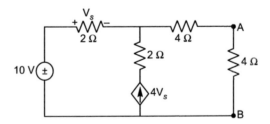

FIGURE 8.76

Solution: Step 1. First open the terminal AB and find $V_{AB} = V_{TH}$.

The 10 V battery is charging by the $4\,V_s$ independent current source

$$2 \times 4V_s - V_s + 10 = 0$$
$$7\,V_s = -10$$

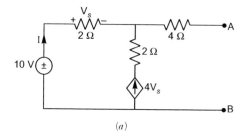

(a)

FIGURE 8.76

$$V_s = -\frac{10}{7} \text{ V} = 2I$$

$$I = -\frac{5}{7}\text{A}.$$

The negative sign shows the direction of current is opposite as we assume,

$$V_{TH} = V_{AB} = -V_s + 10 = \frac{10}{7} + 10 = \frac{80}{7} \text{ V}.$$

Step 2. Now short the terminal AB and calculate I_{SC}.
By applying KVL,

$$-10 + 2I + 4(I + 4V_s) = 0 \qquad (1)$$

and

$$V_s = 2I. \qquad (2)$$

(b)

FIGURE 8.76

Now by Equations (1) and (2)

$$-10 + 2I + 4I + 16 \times 2I = 0$$

$$I = \frac{10}{38}.$$

Now

$$I_{SC} = I + 4 \times 2I = 9I = 9 \times \frac{10}{38} = \frac{90}{38}$$

and we know

$$R_{TH} = \frac{V_{TH}}{I_{SC}} = \frac{80}{7} \times \frac{38}{90} = \frac{304}{63} \; \Omega.$$

The Thevenin equivalent circuit of the given figure is shown in Figure 8.76(c).

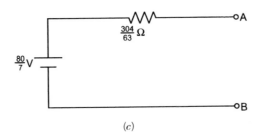

(c)

FIGURE 8.76 Thevenin equivalent circuit.

Problem 8.12. *For the circuit shown in Figure 8.77 obtain the voltage across each current source using the Superposition Theorem.*

Solution: Case I: When 2A is acting and other sources are set to zero, the result is shown in Figure 8.77(a).

Converting the current source into an equivalent voltage source, the circuit becomes as shown in Figure 8.77(b).

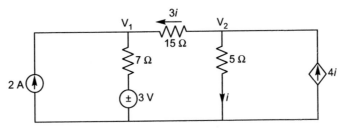

FIGURE 8.77

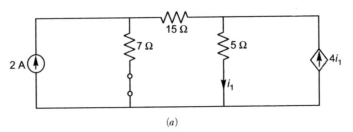

(a)

FIGURE 8.77

Applying KVL,

$$14 = -7 \times 3i_1 - 15 \times 3i_1 + 5i_1$$

$$i_1 = -\frac{14}{61}\text{A}.$$

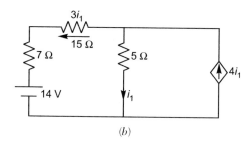

(b)

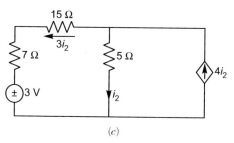

(c)

FIGURE 8.77

Case II: When $3\,\text{V}$ is acting alone and the other source is set to zero, as shown in Figure 8.77(c), applying KVL,

$$3 = -7 \times 3i_2 - 15 \times 3i_2 + 5i_2 \text{ or } i_2 = -\frac{3}{61} \text{ Amp.}$$

Therefore, using the Superposition Theorem,

$$i = i_1 + i_2 = -\frac{14}{61} - \frac{3}{61} = -\frac{17}{61}.$$

Now the voltage drop across the 2 A current source $= 7(3i + 2) + 3$.

$$7\left(-\frac{17 \times 3}{61} + 2\right) + 3 = 11.15\,\text{V}$$

and the voltage across the $4i$ current source $= 5i = 5\left(-\frac{17}{61}\right) = -1.39\,\text{V}.$

Problem 8.13. *Using the Superposition theorem, calculate the current through* $(2 + 3J)$ Ω *impedance of the circuit shown in Figure 8.78.*

Solution: Step 1. The 30 V battery is operating and the 20 V battery is set to zero, as shown in Figure 8.78(a), 6 Ω and 4 Ω resistors are connected in parallel, therefore, Figure 8.78(a) can be reduced as shown in Figure 8.78(b).

$$I = \frac{30}{5 + \left[J5||\{(2 + J3) + 2.4\}\right]}$$

$$= \frac{30}{5 + \left[J5||(4.4 + J3)\right]} = \frac{30(J5 + 4.4 + J3)}{5 \times (J5 + 4.4 + J3) + J5(4.4 + J3)}$$

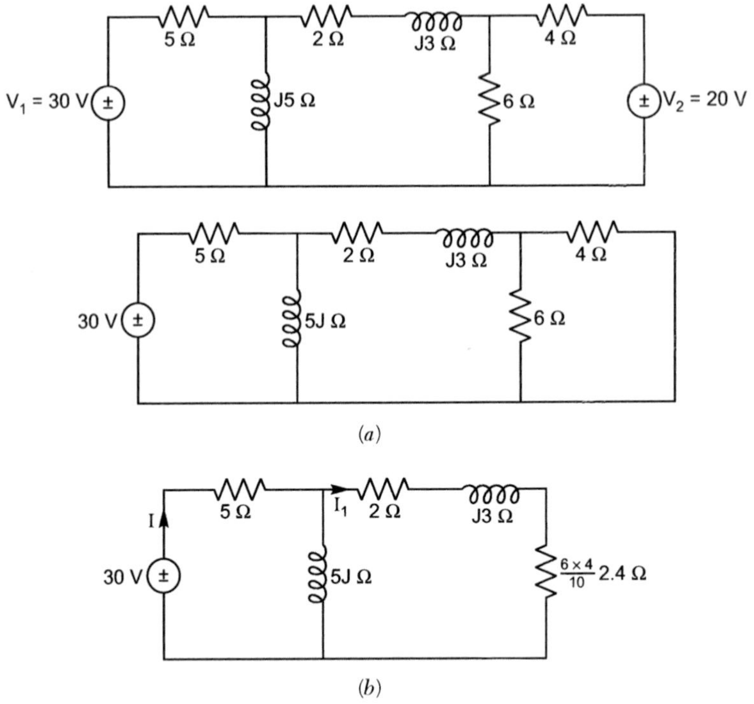

(a)

(b)

FIGURE 8.78

By the current division formula,

$$I_1 = \frac{J5}{J5 + 4.4 + J3} \times I = \frac{30 \times J5}{5(J8 + 4.4) + J5(4.4 + J3)}$$

$$I_1 = \frac{J30}{4.4 + J8 + J4.4 - 3} = \frac{J30}{1.4 + J12.4} = \frac{30\angle90°}{12.48\angle83.6°} = 2.40\angle6.4°.$$

Step 2. The 20 battery is operating and the 30 V battery is set to zero (short circuit), as shown in Figure 8.78(c).

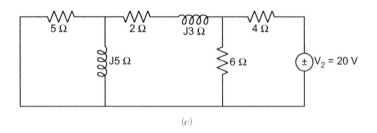

(c)

FIGURE 8.78

5 Ω and 5J are connected in parallel, therefore, Figure 8.78(c) can be reduced as shown in Figure 8.78(d).

$$I = \cfrac{20}{[\{(2.5 + 2.5J) + (2 + J3)\}116] + 4}$$

$$= \cfrac{20}{4 + \{(4.5 + 5.5J)116\}}$$

$$I = \frac{20(10.5 + J5.5)}{4.2 + J22 + 27 + J33} = \frac{20(10.5 + J5.5)}{69 + J55}$$

$$I_2 = \frac{6}{6 + (4.5 + J5.5)} \times I = \frac{120}{69 + J55} = 1.36\angle-37.6°$$

W
2 Ω   J3 Ω I_2 4 Ω

$\left(\frac{5 + 50}{2}\right)$ 6 Ω $\pm$ V_2 = 20 V

(d)

FIGURE 8.78

Therefore, the resultant current $I = I_1 - I_2$

$$= 2.40\angle6.4° - 1.36\angle - 37.6°$$
$$= 2.39 + J0.27 - 1.06 + J0.85$$
$$= 1.33 + J1.12 = 1.74\angle40°\,A.$$

Problem 8.14. *Find the Thevenin's equivalent for the network shown in Figure 8.79 at the right of terminals ab and hence find the source current.*

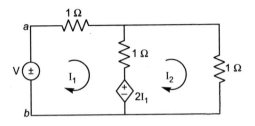

FIGURE 8.79

Solution: Step 1. Open the terminal $a - b$. Now the circuit shown in the figure does not have any independent source, hence, $V_{TH} = 0$.

Now with the current source of 1 Amp.,

$$I_1 = 1 \text{ Amp.}$$

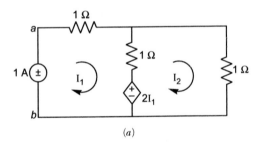

(a)

FIGURE 8.79

Now applying KVL in mesh 2,

$$-2I_1 + (I_2 - 1) + I_2 = 0$$
$$I_2 = 1.5 \text{ Amp.}$$
$$V_{ab} = 1 \times I_1 + I_2 \times 1$$
$$V_{ab} = 2.5 \text{ V}$$
$$R_{TH} = \frac{V_{ab}}{I_{SC}} = \frac{2.5}{1} = 2.5 \,\Omega.$$

Therefore, the Thevenin's equivalent circuit is as shown in Figure 8.79(b).

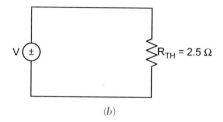

(b)

FIGURE 8.79

Problem 8.15. *In the single source network shown in Figure 8.80, the voltage source $100\angle45°$ causes a current I in the $5\,\Omega$ branch. Find I and then verify the Reciprocity Theorem for this circuit.*

Solution: Case I: Applying mesh analysis we get,

$$I_1(10 + J5) - J5I_2 = 100\angle45°$$
$$-I_15J + 10I_2 + J5I_3 = 0$$
$$5JI_2 + (5 - 5J)I_3 = 0.$$

After solving these three equations we get $I_3 = I = 2.16\angle57.5°$ Amp.

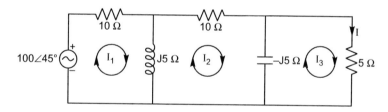

FIGURE 8.80

Case II: Now the source shifted at point 2.

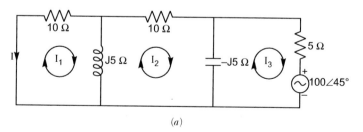

(a)

FIGURE 8.80

Again applying mesh analysis we get,

$$I_1(10 + J5) + J5I_2 = 0$$
$$-I_1J + I_2 \times 10 + J5I_3 = 0$$
$$I_2J5 + I_2(5 - 5J) = 100\angle 45°.$$

After solving these three equations, we get $I = -I_1 = 2.16\angle 57.5°$ Amp.

Problem 8.16. *For the network shown in Figure 8.81 determine voltage across the capacitor using Thevenin's Theorem.*

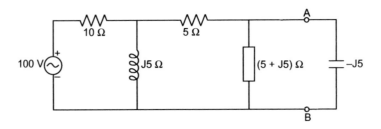

FIGURE 8.81

Solution: Step 1. Open the terminals AB and calculate $V_{AB} = V_{TH}$

$$Z_{eq} = 10 + J5 || (10 + J5)$$
$$= 10 + \frac{J5 \times (10 + J5)}{10(1 + J)} = 10 + \frac{5}{2} \frac{(2J + J^2)}{(1 + J)}$$
$$Z_{eq} = \frac{37 + J}{4}$$
$$I = \frac{V}{Z_{eq}}$$
$$= \frac{100}{37 + J} \times 4$$
$$I_1 = \frac{J5}{J5 + 5 + 5 + 5J} \times \frac{400}{37 + J}$$
$$I_1 = \frac{J5 \times 400}{10(1 + J)(37 + J)} = \frac{200J}{37 + J + 37J + J^2}$$
$$I_1 = \frac{200J}{36 + 38J} = \frac{20}{137}(18J + 19)$$

$$V_{AB} = V_{TH} = (5 + 5J)I_1 = \frac{100}{137}(1 + J)(18J + 19)$$

$$V_{TH} = \frac{100}{137}(1 + 37J).$$

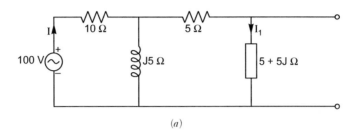

(a)

FIGURE 8.81

Step 2. Calculate Z_{AB} (equivalent impedance across AB),

$$Z_{AB} = \left[\frac{10 \times J5}{10 + 5J} + 5\right] \| (5 + 5J) = \left[\frac{10J}{2 + J} + 5\right] \| (5 + 5J)$$

$$= \frac{\left(\frac{10+15J}{2+J}\right)(5 + 5J)}{\frac{10+15J}{2+J} + (5 + 5J)} = 3 + \frac{7}{3}J$$

$$I = \frac{V_{TH}}{Z_{TH} + Z_L}$$

$$= \frac{\frac{100}{137}(1 + 37J)}{3 + \frac{7}{3}J - J5}$$

$$= \frac{100}{137}\frac{(1 + 37J)}{\left(3 - \frac{8}{3}J\right)}.$$

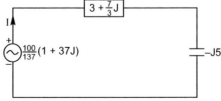

(b) Thevenin equivalent circuit.

FIGURE 8.81

The voltage across the capacitor,

$$V_C = I \times -J5 = \frac{100}{137} \times \frac{(1+37J)}{(3-\frac{8}{3}J)} \times -J5$$

$$V_C = \frac{500}{137}\left(\frac{-J+37}{3-\frac{8}{3}J}\right)$$

$$V_C = 3.6496 \times \frac{37.01\angle-1.54°}{4.01\angle-41.63°}$$

$$V_C = 33.68 \angle 40.1°.$$

Problem 8.17. *In the network shown in Figure 8.82 two voltage sources act on the load impedance connected to the terminals AB. If the load is variable in both reactance and resistance, for what value will load Z_L receive maximum power?*

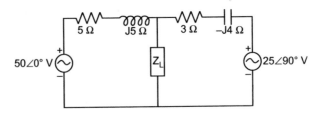

FIGURE 8.82

Solution: For maximum power,

$$Z_L = Z^*_{TH}.$$

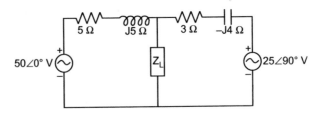

(a)

FIGURE 8.82

For Z_{TH},

$$Z_{AB} = Z_{TH} = \frac{5(1+J)(3-J4)}{5(1+J)+(3-J4)}$$

$$= \frac{5(3-J4+3J+4)}{8+J} = \frac{5(7-J)}{8+J} \times \left(\frac{8-J}{8-J}\right)$$

$$= \frac{5(56-7J-8J+1)}{65} = \frac{1}{13}(57-15J)$$

$$Z_L = Z_{TH}^* = \frac{1}{13}(57+15J).$$

Problem 8.18. *For the circuit shown in Figure 8.83, using Millman's Theorem, find the current in the load impedance, $Z_L = (2+4J)\,\Omega$.*

Solution: Equivalent voltage,

$$E = \frac{E_1 Y_1 + E_2 Y_2 + E_3 Y_3}{Y_1 + Y_2 + Y_3} \left[Y = \frac{1}{Z}\right]$$

$$= \frac{1 \times 1 + 5 \times 1 + 25 \times 0.2}{1+1+0.2} = 5\,V$$

$$Y_{eq} = Y_1 + Y_2 + Y_3 = 2.2$$

$$Z_{eq} = \frac{1}{Y_{eq}} = \frac{1}{2.2} = 0.4545\,\Omega.$$

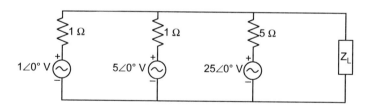

FIGURE 8.83

The equivalent circuit is given in Figure 8.83(a),

$$I = \frac{E}{Z_{eq} + Z_L} = \frac{5}{0.4545 + 2 + 4J}$$

$$I = 1.06\angle{-58.46°}A.$$

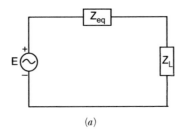

FIGURE 8.83

Problem 8.19. *Determine current I in Figure 8.84 using the Superposition Theorem.*

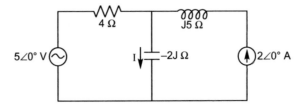

FIGURE 8.84

Solution: Case I: When the voltage source is operating and the current source is set to zero, as shown in Figure 8.84(*a*),

$$I = \frac{5}{4 - 2J}$$
$$= 1 + 0.5J.$$

Case II: Now the current source is operating and the voltage source is set to zero as shown in Figure 8.84(*b*),

$$I' = \frac{4}{4 - 2J} \times 2\angle 0° = 1.6 + J0.8.$$

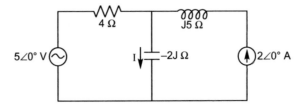

(a)   (b)

FIGURE 8.84

The direction of currents $(I \& I')$ are in the same direction,

$$I = I + I' = 1 + J0.5 + 16 + J0.8$$
$$I = 2.6 + J1.3 = 2.9\angle 26.6°\,A.$$

Problem 8.20. *Given the network of Figure 8.85, find Norton's equivalent circuit to the left side of terminal ab.*

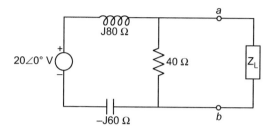

FIGURE 8.85

Solution: For I_N Short the terminal *ab* and calculate $I_{SC} = I_N$.
In this case the 40 Ω resistor is shorted as shown in Figure 8.85(a),

$$I_N = \frac{20}{J80 - J60} = 1\angle -90°\,A$$

FIGURE 8.85

For Z_N, Z_{ab} (equivalent resistance across $a - b$) $= Z_N$

$$= \frac{20J \times 40}{20J + 40}$$
$$= 8 + J16$$
$$Z_N = 17.89\angle 63.43°\ \Omega.$$

Norton's equivalent is shown in Figure 8.85(b).

Problem 8.21. *Find the current in the* 24 Ω *resistor of the circuit of Figure 8.86 by Thevenin's Theorem.*

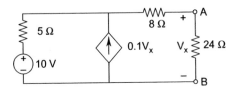

FIGURE 8.86

Solution: For V_{TH}, open the terminal AB,

$$V_{TH} = V_x = 10 - 5(-0.1\,V_x)$$
$$V_x = 20\,V = V_{TH}.$$

Now short the terminal AB, $V_x = 0$

$$I_{SC} = \frac{10}{5+8} = \frac{10}{13}\,A$$
$$R_{TH} = \frac{V_{TH}}{I_{SC}} = \frac{20 \times 13}{10} = 26\,\Omega$$
$$I_L = \frac{V_{TH}}{R_{TH} + R_L} = \frac{20}{26 + 24} = 0.4\,A.$$

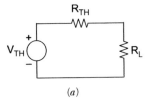

(a)

FIGURE 8.86

Problem 8.22. *For the circuit shown in Figure 8.87 obtain the voltage across each current source using the Superposition Theorem.*

Solution: Case I. When the current source 2 A is working and the voltage source is set to zero (short circuit) converting the current source into an equivalent voltage source, the circuit results in Figure 8.87(b).

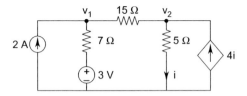

FIGURE 8.87

By KVL,

$$14 = (15 + 7) \times -3i_1 + 5i_1$$
$$14 = -61i_1$$
$$i_1 = -\frac{14}{61}.$$

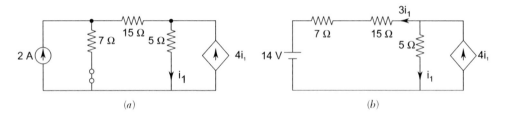

(a)   (b)

FIGURE 8.87

Case II. When a 3 V source is working alone as shown in Figure 8.87(c), by KVL,

$$3 = 22 \times -3i_2 + 5i_2$$
$$i_2 = -\frac{3}{61}.$$

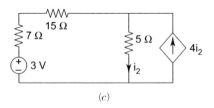

(c)

FIGURE 8.87

By the Superposition Theorem,

$$i = i_1 + i_2 = -\frac{14}{61} - \frac{3}{61} = -\frac{17}{61}.$$

The voltage across the 2 A current source $= 7(3i + 2) + 3$

$$= 7\left(-\frac{51}{61} + 2\right) + 3 = 11.15\,\text{V}.$$

The voltage across the $4i$ current source $= 5i = -\frac{85}{61} = -1.39\,\text{V}.$

Problem 8.23. *Using the Superposition Theorem, calculate the current through* $(2 + 3\text{J})\,\Omega$ *impedance of the circuit shown in Figure 8.88.*

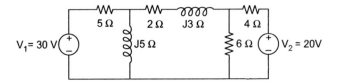

FIGURE 8.88

Solution: Case I. When the 30 V source is working alone as shown in Figure 8.88(a),

$$I_1 = \frac{30}{5 + [\text{J}5.11\{(2 + \text{J}3) + 2.4\}]} = \frac{30(\text{J}5 + 4.4 + \text{J}3)}{5 + (\text{J}5 + 4.4 + \text{J}3) + \text{J}5(4.4 + \text{J}3)}$$

$$I_1' = \frac{\text{J}5}{\text{J}5 + 2 + \text{J}3 + 2.4} \times I_1$$

$$= \frac{30(\text{J}5)}{5(\text{J}5 + 4.4 + \text{J}3) + \text{J}5(4.4 + \text{J}3)}$$

$$I_1' = \frac{\text{J}30}{1.4 + \text{J}12.4} = 2.40\angle 6.4^\circ.$$

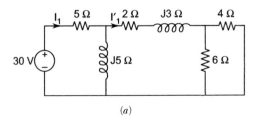

(a)

FIGURE 8.88

Case II. When the 20 V source is working and the other 30 V source is set to zero as shown in Figure 8.88(b),

$$I_2 = \frac{20}{4 + [6 \| \{(2 + 3J) + (5 \| J5)\}]}$$

$$I_2 = \frac{20}{4 + [6 \| (4.5 + J5.5)]} = \frac{20(10.5 + J5.5)}{69 + J55}$$

$$I_2' = \frac{6}{6 + (4.5 + J5.5)} \times I_2 = \frac{120}{69 + J55} = 1.36\angle{-37.6°}.$$

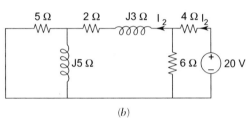

(b)

FIGURE 8.88

Now by the Superposition Theorem,

$$I = I_1' - I_2'$$
$$= 2.4\angle6.4° - 1.36\angle{-37.6°} = 1.33 + J1.12 = 1.74\angle40° \text{ A}.$$

Problem 8.24. *Find the Thevenin's equivalent for the network shown in Figure 8.89 at the right of terminal AB and, hence, find the source current.*

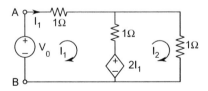

FIGURE 8.89

Solution: For V_{TH}, open the terminal AB as shown in Figure 8.89(a).

$V_{AB} = V_{TH} = 0$ because the circuit does not have any independent source.

For R_{TH}, in this case, a voltage source V_0 is applied at terminal AB and I_1 current is flowing due to V_0 as shown in Figure 8.89(b), and $R_{TH} = \frac{V_0}{I_1}$.

By KVL,

$$V_0 = 1 \times I_1 + 1 \cdot (I_1 - I_2) + 2I_1$$
$$V_0 = 4I_1 - I_2 \qquad (1)$$

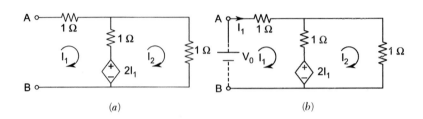

(a)   (b)

FIGURE 8.89

and

$$2I_1 = (I_2 - I_1) + I_2$$
$$3I_1 = 2I_2. \qquad (2)$$

By Equations (1) and (2),

$$V_0 = 4I_1 - \frac{3}{2}I_1 = \frac{5}{2}I_1$$

$$R_{TH} = \frac{5}{2}\ \Omega.$$

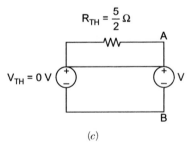

(c)

FIGURE 8.89

Therefore, Thevenin's equivalent circuit is shown in Figure 8.89(c). Hence, the source current

$$I = \frac{V}{R_{TH}} = \frac{2\,V}{5}.$$

QUESTIONS FOR DISCUSSION

1. Explain the Superposition Theorem for A.C.
2. Can we apply the Superposition Theorem for A.C. and D.C.?
3. What is linearity?
4. Explain the Maximum Power Transfer Theorem for A.C.
5. What is Tellegen's Theorem?
6. What is Millman's Theorem?
7. What is the difference between the Thevenin theorem and the Norton theorem?
8. Tell us the procedure to calculate R_{TH} in the Thevenin Theorem.
9. What is the procedure to calculate V_{TH} in the Thevenin Theorem?
10. Can we have any relation between R_{TH}, V_{TH} and I_N?
11. What is reciprocity?

OBJECTIVE QUESTIONS

1. In the lattice network, find the value of R for the maximum power transfer to the load
   (a) 5 Ω   (b) 6.5 Ω   (c) 8 Ω   (d) 9 Ω.

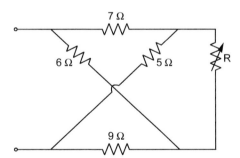

FIGURE 8.90

2. In Figure 8.87, if we connect a source of 2 V with an internal resistance of 1 Ω at A′A with a positive terminal at A, then the current through R is
   (a) 2 A   (b) 1.66 A   (c) 1 A   (d) 0.625 A.

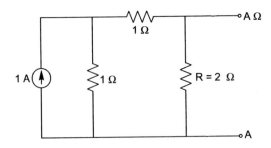

FIGURE 8.91

3. In the circuit shown the value of I is
   (*a*) 1 A   (*b*) 2 A   (*c*) 4 A   (*d*) 8 A.

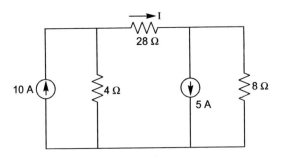

FIGURE 8.92

4. When all the resistances in the circuit are of one Ω each, the equivalent resistance across the points A and B will be
   (*a*) 1 Ω   (*b*) 0.5 Ω   (*c*) 2 Ω   (*d*) 1.5 Ω.

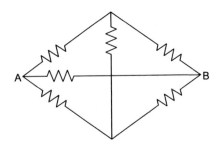

FIGURE 8.93

5. The equivalent circuit of the following circuit is

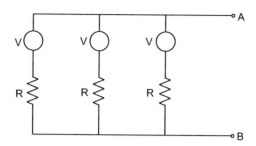

FIGURE 8.94

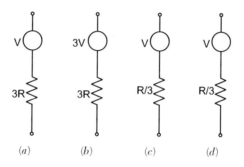

FIGURE 8.94

6. In a linear network, the ratio of voltage excitation to current response is unaltered when the position of excitation and response are inter-changed. This is the
    (*a*) Principle of duality   (*b*) Reciprocity theorem
    (*c*) Equivalence theorem   (*d*) Principle of superposition.
7. If all the elements in a particular network are linear, then the Superposition Theorem would hold when the excitation is
    (*a*) D.C. only   (*b*) A.C. only
    (*c*) either A.C. or D.C.   (*d*) an impulse.
8. In the circuit shown in the figure, the voltage across the 2 Ω resistor is
    (*a*) 6 V   (*b*) 4 V   (*c*) 2 V   (*d*) zero.
9. The principles of homogeneity and superposition are applied to
    (*a*) linear time-variant systems
    (*b*) non-linear time-variant systems
    (*c*) linear time-invariant systems
    (*d*) non-linear time-invariant systems.

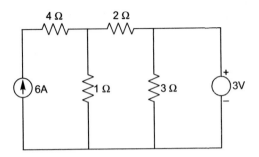

FIGURE 8.95

10. The value of the current I flowing in the 1 Ω resistor in the circuit shown in the figure will be
    (a) 10 A   (b) 6 A   (c) 5 A   (d) zero.

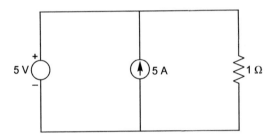

FIGURE 8.96

11. In the circuit shown in the given figure the current I through R_L is
    (a) 2 A   (b) Zero   (c) −2 A   (d) −6 A.

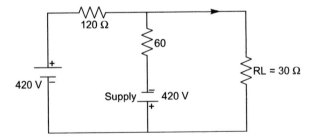

FIGURE 8.97

12. What is R_{TH} in the given circuit
    (a) 0 Ω   (b) 1 Ω   (c) 3 Ω   (d) $\frac{1}{3}$ Ω.

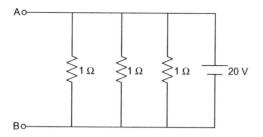

FIGURE 8.98

UNSOLVED PROBLEMS

1. Find, using Thevenin's theorem, the current in the 5 Ω resistor connected across AB in the network shown in Figure 8.99.

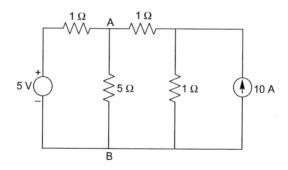

FIGURE 8.99

2. The Thevenin impedance across the terminal AB of the given network is what?

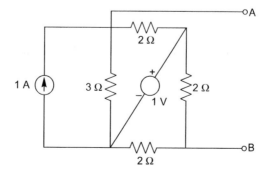

FIGURE 8.100

3. State and explain
    (*a*) Tellegen's Theorem
    (*b*) Reciprocity Theorem
    (*c*) Millman's Theorem
    (*d*) Maximum Power Transfer Theorem
4. State and prove Thevenin's Theorem.
5. State and prove the Maximum Power Transfer Theorem for A.C. circuits.
6. For the network shown in Figure 8.101, determine, using Thevenin's Theorem, voltage across the capacitor.

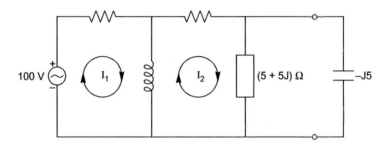

FIGURE 8.101

7. Obtain Thevenin and Norton equivalent circuits at terminal AB of the circuit shown in Figure 8.102.

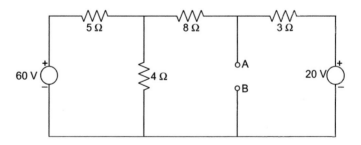

FIGURE 8.102

8. Find the load resistance R_L for the maximum power transfer in the circuit shown in Figure 8.103. Calculate the maximum power also.

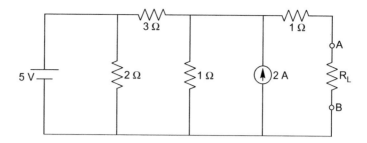

FIGURE 8.103

9. Find both Thevenin and Norton's equivalent circuits for the network shown in Figure 8.104.

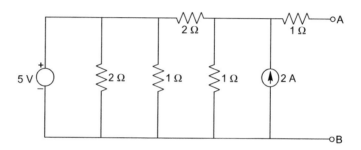

FIGURE 8.104

10. Using the Superposition Theorem find the currents in the various branches of the network shown in Figure 8.105.

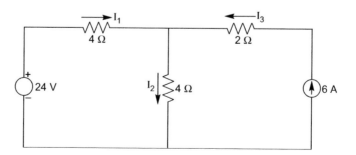

FIGURE 8.105

11. Calculate the load current I in the circuit in Figure 8.106 by Millman, Thevenin, and Norton Theorems.

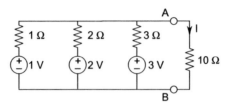

FIGURE 8.106

12. For the Maximum Power Transfer Theorem in the A.C. circuit show that $R_L = R_{TH}$ and $X_L = -X_{TH}$.

NETWORK FUNCTIONS; POLES AND ZEROS

Chapter **9**

9.1 INTRODUCTION

In this chapter we introduce the concept of *transfer functions* relating currents and voltages in different sections of a network. These functions are mathematically similar to the transform impedance or admittance functions and are included in the broader category of functions called *network functions*.

Terminal Pairs or Ports. Any network may be represented schematically by a rectangle or box as shown in Figure 9.1. A network may be used for a variety of purposes. Thus consider its use as a load connected to some other network. In order to connect it to the active network, there must be available two terminals of this passive network. Figure 9.1 shows a network with one pair of terminals 1-1′ or with one port. Such a network may be called a *one-port network or one terminal-pair network*. When such a one-port network is connected to an energy source or an active network at its pair of terminals, the energy source provides the driving force for this one-port network and the pair of terminals constitute the *driving-point* of the network. One pair of terminals is known as a port.

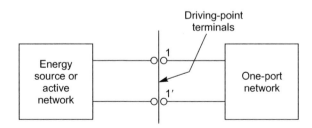

FIGURE 9.1 **One-port network driven by a source.**

Figure 9.2 shows a two-port network or two terminal-pair network. In this case, port 1 is connected to the driving force or the input and is called the *input port* or the *driving port*. On the other hand, port 2 (*i.e.*, terminal-pair 2-2′), is connected to the load and is called the *output port* or the *load port*.

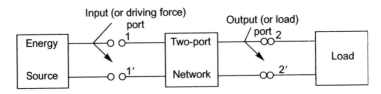

FIGURE 9.2 **Two-port network driven by an energy source.**

Figure 9.3 shows a general *n*-port network. Here the driving force (energy source) may be connected to one or more ports and other networks may be connected at the remaining ports.

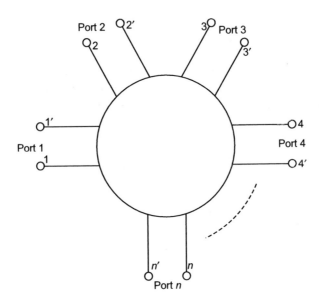

FIGURE 9.3 **General *n*-port network.**

9.2 NETWORK FUNCTIONS FOR A ONE-PORT NETWORK

For a one-port network in the zero or relaxed state (*i.e.*, with no initial conditions) containing no internal voltage or current sources (except controlled sources), the transform impedance $Z(s)$ at its port is defined as the ratio of the transform voltage $V(s)$ to the transform current $I(s)$ at the same port.

Thus, we may write,

$$Z(s) = \frac{V(s)}{I(s)} \quad [\text{Driving-point impedance}]. \tag{9.1}$$

The transform admittance $Y(s)$ at the port is reciprocal of the transform impedance. Thus,

$$Y(s) = \frac{1}{Z(s)} = \frac{I(s)}{V(s)} \quad [\text{Driving-point admittance}]. \tag{9.2}$$

9.3 NETWORK FUNCTIONS FOR A TWO-PORT NETWORK

For a two-port network in the zero or relaxed state (*i.e.*, with no initial conditions) containing no internal voltage of current sources except controlled sources), the transform impedance $Z(s)$ at any port, say port 1, is defined as the ratio of transform voltage $V(s)$ at this port to the transform current $I(s)$ at either this port or the other port. Thus, without specifying the port at which current $I(s)$ is considered, the transform impedance is given by,

$$Z(s) = \frac{V(s)}{I(s)}. \tag{9.3}$$

Transform admittance $Y(s)$ at any port being reciprocal of transform impedance is given by,

$$Y(s) = \frac{1}{Z(s)} = \frac{I(s)}{V(s)} \tag{9.4}$$

If $V(s)$ and $I(s)$ are at the same port, then the transform impedance and transform admittance are specified as *driving-point impedance and driving-point admittance, respectively.*

Thus, for a two-port network in a zero state with no internal energy sources, the *driving-point impedance* is defined as the ratio of the transform voltage at any port to the transform current at the same port.

Driving-point admittance is the reciprocal of the driving-point impedance at the same port.

Figure 9.4 gives a two-port network with reference directions of port voltages and port currents. The reference direction for the current is taken as that into the network.

FIGURE 9.4   A two-port network.

With reference to the two-port network of Figure 9.4, the driving-point impedance at port 1 is given by,

$$Z_{11}(s) = \frac{V_1(s)}{I_1(s)}.$$ (9.5)

While the driving-point admittance at port 1 is given by,

$$Y_{11}(s) = \frac{I_1(s)}{V_1(s)}.$$ (9.6)

Transfer functions. Transfer functions, in general, relate a quantity at one part to a quantity at another port. Thus, the transfer functions which relate voltages and currents may take the following forms:

(a) *voltage transfer ratio.* This is the ratio of voltage at one-port to voltage at another port.

$$\frac{V_2(s)}{V_1(s)}$$

(b) *current transfer ratio.* This is the ratio of the current at one port to the current at another port.

$$\frac{I_2(s)}{I_1(s)}$$

(c) *transfer admittance.* This is the ratio of current at one port to voltage at another port.

$$\frac{I_2(s)}{V_1(s)}$$

(d) *transfer impedance.* This is the ratio of voltage at one port to current at another port.

$$\frac{V_2(s)}{I_1(s)}$$

Conventionally, transfer function is defined as the ratio of an output quantity to an input quantity. [The numerator term is the output quantity and the denominator term is the input quantity.]

With reference to the two-port network of Figure 9.4, the two output quantities are $V_2(s)$ and $I_2(s)$, while the two input quantities are $V_1(s)$ and $I_1(s)$. Thus, for the two-port network, the four transfer functions are as below:

$$
\left.
\begin{aligned}
Z_{21}(s) &= \frac{V_2(s)}{I_1(s)} \\[6pt]
Y_{21}(s) &= \frac{I_2(s)}{V_1(s)} \\[6pt]
G_{21}(s) &= \frac{V_2(s)}{V_1(s)} \\[6pt]
\alpha_{21}(s) &= \frac{I_2(s)}{I_1(s)}
\end{aligned}
\right\}
\tag{9.7}
$$

The transfer functions listed above have subscripts 2 and 1 in the order of *effect cause*. Thus, $Z_{21}(s)$ implies that the cause namely, $I_1(s)$, has resulted in the effect, namely, $V_2(s)$.

In the following examples, we work out network functions for a few simple networks.

Example 9.1. *Find the transform impedance $Z(s)$ of the given one-port network.*

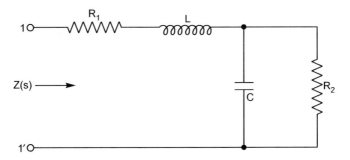

FIGURE 9.5

Solution: The transform network corresponding to the given network is shown in Figure 9.5(a).

Transform admittances Cs and $\frac{1}{R_2}$ may be combined together and the resulting series network is shown in Figure 9.5(b).

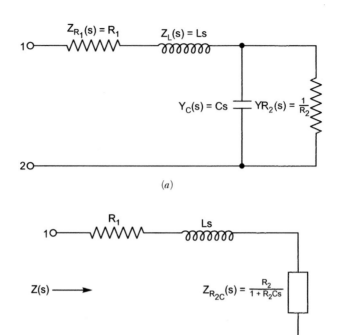

FIGURE 9.5

Hence, the transform impedance $Z(s)$ at the given port is,

$$Z(s) = R_1 + Ls + \frac{R_2}{1 + R_2Cs}$$
$$= \frac{R_2LCs^2 + s(L + R_1R_2C) + (R_1 + R_2)}{1 + R_2Cs}$$

or

$$Z(s) = L\frac{s^2 + s\frac{L+R_1R_2C}{R_2LC} + \frac{R_1+R_2}{R_2LC}}{s + \frac{1}{R_2C}}.$$

Example 9.2. *Find the transform impedance $Z(s)$ of the given one-port network.*

Solution: The transform network corresponding to the given network is shown in Figure 9.6(a).

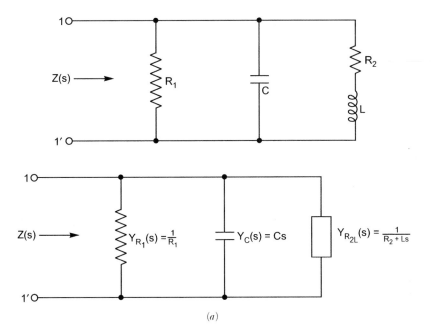

FIGURE 9.6

Combining the three admittances shown in Figure 9.6(*a*), the total admittance at the port is

$$Y(s) = \frac{1}{R_1} + Cs + \frac{1}{R_2 + Ls}.$$

$$= \frac{C}{R_1} \cdot \frac{s^2 + s\frac{L+CR_1R_2}{CL} + \frac{R_1R_2}{CL}}{s + \frac{R_2}{L}} \quad \text{and} \quad Z(s) = \frac{1}{Y(s)}.$$

Example 9.3. *The given two-port network has input voltage* $V_1(s)$ *and output voltage* $V_2(s)$. *Find the transfer functions* $G_{21}(s)$ *and* $Z_{21}(s)$ *and the driving impedance* $Z_{11}(s)$.

Solution: The voltage equations are:

$$V_1(s) = I_1(s) \cdot R + Ls \cdot I_1(s) \tag{1}$$

$$V_2(s) = I_1(s)Ls. \tag{2}$$

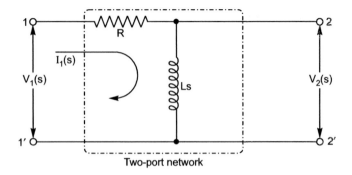

FIGURE 9.7

Hence,

$$G_{21}(s) = \frac{V_2(s)}{V_1(s)} = \frac{Ls \cdot I_1(s)}{(R + Ls)I_1(s)} = \frac{Ls}{Ls + R} = \frac{s}{s + \frac{R}{L}} \qquad (3)$$

$$Z_{21}(s) = \frac{V_2(s)}{I_1(s)} = \frac{I_1(s) \cdot Ls}{I_1(s)} = Ls \qquad (4)$$

$$Z_{11}(s) = \frac{V_1(s)}{I_1(s)} = \frac{I_1(s)(Ls + R)}{I_1(s)} = Ls + R. \qquad (5)$$

Example 9.4. *The given two-port network has input voltage $V_1(s)$ and out-put voltage $V_2(s)$. Find the transfer functions $G_{21}(s)$ and $Z_{21}(s)$ and the driving-point admittance $Y_{11}(s)$.*

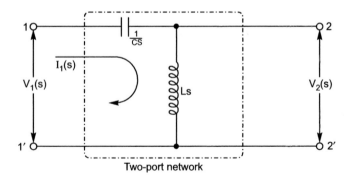

FIGURE 9.8

Solution: The voltage transform equations are:

$$V_1(s) = I_1(s)\left[\frac{1}{Cs} + Ls\right] \tag{1}$$

$$V_2(s) = I_1 s \cdot Ls. \tag{2}$$

Hence, the transfer voltage ratio is,

$$G_{21}(s) = \frac{V_2(s)}{V_1(s)} = \frac{Ls}{Ls + \frac{1}{Cs}} = \frac{s^2}{s^2 + \frac{1}{LC}} \tag{3}$$

The transfer impedance is

$$Z_{21}(s) = \frac{V_2(s)}{I_1(s)} = \frac{I_1(s) \cdot Ls}{I_1(s)} = Ls \tag{4}$$

$$Y_{11}(s) = \frac{I_1(s)}{V_1(s)} = \frac{I_1(s)}{I_1(s)[Ls + \frac{1}{Cs}]}$$

$$= \frac{1}{L} \cdot \frac{s}{s^2 + \frac{1}{LC}}. \tag{5}$$

Example 9.5. *For the given network, find the transfer functions* $G_{21}(s)$, $Z_{21}(s)$ *and driving-point impedance* $Z_{11}(s)$.

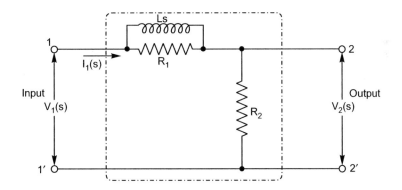

FIGURE 9.9

Solution: Transform impedances R_1 and Ls may be combined into an equivalent transform impedance $Z_{R_1L}(s)$ given by,

$$Z_{R_1L}(s) = \frac{1}{\frac{1}{R_1} + \frac{1}{Ls}} = \frac{R_1 Ls}{R_1 + Ls}. \tag{1}$$

Now

$$V_2(s) = I_1(s \cdot R_2) \tag{2}$$

and

$$V_1(s) = I_1(s)\left[R_2 + \frac{R_1 Ls}{R_1 + Ls}\right]. \tag{3}$$

Hence, the transfer functions are as below:

$$G_{21}(s) = \frac{V_2(s)}{V_1(s)} = \frac{R_2}{R_2 + \frac{R_1 Ls}{R_1 + Ls}} = \frac{Ls + R_1}{Ls(\frac{R_1 + R_2}{R_2}) + R_1}$$

or

$$G_{21}(s) = \frac{s + \frac{R_1}{L}}{s \cdot \frac{R_1 + R_2}{R_2} + \frac{R_1}{L}} \tag{4}$$

$$Z_{12}(s) = \frac{V_2(s)}{I_1(s)} = \frac{I_1(s)R_2}{I_1(s)} = R_2 \tag{5}$$

$$Z_{11}(s) = \frac{V_1(s)}{I_1(s)} = R_2 + \frac{R_1 Ls}{R_1 + Ls}$$

$$= \frac{s + \frac{R_1 R_2}{L(R_1 + R_2)}}{s + (\frac{R_1}{L})}(R_1 + R_2). \tag{6}$$

Example 9.6. *The given network is driven by a current source and is terminated by resistor R_2 at port 2. For this terminated two-port network, calculate (a) transfer functions $G_{21}(s), a_{21}(s), Z_{21}(s),$ and $Y_{21}(s)$ and (b) driving-point impedance $Z_{11}(s)$.*

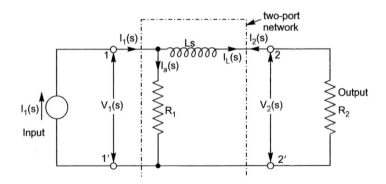

FIGURE 9.10

Solution: The given network is a current divider network. Hence, we write the following equations for the various currents involved in the network,

$$I_1(s) = I_a(s) + I_l(s) = V_1(s) \cdot \frac{1}{R_1} + V_1(s) + \frac{1}{Ls + R_2} \tag{1}$$

$$I_l(s) = \frac{\frac{1}{Ls+R_2}}{\frac{1}{R_1} + \frac{1}{Ls+R_2}} = I_1(s) \cdot \frac{R_1}{L} \cdot \frac{1}{s + \frac{R_1+R_2}{L}}. \tag{2}$$

But

$$I_2(s) = -I'_l(s). \tag{3}$$

Hence,

$$\alpha_{21}(s) = \frac{I_2(s)}{I_1(s)} = -\frac{R_1}{L} \cdot \frac{1}{s + \frac{R_1+R_2}{L}} \tag{4}$$

$$G_{21}(s) = \frac{V_2(s)}{V_1(s)} = \frac{I_l(s)R_2}{I_l(s)(R_2 + Ls)} = \frac{R_2}{L} \cdot \frac{1}{s + \frac{R_2}{L}} \tag{5}$$

$$Z_{21}(s) = \frac{V_2(s)}{I_1(s)} = \frac{I_l(s)R_2}{I_1(s)} = \frac{R_1R_2}{L} \cdot \frac{1}{s + \frac{R_1+R_2}{L}} \tag{6}$$

$$Y_{21}(s) = \frac{I_2(s)}{V_1(s)} = \frac{-I_l(s)}{I_l(s)[Ls + R_2]} = \frac{-1}{L} \cdot \frac{1}{s + \frac{R_2}{L}} \tag{7}$$

$$Z_{11}(s) = \frac{V_1(s)}{I_1(s)} = \frac{V_1(s)}{V_1(s)\left[\frac{1}{R_1} + \frac{1}{Ls+R_2}\right]}$$

$$= \frac{R_1}{L} \cdot \frac{s + \frac{R_2}{L}}{s + \frac{R_1+R_2}{L}}. \tag{8}$$

9.4 NETWORK FUNCTIONS AS A QUOTIENT OF POLYNOMIALS IN S

We find that the expression for network function may, in general, be put in the form of a ratio of polynomials in s. Thus,

$$\text{the network function} = \frac{P(s)}{Q(s)}$$

$$= \frac{a_0 s^n + a_1 s^{n-1} + a_2 s^{n-2} + \cdots + a_{n-1}s + a_n}{b_0 s^m + b_1 s^{m-1} + b_2 s^{m-2} + \cdots + b_{m-1}s + b_m}. \tag{9.8}$$

This ratio is a rational function of s when n and m are integers. In Equation (9.8), n is the degree of the polynomial in the numerator while m is the degree of the polynomial in the denominator,

$$I_k(s) = \frac{\Delta_{ik}}{\Delta} V_j(s). \tag{9.9}$$

Hence, the network function

$$\frac{I_k(s)}{V_j(s)} = \frac{\Delta_{jk}}{\Delta} = Z. \tag{9.10}$$

Hence, Δ is the determinant of impedance matrix $[Z]$ and is of order l, whereas Δ_{jk} is a cofactor of Δ with row j and column k deleted and is, therefore, a determinant of order $(l-1)$.

Now in determinant Δ, every element in its most general form is of the following nature:

$$Z_{jk} = R_{jk} + L_{jk} \cdot s + \frac{1}{C_{jk} \cdot s} \left[\begin{array}{l} Z_L = J\omega L = sL \\ Z_C = \frac{1}{J\omega C} = \frac{1}{sC} \end{array} \right]. \tag{9.11}$$

Hence, the expression for Δ_{jk} and Δ will contain products, sums, and quotients of terms of this type. Hence, any network function $I_k(s)/V_j(s)$, being the ratio of determinants Δ_{jk} and Δ, will always be in the form of a rational quotient of polynomials of the type given by Equation (9.8). This is true for all networks.

9.5 POLES AND ZEROS OF NETWORK FUNCTIONS

We have seen that every network function may be expressed in the form of a quotient of polynomials in s. Thus, let the network function be given by,

$$N(s) = \frac{P(s)}{Q(s)} = \frac{a_0 s^n + a_1 \cdot s^{n-1} + \cdots + a_{n-1} \cdot s + a_n}{b_0 s^m + b_1 \cdot s^{m-1} + \cdots + b_{m-1} \cdot s + b_m}, \tag{9.12}$$

where coefficients a_0 to a_n and b_0 to b_m are real and positive for a network containing passive elements only and containing no controlled sources. Now the polynomial in the numerator is of degree n. Hence, equation $P(s) = 0$ has n roots. Let these roots be $Z_1, Z_2, \ldots, Z_n$. Then numerator $P(s)$ may be written as the product of n linear factors $(s - Z_1), (s - Z_2), \ldots, (s - Z_n)$. Similarly, the polynomial in the denominator is of degree m. Hence, equation

$Q(s) = 0$ has m roots. Let these roots be $p_1, p_2, \ldots p_m$, then quantity $Q(s)$ may be written as the product of m linear factors $(s - p_1), (s - p_2), \ldots (s - p_m)$. Hence, Equation (9.12) giving the network function $N(s)$ may be written as,

$$N(s) = K\frac{(s - Z_1)(s - Z_2)\cdots(s - Z_n)}{(s - p_1)(s - p_2)\cdots(s - p_m)}, \tag{9.13}$$

where K is a constant called the *scale factor*.

From the study of Equation (9.13) we draw the following conclusions:

(*i*) When the variable s has value equal to any of the roots $Z_1, Z_2, \ldots, Z_n$, the network function $N(s)$ becomes zero. Hence, these complex frequencies $Z_1, Z_2, \ldots, Z_n$ are called the *zeros* of the network function or we can say roots of the numerator polynomials are known as zeros.

(*ii*) When the variable s has any of the values $p_1, p_2 \ldots p_m$, the network function $N(s)$ becomes infinite. Hence, these complex frequencies $p_1, p_2, \ldots, p_m$ are called the *poles* of the network function or we can say, the roots of the denominator polynomials are known as poles.

Factors $(s - Z_1), (s - Z_2)$, etc., in the numerator of Equation (9.13) giving network function $N(s)$ are called the *zero factors*, while factors $(s - p_1)$, $(s - p_2)$, etc., in the denominator of Equation (9.13) giving the network function $N(s)$ are called the *pole factors*.

From Equation (9.13) we conclude that any network function is completely known if we know its zeros, poles, and the scale factor K. Accordingly poles and zeros are very useful quantities in describing the behavior of any network.

If a pole or zero is not repeated, it is said to be *simple* or *distinct*. On the other hand, if a pole or zero is repeated once, it is said to be *double*. Similarly, if a pole or zero is repeated twice, it is said to be *triple*. In a general case, when r poles or zeros have the same value, that pole or zero is said to be of *multiplicity r*.

Poles and zeros of infinity and at zero. Multiple poles and zeros are taken to be present at s equal to *infinity*, the degree or multiplicity depending upon the values of n and m. In Equation (9.12) when $n > m$, the poles at infinity are of degree or multiplicity $(n - m)$, so that the total number of poles also becomes n. Out of these n poles, m poles exist at finite values of complex frequency s while $(n - m)$ poles exist at $s = \infty$. Similarly when $n < m$, the zeros at infinity are of degree or multiplicity $(m - n)$ so that the total number of zero also becomes m. Out of these m zeros, n zeros exist at finite values of s while $(m - n)$ zeros exist at $s = \infty$.

Total number of zeros equals total number of poles. For any rational network function, counting poles and zero at zero and infinity, in addition to the poles and zeros at finite values of s, the total number of zeros is equal to the total number of poles. Thus, consider the following network function,

$$N(s) = \frac{s(s+2)}{(s+3)(s+1+j1)(s+1-j1)}. \tag{9.14}$$

Here the network function has a zero at $s = 0$, and a zero at $s = -2$. Thus, in all it has 2 zeros. The function has a pole each at $s = -3, s = -1 - j1$, and $s = -1 + j1$. Thus, the number of poles equals 3. Hence, there is included a zero at infinity in order to make the number of zeros equal the number of poles.

Figure 9.11 shows the plot of poles and zeros of network function N(s) as given by Equation (9.14). Symbol O is generally used to indicate a zero while symbol X is used to indicate a pole.

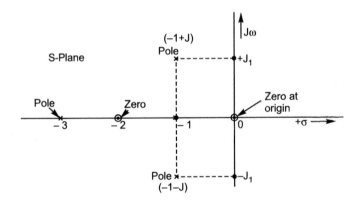

FIGURE 9.11 Location of poles and zeros in the network function of Equation (9.14).

Three-dimensional plot of a network function. At poles, a network function becomes infinite while at zeros, it becomes zero.

Accordingly poles and zeros constitute *critical frequencies*. At every other frequency, the network function has a finite non-zero value. Figure 9.12 shows a three-dimensional plot of the magnitude of the network function given by Equation (9.14) as a function of complex frequency $s(=\sigma + j\omega)$ for only one quadrant in the s-plane.

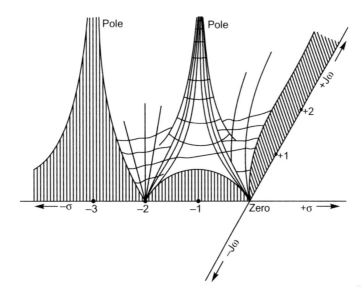

FIGURE 9.12   Magnitude of the network function of Equation (9.14) plotted against the complex frequency.

9.6   SIGNIFICANCE OF POLES AND ZEROS IN NETWORK FUNCTIONS

Poles and zeros provide considerable useful information in network functions. Thus consider the following cases:

(a) *Driving-point Impedance.* Driving-point impedance $Z_{11}(s)$ equals $V_1(s)/I_1(s)$. Hence, a pole of $Z_{11}(s)$ implies zero current for a finite value of driving voltage (*i.e.*, it signifies an open circuit). On the other hand, a zero of $Z_{11}(s)$ implies zero voltage $V_1(s)$ for a finite value of driving-point current $I_1(s)$ (*i.e.*, it signifies a *short circuit*).

(b) *Driving-point Admittance.* Driving-point admittance $Y_{11}(s)$ equals $I_1(s)/V_1(s)$. Hence, a pole of $Y_{11}(s)$ implies zero voltage $V_1(s)$ for a finite value of driving current (*i.e.*, it signifies a short circuit). On the other hand, a zero of $Y_{11}(s)$ implies zero current $I_1(s)$ for a finite value of driving voltage $V_1(s)$ (*i.e.*, it signifies an open circuit).

(c) *Voltage Transfer Ratio* $G_{21}(s)$. Voltage transfer ratio $G_{21}(s)$ is given by,

$$G_{21}(s) = \frac{V_2(s)}{V_1(s)}. \tag{9.15}$$

Equation (9.15) may be written as,

$$V_2(s) = V_1(s) \cdot G_{21}(s). \tag{9.16}$$

In most of the network problems, the input voltage $V_1(s)$ or $v_1(t)$ in the time domain is given and we are required to compute the output voltage or the response $V_2(s)$ or $v_2(t)$. In order to find $V_2(s)$, we must evaluate $G_{21}(s)V_1(s)$. This $G_{21}(s)$ may be computed knowing the network elements. The expression for $G_{21}(s), V_1(s)$, is obtained in the form of a ratio of polynomials in s. This expression may be expanded by partial fractions. Then the denominator of each partial fraction term gives a pole of either $G_{21}(s)$ or $V_1(s)$.

Thus, with no repeated roots, we have

$$G_{21}(s) \cdot V_1(s) = \sum_{j=1}^{u} \frac{K_j}{s - p_j} + \sum_{k=1}^{v} \frac{K_k}{s - p_k}, \tag{9.17}$$

where u and v are the number of poles of $G_{21}(s)$ and $V_1(s)$, respectively.

The inverse Laplace transformation Equation (9.17) yields,

$$v_2(t) = \mathcal{L}^{-1}[G_{21}(s) \cdot V_1(s)].$$

We conclude, therefore, that the poles *determine the time variation of the response*. Zeros on the other hand, *determine the magnitudes of coefficients* K_j *and* K_k *in the partial fraction expansion and thus determine the magnitude of the response.*

(*d*) *Other Network Functions*. The significance of poles and zeros in other transfer functions, namely $Y_{21}(s), Z_{21}(s)$, and $\alpha_{21}(s)$, is the same as discussed above for voltage transfer ratio $G_{21}(s)$. In each of these transfer functions, poles determine the time variation of the response whereas zeros determine the magnitudes.

9.7   RESTRICTIONS ON LOCATIONS OF POLES AND ZEROS IN DRIVING-POINT FUNCTIONS

The restrictions which get imposed on the location of poles and zeros in driving-point functions are discussed below.

(*i*) *Coefficients in the polynomials* P(*s*) *and* Q(*s*) *of* $N(s) = \frac{P(s)}{Q(s)}$ *must be real and positive.*

We saw earlier that any network function may be put in the form of a quotient of polynomials of s as below,

$$N(s) = \frac{P(s)}{Q(s)}$$

$$= \frac{a_0 s^n + a_1 s^{n-1} + \cdots + a_n}{b_0 s^m + b_1 s^{m-1} + \cdots + b_m}, \qquad (9.18)$$

where coefficients a's and b's are real positive. Hence, when s is real, $P(s)$ and $Q(s)$ will also be real. Such functions are called *real functions*.

(*ii*) *Poles and zeros, if complex, must occur in conjugate pairs.* Let polynomial $P(s)$ in Equation (9.19) be a real function. Now let one of the zeros of $P(s)$ be complex. Let it be equal to $(-a - jb)$. Then its conjugate, namely $(-a + jb)$, must also form a zero of the same polynomial $P(s)$, failing which $P(s)$ will have some complex coefficients. This may be established as below.

Let $P(s)$ contain only one pair of conjugate zeros, namely $(-a - jb)$ and $(-a + jb)$. Then $P(s)$ is given by

$$P(s) = (s + a + jb)(s + a - jb)$$

$$= (s + a)^2 + (b)^2 \qquad (9.19)$$

or

$$P(s) = s^3 + 2as + (a^2 + b^2). \qquad (9.20)$$

From Equation (9.20), it is evident that $P(s)$ is real since a and b are real.

Next consider two complex zeros which are not conjugate. Let these be $(-a - jb)$ and $(-c - jd)$. Then the product $(s + a + jb)(s + c + jd)$, where $a \neq c$ and $b \neq -d$, will not have all real coefficients. Hence, such zeros are inadmissible in network functions.

Thus, we conclude that in a real network function, the zeros must either be real or must occur in conjugate pairs.

(*iii*) *The real part of all poles and zeros must be either negative or zero.* From Equation (9.20) we further observe that in order to make $P(s)$ a real function, the second term, namely $2as$ must be positive. Hence, 'a' must be positive. Hence, the real part of all poles and zeros, namely $(-a)$, must be negative.

This restriction that the real part of the pole or zero must be negative or zero, arises from the assumption that any network constituted by passive elements only, must be stable (*i.e.*, the excitation due to any initial condition in any elements results in bounded output). By bounded output is meant an output which remains finite no matter how large time is made.

Thus, when a voltage source V(s) is applied to a one-port network containing passive elements only, the resulting response is a current I(s) at the input. Conversely, a current source I(s) as the excitation results in a voltage V(s) across the port as the response. The network must be stable with either excitation current I(s) or excitation voltage V(s). Hence, the conclusions drawn, for say zeros, will apply for poles as well and *vice versa*. Now let us consider the driving-point impedance Z(s). Let the denominator in the expression for Z(s) contain the factor $(s - s_p)$, where $s_p = \sigma_p + j\omega_p$. This factor $(s - s_p)$ will cause a time-domain response of the following form:

$$\text{Response} = K_p e^{S_p t}$$
$$= K_p \cdot e^{\sigma_p t} \cdot e^{j\omega_h t}. \tag{9.21}$$

Since the network function is real, the denominator must also contain factor $(s - s_p')$, where $s_p' = \sigma_p - j\omega_p$. The corresponding time-domain response is given by,

$$\text{Response} = K_p' \cdot e^{S_p' t}$$
$$= K_p' \cdot e^{\sigma_p t} \cdot e^{-j\omega_p t}. \tag{9.22}$$

Combining the two responses, the combined response is given by,

$$\text{Total response} = e^{\sigma_p t} \left[K_p \cdot e^{j\omega_p t} + K_p' \cdot e^{-j\omega_p t} \right] \tag{9.23}$$
$$= K_0 e^{\sigma_p t} \cdot \sin(\omega_p t + \theta_p). \tag{9.24}$$

From Equation (9.24) we see that for the response to be bounded, all that is necessary is to have $\sigma_p < 0$.

Thus, we conclude that for a driving-point impedance (or admittance) to be bounded, the poles (and zeros) should have negative or zero real parts. Thus, in the s-plane all poles and zeros must lie in the *left half of the s-plane* and none can lie in the *right half of the s-plane*. Of course, poles and zeros can lie on the boundary dividing the left half and the right half.

(*iv*) *If the real part of any pole or zero is zero, then that pole or zero must be simple (not repeated).* We have seen that the real part of a complex pole or zero must be either negative or zero. Thus, the real parts of poles and zeros may have zero value (*i.e.*, the poles and zeros may lie on the boundary dividing the left half of the s-plane and right of the s-plane). But the restriction offered is that such poles and zeros must be simple (not multiple or repeated). This restriction crops us because if the poles or zeros are of order n, then the time-domain response contains term t^{n-1} and this term becomes infinite as time t increases to infinity.

Multiple poles and zeros may, however, exist at locations in the left half of the s-plane. If there exist n such multiple poles at any location, then such poles result in response of the form $i^{n-1} \cdot e^{-\sigma t}$ which constitutes a bounded response since,

$$\lim_{t \to \infty} t^{n-1} . e^{-\sigma t} = 0 \qquad (9.25)$$

by L'Hospital's rule (for finite value of n).

(v) *Polynomial* P(s) *or* Q(s) *cannot have missing terms between those of highest and lowest order unless all even order terms or all odd order terms are missing.* From the earlier discussion we conclude that the polynomials P (s) and Q (s) in the expression for driving-point function are constituted by the products of the following factors: (i) constant k (ii) ($s + a$), where a is either zero or real and positive (iii) ($s^2 + bs + c$) and (iv) ($s^2 + d$) where b, c, and d are real and positive. Hence, polynomials P(s) and Q(s) formed by the product of three factors will also have all coefficients which are real and positive since the negative sign cannot appear through any mechanism. Further, since the negative sign cannot be introduced, there exists no chance of cancellation of a positive term with a negative term. Thus, there is no possibility of a coefficient becoming zero through the mechanism of cancellation of a positive term with a negative term. Thus, polynomial P(s) or Q(s) as given by Equation (9.18) has all coefficients *non-zero* except under the following two conditions:

(a) polynomials composed of factors like ($s^2 + d$) only. In that case the polynomial contains only even order terms and is free of odd order terms.

(b) polynomials having a simple (not multiple) zero at the origin and all other factors of the type ($s^2 + d$). In this case, polynomials formed by terms of type ($s^2 + d$) is an even polynomial but on multiplication by a single s, the polynomial becomes an odd polynomial (*i.e.*, coefficients of all even order are zero).

We thus conclude that in a polynomial P(s) or Q(s) representing a driving-point function, no coefficient may be zero (*i.e.*, no term in the polynomial may be missing unless all the even order or all the odd order terms are missing).

(vi) *The degrees of* P(s) *and* Q(s) *may differ by zero or one only.* Consider the behavior of a one-port network at very high and at very low frequencies,

$$m - n = \pm 1. \qquad (9.26)$$

(a) *At very high frequencies.* At a very high frequencies, the impedance of an inductor, $Z_L(s) = Ls$, is very large compared to the impedance of any other element and thus dominates the impedance function of the network. Hence, the impedance function of the network may be represented by equivalent inductance L_{eq} as shown in Figure 9.13(a). Impedance $Z_c(s) = \frac{1}{Cs}$ of C is extremely small and may be neglected. Hence, if the network does not contain L, then either the resistor will dominate the impedance function of the network or the equivalent network would be short circuit since impedance $\frac{1}{sC}$ of C approaches zero as frequency s approaches infinity. Thus, in this case, the impedance function may be represented by R_{eq} as shown in Figure 9.13(b), where R_{eq} may be finite or zero.

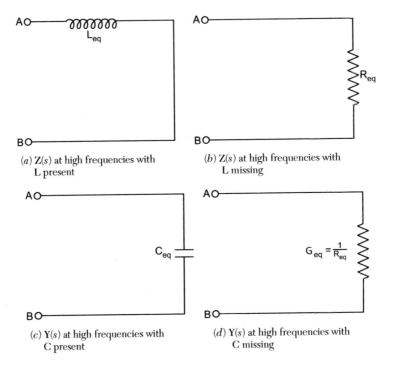

(a) Z(s) at high frequencies with
L present

(b) Z(s) at high frequencies with
L missing

(c) Y(s) at high frequencies with
C present

(d) Y(s) at high frequencies with
C missing

FIGURE 9.13   **High-frequency equivalent networks for a one-port network.**

At very high frequencies, the admittance of a capacitor $Y_c(s) = Cs$, is very large compared to admittance of any other element and will thus dominate the admittance function of the network. Hence, the admittance function of the network may be represented by equivalent capacitance C_{eq} as shown in Figure 9.13(c).

Admittance of L, $Y_L(s) = \frac{1}{Ls}$, is extremely small and may be neglected. Hence, if the network does not contain C, then either the resistor will dominate the admittance function of the network or the equivalent network would be open circuit, since the admittance $\frac{1}{Ls}$ of L approaches zero as frequency s

approaches infinity. Thus, in this case, the admittance function may be represented by G_{eq} as shown in Figure 9.13(d), where G_{eq} may be finite or zero.

We thus see that equivalent representation of a one-port network at high frequencies involves at most one kind of element and will assume one of the forms shown in Figure 9.13.

(b) *At low frequencies.* Conclusions similar to the above may be drawn for impedance and admittance functions of a one-port network at low frequencies. Thus, at low frequencies also, *i.e.*, when s approaches zero, the equivalent representation of a one-port network involves at most one kind of element.

(vii) *Requisite conditions for a driving-point voltage to assume equivalent representation of Figure 9.13.* We know explore the conditions under which the driving-point function of a general network assumes the equivalent representation of Figure 9.13.

From Equation 9.12, the value of the driving-point function $N(s)$ when $S \to \infty$ is given by,

$$\lim_{s \to \infty} N(s) = \lim_{s \to \infty} \frac{a_0 s^n + a_1 \cdot s^{n-1} + \cdots + a_{n-1} \cdot s + a_n}{b_0 s^m + b_1 \cdot s^{m-1} + \cdots + b_{m-1} \cdot s + a_m} \tag{9.27}$$

$$= \lim_{s \to \infty} \frac{a_0 \cdot s^n}{b_0 \cdot s^m} \tag{9.28}$$

$$= \lim_{s \to \infty} \frac{a_0}{b_0} \cdot s^{n-m}. \tag{9.29}$$

But we have already established that a one-port network assumes one of the three forms shown in Figure 9.13 at high frequencies (*i.e.*, when $s \to \infty$). In order that the driving-point function $N(s)$ of a general network with s tending to infinity assumes one of the three forms shown in Figure 9.13, it is necessary that $\lim_{s \to \infty} N(s)$ assumes one of the following three forms: s, 1, or 1/s.

From Equation (9.29) we see that this condition is satisfied when

$$|n - m| \geq 1 \tag{9.30}$$

or

$$(n - m) = -1, 0 \text{ or } 1. \tag{9.31}$$

Here n and m are the degrees of polynomials $P(s)$ and $Q(s)$, respectively.

Hence, we conclude that the degrees of the polynomials $P(s)$ in the numerator and $Q(s)$ in the denominator for a driving-point function may differ by one at most.

(viii) *Lowest degree terms in* $P(s)$ *and* $Q(s)$ *may differ in degree at most by one.* At low frequencies (*i.e.*, as s approaches zero) in $P(s)$ and $Q(s)$, the high degree terms such as s^n, s^{m-1}, s^m, s^{m-1}, etc., are small and may be neglected. Hence, network function $N(s)$ may be put as,

$$N(s) = \frac{\cdots + a_{n-1}s + a_n}{\cdots + b_{m-1}s + b_n}.$$ (9.32)

In order that the driving-point function $N(s)$ of a network, with s tending to zero assumes one of the four forms shown in Figure 9.13, it is necessary that $\lim_{s \to 0} N(s)$ assume one of the following 3 forms: s, 1, or $\frac{1}{s}$. This condition may be satisfied in only three possible ways. We will now explore these three possibilities. As s approaches zero, all high order terms except $a_{n-1}s$, a_n, $b_{m-1}s$, and b_m become zero.

Let $N(s)$ be a driving-point impedance. Then the three possibilities are as below.

(a) $a_n = 0$ and $b_m \neq 0$, then

$$\lim_{s \to \infty} N(s) = \lim_{s \to 0} \frac{a_{n-1}}{b_{m-1} + \frac{b_m}{s}}$$ (9.33)

$$= \lim_{s \to 0} \frac{a_{n-1}}{b_{m-1}} \cdot \frac{1}{1 + \frac{b_m}{b_{m-1} \cdot s}}$$

$$= \lim_{s \to 0} \frac{a_{n-1}}{b_m} s = Ls.$$ (9.34)

Thus, impedance $N(s)$ represents an inductor of value $\left(\frac{a_{n-1}}{b_m}\right)$.

(b) $b_m = 0$ and $a_n \neq 0$, then

$$\lim_{s \to 0} N(s) = \lim_{s \to 0} \frac{a_{n-1} \cdot s + a_n}{b_{m-1} \cdot s}$$ (9.35)

$$= \lim_{s \to 0} \frac{a_{n-1}}{b_{m-1}} \cdot \left[1 + \frac{n_n}{a_{n-1} \cdot s}\right]$$ (9.36)

$$= \lim_{s \to 0} \frac{a_n}{b_{m-1}} \cdot \frac{1}{s} = \frac{1}{sC}.$$ (9.37)

Thus, impedance $N(s)$ represents a capacitor of value $\left(\frac{b_{m-1}}{a_n}\right)$.

(c) $a_n \neq 0$ and $b_m \neq 0$, then

$$\lim_{s \to 0} N(s) = \lim_{s \to 0} \frac{a_{n-1} + \frac{a_n}{s}}{b_{m-1} + \frac{b_m}{s}}$$ (9.38)

$$= \frac{a_n}{b_m}.$$ (9.39)

Thus, driving-point impedance N(s) becomes a constant which represents the impedance of a resistor of value $\left(\frac{a_n}{b_m}\right)$.

Next, let N(s) be the driving-point admittance function.

Then the three possibilities are as below:

(a) $a_n = 0$ and $b_m \neq 0$, then in accordance with Equation (9.34), admittance N(s) when s approaches zero equals

$$\lim_{s \to 0} \frac{a_{n-1}}{b_m} s = Cs[Y_c = Cs].$$

This represents a capacitor of value $\left(\frac{a_{n-1}}{b_m}\right)$.

(b) $b_m = 0$ and $a_n \neq 0$, then in accordance with Equation (9.37), admittance N(s) when s approaches zero equals

$$\lim_{s \to 0} \frac{a_n}{b_{m-1}} \cdot \frac{1}{s} = \frac{1}{Ls}\left[Y_L = \frac{1}{Ls}\right].$$

This represents an inductor of value $\left(\frac{b_{m-1}}{a_n}\right)$.

(c) $a_n \neq 0$ and $b_m \neq 0$, then in accordance with Equation (9.38), N(s) represents a resistor of value $\left(\frac{b_m}{a_n}\right)$.

In case (a) above (i.e., when $a_n = 0$ and $b_m \neq 0$) the term of the lowest degree in P(s) is $a_{n-1} \cdot s$, while the term of the lowest degree in Q(s) is b_m. Thus the term of the lowest degree in the numerator is one degree higher than the term of the lowest degree in the denominator.

Similarly, we see that in case (b), the degree of the lowest degree term in the numerator is one lower than the degree of the lowest degree term in the denominator.

Finally in case (c), the terms of the lowest degree in the numerator and denominator are of the same degree.

We thus conclude that in any driving-point function N(s), the terms of the lowest degree in the numerator and denominator *may differ in degree at most by one*.

Items (i) to (vii) above give the necessary conditions for driving-point functions (with common factors in numerator P(s) and denominator Q(s) cancelled).

Example 9.7. *Verify whether the following expression for driving-point impedance Z(s) is suitable for representing a passive one-port network,*

$$Z(s) = \frac{s^4 + 2s^3 - 2s + 1}{s^3 + s^2 - 2s + 12}.$$

Solution: This function is not suitable for representing the impedance of a one-port network for the following reasons:

(a) In the numerator one coefficient is negative.
(b) In the denominator one coefficient is negative.
(c) In the numerator one coefficient is missing.
(d) The denominator on factorization equals $(s + 3)(s^2 - 2s + 4)$.

This has 2 poles of Z(s) in the right half of the s-plane.
This is known as the *inspection test*.

Example 9.8. *Verify whether the following expression for driving-point impedance is suitable for representing a passive one-port network,*

$$Z(s) = \frac{s^2 + s + 2}{s^4 + 5s^3 + 6s^2}.$$

Solution: This function is not suitable for representing the impedance of a one-port network for the following reasons:

(a) The degree of the numerator is 2 while that of the denominator is 4; a difference of 2 which is in excess of the maximum permitted difference of 1.
(b) The denominator equals $s^2(s + 2)(s + 3)$. Thus, there is a double pole at the origin in the s-plane which is not permitted.

Example 9.9. *Verify whether the following expression for the driving-point impedance Z(s) is suitable for representing a passive one-port network,*

$$Z(s) = \frac{s^2(s^2 - 2s + 6)}{(s + 3)}.$$

Solution: This function is not suitable for representing the driving-point impedance of a one-port network for the following reasons:

(a) The term of the lowest degree in the numerator is of degree 2 while that in the denominator is zero; a difference of 2 which is in excess of the maximum permitted difference of one.
(b) The numerator has a factor s^2 indicating a double zero at the origin in the s-plane which is not permitted.
(c) The factor $(s^2 - 2s + 6)$ yields two poles of Z(s) in the right half of the s-plane.
(d) The degree of numerator is 4 while that of denominator is one; a difference of 3 which is in excess of the maximum permitted difference of one.

9.8 RESTRICTIONS ON LOCATIONS OF POLES AND ZEROS FOR TRANSFER FUNCTIONS

Now we consider the restrictions on locations of poles and zeros for transfer functions Z_{21}, Y_{21}, G_{21}, and α_{21}. Some of these restrictions may be established using the same technique as used for driving-point functions. Here again we refer to the two-port network of Figure 9.2. Let 1-1′ be the input port and 2-2′ be the output port. We connect the energy source at port 1-1′ and record the output voltage or current at port 2-2′.

The following restrictions are then found to apply to poles and zeros of various transfer functions:

(*i*) *The real part of poles must be negative or zero (but simple, not repeated).* Now if the network is passive, application of a step voltage or current will result in an output response which is bounded. For the response to remain bounded we had earlier found that for a one-port network, the restriction is that the real part of the poles be negative or zero. In case this negative part is zero (including the origin), the poles must be simple (not multiple). The same restriction applies to the poles of the transfer function of a two-port network. For a one-port network, these restrictions apply to zeros also because for a one-port network $Y_{11}(s) = \frac{1}{Z_{11}(s)}$.

For a two-port network, however, in general,

$$Z_{12}(s) \neq \frac{1}{Y_{12}(s)} \quad \text{and} \quad G_{12} \neq \frac{1}{\alpha_{12}(s)}.$$

Hence, these restrictions do not apply to zeros of the transfer functions of a two-port network. We thus summarize that for a transfer function:

(*a*) the poles must lie in the left half of the complex s-plane including the boundary (if simple).

(*b*) the zeros may, however, lie in the right half plane provided they are conjugate if complex.

Transfer functions with zeros in the left half plane are called *minimum-phase transfer functions* while transfer functions with zeros in the right half plane are called *non-maximum phase transfer functions*.

(*ii*) *For $G_{21}(s)$ and $\alpha_{21}(s)$, the maximum degree of $P(s)$ equals the degree of $Q(s)$.* Next we establish the restriction regarding the difference in degrees $P(s)$ and $Q(s)$ for the transfer function $N(s) = P(s)/Q(s)$. We establish this for a simple network, namely, *T*- and π-networks shown in Figure 9.14. We use the *T*-network of Figure 9.14(*a*) to establish properties of $G_{21}(s)$ and $Z_{21}(s)$ and use the π-network of Figure 9.14(*b*) to establish properties of $\alpha_{21}(s)$ and $Y_{21}(s)$.

Now in Figure 9.14(a) with output terminals 2-2' open, Z_3 may be shorted without affecting G_{21}. Then for the T-network of Figure 9.14(a) for input voltage V_1 at port 1,

$$G_{21}(s) = \frac{V_2(s)}{V_1(s)} = \frac{Z_2(s)}{Z_1(s) + Z_2(s)}. \tag{9.40}$$

Similarly in Figure 9.14(b) with output terminals shorted $Y_3(s)$ may be opened without affecting $\alpha_{21}(s)$. Then with input current I_1 at port 1-1', we have

$$\alpha_{21}(s) = \frac{I_2(s)}{I_1(s)} = \frac{Y_1(s)}{Y_1(s) + Y_2(s)}. \tag{9.41}$$

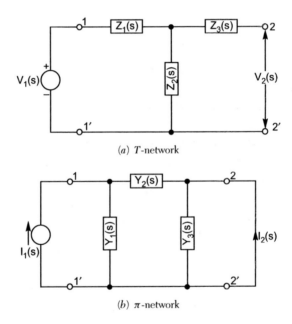

(a) T-network

(b) π-network

FIGURE 9.14   T- and π-network structures representing a general two-port network.

Equations (9.40) and (9.41) are duals. Hence, the results obtained for any one apply equally well to the other. For simplicity we deal with the T-network of Figure 9.14(a) and the results so obtained for $G_{21}(s)$ will then apply to $\alpha_{21}(s)$ as well.

In the last article we established that for a one-port network as s approaches infinity, the entire network reduces essentially to an equivalent network in which only one kind of elements dominated.

Hence, in the T-network of Figure 9.14(a) we consider Z_1, Z_2 and Z_3 to be constituted by one kind of element only. But there are only two sub-networks, namely, Z_1 and Z_2 in the T-network of Figure 9.14(a) with output open circuited. Hence, the following two possibilities exist:

(a) $Z_1(s)$ and $Z_2(s)$ are of the same kind. Then $G_{21}(s)$ becomes a constant and the network containing $Z_1(s)$ and $Z_2(s)$ serves simply as a potential divider independent of time. But since we are considering only the high-frequency circuit of general sub-networks $Z_1(s)$ and $Z_2(s)$ (i.e., equivalent network when s approaches infinity) the voltage ratio $G_{21}(s)$ becomes constant only under the condition that the degree of P(s) equals the degree of Q(s).

(b) $Z_1(s)$ and $Z_2(s)$ are different kinds. In this case it may be shown that, whatever be the nature of $Z_1(s)$ and $Z_2(s)$, whether it is an inductor, capacitor, or resistor the degree of the numerator is equal to or less than the degree of the denominator and it can never exceed the degree of the denominator.

The same is true for the current ratio $\alpha_{21}(s)$. Thus, we conclude that in the expression for the voltage transfer ratio $G_{21}(s)$ or the current transfer ratio $\alpha_{21}(s)$, the degree of the numerator may at the most equal the degree of the denominator.

(iii) For $Z_{21}(s)$ and $Y_{21}(s)$, the maximum degree of P(s) equals the degree of Q(s) plus one. In order to study the properties of $Z_{21}(s)$ we make use of the T-network of Figure 9.14(a) with a current source $I_1(s)$ connected to an input port 1-1' as shown in Figure 9.15(a). Similarly, in order to study the properties of $Y_{21}(s)$, we make use of the π-network of Figure 9.15(b) with a voltage source $V_1(s)$ connected to the input port 1-1' as shown in Figure 9.15(b).

Now in Figure 9.15(a) with output terminals 2-2' open, Z_3 may be shorted without affecting $Z_{12}(s)$. Then for the T-network of Figure 9.15(a) for the input current $I_1(s)$ at port 1, we get

$$Z_{21}(s) = \frac{V_2(s)}{I_1(s)} = \frac{Z_2(s) \cdot I_1(s)}{I_1(s)} = Z_2(s). \tag{9.42}$$

Similarly, in Figure 9.15(b) with output terminals 2-2' short circuited, $Y_3(s)$ may be opened without affecting $Y_{21}(s)$. Then for the π-network of Figure 9.15(b) with the input voltage $V_1(s)$ at port 1, we get

$$Y_{21}(s) = \frac{I_2(s)}{V_1(s)} = \frac{-V_1(s) \cdot Y_2(s)}{V_1(s)} = -Y_2(s). \tag{9.43}$$

Equations (9.42) and (9.41) are duals. Hence, the results obtained for any one, say $G_21(s)$, obtained using the T-network of Figure 9.15(a) apply equally well to $Y_{21}(s)$ obtained using the π-network of Figure 9.15(b).

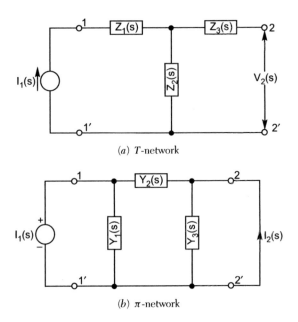

(a) T-network

(b) π-network

FIGURE 9.15   T- and π-network structures representing a general two-port network.

Now from Equation (9.42) we find that the value of impedance $Z_2(s)$ maximizes the degree of $P(s)$ in $Z_{21}(s)$ in impedance Ls of an inductor L. This is shown in Figure 9.16(a). Equation (9.42) then yields,

$$Z_{21}(s) - Z_2(s) = Ls. \tag{9.44}$$

Similarly from Equation (9.43) we find that the value of admittance $Y_2(s)$ which maximizes the degree of numerator $P(s)$ in $Y_{21}(s)$ is admittance Cs of a capacitor C. This is shown in Figure 9.16(b). Equation (9.43) then yields,

$$Y_{21}(s) = -Y_{12}(s) = -Cs. \tag{9.45}$$

Equations (9.44) and‘(9.45) represent $Z_{21}(s)$ and $Y_{21}(s)$, respectively, as s approaches infinity. Hence, we conclude that for $Z_{21}(s)$ and $Y_{21}(s)$, the maximum degree of $P(s)$ equals the degree of $Q(s)$ plus one.

(iv) *The degree of* $P(s)$ *may be as small as zero and independent of the degree of* $Q(s)$. Again, we refer to Equations (9.42) and (9.43) giving the expressions for $Z_{21}(s)$ and $Y_{21}(s)$ for a *T*-network and π-network equivalent circuits for a general two-port network.

From Equation (9.42) we find that the value of impedance $Z_2(s)$ which minimizes the degree of numerator $P(s)$ in the expression for $Z_{21}(s)$ is the impedance R of a resistor.

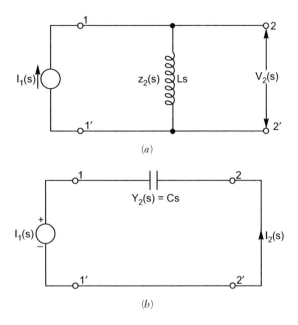

(a)

(b)

FIGURE 9.16 **Special cases of a two-port network which maximize the degree of $P(s)$ in transfer functions $Z_{21}(s)$ and Y_{21} (s).**

Thus,

$$Z_{21}(s) = R. \tag{9.46}$$

Similarly, from Equation (9.43), we find that the value of admittance $Y_{21}(s)$, which minimizes the degree of numerator $P(s)$ in the expression for $Y_{21}(s)$, is admittance $\frac{1}{R}$ of resistor R.

Thus, the degree of numerator $P(s)$ in the expression for transfer function $Z_{21}(s)$ and $Y_{21}(s)$ is zero irrespective of the degree of denominator $Q(s)$.

It may be shown that the same is true for other transfer functions, namely, $G_{21}(s)$ and $\alpha_{21}(s)$.

(v) Polynomial $P(s)$ may have terms missing between the terms of lowest and highest degree and flow of the coefficients may be negative. In order to establish this, we consider the symmetrical lattice network of Figure 9.17 as representing the general two-port network.

We apply current $I_1(s)$ at the input port 1. Then from Figure 9.17(*b*), we get

$$V_2(s) = V_1(s) \left[\frac{Z_b(s)}{Z_a(s) + Z_b(s)} - \frac{Z_a(s)}{Z_a(s) + Z_b(s)} \right]. \tag{9.47}$$

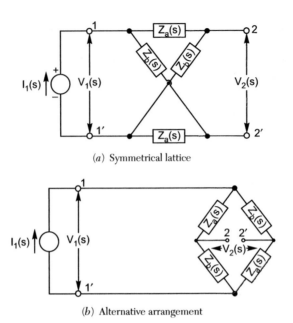

(a) Symmetrical lattice

(b) Alternative arrangement

FIGURE 9.17 **Symmetrical lattice representing a general two-port network.**

But

$$V_1(s) = I_1(s) \cdot \frac{1}{2}[Z_a(s) + Z_b(s)]. \tag{9.48}$$

Hence, from Equations (9.47) and (9.48), we get

$$Z_{21}(s) = \frac{V_2(s)}{I_1(s)} = \frac{1}{2}[Z_b(s) - Z_a(s)]. \tag{9.49}$$

Now each of the sub-networks $Z_a(s)$ and $Z_b(s)$ may be expressed as a ratio of polynomials of s. Thus, let

$$Z_a(s) = \frac{n_a}{d_a} \tag{9.50}$$

and

$$Z_b(s) = \frac{n_b}{d_b}. \tag{9.51}$$

Then

$$
\begin{aligned}
Z_{21}(s) &= \frac{1}{2}\left[\frac{n_b}{d_b} - \frac{n_a}{d_a}\right] \\
&= \frac{1}{2}\left[\frac{n_b \cdot d_a - n_a \cdot d_b}{d_a \cdot d_b}\right].
\end{aligned} \tag{9.52}
$$

From Equation (9.52) we conclude that the lowest possible degree in the numerator of $Z_{21}(s)$ may be zero. This situation corresponds to cancellation of all terms in s leaving only a constant term. This is the same property as (iv) established earlier using the T-network form of the equivalent circuit.

We also see from Equation (9.52) that in the numerator polynomial of $Z_{21}(s)$, any term may be zero or negative. Similar remarks apply to other transfer functions.

(vi) *The coefficients in the polynomial* $P(s)$ *and* $Q(s)$ *of* $N(s) = P(s)/Q(s)$ *must be real and those of* $Q(s)$ *must be positive.* In similarity with the driving-point functions the coefficients in the polynomials $P(s)$ and $Q(s)$ for transfer function $N(s)$ must be real. Further, the coefficients in denominator polynomial $Q(s)$ must be positive. However, the numerator polynomial $P(s)$ may have negative coefficients as seen from Equation (9.52).

(vii) *Poles and zeros must be conjugate if imaginary or complex.* This restriction for transfer functions is the same as in the case of driving-point functions.

$(viii)$ *The polynomial* $Q(s)$ *may not have any missing terms between the lowest and highest degree terms unless all even or all odd terms are missing.* This restriction on denominator polynomial $Q(s)$ is the same as in the case of the driving-point functions.

The restrictions on the location of poles and zeros for transfer functions established above are summarized in Table 9.1.

TABLE 9.1   Restrictions on Pole and Zero Location in Transfer Functions (With Common Factors in $P(s)$ and $Q(s)$ cancelled)

(i)	(a) The coefficients in the polynomials $P(s)$ and $Q(s)$ of $N(s) = P(s)/Q(s)$ must be real.
	(b) Coefficients in $Q(s)$ must be positive but some of the coefficients in $P(s)$ may be negative.
(ii)	Poles and zeros, if imaginary or complex, must be conjugate.
(iii)	(a) The real parts of poles must be zero or negative.
	(b) If the real part is zero (including the origin), then that pole must be simple (not multiple).
(iv)	The polynomial $Q(s)$ may not have any missing terms between the highest and the lowest degree terms, unless all even or all odd terms are missing.
(v)	The polynomial $P(s)$ may have terms missing between the lowest and highest degree terms.
(vi)	The degree of $P(s)$ may be as small as zero, independent of the degree of $Q(s)$.
(vii)	For $G_{21}(s)$ and $\alpha_{21}(s)$, the maximum degree of $P(s)$ equals the degree of $Q(s)$.
$(viii)$	For $Z_{21}(s)$ and $Y_{21}(s)$, the maximum degree of $P(s)$ equals the degree of $Q(s)$ plus one.

9.9   TIME-DOMAIN RESPONSE FROM A POLE AND ZERO PLOT

The pole and zero plot of a given network function gives useful information about the critical frequencies at which the network function is abnormal (*i.e.*, either zero or infinite). At the same time, the time-domain response of the given network may be assessed from the pole and zero plot of a network function and from the knowledge of the network sources. Thus, let a driving voltage V(s) be applied to a network having driving-point admittance Y(s). Then the resulting transform current I(s) is given by,

$$I(s) = Y(s) \cdot V(s) = \frac{P(s)}{Q(s)}. \tag{9.53}$$

The ratio $P(s)/Q(s)$ may be put as,

$$\frac{P(s)}{Q(s)} = K \cdot \frac{(s - z_1)(s - z_2) \cdots (s - z_n)}{(s - p_1)(s - p_2) \cdots (s - p_m)}. \tag{9.54}$$

We saw earlier that the poles of this function determine the time-domain behavior of the current $i(t)$. Further, the zeros together with the poles and the scale factor K determine the magnitude of each of the terms of $i(t)$. This is elaborated on here.

Let the damping ratio be ζ and the undamped natural frequency be ω_n for the given circuit. Then the poles and zeros of Equation (9.54) expressed in terms of ζ and ω_n will have the following forms for different value of ζ,

$$S_1 S_2 = -\zeta \omega_n \pm j\omega_n \sqrt{1 - \zeta^2} \qquad \text{for } \zeta < 1 \tag{9.55}$$

$$S_1, S_2 = -\zeta \omega_n \pm \omega_n \sqrt{\zeta^2 - 1} \qquad \text{for } \zeta > 1 \tag{9.56}$$

$$S_1, S_2 = -\omega_n \qquad \text{for } \zeta = 1 \tag{9.57}$$

$$S_1, S_2 = \pm j\omega_n \qquad \text{for } \zeta = 0. \tag{9.58}$$

Equation (9.55) gives the location of poles in the s-plane. For these poles, the time-domain response in terms of ζ and ω_n is given by the following equation:

$$i(t) = K_1 \cdot e^{\left(-\zeta \omega_n + j\omega_n \sqrt{1 - \zeta^2}\right)t} + K_2 \cdot e^{\left(-\zeta \omega_n - j\omega_n \sqrt{1 - \zeta^2}\right)t} \tag{9.59}$$

$$= K_4 \cdot e^{-\zeta \omega_n^t} \cdot \sin \omega_n \sqrt{1 - \zeta^2} t. \tag{9.60}$$

We also know that the contours of root s on the complex s-plane have the following properties for different values of parameters.

(i) For constant ω_n, the contours are circles of radius ω_n as shown in Figure 9.18(a).

(ii) For constant damping ratio, ζ, contours are straight lines through the origin at angle $\theta = \cos^{-1}\zeta$ with a negative real axis as shown in Figure 9.18(b).

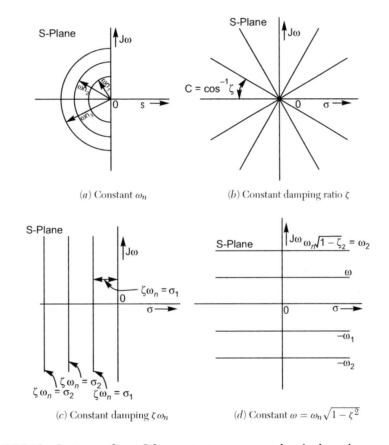

(a) Constant ω_n

(b) Constant damping ratio ζ

(c) Constant damping $\zeta\omega_n$

(d) Constant $\omega = \omega_n\sqrt{1-\zeta^2}$

FIGURE 9.18 **Contours of root S for constant parameter values in the s-plane.**

(iii) For constant damping $\zeta\omega_n$, contours are straight lines parallel to the $j\omega$ axis in the s-plane as shown in Figure 9.18(c).

(iv) For constant frequency $\omega_n\sqrt{1-\zeta^2}$, contours are straight lines parallel to the real axis in the s-plane as shown in Figure 9.18(d). $\omega_n\sqrt{1-\zeta^2}$ is known as damping frequency.

In order to grasp the significance of these contours, let us consider the poles shown in Figure 9.13. No zeros are included here for the sake of clarity. These sets of poles are consistent with the restrictions on the location of poles and zeros. Thus, poles p_1 and p_1' constitute one conjugate pair. Similarly, poles p_2 and p_2' constitute another pair, and p_3 and p_4 form the remaining poles.

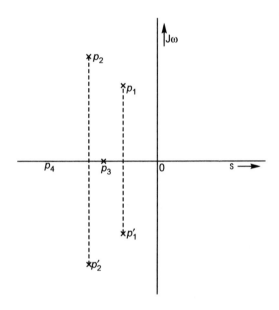

FIGURE 9.19   An array of poles.

Now the pair of poles, p_1 and p_1', signify the time-domain response of the type given by Equation (9.60). This response is in the form of a damped sinusoidal oscillation of frequency $\omega_n\sqrt{1-\zeta^2}$ which is the height of pole p_1 from σ-axis. Similarly, the pair of poles, p_2 and p_2' signify the time-domain response in the form of another damped oscillation of frequency equal to the height of the pole p_2 above the σ-axis. Evidently for the locations for poles chosen in Figure 9.19, the frequency of oscillation corresponding to pole pair $p_1 - p_1'$ is lower than that corresponding to pole pair $p_2 - p_2'$. The damping (*i.e.*, the rate of decrease of amplitude) is given by $e^{-\zeta\omega_n}$ where $\zeta\omega_n$ (or σ) is the distance of the pole from the $j\omega$ axis. Thus, the rate of damping for pole pair $p_2 - p_2'$ is more than that for pole pair $p_1 - p_1'$. The natural frequency ω_n, being equal to the radial distance of the pole from the origin in the s-plane, is more for pole pair $p_2 - p_2'$ than for pole pair $p_1 - p_1'$. The actual frequency of oscillation $\omega \, (= \omega_n \sqrt{1 - \zeta^2})$ depends both on ω_n and the damping ratio ζ and equals the distance of the pole from the σ-axis.

Poles p_3 and p_4 correspond to the zero value of frequency $\omega_n \sqrt{1 - \zeta^2}$ in Equation (9.60) and each results in the time-domain response in the form of exponential decay of the form $e^{-\sigma t} (= e^{-\zeta \omega_n t})$. Pole p_4 has a larger value of negative σ than pole p_3. Hence, damping for pole p_4 is greater than that for pole p_3. Stated otherwise pole p_3 has a greater time constant than pole p_4.

The time-domain response of each of the four poles (or pole pairs) for the poles of Figure 9.19 is shown in Figure 9.20 for the arbitrary amplitude of each response.

The total time-domain response corresponding to the poles in Figure 9.19 is obtained by adding these individual resonses. Thus, the total response is,

$$i(t) = K1 \cdot e^{p_1 t} + K_2 \cdot e^{p_1' t} + K_3 \cdot e^{p_2' t} + K_4 \cdot e^{p_2' t} + K_5 \cdot e^{p_3 t} + K_6 \cdot e^{p_4 t}.$$
(9.61)

The first two terms form a conjugate pair and these two together give a damped sinusoidal response. Similarly, the third and fourth terms combined together give another damped sinusoidal response.

Poles lying on the σ-axis result in an exponential response. As the pole moves more and more along the $-\sigma$ axis, the response becomes more and more damped (*i.e.*, the time constant of the exponential response goes on decreasing as shown by responses (I), (II), and (III) shown in Figure 9.21). For the pole at the origin in the s-plane, the response is a step function as shown.

As the pole moves along the $j\omega$ axis (and the conjugate pole moves along the $-j\omega$ axis), the frequency of the sinusoidal response goes on increasing. Thus, for poles (IV), (VII), and (X) located on the $+j\omega$ axis, the frequency increases progressively as we move from (IV) through (VII) to (X). But since $\sigma = 0$, the amplitude of response remains time invariant. For other locations of poles, namely (V), (VI), (VIII), (IX) (XI), and (XII), the response is a damped sinusoid, and the order of damping and the frequency of sinusoid depending on the location of the pole.

Magnitude of terms in time-domain response. Equation (9.54) gives the network function $I(s)$ in the frequency domain. In order to find the corresponding time-domain response, we expand the expression on the right-hand side of Equation (9.54) by partial fraction. Thus, we get,

$$I(s) = \frac{K_1}{s - p_1} + \frac{K_2}{s - p_2} + \cdots + \frac{K_m}{s - p_m}.$$
(9.62)

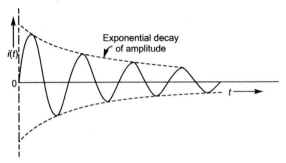

(a) Time-domain response for pole pair $p_1|p_1'$.
(Frequency $\omega_n\sqrt{1-\zeta^2}$ lower damping $\zeta\omega_n$ lower.)

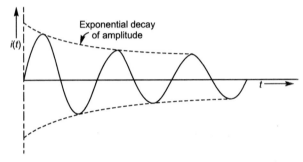

(b) Time-domain response for pole pair $p_2 - p_2'$.
(Frequency $\omega_n\sqrt{1-\zeta^2}$ higher Damping $\zeta\omega_n$ higher.)

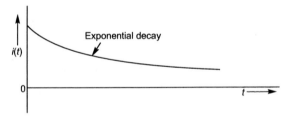

(c) Time-domain response for pole p_3.
(Damping $\sigma(=\zeta\omega_n)$ lower.)

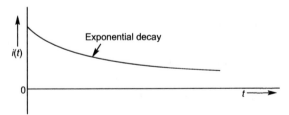

(d) Time-domain response for pole p_4.
(Damping $\sigma(=\zeta\omega_n)$ higher.)

FIGURE 9.20   Time-domain response corresponding to poles (or pole pairs) of Figure 9.19.

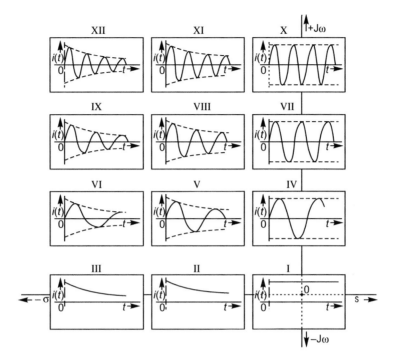

FIGURE 9.21   **Time-domain response as dependent on the location of the pole.**

Any of the coefficients may be found by the Heaviside method. Thus, the coefficient K_l of the lth factor is given by,

$$K_l = K \left[\frac{(s - z_1)(s - z_2) \cdots (s - z_n)}{(s - p_1)(s - p_2) \cdots (s - p_m)} \cdot (s - p_l) \right]_{s = +p_l}. \tag{9.63}$$

On substituting p_l for s in Equation (9.63), we get

$$K_l = K \frac{(p_l - z_1)(p_l - z_2) \cdots (p_l - z_n)}{(p_l - p_1)(p_l - p_2) \cdots (p_l - z_m)}. \tag{9.64}$$

This equation consists of factors of the form $(p_l - z_x)$ and $(p_i - p_y)$, where $p_l, p_y,$ and Z_x are all known complex numbers. Now $(p_l - z_x)$ being the difference of two complex numbers is another complex number and may be written as,

$$(p_l - z_x) = M_{xl} \cdot e^{j\phi}xl, \tag{9.65}$$

where M_{xl} is the magnitude of the phasor $(p_l - z_x)$ and ϕ_{xl} is the phase angle of the same phasor. Similarly, each factor $(p_l - z_y)$ in the denominator of Equation (9.64) being a difference of two complex numbers is another complex number and may be written as,

$$(p_l - p_y) = Q_{yl} \cdot e^{j\theta} yl, \tag{9.66}$$

where Q_{yl} is the magnitude of the phasor $(p_l - p_y)$ and θ_{yl} is the phase angle of the same phasor.

Figure 9.22 shows the difference of two complex quantities p_l and Z_x in the s-plane. It may be readily seen that the difference $(p_l - z_x)$ is nothing but a phasor AB directed from Z_x to p_l. The magnitude M_{xl} of this phasor is the distance from Z_x to p_l, while the phase along ϕ_{xl} is the angle of the line from Z_x to p_l measured with respect to the $\sigma = 0$ line. Similar remarks apply to phasor Q_{yl}.

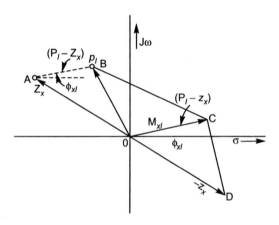

FIGURE 9.22 Subtraction of quantity z_x from p_l.

Thus each factor in Equation (9.64) may be expressed in terms of its magnitude and phase angle. Hence, Equation (9.64) may be written as,

$$K_l = K \cdot \frac{M_{1l} \cdot M_{2l} \cdot M_{3l} \cdots M_{nl}}{Q_{1l} \cdot Q_{2l} \cdot Q_{3l} \cdots Q_{ml}} \tag{9.67}$$
$$xe^{j}[(\phi_{1l} + \phi_{2l} + \cdots + \phi_{nl}) - (\theta_{1l} + \theta_{2l} + \cdots + \theta_{ml})].$$

Equation (9.67) gives the magnitude and phase angle of the coefficient K_l. Thus, all the coefficients from K_1 to K_m may be evaluated by this technique.

Magnitudes $M_{1l}, M_{2l} \ldots$ and $Q_{1l}, Q_{2l} \ldots$ may be easily measured from a plot of poles and zeros on the s-plane.

Similarly, the phase angles $\phi_{1l}, \phi_{2l} \ldots$ and $\theta_{1l}, \theta_{2l} \ldots$ may also be measured from the plot of poles and zeros on the s-plane.

Thus, the method of evaluating a network function say $I(s)$ as given by Equation (9.62) using the poles and zeros consists of the following steps:

(*i*)   Plot to scale on the s-plane the poles and zeros of the network function $I(s) = P(s)/Q(x)$.

(*ii*) Measure or calculate the distances $Q_{1l}, Q_{2l}, \ldots Q_{ml}$ of a given pole, say p_l, from each of the other finite poles.

(*iii*) Similarly, measure or calculate the distances $M_{1l}, M_{2l}, \ldots M_{nl}$ of the same pole p_l from each of the zeros.

(*iv*) Measure or calculate the angle of the line joining that pole p_l to each of the other finite poles and the zeros, giving

$$\theta_{1l}, \theta_{2l}, \ldots, \theta_{ml} \quad \text{and} \quad \phi_{1l}, \phi_{2l}, \ldots, \phi_{nl}.$$

(*v*)   Substitute these quantities in Equation (9.67) to evaluate K_l.

(*vi*) Similarly, evaluate each of the coefficients from K_1 to K_m and substitute these values in Equation (9.62) to evaluate the given network function $I(s)$.

Example 9.10. *Let the transform current* $I(s)$ *in a network be given by the following equation*

$$I(s) = \frac{2s}{(s+1)(s+2)}.$$

Plot the poles and zeros in the s-plane and hence obtain the time-domain response.

Solution: Evidently the function has poles at -1 and -2 and a zero at the origin. The plot of poles and zeros is as shown in Figure 9.23.

Zeros $s = 0$
Poles $s = -1, -2$.

The function can be easily expanded by partial fractions and then the time-domain response $i(t)$ may be found. But the time-domain value $i(t)$ can also be evaluated using the plot of poles and zeros on the s-plane.

Since factors $(s+1)$ and $(s+2)$ appear in the denominator polynomial, the time-domain response will be,

$$i(t) = K_1 \cdot e^{-t} + K_2 \cdot e^{-2t}. \tag{1}$$

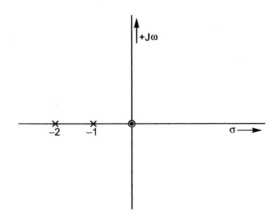

FIGURE 9.23

The coefficients K_1 and K_2 may be evaluated from the pole-zero plot. Thus, for the plot of poles and zeros as given in Figure 9.23, we get

$$M_{01} = 1 \quad \text{and} \quad \phi_{01} = 180°$$
$$Q_{21} = 1 \quad \text{and} \quad \theta_{21} = 0°.$$

Hence,

$$K_1 = F \cdot \frac{M_{01} \cdot e^{j\phi_{01}}}{Q_{21} \cdot e^{j\theta_{21}}} = 2 \cdot \frac{e^{j180°}}{e^{j0°}}$$
$$= 2e^{j180°} = -2.$$

Similarly,

$$M_{02} = 2 \quad \text{and} \quad \phi_{02} = 180°$$
$$Q_{12} = 1 \quad \text{and} \quad \theta_{12} = 180°.$$
$$K_2 = F \cdot \frac{M_{02} \cdot e^{j\phi_{02}}}{Q_{12} \cdot e^{j\theta_{12}}} = 2\frac{2e^{j180°}}{1e^{j180°}} = 4.$$

Substituting the values of coefficients K_1 and K_2 into Equation (1), we get

$$i(l) = -2e^{-t} + 4 \cdot e^{-2t}.$$

9.10 ROUTH CRITERION

The Routh stability criterion provides a convenient method of determining a control system's stability. It determines the number of characteristic roots within the unstable right half of the s-plane, and the number of characteristic roots in the stable left half, and the number of roots on the imaginary axis. It does not locate the roots. The method may also be used to establish limiting values for a variable factor beyond which the system would become unstable.

The characteristic equation being tested for stability is generally of the form

$$a_n s^n + a_{n-1} s^{n-1} + \cdots a_1 s + a_0 = 0$$

The Routh array is constructed as follows

$$
\begin{array}{cccc}
s^n & a_n & a_{n-2} & a_{n-4} \\
s^{n-1} & a_{n-1} & a_{n-3} & a_{n-5} \\
s^{n-2} & b_1 & b_2 & b_3 \\
s^{n-3} & c_1 & c_2 & c_3 \\
\vdots & \vdots & \vdots & \vdots \\
s^1 & Y_1 & Y_2 & \\
s^0 & Z_1 & & \\
\end{array}
$$

$$b_1 = (a_{n-1} \cdot a_{n-2} - a_n \cdot a_{n-3})/a_{n-1}$$
$$b_2 = (a_{n-1} \cdot a_{n-4} - a_n \cdot a_{n-5})/a_{n-1}.$$

The numerator in each case is formed from the elements in the two rows above in column 1 (pivot column) and in the column to the right of the element being calculated. The calculated element is made 0 if the row is too short to complete the calculation. The last row will have just one element,

$$c_1 = (b_1 \cdot a_{n-3} - a_{n-1} \cdot b_2)/b_1 \quad c_2 = (b_1 \cdot a_{n-5} - a_{n-1} \cdot b_3)/b_1 \; c_3 = 0.$$

ROUTH STABILITY CRITERIA

For system stability, the primary requirement is that all of the roots of the characteristic equation have negative real parts.

All of the roots of the characteristic equation have negative real parts only if the elements in column 1 of the Routh array are the same sign.

The number of sign changes in column 1 is equal to the number of roots of the characteristic equation with positive real parts.

Example 9.11. *Test the stability of a system having a characteristic equation,*

$$F(s) = s^3 + 6s^2 + 12s^3 + 8 = 0.$$

Solution: The Routh array is constructed as follows:

$$
\begin{array}{cccc}
s^3 & 1 & 12 & 0 \\
s^2 & 6 & 8 & 0 \\
s^1 & \dfrac{64}{6} & 0 & \\
s^0 & 0 & &
\end{array}
$$

Column 1 (pivot column) includes no changes of sign and therefore the roots of the characteristic equation have only real parts and the system is stable.

SPECIAL CASES FOR RESOLVING THE ROUTH STABILITY ARRAY

1. If a zero appears in the first column 1 of any row, marginal stability or instability is indicated. The normal method of constructing the array cannot be continued because the divisor would be zero. A convenient method of resolving this method is to simply replace the zero by a small number ε and continue as normal. The limit, $\varepsilon > 0$, is then determined and the first column is checked for sign changes.

Example 9.12. *Test the stability of a system having a characteristic equation,*

$$F(s) = s^5 + 2s^4 + 2s^3 + 4s^2 + s + 1.$$

The Routh array is constructed as follows

$$
\begin{array}{cccc}
s^5 & 1 & 2 & 1 \\
s^4 & 2 & 4 & 1 \\
s^3 & \varepsilon & 0.5 & \\
s^2 & -\dfrac{1}{\varepsilon} & 1 & \\
s^1 & 0.5 & & \\
s^0 & 1 & &
\end{array}
$$

There are two sign changes in column 1 and there are therefore 2 positive roots and the system is unstable.

2. If all of the elements in a row are zero (two rows are proportional), this indicates that the characteristic polynomial is divided exactly by the polynomial one row above the all zero row (always an even-ordered polynomial). Call this polynomial $N(s)$.

This also indicates the presence of a divisor polynomial N(s) whose roots are all symmetrically arranged about the origin, *i.e.*, they are of the form

$$s = \pm \alpha \ldots \quad \text{or} \quad s = \pm J\omega \quad \text{or} \quad s = -\alpha + J\omega \quad \text{and} \quad s = +\alpha + J\omega.$$

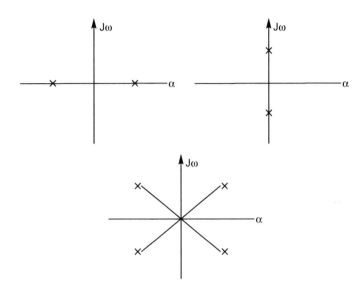

FIGURE 9.24

An all zero row will always be associated with an odd power of s.

In order to complete the array the previous row is differentiated with respect to s and the array is completed in the normal way.

When assessing this modified array the number of sign changes in the first column (before the all zero row) indicates the number of roots of the remainder polynomial with positive real parts. From the all zero row down, each change of sign in column 1 indicates the number of roots in the divisor polynomial with positive real roots and as the roots are symmetrical this would indicate the number of roots in the right half of the s-plane and the number of roots in the left-half of the s-plane. Roots not accounted for in this way (*i.e.*, no sign change) must lie on the imaginary axis.

Example 9.13. *Consider a closed loop control system with negative feedback which has an open loop transfer function,*

$$G(s) = K/s(s + 1) \cdot (s^2 + s + 1).$$

Solution: The closed loop characteristic equation $= F(s) = s^4 + 2s^3 + 2s^2 + s + K = 0$.

The Routh array is constructed as follows:

$$
\begin{array}{cccc}
s^4 & 1 & 2 & K \\
s^3 & 2 & 1 & \\
s^2 & \dfrac{3}{2} & K & \\
s^1 & \dfrac{\left(\frac{3}{2} - 2K\right)}{\frac{3}{2}} & & \\
s^0 & K. & &
\end{array}
$$

In this array row 3 becomes an all zero row if $K = \frac{3}{4}$ and the divisor polynomial of row 2 $= \left(\frac{3}{2}\right)s^2 + \left(\frac{3}{4}\right) = 0 = 2s^2 + 1$.

By dividing $F(s)$ by $(2s^2 + 1)$ the equation is obtained

$$
F(s) = \left(\frac{1}{2}s^2 + s + \frac{3}{4}\right)N(s) = (2s^2 + 1).
$$

The array is completed when $K = \frac{3}{4}$ by differentiating $N(s)$ with respect to s.

The coefficients of $N'(s)$ are used to replace the zero coefficients in row 3,

$$
\begin{array}{cccc}
s^4 & 1 & 2 & K \\
s^3 & 2 & 1 & \\
s^2 & \dfrac{3}{2} & \dfrac{3}{4} & \\
s^1 & 4 & & \\
s^0 & \dfrac{3}{4}. & &
\end{array}
$$

There are no sign changes up to/including row 2 indicating the roots of the remainder polynomial are in the right half of the plane.

As there are no changes of sign from row 2 down the roots of the divisor polynomial must lie on the imaginary axis.

To locate these roots set s in $N(s)$ to $j\omega$, i.e., $2(j\omega)^2 + 1 = 0$, therefore

$$
\omega = \frac{1}{\sqrt{2}} \text{ rads/unit time.}
$$

It is sometimes required to find a range of values of a parameter for which the system is stable. This can be achieved by use of the Routh criteria using the method illustrated by the following example.

The system characteristic equation $= F(s) = s^3 + 3s^2 + 3s + K = 0$

$$
\begin{array}{cccc}
s^3 & 1 & 3 & 0 \\
s^2 & 3 & K & \\
s^1 & \dfrac{9-K}{3} & & \\
s^0 & K. & &
\end{array}
$$

In order for the system to be stable there should be no sign change in column 1. To achieve this K must be greater than 0 and K must be less than 9. Therefore, for system stability ...$0 < K < 9$.

SOLVED PROBLEMS

Problem 9.1. *The pole-zero configuration of a transfer function is given below (Figure 9.25). The value of the transfer function at $s = 1$ is found to be 5. Determine the transfer function and gain factor K.*

Solution: The poles are located at $s = -2$ and $s = -4$; the zeros are located at $s = 0$ and $s = -3$.

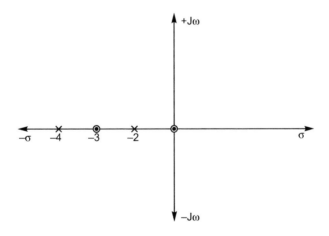

FIGURE 9.25

Thus, the transfer function is

$$
G(s) = \frac{Ks(s + 3)}{(s + 2)(s + 4)}.
$$

It is given that at $s = 1$, the value of $G(s)$ is 5.

$$G(1) = 5 = \frac{K \times 1 \times 4}{3 \times 5}$$

$$K = \frac{75}{4} = 18.75$$

Now

$$G(s) = \frac{18.75s(s + 3)}{(s + 2)(s + 4)}.$$

Problem 9.2. *For the circuit shown in Figure 9.26, find the transfer function.*

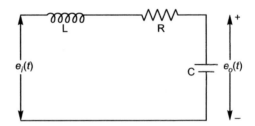

FIGURE 9.26

Solution: Transfer function

$$G(s) = \left.\frac{e_0(s)}{e_i(s)}\right| \quad \text{all initial conditions are zero.}$$

$$ei(t) = L\frac{di}{dt} + i(t)R + \frac{1}{C}\int i\,dt.$$

by the Laplace transform,

$$e_0(s) = \frac{1}{Cs} \quad \text{and} \quad e_0(t) = \frac{1}{C}\int i\,dt$$

$$e_i(s) = \left(R + Ls + \frac{1}{Cs}\right)$$

$$G(s) = \frac{1}{Cs} \times \frac{1}{(R + Ls + \frac{1}{Cs})}$$

$$G(s) = \frac{1}{RCs + LCs^2 + 1}.$$

Problem 9.3. *A system has the following transfer function*

$$\frac{C(s)}{R(s)} = \frac{(s+1)(s+3)}{(s+2)(s^2+8s+32)}.$$

Determine the time response C(t) of the system for a unit step input.

Solution: $r(t) = u(t)$

$$R(s) = \frac{1}{s}.$$

Now,

$$C(s) = \frac{(s+1)(s+3)}{s(s+2)(s^2+8s+32)}$$

$$C(s) = \frac{A}{s} + \frac{B}{s+2} + \frac{Cs+D}{s^2+8s+32}.$$

After solving,

$$A = \frac{3}{64}, \quad B = \frac{1}{40}$$

$$C = -\frac{23}{320}, \quad D = 0$$

$$C(s) = \frac{3}{64} \cdot \frac{1}{s} + \frac{1}{40} \cdot \left(\frac{1}{s+2}\right) - \frac{23}{320}\left[\frac{s}{(s+4)^2+4^2}\right]$$

$$C(s) = \frac{3}{64}\left(\frac{1}{s}\right) + \frac{1}{40}\left(\frac{1}{s+2}\right) - \frac{23}{320}\left[\frac{s+4}{(s+4)^2+(4)^2}\right]$$

$$\qquad + \frac{23}{320}\left[\frac{4}{(s+4)^2+4^2}\right].$$

Taking the Laplace inverse

$$C(t) = \frac{3}{64} + \frac{1}{40}e^{-2t} - \frac{23}{320}e^{-4t}\cos 4t + \frac{23}{80}e^{-4t}\sin 4t.$$

Problem 9.4. *The transfer function of a system is given below:*

$$G(s) = \frac{(s+3)}{s(s^2+2)}.$$

Determine the response for the unit step and ramp function.

Solution: For the unit step input,

$$R(t) = 1 \quad \text{and} \quad R(s) = \frac{1}{s}$$

$$G(s) = \frac{s+3}{s(s^2+2)} = \frac{C(s)}{R(s)}$$

$$C(s) = \frac{s+3}{s(s^2+2)} R(s) = \frac{s+3}{s(s^2+2)} \times \frac{1}{s}.$$

Now by the partial fraction method,

$$C(s) = \frac{A}{s} + \frac{B}{s^2} + \frac{Cs+D}{(s^2+2)}$$

$$C(s) = \frac{A(s^3+2s) + B(s^2+2) + (Cs^3 + Ds^2)}{s^2(s^2+2)}$$

$$= \frac{s^3(A+C) + s^2(B+D) + s(2A) + 2B}{s^2(s^2+2)}.$$

After comparing, $A + C = 0, B + D = 0, 2A = 1, 2B = 3$, we get

$$A = \frac{1}{2}, \quad B = \frac{3}{2}, \quad C = -\frac{1}{2}, \quad D = -\frac{3}{2}$$

$$C(s) = \frac{1}{2s} + \frac{3}{2s^2} - \frac{1}{2}\left(\frac{s+3}{s^2+2}\right)$$

$$C(s) = \frac{1}{2s} + \frac{3}{2s^2} - \frac{1}{2}\left(\frac{s}{s^2+2}\right) - \frac{3}{4}\left(\frac{2}{s^2+2}\right)$$

$$C(s) = \frac{1}{2s} + \frac{3}{2s^2} - \frac{1}{2}\left(\frac{s}{s^2+2}\right) - \frac{3}{4}\left(\frac{2}{s^2+2}\right).$$

Taking the Laplace Inverse,

$$C(t) = \frac{1}{2} + \frac{3}{2}t - \frac{1}{2}\cos\sqrt{2}t - \frac{3}{4}\sin\sqrt{2}t.$$

For the ramp function, $R(s) = \frac{1}{s^2}$,

$$C(s) = \frac{(s+3)}{s^3(s^2+2)} = \frac{A}{s} + \frac{B}{s^2} + \frac{C}{s^3} + \frac{Ds+E}{s^2+2},$$

$$C(s) = -\frac{3}{4s} + \frac{1}{2s^2} + \frac{3}{2s^3} + \frac{1}{4}\left(\frac{1}{s^2+2}\right).$$

Taking the Laplace inverse,

$$C(t) = -\frac{3}{4} + \frac{1}{2}t + \frac{3}{4}t^2 + \frac{1}{8}\sin\sqrt{2}t.$$

Problem 9.5. *The transfer function of a system is given below:*

$$G(s) = \frac{(s^2 + 3)}{s(s + 4)(s^2 + 4)}.$$

Determine the impulse response of the system.

Solution: For the impulse response, the input is the impulse signal $\delta(t)$ and we know

$$\mathcal{L}[\delta(t)] = 1 \quad \text{or} \quad R(s) = 1$$

$$C(s) = R(s)G(s) = \frac{1 \times (s^2 + 3)}{s(s + 4)(s^2 + 4)}$$

$$C(s) = \frac{A}{s} + \frac{B}{s + 4} + \frac{Cs + D}{s^2 + 4}$$

$$= \frac{A(s + 4)(s^2 + 4) + Bs(s^2 + 4) + (Cs + D)s(s + 4)}{s(s + 4)(s^2 + 4)}$$

$$= \frac{s^3(A + B + C) + s^2(4A + D + 4C) + s(4A + 4B + 4D) + 16}{s(s + 4)(s^2 + 4)}.$$

After comparing we get,

$$A = \frac{3}{16}, \quad B = -\frac{19}{80}, \quad C = \frac{1}{20} = D$$

$$C(s) = \frac{3}{16s} - \frac{19}{80(s + 4)} + \frac{1}{20}\left(\frac{s + 1}{s^2 + 4}\right)$$

$$C(s) = \frac{3}{16s} - \frac{19}{80(s + 4)} + \frac{1}{20}\left(\frac{s}{s^2 + 4}\right) + \frac{1}{80}\left(\frac{4}{s^2 + 4}\right).$$

Taking the Laplace inverse,

$$C(t) = \frac{3}{16} - \frac{19}{80}e^{-4t} + \frac{1}{20}\cos 2t + \frac{1}{80}\sin 2t.$$

Problem 9.6. *Determine the transfer of the network shown in Figure 9.27.*

Solution: The s-domain representation of Figure 9.27 is shown in Figure 9.27(*a*). Applying mesh analysis,

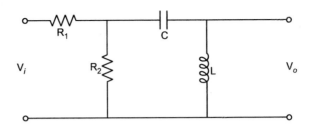

FIGURE 9.27

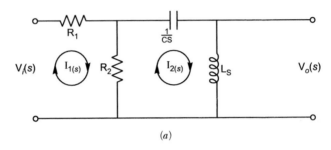

(a)

FIGURE 9.27

For mesh 1,

$$V_i(s) = R_1 I_1(s) + R_2[I_1(s) - I_2(s)]$$
$$V_i(s) = I_1(s)[R_1 + R_2] - I_2(s)R_2. \tag{1}$$

For mesh 2, $R_2[I_2(s) - I_1(s)] + I_2(s) \times (\frac{1}{Cs} + Ls)$

$$I_1(s) = \frac{I_2(s)\left(R_2 + Ls + \frac{1}{Cs}\right)}{R_2} \tag{2}$$

$$V_0(s) = Ls I_2(s). \tag{3}$$

By Equations (1) and (2),

$$V_i(s) = I_2(s)\left[\left\{\frac{R_2 + Ls + \frac{1}{Cs}}{R_2}\right\}(R_2 + R_1) - R_2\right]$$

$$V_i(s) = I_2(s)\left[\frac{R_2Ls + \frac{R_2}{Cs} + R_1R_2 + R_1Ls + \frac{R_1}{Cs}}{R_2}\right]. \tag{4}$$

The transfer function $= \frac{V_0(s)}{V_i(s)}$ and by Equations (3) and (4),

$$= \frac{CLs^2 R_2}{R_2LCs^2 + R_2 + R_1R_2Cs + R_1LCs^2 + R_1}.$$

Problem 9.7. *Derive the transfer function of the system shown in Figure 9.28, A being the amplifier gain.*

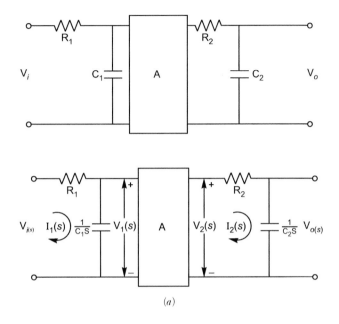

(a)

FIGURE 9.28

Solution: The *s*-domain representation of Figure 9.28 is shown in Figure 9.28(a).

$$V_i(s) = I_1(s)\left(R_1 + \frac{1}{C_1 s}\right) \tag{1}$$

$$\frac{v_2(s)}{v_1(s)} = A \tag{2}$$

$$-v_2(s) + I_2(s)\left(R_2 + \frac{1}{C_2 s}\right) = 0$$

$$v_2(s) = I_2(s)\left[R_2 + \frac{1}{C_2 s}\right] \tag{3}$$

$$V_0(s) = \frac{I_2(s)}{C_2 s} \tag{4}$$

and

$$v_1(s) = \frac{I_1(s)}{C_1 s}. \tag{5}$$

Putting the values of $v_1(s)$ and $v_2(s)$ from Equation (5) and (3) into Equation (2)

$$\frac{I_2(s)\left[R_2 + \frac{1}{C_2 s}\right]}{\frac{I_1(s)}{C_1 s}} = A$$

$$I_2(s)\left[\frac{R_2 C_2 s + 1}{C_2 s}\right] = \frac{A I_1(s)}{s}$$

$$I_1(s) = \frac{C_1(R_2 C_2 s + 1)I_2(s)}{C_2 A}. \tag{6}$$

Now the transfer function by Equations (4) and (1),

$$\frac{V_0(s)}{V_i(s)} = \frac{I_2(s)}{C_2 s} \times \frac{1}{I_1(s)(R_1 + \frac{1}{C_1 s})}$$

$$= \frac{I_2(s)C_1 s C_2 A}{C_2 s C_1(R_2 C_2 s + 1)I_2(s)(R_1 C_1 s + 1)}$$

$$= \frac{A}{(R_1 C_1 s + 1)(R_2 C_2 s + 1)}.$$

Problem 9.8. *A system is represented by a relation given below:*

$$X(s) = R(s)\frac{100}{s^2 + 2s + 50} \quad \text{if } r(t) = 1.$$

Find the value of $X(0)$ *and* $X(\infty)$.

Solution: $r(t) = 1$

$$R(s) = \frac{1}{s}$$

$$X(s) = \frac{100}{s(s^2 + 2s + 50)}.$$

By the initial value theorem,

$$X(0) = \lim_{s \to \infty} sX(s) = \lim_{s \to \infty} \frac{100}{s^2 + 2s + 50}$$

$$X(0) = 0.$$

By the final value theorem,

$$X(\infty) = \lim_{s \to 0} sX(s) = \lim_{s \to 0} \frac{100}{(s^2 + 2s + 50)}$$

$$= \frac{100}{50} = 2.$$

Problem 9.9. *Determine the stability of a system having the following characteristic equation,*

$$s^6 + s^5 + 5s^4 + 3s^3 + 2s^2 - 4s - 8 = 0.$$

Solution: The Routh table is

$$
\begin{array}{c|cccc}
s^6 & 1 & 5 & 2 & -8 \\
s^5 & 1 & 3 & -4 & 0 \\
s^4 & 2 & 6 & -8 & 0 \\
s^3 & 0 & 0 & 0 & 0 \\
s^2 & & & & \\
s^1 & & & & \\
s^0 & & & & \\
\end{array}
$$

The fourth row vanishes and the application of Routh criterion fails.

$$A(s) = 2s^4 + 6s^2 - 8$$

$$\frac{dA(s)}{ds} = 8s^3 + 12s - 0$$

Now, the coefficients of the fourth row are 8, 12, and 0, and the modified Routh array is given below:

$$
\begin{array}{c|cccc}
s^6 & 1 & 5 & 2 & -8 \\
s^5 & 1 & 3 & -4 & 0 \\
s^4 & 2 & 6 & -8 & 0 \\
s^3 & 8 & 12 & 0 & 0 \\
s^2 & 3 & -8 & 0 & 0 \\
s^1 & \dfrac{100}{3} & 0 & 0 & 0 \\
s^0 & -8 & 0 & 0 & 0 \\
\end{array}
$$

There is one sign change in the first column, therefore, the system is unstable because the system has one root with a positive real part.

Problem 9.10. *Determine the stability of a system having the following characteristic equation,*

(a) $s^4 + 10s^3 + 35s^2 + 50s + 24$
(b) $s^3 + 2s^2 + 2s + 40 = 0$
(c) $a_0 s^3 + a_1 s^2 + a_2 s + a_3 = 0.$

Solution: (a) $s^4 + 10s^3 + 35s^2 + 50s + 24$

$$
\begin{array}{llll}
s^4 & 1 & 35 & 24 \\
s^3 & 10 & 10 & \\
s^2 & 30 & 24 & \\
s^1 & 42 & & \\
s^0 & 24 & &
\end{array}
$$

There is no sign change in the first column, therefore, the system is stable.

(b) $s^3 + 2s^2 + 2s + 40 = 0$

$$
\begin{array}{lll}
s^3 & 1 & 2 \\
s^2 & 2 & 40 \\
s^1 & -18 & \\
s^0 & 40 &
\end{array}
$$

There are two sign changes (2 to -18 and -18 to 40) in the first column, therefore, two roots lie in the right half of the s-plane, hence, the system is unstable.

(c) $a_0 s^3 + a_1 s^2 + a_2 s + a_3 = 0$

$$
\begin{array}{lcc}
s^3 & a_0 & a_2 \\
s^2 & a_1 & a_3 \\
s^1 & \dfrac{a_1 a_2 - a_0 a_3}{a_1} & 0 \\
s^0 & a_3 & 0
\end{array}
$$

The system is stable if $a_1 a_2 - a_0 a_3 > 0 a_2 > \frac{a_0 a_3}{a_1}$ and a_3 should be positive.

Problem 9.11. *Determine the stability of a system having the following characteristic equation,*

(a) $s^5 + 2s^3 + 4s = 0$
(b) $s^6 + 2s^4 + 2s^2 + 1 = 0$.

Solution: (a) $s^5 + 2s^3 + 4s = 0$.

All terms are odd.

Now

$$
\frac{dA(s)}{ds} = 5s^4 + 6s^2 + 4.
$$

Now the Routh table is

$$
\begin{array}{llll}
s^5 & 1 & 2 & 4 \\
s^4 & 5 & 6 & 4 \\
s^3 & 0.8 & 3.2 & 0 \\
s^2 & -14 & 4 & 0 \\
s^1 & 3.42 & 0 & \\
0 & 4 & &
\end{array}
$$

There are two sign changes (0.8 to -14 and -14 to 3.42) in the first column, therefore, two roots lie in the right half of the s-plane. Hence, the system is unstable.

(b) $A(s) = s^6 + 2s^4 + 2s^2 + 1$

$$\frac{dA(s)}{ds} = 6s^5 + 8s^3 + 4s$$

$$
\begin{array}{lcccc}
s^6 & 1 & 2 & 2 & 1 \\
s^5 & 6 & 8 & 4 & 0 \\
s^4 & \dfrac{2}{3} & \dfrac{4}{3} & 1 & 0 \\
s^3 & -4 & -5 & 0 & \\
s^2 & \dfrac{13}{6} & 1 & 0 & \\
s^1 & -\dfrac{21}{6} & 0 & & \\
s^0 & 1 & & &
\end{array}
$$

There are four sign changes ($\frac{2}{3}$ to -4, -4 to $\frac{13}{6}$, $\frac{13}{6}$ to $-\frac{21}{6}$, $-\frac{21}{6}$ to 1). This system has six roots and four roots lie in the right half of the s-plane. Hence, the system is unstable.

Problem 9.12. *Show that the following system is unstable,* $F(s) = s^4 + s^3 + 2s^2 + 2s + 3 = 0$.

Solution:

$$
\begin{array}{lccc}
s^4 & 1 & 2 & 3 \\
s^3 & 1 & 2 & 0 \\
s^2 & \in & 3 & \\
s^1 & 2 - \dfrac{\in 3}{\in} & 0 & \\
s^0 & 3 & 0 &
\end{array}
$$

The value of

$$2 - \frac{3}{\in} = -\infty, \quad \text{hence, two} \in \to 0$$

sign changes ($\in$ to $-\infty$, $-\infty$ to 3). Therefore, the system is unstable.

Problem 9.13. (i) *For the given network function, draw the pole-zero diagram and hence obtain the time-domain response i(t),*

$$I(s) = \frac{s^2 + 4s + 3}{s^2 + 2s}.$$

(ii) *Determine the number of roots with positive real parts and negative real parts for the following polynomial equation using Routh Criterion:*

$$Q(s) = s^4 + 2s^3 + 8s^2 + 10s + 15.$$

Solution:

(i)
$$I(s) = \frac{s^2 + 4s + 3}{s(s + 2)} = \frac{(s + 1)(s + 3)}{s(s + 2)}.$$

The poles are at $s = 0, s = -2$, and zeros are at $s = -1, s = -3$ as shown in Figure 9.29.

$$I(s) = 1 + \frac{2s + 3}{s(s + 2)} = 1 + \frac{A}{s} + \frac{B}{s + 2}$$

$$I'(s) = \frac{A}{s} + \frac{B}{s + 2}.$$

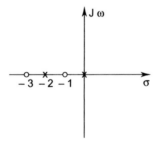

FIGURE 9.29

After solving,

$$A = \frac{3}{2}, \quad B = \frac{1}{2}$$

$$I(s) = 1 + \frac{3}{2s} + \frac{1}{2(s + 2)}.$$

Taking the Laplace inverse,

$$I(t) = \delta(t) + \frac{3}{2}u(t) + \frac{1}{2}e^{-2t} \text{ Amp}.$$

(*ii*) Routh criterion,

$$
\begin{array}{cccc}
s^4 & 1 & 8 & 15 \\
s^3 & 2 & 10 & 0 \\
s^2 & \dfrac{2 \times 8 - 10 \times 1}{2} = 3 & \dfrac{2 \times 15 - 1 \times 0}{2} = 15. & \\
s & 0 & 0 &
\end{array}
$$

The Routh criterion fails in the fourth row as all the terms are zero in this row. The auxiliary equation is

$$A(s) = 3s^2 + 15$$

$$\frac{dA(s)}{ds} = 6s + 0.$$

The remaining portion of tabulation is

$$
\begin{array}{ccc}
s^1 & 6 & 0 \\
s^0 & 15 &
\end{array}
$$

since there are no sign changes in the first column of the entire Routh's tabulation.

Problem 9.14. *Find the open circuit transfer impedance $\frac{V_2}{I_1}$ and the open circuit voltage ratio $\frac{V_2}{V_1}$ for the ladder network shown in Figure 9.30.*

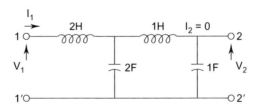

FIGURE 9.30

Solution: The *s*-domain representation of Figure 9.30 is shown in Figure 9.30(*a*). Applying KVL,

$$V_1 = 2sI_1 + \frac{1}{2s}(I_1 - I_3) \tag{i}$$

$$V_2 = \frac{1}{s}I_3 \tag{ii}$$

$$I_3 = \frac{\frac{1}{2s} \times I_1}{\frac{1}{2s} + s + \frac{1}{s}}$$

$$= \frac{I_1}{2s^2 + 3}. \hspace{3cm} (iii)$$

From Equations (ii) and (iii),

$$\frac{V_2}{I_1} = \frac{1}{s(2s^2 + 3)}.$$

From Equations (i) and (iii),

$$V_1 = \left[\left(2s + \frac{1}{2s}\right)(2s^2 + 3) - \frac{1}{2s}\right] \times I_3 = \left(\frac{4s^4 + 7s^2 + 1}{s}\right) I_3. \hspace{1cm} (iv)$$

From Equations (ii) and (iv),

$$\frac{V_2}{V_1} = \frac{1}{4s^4 + 7s^2 + 1}.$$

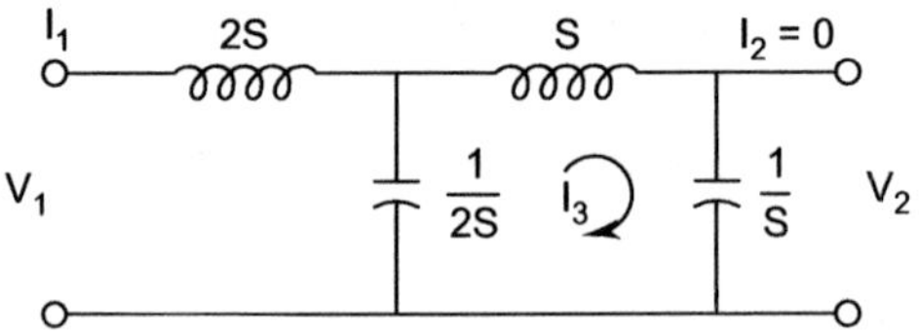

(a)

FIGURE 9.30

QUESTIONS FOR DISCUSSION

1. What is a zero?
2. What is a pole?
3. Define stability.
4. What is convolution integral?
5. Which system is more stable,

    (a) $F(s) = \dfrac{1}{s+2}$ or   (b) $F(s) = \dfrac{1}{s+4}$?

6. Which function is a stable function? Explain.

    (a) $F(s) = \dfrac{1}{(s+1)(s-2)}$ (b) $F(s) = \dfrac{1}{(s+1)(s+2)}.$

OBJECTIVE QUESTIONS

1. The output of a linear time-invariant control system is $C(t)$ for a input $r(t)$. If $r(t)$ is modified by passing it through a block whose transfer function is e^{-s} and then applied to the system, the modified output of the system would be
   (a) $C(t)/1 + e^t$ (b) $C(t)/1 + e^{-t}$
   (c) $C(t-1)u(t-1)$ (d) $C(t)u(t-1)$.

2. A linear time-invariant system initially at rest, when subjected to a unit step input, gives a response $Y(t) = te^{-t} t > 0$. The transfer function of the system is
   (a) $\dfrac{1}{(s+1)^2}$ (b) $\dfrac{1}{s}(s+1)^2$ (c) $\dfrac{s}{(s+1)^2}$ (d) $\dfrac{1}{s}(s+1)$.

3. The system response can be tested better with
   (a) sinusoidal input signal   (b) unit impulse input signal (IES-92)
   (c) ramp input signal   (d) exponentially decaying signal.

4. For a system to be stable
   (a) All poles and zeros of the transfer function must be in the right half of the s-plane
   (b) All poles and zeros of the transfer function must be on the imaginary axis
   (c) All the poles of the transfer function must be in the left half of the s-plane
   (d) All the zeros of the transfer function must be in the left half of the s-plane.

UNSOLVED PROBLEMS

1. Show that the following system is stable or unstable:
   (a) $s^5 + 4s^4 + 3s^3 + s^2 + 2s + 5$ (b) $s^4 + 5s^2 + 2$
   (c) $2s^4 + s^3 + 3s^2 + s + 8$ (d) $s^3 + 5s$.

2. Define stability and find the range of 'a' for following functions:
   (a) $s^4 + s^3 + as^2 + s + 1$ (b) $2s^3 + s^2 + 2s + a$
   (c) $s^5 + 3s^3 + as$ (d) $s^4 + 2s^2 + a$.

3. Determine the stability of a system having the following characteristic equations:
   (a) $2s^6 + 3s^5 + s^4 + 2s^3 + s^2 + s + 5$
   (b) $s^5 - 2s^4 + s^3 + 2s^2 + s + 1$.

10

TWO-PORT
NETWORKS

10.1 INTRODUCTION

A pair of terminals at which a signal may enter or leave a network is called a *port* and a network having only one such pair of terminals is called a one-*port* network or simply a one-port. Similarly, a network having two ports, is known as a two-port network. When more than one pair of terminals is present, the network is known as a multi-port network.

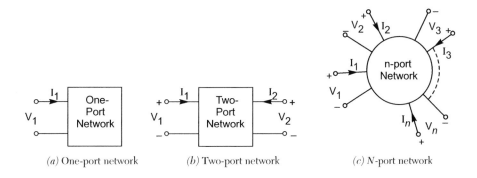

(a) One-port network   (b) Two-port network   (c) N-port network

FIGURE 10.1 Different type of networks.

A general network having two pair of terminals, one as the input terminal and the other as the output terminal, is a very important in various electrical and electronics systems. In such a network an electrical signal enters the input signal and leaves from the output terminals. Two ports containing no sources in their branches are called passive ports. If there is any source (voltage or current) in any branch they are called active ports.

A two-port network is shown in Figure 10.1(b). It has four variables (V_1, I_1, V_2, and I_2). The first two variables (V_1 and I_1) are input variables and (V_2 and I_2) are output variables. Only two of the four variables are independent and the other two are dependent. The dependence of two of the four variables on the other two is described in a number of ways depending on which of the variables are chosen to be the independent variables. There are six possible sets of equations between V_1, I_1, V_2, and I_2.

TABLE 10.1   Two-Port Parameters

S. No.	Name	Function		Matrix form
		Express	**In terms of**	
1.	Open circuit Impedance or [Z] Parameter	V_1, V_2	I_1, I_2	$\begin{bmatrix} V_1 \\ V_2 \end{bmatrix} = \begin{bmatrix} Z_{11} & Z_{12} \\ Z_{21} & Z_{22} \end{bmatrix} \begin{bmatrix} I_1 \\ I_2 \end{bmatrix}$
2.	Short circuit admittance or [Y] Parameter	I_1, I_2	V_1, V_2	$\begin{bmatrix} I_1 \\ I_2 \end{bmatrix} = \begin{bmatrix} Y_{11} & Y_{12} \\ Y_{21} & Y_{22} \end{bmatrix} \begin{bmatrix} V_1 \\ V_2 \end{bmatrix}$
3.	ABCD or Transmission Parameter	V_1, I_1	V_2, I_2	$\begin{bmatrix} V_1 \\ I_1 \end{bmatrix} = \begin{bmatrix} A & B \\ C & D \end{bmatrix} \begin{bmatrix} V_2 \\ -I_2 \end{bmatrix}$
4.	Inverse Transmission Parameter	V_2, I_2	V_1, I_1	$\begin{bmatrix} V_2 \\ I_2 \end{bmatrix} = \begin{bmatrix} A' & B' \\ C' & D' \end{bmatrix} \begin{bmatrix} V_1 \\ -I_1 \end{bmatrix}$
5.	Hybrid or [h] Parameter	V_1, I_2	I_1, V_2	$\begin{bmatrix} V_1 \\ I_2 \end{bmatrix} = \begin{bmatrix} h_{11} & h_{12} \\ h_{21} & h_{22} \end{bmatrix} \begin{bmatrix} I_1 \\ V_2 \end{bmatrix}$
6.	Inverse hybrid or [g] Parameter	I_1, V_2	V_1, I_2	$\begin{bmatrix} I_1 \\ V_2 \end{bmatrix} = \begin{bmatrix} g_{11} & g_{12} \\ g_{21} & g_{22} \end{bmatrix} \begin{bmatrix} V_1 \\ I_2 \end{bmatrix}$

It is natural to ask why we need six different ways of evaluation and the answer is two-fold. Firstly, some parameter sets are more easily measured when their numerical values lie in a given range. Second, a particular problem may be solved more easily in terms of one set than the others.

10.2   OPEN CIRCUIT IMPEDANCE OR [Z] PARAMETERS

$$\begin{bmatrix} V_1 \\ V_2 \end{bmatrix} = \begin{bmatrix} Z_{11} & Z_{12} \\ Z_{21} & Z_{22} \end{bmatrix} \begin{bmatrix} I_1 \\ I_2 \end{bmatrix} \text{ or } [V] = [Z][I]$$

$$V_1 = Z_{11}I_1 + Z_{12}I_2 \tag{10.1}$$

$$V_2 = Z_{21}I_1 + Z_{22}I_2 \tag{10.2}$$

Now if $I_1 = 0$ means the input port is an open circuit

$$Z_{12} = \frac{V_1}{I_2} \bigg|_{I_1=0}$$

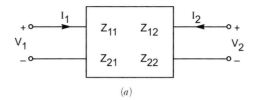

(a)

and

$$Z_{22} = \frac{V_2}{I_2} \bigg|_{I_1=0}$$

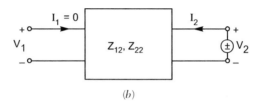

(b)

if

$I_2 = 0$ means the output port is an open circuit

$$Z_{11} = \frac{V_1}{I_1} \bigg|_{I_2=0}$$

and

$$Z_{21} = \frac{V_2}{I_1} \bigg|_{I_2=0}.$$

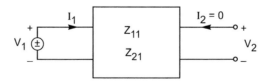

FIGURE 10.2 **Determination of Z-parameters.**

[Z] parameters are known as open circuit impedance parameters because, in both the cases, one port is open (*i.e.*, $I_1 = 0$ and $I_2 = 0$)

Z_{11} and Z_{22} are driving-point impedance at port 1 and port 2. In the case of driving-point function the data should be from the same port.

Z_{12} and Z_{21} are transfer impedance because in Z_{12} and Z_{21}; one data is from port 1 and the other data is from port 2,

$$Z_{12} = \frac{V_1}{I_2} \; [V_1 \text{ from port 1 and } I_2 \text{ from port 2}] = \text{reverse transfer impedance}$$

$$Z_{21} = \frac{V_2}{I_1} \; [I_1 \text{ from port 1 and } V_2 \text{ from port 2}] = \text{forward transfer impedance.}$$

It is seen that dimensionally each parameter is an impedance (ratio voltage and current), therefore, the impedance parameters may be either calculated or measured by first opening port 2 and then determining the ratios $(\frac{V_1}{I_1})$ and $(\frac{V_2}{I_1})$, and then opening port 1 and determining the ratios $(\frac{V_1}{I_2})$ and $(\frac{V_2}{I_2})$. But in this chapter we will solve all problems using the comparing method.

10.2.1 Comparing method

Step 1. Apply KVL in both the loops (in the case of two loops) and get the expression in terms of (V_1, I_1, I_2) and (V_2, I_1, I_2) for loop one and two.

Step 2. Write the desired standard parameters equation and compare the above two equation with the standard equation and determine the parameters. We will rearrange the above two equations according to standard equations and then compare for the different type of parameters.

Example 10.1. *Determine the Z parameters of the given network as shown in Figure 10.3.*

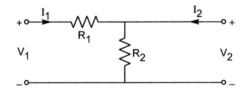

FIGURE 10.3

Solution: Applying KVL in both the loops,

$$V_1 = I_1 R_1 + (I_1 + I_2)R_2$$

or

$$V_1 = I_1(R_1 + R_2) + I_2 R_2 \tag{1}$$

$$V_2 = (I_1 + I_2)R_2 = I_1 R_2 + I_2 R_2. \tag{2}$$

Now the standard equation for the Z parameters

$$V_1 = Z_{11}I_1 + Z_{12}I_2 \tag{3}$$

$$V_2 = Z_{21}I_1 + Z_{22}I_2. \tag{4}$$

Now comparing Equations (3) and (4) with Equations (1) and (2), we get

$$Z_{11} = (R_1 + R_2), \quad Z_{12} = R_2, \quad Z_{21} = R_2, \quad Z_{22} = R_2.$$

Equation (10.1) is the sum of the two terms $Z_{11}I_1$ and $Z_{12}I_2$. The first term $Z_{11}I_1$ is obtained by multiplying the impedance Z_{11} with the input current I_1. The second term $Z_{12}I_2$ is dependent upon I_2 which does not flow in the input circuit. Hence, this term can be replaced by a current controlled voltage source. Similarly, Equation (10.2) shows that port 2 can be represented by an impedance Z_{22} and a current controlled voltage source $Z_{21}I_1$ as shown in Figure 10.4.

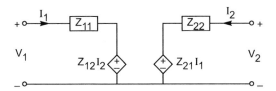

FIGURE 10.4 **Equivalent circuit of a two-port network in terms of Z-parameters.**

10.3 SHORT CIRCUIT ADMITTANCE OR [Y] PARAMETERS

$$\begin{bmatrix} I_1 \\ I_2 \end{bmatrix} = \begin{bmatrix} Y_{11} & Y_{12} \\ Y_{21} & Y_{22} \end{bmatrix} \begin{bmatrix} V_1 \\ V_2 \end{bmatrix}$$

$$I_1 = Y_{11}V_1 + Y_{12}V_2 \tag{10.3}$$

$$I_2 = Y_{21}V_1 + Y_{22}V_2 \tag{10.4}$$

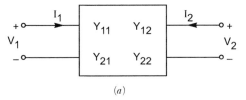

(a)

Now if $V_1 = 0$ the input port is short circuit

$$Y_{12} = \frac{I_1}{V_2} \bigg|_{V_1=0}$$

and

$$Y_{22} = \frac{I_2}{V_2} \bigg|_{V_1=0} .$$

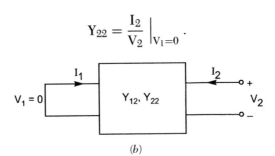

(b)

If

$V_2 = 0$ the output port is short circuit

$$Y_{11} = \frac{I_1}{V_1} \bigg|_{V_2=0}$$

$$Y_{21} = \frac{I_2}{V_1} \bigg|_{V_2=0} .$$

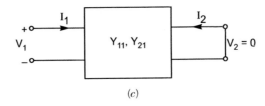

(c)

FIGURE 10.5   Determination of Y-parameters.

[Y] parameters are known as short circuit admittance parameters because in both the cases, one-port is short (*i.e.*, $V_1 = 0$ and $V_2 = 0$). The constants Y_{11}, Y_{12}, Y_{21}, and Y_{22} are called the admittance or Y-parameters of the two-port network. Equations (10.3) and (10.4) are called the standard equation or defining equations of the Y-parameters.

Y_{11} and Y_{22} are driving-point admittance at port 1 and port 2.

Y_{12} and Y_{21} are known as transfer admittance.

Y_{11} is known as short circuit input admittance.

Y_{12} is known as short circuit reverse transfer admittance.

Y_{21} is known as short circuit forward transfer admittance.

Similarly, Y_{22} is known as short circuit output admittance.

It is seen that dimensionally each parameter is admittance (ratio of current and voltage). Also, each parameter is obtained by short circuiting either the input or output port. Therefore, these parameters are called the short circuit admittance parameters. The admittance parameters may be either calculated or measured by first shorting port 1 and then determining the ratios ($\frac{I_1}{V_2}$) and ($\frac{I_2}{V_2}$), and then shorting port 2 and determining the ratios ($\frac{I_1}{V_1}$) and ($\frac{I_2}{V_1}$). But in this chapter we will solve all problems using the comparing method discussed in the last section.

Example 10.2. *Determine the Z-and Y-parameters of the given network as shown in Figure 10.6.*

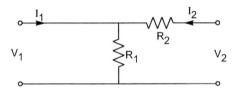

FIGURE 10.6

Solution: Applying KVL,

$$V_1 = (I_1 + I_2)R_1$$

or

$$V_1 = I_1 R_1 + I_2 R_1 \tag{1}$$
$$V_2 = R_2 I_2 + (I_1 + I_2)R_1$$

or

$$V_2 = I_1(R_1) + I_2(R_1 + R_2). \tag{2}$$

Now comparing Equations (1) and (2) with the standard equation of Z-parameters, we get

$$Z_{11} = R_1, \quad Z_{12} = R_1, \quad Z_{21} = R_1, \quad Z_{22} = R_1 + R_2.$$

Now the standard equation for Y-parameters as

$$I_1 = Y_{11}V_1 + Y_{12}V_2 \tag{3}$$
$$I_2 = Y_{21}V_1 + Y_{22}V_2. \tag{4}$$

Now by Equations (1) and (2),

$$I_1 = \left(\frac{R_1 + R_2}{R_1 \cdot R_2}\right) V_1 - \frac{V_2}{R_2} \tag{5}$$

$$[\text{equation (1)} \times (R_1 + R_2) - \text{Equation (2)} \times R_1]$$

and

$$I_2 = -\frac{V_1}{R_2} + \frac{V_2}{R_2}. \tag{6} \ [\text{Equation (2)} - \text{Equation (1)}]$$

Now comparing Equations (5) and (6) with (3) and (4),

$$Y_{11} = \left(\frac{R_1 + R_2}{R_1 \cdot R_2}\right), \quad Y_{12} = -\frac{1}{R_2}, \quad Y_{21} = -\frac{1}{R_2}, \quad Y_{22} = \frac{1}{R_2}.$$

A circuit model using Y-parameters can be developed from Equations (10.3) and (10.4) as shown in Figure 10.7.

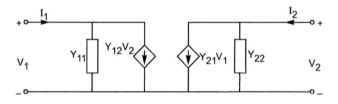

FIGURE 10.7   **A two-controlled source model of Y-parameters.**

10.4   ABCD OR TRANSMISSION PARAMETERS

If we select V_2 and I_2 as independent variables, the two-port equations may be written as

$$\begin{bmatrix} V_1 \\ I_1 \end{bmatrix} = \begin{bmatrix} A & B \\ C & D \end{bmatrix} \begin{bmatrix} V_2 \\ -I_2 \end{bmatrix}$$

$$V_1 = AV_2 + B(-I_2) \tag{10.5}$$

$$I_1 = CV_2 + D(-I_2). \tag{10.6}$$

Now if $V_2 = 0$, the output port is short circuit,

$$B = \frac{V_1}{-I_2} \Big|_{V_2=0}$$

$$D = \frac{I_1}{-I_2} \Big|_{V_2=0}.$$

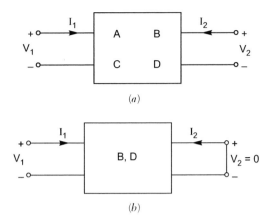

(a)

(b)

If

$$I_2 = 0, \text{output port is open circuit,}$$

$$A = \frac{V_1}{V_2}\bigg|_{I_2=0}$$

$$C = \frac{I_1}{V_2}\bigg|_{I_2=0}.$$

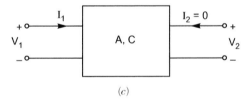

(c)

FIGURE 10.8   Determination of ABCD parameters.

As per convention, the currents entering the upper terminals of the ports are taken as positive. In a power transmission network the output current I_2 is coming out of the upper terminal of the output port. Hence, the reference direction of the output current I_2 is opposite to that shown in the figure.

The ABCD parameters are defined as follows:

$$A = \frac{V_1}{V_2}\bigg|_{I_2=0} = \text{open circuit reverse voltage transfer ratio}$$

$$B = \frac{V_1}{-I_2}\bigg|_{V_2=0} = \text{short circuit reverse transfer impedance}$$

$$C = \frac{I_1}{V_2}\Big|_{I_2=0} = \text{open circuit reverse transfer admittance}$$

$$D = \frac{I_1}{-I_2}\Big|_{V_2=0} = \text{short circuit reverse current transfer ratio.}$$

It is seen that all the four parameters are transfer functions. The ABCD parameters are widely used in transmission line calculations.

In transmission line theory, the input port is called the sending input and the output port is called the receiving end.

Example 10.3. *Determine the* ABCD *parameters of the given network as shown in Figure 10.9.*

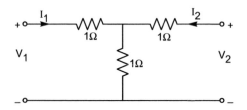

FIGURE 10.9

Solution: Loops equations are

$$V_1 = I_1 \cdot 1 + (I_1 + I_2) \cdot 1$$

or

$$V_1 = 2I_1 + I_2 \tag{1}$$
$$V_2 = I_2 \cdot 1 + (I_1 + I_2) \cdot 1$$

or

$$V_2 = I_1 + 2I_2. \tag{2}$$

Now the standard equations for ABCD parameters

$$V_1 = AV_2 - BI_2 \tag{3}$$
$$I_1 = CV_2 - DI_2. \tag{4}$$

Rearrange Equation (2) into the form of Equation (4),

$$I_1 = V_2 - 2I_2. \tag{from (2a)}$$

Putting the value of I_1 from Equation (2a) into Equation (1),

$$V_1 = 2(V_2 - 2I_2) + I_2 = 2V_2 - 3I_2. \tag{5}$$

Now comparing Equations (5) and (2a) with Equations (3) and (4),

$$A = 2, B = 3, C = 1, D = 2.$$

Example 10.4. *Determine the Z- and ABCD parameters of the given network as shown in Figure 10.10.*

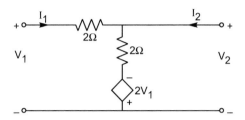

FIGURE 10.10

Solution: The two loop equations are

$$3V_1 = 2I_1 + 2(I_1 + I_2)$$
$$V_1 = 4I_1 + 2I_2$$
$$V_1 = \frac{4}{3}I_1 + \frac{2}{3}I_2 \tag{1}$$
$$V_2 = 2(I_1 + I_2) - 2V_1 = 2I_1 + 2I_2 - \frac{8}{3}I_1 - \frac{4}{3}I_2$$

or

$$V_2 = -\frac{2}{3}I_1 + \frac{2}{3}I_2. \tag{2}$$

Comparing Equations (1) and (2) with the standard equation of Z-parameters (10.1) and (10.2),

$$Z_{11} = \frac{4}{3}, \quad Z_{12} = \frac{2}{3}, \quad Z_{21} = -\frac{2}{3}, \quad Z_{22} = \frac{2}{3}.$$

Now the standard equations for the ABCD parameters are

$$V_1 = AV_2 - BI_2 \qquad \text{(from 10.5)}$$
$$I_1 = CV_2 - DI_2. \qquad \text{(from 10.6)}$$

Now rearrange Equation (2) into the form of Equation (10.6),

$$I_1 = -\frac{3}{2}V_2 + I_2. \tag{2a}$$

Putting the value of I_1 from Equation $(2a)$ into Equation (1),

$$V_1 = \frac{4}{3}\left(-\frac{3}{2}V_2 + I_2\right) + \frac{2}{3}I_2$$

or

$$V_1 = -2V_2 + 2I_2. \tag{3}$$

Comparing Equations (3) and $(2a)$ with Equations (10.5) and (10.6),

$$A = -2, \quad B = -2, \quad C = -\frac{3}{2}, \quad D = -1.$$

10.5   HYBRID OR h-PARAMETERS

If I_1 and V_2 are taken as the independent variables, the two-port equations may be written as

$$V_1 = h_{11}I_1 + h_{12}V_2 \tag{10.7}$$
$$I_2 = h_{21}I_1 + h_{22}V_2. \tag{10.8}$$

The constants h_{11}, h_{12}, h_{21}, and h_{22} are called the hybrid (or h) parameters of the two-port network.

If

$V_2 = 0$ the output port is short circuit.

$$h_{11} = \frac{V_1}{I_1}\bigg|_{V_2=0} = \text{short circuit input impedance.}$$

$$h_{21} = \frac{I_2}{I_1}\bigg|_{V_2=0} = \text{short circuit forward current ratio.}$$

If

$I_1 = 0$ the input-port is open circuit.

$$h_{12} = \frac{V_1}{V_2}\bigg|_{I_1=0} = \text{open circuit reverse voltage transfer ratio}$$

$$h_{22} = \frac{I_2}{V_2}\bigg|_{I_1=0} = \text{open circuit output admittance.}$$

From the definitions of h-parameters, it is seen that h_{11} has the dimensions of impedance, h_{12} and h_{21} are dimensionless, and h_{22} has the dimensions of admittance. Thus, these parameters are dimensionally homogeneous. That is why these parameters are called hybrid parameters. These parameters are most suitable for transistor testing because of the following reasons:

1. The output impedance of a transistor is very high and shorting of port 2-2′ makes it easier to measure h_{11} and h_{21}.
2. The input impedance is low and the open circuit test can be performed makes it conveniently to measure h_{12} and h_{22}.

In transistor applications, the symbols $h_{11}, h_{12}, h_{21}, h_{22}$ are replaced by h_i, h_r, h_f, h_o, respectively, to denote input, reverse, forward, and output.

The standard Equations (10.7 and 10.8) may be used to determine the equivalent circuit of a two-port network in terms of the h-parameter. This circuit is shown in Figure 10.11.

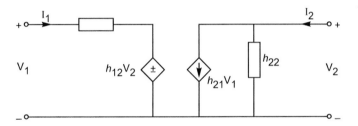

FIGURE 10.11   Equivalent circuit of a two-port network in terms of h-parameters.

Example 10.5. *Find the Z- and h-parameters of the given network as shown in Figure 10.12.*

Solution: By using KCL, the current shown in each branch is as shown in Figure 10.12(a).

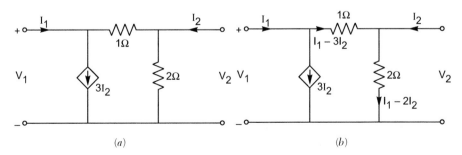

(a)   (b)

FIGURE 10.12

By applying KVL, we get

$$V_1 = 1 \cdot (I_1 - 3I_2) + 2(I_1 - 2I_2)$$

or

$$V_1 = 3I_1 - 7I_2 \tag{1}$$
$$V_2 = 2(I_1 - 2I_2) = 2I_1 - 4I_2. \tag{2}$$

By comparing Equations (1) and (2) with the standard equation of Z-parameters (10.1 and 10.2), we get

$$Z_{11} = 3, \quad Z_{12} = -7, \quad Z_{21} = 2, \quad Z_{22} = -4.$$

Now the standard equations for the h-parameters are

$$V_1 = h_{11}I_1 + h_{12}V_2 \qquad \text{(from 10.7)}$$
$$I_2 = h_{21}I_2 + h_{22}V_2. \qquad \text{(from 10.8)}$$

Rearrange Equation (2) into the form of Equation (10.8),

$$I_2 = \frac{I_1}{2} - \frac{V_2}{4}. \tag{2a}$$

Putting the value of I_2 from Equation (2a) into Equation (1),

$$V_1 = 3I_1 - 7\left(\frac{I_1}{2} - \frac{V_2}{4}\right) = -\frac{I_1}{2} + \frac{7}{4}V_2. \tag{3}$$

Comparing Equations (3) and (2a) with Equations (10.7) and (10.8) we get,

$$h_{11} = -\frac{1}{2}\,\Omega, \quad h_{12} = \frac{7}{4}, \quad h_{21} = \frac{1}{2}, \quad h_{22} = -\frac{1}{2}\,\mho.$$

10.6   INVERSE HYBRID (OR g) PARAMETERS

If V_1 and I_2 are chosen as independent variables, the two-port network equations may be written as

$$I_1 = g_{11}V_1 + g_{12}I_2 \tag{10.9}$$
$$V_2 = g_{21}V_1 + g_{22}I_2. \tag{10.10}$$

In matrix form, these equations are written as

$$\begin{bmatrix} I_1 \\ V_2 \end{bmatrix} = \begin{bmatrix} g_{11} & g_{12} \\ g_{21} & g_{22} \end{bmatrix} \begin{bmatrix} V_1 \\ I_2 \end{bmatrix}.$$

The constants g_{11}, g_{12}, g_{21}, and g_{22} are known as inverse hybrid parameters or g-parameters. The g-parameters are defined as follows by using Equations (10.9) and (10.10).

If $I_2 = 0$ the output port is open circuit.

$$g_{11} = \left. \frac{I_1}{V_1} \right|_{I_2=0} = \text{ open circuit input admittance.}$$

$$g_{21} = \left. \frac{V_2}{V_1} \right|_{I_2=0} = \text{ open circuit forward voltage gain.}$$

If

$V_1 = 0$ the input port is short circuit.

$$g_{12} = \left. \frac{I_1}{I_2} \right|_{V_1=0} = \text{ short circuit reverse current gain.}$$

$$g_{22} = \left. \frac{V_2}{I_2} \right|_{V_1=0} = \text{ short circuit output impedance.}$$

From the definitions of the g-parameters, it is seen that g_{11} has the dimensions of admittance, g_{21} and g_{12} are dimensionless, and g_{22} has the dimensions of impedance.

By Equations (10.9) and (10.10), the equivalent circuit of a two-port network is as shown in the figure.

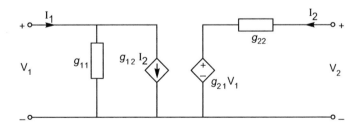

FIGURE 10.13   **Equivalent circuit of a two-port network in terms of g-parameters.**

Example 10.6. *Determine the Z- and g-parameters of the given network as shown in Figure 10.14.*

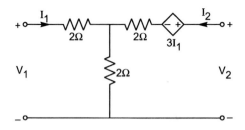

FIGURE 10.14

Solution: Two loop equations are

$$V_1 = 2I_1 + 2(I_1 + I_2)$$

or

$$V_1 = 4I_1 + 2I_2 \tag{1}$$
$$V_2 = 3I_1 + 2I_2 + 2(I_1 + I_2)$$
$$V_2 = 5I_1 + 4I_2. \tag{2}$$

Comparing Equations (1) and (2) with the standard equation of the Z-parameters (10.1) and (10.2), we get

$$Z_{11} = 4, \quad Z_{12} = 2, \quad Z_{21} = 5, \quad Z_{22} = 4.$$

Now the standard equations of the g-parameters,

$$I_1 = g_{11}V_1 + g_{12}I_2 \qquad \text{(from 10.9)}$$
$$V_2 = g_{21}V_1 + g_{22}I_2. \qquad \text{(from 10.10)}$$

Rearranging Equation (1) into the form of Equation (10.9),

$$I_1 = \frac{V_1}{4} - \frac{I_2}{2}. \tag{1a}$$

Putting the values of I_1 from Equation (1a) into Equation (2),

$$V_2 = 5\left[\frac{V_1}{4} - \frac{I_2}{2}\right] + 4I_2$$
$$V_2 = \frac{5}{4}V_1 + \frac{3}{2}I_2. \tag{3}$$

Comparing Equations (1a) and (3) with Equations (10.9) and (10.10), we get

$$g_{11} = \frac{1}{4}, \quad g_{12} = -\frac{1}{2}, \quad g_{21} = \frac{5}{4}, \quad g_{22} = \frac{3}{2}.$$

Inverse transmission (or $A'B'C'D'$) parameters

If V_1 and I_1 are chosen as independent variables the two-port network equations may be written as

$$V_2 = A'V_1 - B'I_1 \tag{10.11}$$

$$I_2 = C'V_1 - D'I_1. \tag{10.12}$$

In matrix form, these equations are written as

$$\begin{bmatrix} V_2 \\ I_2 \end{bmatrix} = \begin{bmatrix} A' & B' \\ C' & D' \end{bmatrix} \begin{bmatrix} V_1 \\ -I_1 \end{bmatrix}.$$

The constants $A', B', C',$ and D' are known as inverse transmission parameters. The inverse transmission parameters are defined as follows:

$$A' = \left.\frac{V_2}{V_1}\right|_{I_1=0} = \text{open circuit forward voltage transfer ratio.}$$

$$B' = \left.\frac{V_2}{-I_1}\right|_{V_1=0} = \text{short circuit forward transfer impedance.}$$

$$C' = \left.\frac{I_2}{V_1}\right|_{I_1=0} = \text{open circuit forward transfer admittance.}$$

$$D' = \left.\frac{I_2}{-I_1}\right|_{V_1=0} = \text{short circuit forward current transfer ratio.}$$

Example 10.7. *Determine the Z- and $A'B'C'D'$ parameters of the given network as shown in Figure 10.15.*

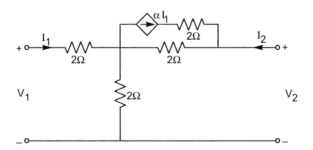

FIGURE 10.15

Solution: The two loops equations are

$$V_1 = 2I_1 + 2(I_1 + I_2)$$

or

$$V_1 = 4I_1 + 2I_2 \tag{1}$$
$$V_2 = 2(I_2 + aI_1) + 2(I_1 + I_2)$$

or

$$V_2 = I_1(2 + 2\alpha) + 4I_2. \tag{2}$$

> **NOTE** There is no role of 2Ω resistor, connected in series with the (αI_1) current source.

After comparison of Equations (1) and (2) with Equations (10.1) and (10.2), we get

$$Z_{11} = 4, \quad Z_{12} = 2, \quad Z_{21} = 2 + 2\alpha, \quad Z_{22} = 4.$$

Now the standard equations for $A'B'C'D'$ parameters are

$$V_2 = A'V_1 - B'I_1 \qquad \text{(from 10.11)}$$
$$I_2 = C'V_1 - D'I_1. \qquad \text{(from 10.12)}$$

Rearranging Equation (1) into the form of Equation (10.12),

$$I_2 = \frac{V_1}{2} - 2I_1. \tag{1a}$$

Putting the value of I_2 from Equation (1a) into Equation (2),

$$V_2 = I_1(2 + 2\alpha) + 4\left[\frac{V_1}{2} - 2I_1\right]$$
$$V_2 = 2V_1 - I_1(6 - 2\alpha). \tag{3}$$

Comparing Equation (3) and (1a) with Equations (10.11) and (10.12),

$$A' = 2, \quad B' = (6 - 2\alpha), \quad C' = \frac{1}{2}, \quad D' = 2.$$

10.7   INTER-RELATIONSHIP OF PARAMETERS

Since all parameters describe the same two-port network, they are inter-related. One set of parameters may be expressed in terms of the other set.

There are the following steps for the transformation:

Step 1. Write the standard equations for both sets.

Step 2. Solve the equations (or rearrange) of the second set for the dependent (unknown) variables of the first set of equations and express the resulting equations in the form of the equations of the first set.

Step 3. Compare the equations obtained in step 2 with those of the first set to obtain the required parameters.

10.7.1 Z-parameters in Terms of Other Parameters

(i) Z-parameters in terms of Y-parameters.

The standard equation of Z- and Y-parameters are

$$V_1 = Z_{11}I_1 + Z_{12}I_2 \qquad \text{(from 10.1)}$$
$$V_2 = Z_{21}I_1 + Z_{22}I_2 \qquad \text{(from 10.2)}$$

and

$$I_1 = Y_{11}V_1 + Y_{12}V_2 \qquad \text{(from 10.3)}$$
$$I_2 = Y_{21}V_1 + Y_{22}V_2. \qquad \text{(from 10.4)}$$

Now rearrange Equations (10.3) and (10.4) into the form of Equations (10.1) and (10.2).

Equation (10.3) $\times Y_{22}$ – Equation (10.4) $\times Y_{12}$,

$$I_1Y_{22} - I_2Y_{12} = V_1(Y_{11}Y_{22} - Y_{21}Y_{12}) = V_1[Y]$$

$$|Y| = \begin{vmatrix} Y_{11} & Y_{12} \\ Y_{21} & Y_{22} \end{vmatrix} = (Y_{11}Y_{22} - Y_{12}Y_{21})$$

$$V_1 = \frac{Y_{22}}{|Y|} \quad I_1 - \frac{Y_{12}}{|Y|}I_2. \qquad (10.13)$$

Similarly,

$$V_2 = -\frac{Y_{21}}{|Y|}I_1 + \frac{Y_{11}}{|Y|}I_2. \qquad (10.14)$$

Now comparing Equations (10.1) and (10.2) with Equations (10.13) and (10.14),

$$Z_{11} = \frac{Y_{22}}{|Y|}, \quad Z_{12} = -\frac{Y_{12}}{|Y|}, \quad Z_{21} = -\frac{Y_{21}}{|Y|}, \quad Z_{22} = \frac{Y_{11}}{|Y|}.$$

(ii) Z-parameters in terms of ABCD parameters.

The standard equation of ABCD parameters,

$$V_1 = AV_2 - BI_2 \qquad \text{(from 10.5)}$$
$$I_1 = CV_2 - DI_2. \qquad \text{(from 10.6)}$$

Now rearrange Equations (10.5) and (10.6) into the form of Equations (10.1 and 10.2),

$$V_2 = \frac{I_1}{C} + \frac{D}{C}I_2 \ [\text{Rearrange Equation 10.6}].\qquad(10.15)$$

Putting the value of V_2 from Equation (10.15) into Equation (10.5),

$$V_1 = A\left[\frac{I_1}{C} + \frac{D}{C}I_2\right]BI_2$$

$$V_1 = \frac{A}{C}I_1 + \left(\frac{AD - BC}{C}\right)I_2 = \frac{A}{C}I_1 + \frac{|T|}{C}I_2\qquad(10.16)$$

$$|T| = \begin{vmatrix} A & B \\ C & D \end{vmatrix} = AD - BC.$$

Now comparing Equations (10.16) and (10.15) with Equations (10.1) and (10.2),

$$Z_{11} = \frac{A}{C}, \quad Z_{12} = \frac{|T|}{C}, \quad Z_{21} = \frac{1}{C}, \quad Z_{22} = \frac{D}{C}.$$

(*iii*) Z-parameters in terms of $A'B'C'D'$ parameters.
The $A'B'C'D'$ parameter equations are

$$V_2 = A'V_1 - B'I_1 \qquad\qquad \text{(from 10.11)}$$
$$I_2 = C'V_1 - D'I_1. \qquad\qquad \text{(from 10.12)}$$

Rearranging Equation (10.12) into the form of Equation (10.1), we get

$$V_1 = \frac{D'}{C'}I_1 + \frac{I_2}{C'}.\qquad(10.17)$$

Putting the value of V_1 from Equation (10.17) into Equation (10.11),

$$V_2 = A'\left[\frac{D'}{C'}I_1 + \frac{I_2}{C'}\right] - B'I_1$$

$$V_2 = \frac{|T'|}{C'}I_1 + \frac{A'}{C'}I_2\qquad(10.18)$$

$$|T'| = \begin{vmatrix} A' & B' \\ C' & D' \end{vmatrix} = A'D' - B'C'.$$

Now comparing Equation (10.17) with (10.1) and (10.18) with (10.2)

$$Z_{11} = \frac{D'}{C'}, \quad Z_{12} = \frac{1}{C'}, \quad Z_{21} = \frac{|T'|}{C'}, \quad Z_{22} = \frac{A'}{C'}.$$

(*iv*) Z-parameters in terms of *h*-parameters.
The *h*-parameters equations are

$$V_1 = h_{11}I_1 + h_{12}V_2 \qquad \text{(from 10.7)}$$

$$I_2 = h_{21}I_1 + h_{22}V_2. \qquad \text{(from 10.8)}$$

The standard equations for Z-parameters are

$$V_1 = Z_{11}I_1 + Z_{12}I_2 \qquad \text{(from 10.1)}$$

$$V_2 = Z_{21}I_1 + Z_{22}I_2. \qquad \text{(from 10.2)}$$

Rearranging Equation (10.8) into the form of Equation (10.2)

$$V_2 = -\frac{h_{21}}{h_{22}}I_1 + \frac{I_2}{h_{22}}. \qquad (10.19)$$

Putting the value of V_2 from Equation (10.19) into Equation (10.7),

$$V_1 = h_{11}I_1 + h_{12}\left[-\frac{h_{21}}{h_{22}}I_1 + \frac{I_2}{h_{22}}\right]$$

$$V_1 = \left(\frac{h_{11} \cdot h_{22} - h_{12} \cdot h_{21}}{h_{22}}\right)I_1 + \frac{h_{12}}{h_{22}}I_2. \qquad (10.20)$$

Comparing Equation (10.20) with (10.1) and (10.19) with (10.2), we get

$$Z_{11} = \frac{|h|}{h_{22}}, \quad Z_{12} = \frac{h_{12}}{h_{22}}, \quad Z_{21} = -\frac{h_{21}}{h_{22}}, \quad Z_{22} = \frac{1}{h_{22}}$$

$$|h| = \begin{vmatrix} h_{11} & h_{12} \\ h_{21} & h_{22} \end{vmatrix} = h_{11}h_{22} - h_{21}h_{12}.$$

(v) Z-parameters in terms of g-parameters.
This is similar to the above case, where we get

$$Z_{11} = \frac{1}{g_{11}}, \quad Z_{12} = -\frac{g_{12}}{g_{11}}, \quad Z_{21} = \frac{g_{21}}{g_{11}}, \quad Z_{22} = \frac{|g|}{g_{11}}$$

$$|g| = \begin{vmatrix} g_{11} & g_{12} \\ g_{21} & g_{22} \end{vmatrix} = g_{11}g_{22} - g_{12}g_{21}.$$

Example 10.8. *Calculate the values of the Z-parameters, if the values of other parameters are as given below:*

(*a*) $A = 2$, $B = -1$, $C = 3$ *and* $D = -2$
(*b*) $h_{11} = 1$, $h_{12} = -2$, $h_{21} = -3$, $h_{22} = 2$
(*c*) $Y_{11} = \frac{1}{3}$, $Y_{12} = \frac{2}{3}$, $Y_{21} = -\frac{1}{3}$, $Y_{22} = \frac{1}{6}$

Solution: (a) $A = 2$, $B = -1$, $C = 3$, $D = -2$.

We know the Z-parameters in terms of the ABCD parameters are

$$Z_{11} = \frac{A}{C} = \frac{2}{3}$$

$$Z_{12} = \frac{|T|}{C} = \frac{AD - BC}{C} = \frac{2 \times -2 - (-1) \times 3}{3} = -\frac{1}{3}$$

$$Z_{21} = \frac{1}{C} = \frac{1}{3}$$

$$Z_{22} = \frac{D}{C} = -\frac{2}{3}.$$

Now the [Z] matrix

$$= \begin{bmatrix} \dfrac{2}{3} & -\dfrac{1}{3} \\ \dfrac{1}{3} & -\dfrac{2}{3} \end{bmatrix}.$$

(b) $h_{11} = 1$, $h_{12} = -2$, $h_{21} = -3$, $h_{22} = 2$

We know the Z-parameters in terms of the h-parameters are

$$Z_{11} = \frac{|h|}{h_{22}} = \frac{h_{11} \cdot h_{22} - h_{21} \cdot h_{12}}{h_{22}} = \frac{1 \times 2 - (-3) \times -2}{2} = -2$$

$$Z_{12} = \frac{h_{12}}{h_{22}} = -1$$

$$Z_{21} = -\frac{h_{21}}{h_{22}} = \frac{3}{2}$$

$$Z_{22} = \frac{1}{h_{22}} = \frac{1}{2}$$

(c) $Y_{11} = \frac{1}{3}$, $Y_{12} = \frac{2}{3}$, $Y_{21} = -\frac{1}{3}$, $Y_{22} = \frac{1}{6}$.

We know the Z-parameters in terms of the Y-parameters are

$$Z_{11} = \frac{Y_{22}}{|Y|} = \frac{Y_{22}}{Y_{11} \cdot Y_{22} - Y_{21} \cdot Y_{12}} = \frac{\frac{1}{6}}{\frac{1}{3} \times \frac{1}{6} + \frac{2}{9}} = \frac{\frac{1}{6}}{\frac{5}{18}} = \frac{3}{5}$$

$$Z_{12} = \frac{Y_{12}}{|Y|} = -\frac{2}{3} \times \frac{18}{5} = -\frac{12}{5} \qquad \left[|Y| = \frac{5}{18} \right]$$

$$Z_{21} = -\frac{Y_{21}}{|Y|} = \frac{1}{3} \times \frac{18}{5} = \frac{6}{5}$$

$$Z_{22} = \frac{Y_{11}}{|Y|} = \frac{1}{3} \times \frac{18}{5} = \frac{6}{5}.$$

10.7.2 Y-parameters in Terms of Other Parameters

(i) Y-parameters in terms of Z-parameters.
  The standard equation of Y- and Z-parameters are

$$I_1 = Y_{11}V_1 + Y_{12}V_2 \qquad \text{(from 10.3)}$$
$$I_2 = Y_{21}V_1 + Y_{22}V_2 \qquad \text{(from 10.4)}$$

and

$$V_1 = Z_{11}I_1 + Z_{12}I_2 \qquad \text{(from 10.1)}$$
$$V_2 = Z_{21}I_1 + Z_{22}I_2. \qquad \text{(from 10.2)}$$

Now rearrange Equations (10.1) and (10.2) into the form of Equations (10.3) and (10.4).
Equation (10.1) × Z_{22} – Equation (10.2) × Z_{12},

$$V_1 Z_{22} - V_2 Z_{12} = I_1(Z_{11} \cdot Z_{22} - Z_{21} \cdot Z_{12})$$

$$I_1 = \frac{Z_{22}}{|Z|}V_1 - \frac{Z_{12}}{|Z|}V_2. \quad (10.21) \ [|Z| = Z_{11} \cdot Z_{22} - Z_{21} \cdot Z_{12}]$$

Similarly,

$$I_2 = -\frac{Z_{21}}{|Z|}V_1 + \frac{Z_{11}}{|Z|}V_2. \qquad (10.22)$$

Comparing Equation (10.21) with (10.3) and (10.22) with (10.4), we get

$$Y_{11} = \frac{Z_{22}}{|Z|}, \quad Y_{12} = -\frac{Z_{12}}{|Z|}, \quad Y_{21} = -\frac{Z_{21}}{|Z|}, \quad Y_{22} = \frac{Z_{11}}{|Z|}.$$

(ii) Y-parameters in terms of ABCD parameters.
The standard equations of ABCD parameters are

$$V_1 = AV_2 - BI_2 \qquad \text{(from 10.5)}$$
$$I_1 = CV_2 - DI_2. \qquad \text{(from 10.6)}$$

Now rearranging Equation (10.5) into the form of Equation (10.4),

$$I_2 = -\frac{V_1}{B} + \frac{A}{B}V_2. \qquad (10.23)$$

Putting the value of I_2 from Equation (10.23) into Equation (10.6),

$$I_1 = CV_2 - D\left[-\frac{V_1}{B} + \frac{A}{B}V_2\right]$$

$$I_1 = \frac{D}{B}V_1 + \frac{BC - AD}{B}V_2. \qquad (10.24)$$

Comparing Equation (10.24) with Equation (10.3) and (10.23) with (10.4), we get

$$Y_{11} = \frac{D}{B}, \quad Y_{12} = \frac{BC - AD}{B} = -\frac{|T|}{B}, \quad Y_{21} = -\frac{1}{B}, \quad Y_{22} = \frac{A}{B}.$$

(*iii*) Y-parameters in terms of A'B'C'D' parameters.
Similar to the above case, we get

$$Y_{11} = \frac{A'}{B'}, \quad Y_{12} = -\frac{1}{B'}, \quad Y_{21} = -\frac{|T'|}{B'}, \quad Y_{22} = \frac{D'}{B'}.$$

(*iv*) Y-parameters in terms of *h*-parameters.
The standard equations of the *h*-parameters are

$$V_1 = h_{11}I_1 + h_{12}V_2 \qquad \text{(from 10.7)}$$
$$I_2 = h_{21}I_2 + h_{22}V_2. \qquad \text{(from 10.8)}$$

Rearranging Equation (10.7) into the form of Equation (10.3),

$$I_1 = \frac{V_1}{h_{11}} - \frac{h_{12}}{h_{11}}V_2. \qquad (10.25)$$

Putting the value of I_1 from Equation (10.25) into Equation (10.8),

$$I_2 = h_{21}\left[\frac{V_1}{h_{11}} - \frac{h_{12}}{h_{11}}V_2\right] - h_{22}V_2$$

or

$$I_2 = \frac{h_{21}}{h_{11}}V_1 + \frac{|h|}{h_{11}}V_2. \qquad (10.26) \; [|h| = h_{11}h_{22} - h_{21}h_{12}]$$

Comparing Equation (10.25) with (10.3) and (10.26) with (10.4), we get

$$Y_{11} = \frac{1}{h_{11}}, \quad Y_{12} = -\frac{h_{12}}{h_{11}}, \quad Y_{21} = \frac{h_{21}}{h_{11}}, \quad Y_{22} = \frac{|h|}{h_{11}}.$$

(v) Y-parameters in terms of *g*-parameters.
Similar to the above case, we get

$$Y_{11} = \frac{|g|}{g_{22}}, \quad Y_{12} = \frac{g_{12}}{g_{22}}, \quad Y_{21} = -\frac{g_{21}}{g_{22}}, \quad Y_{22} = \frac{1}{g_{22}}$$

$$|g| = \begin{bmatrix} g_{11} & g_{12} \\ g_{21} & g_{22} \end{bmatrix} = g_{11}g_{22} - g_{12}g_{21}.$$

Example 10.9. *Calculate the values of the Y-parameters, if the values of other parameters are as given below:*

   (*a*) A = 2, B = −1, C = 3 *and* D = −2
   (*b*) $h_{11} = 1, h_{12} = -2, h_{21} = -3, h_{22} = 2$
   (*c*) $Z_{11} = \frac{1}{3}, Z_{12} = \frac{2}{3}, Z_{21} = -\frac{1}{3}, Z_{22} = \frac{1}{6}$
   (*d*) A′ = −1, B′ = 2, C′ = −2, D′ = 3.

Solution: (a) We know the Y-parameters in terms of the transmission parameters given A = 2, B = −1, C = 3, D = −2 are

$$Y_{11} = \frac{D}{B} = \frac{-2}{-1} = 2$$

$$Y_{12} = -\frac{[T]}{B} = -\frac{(2 \times -2 - (-1) \times 3)}{-1} = -1$$

$$Y_{21} = -\frac{1}{B} = 1$$

$$Y_{22} = \frac{A}{B} = -2.$$

(b) The Y-parameters in terms of hybrid (h) parameters are

$$Y_{11} = \frac{1}{h_{11}} = 1$$

$$Y_{12} = -\frac{h_{12}}{h_{11}} = 2$$

$$Y_{21} = \frac{h_{21}}{h_{11}} = -3$$

$$Y_{22} = \frac{|h|}{h_{11}} = -4. \qquad\qquad [|h| = h_{11}h_{22} - h_{21}h_{12} = -4]$$

(c) $Z_{11} = \frac{1}{3}, \quad Z_{12} = \frac{2}{3}, \quad Z_{21} = -\frac{1}{3}, \quad Z_{22} = \frac{1}{6}.$
We know the Y-parameters in terms of the Z-parameters are

$$Y_{11} = \frac{Z_{22}}{|Z|} = \frac{\frac{1}{6}}{\frac{1}{3} \times \frac{1}{6} + \frac{2}{3} \times \frac{1}{3}} = \frac{\frac{1}{6}}{\frac{5}{18}} = \frac{3}{5} [|Z| = Z_{11}Z_{22} - Z_{21}Z_{12}]$$

$$Y_{12} = -\frac{Z_{12}}{|Z|} = -\frac{2}{3} \times \frac{18}{5} = -\frac{12}{5}$$

$$Y_{21} = -\frac{Z_{21}}{|Z|} = \frac{1}{3} \times \frac{18}{5} = \frac{6}{5}$$

$$Y_{22} = \frac{Z_{11}}{|Z|} = \frac{1}{3} \times \frac{18}{5} = \frac{6}{5}.$$

(d) $A' = -1$, $B' = 2$, $C' = -2$, $D' = 3$.

$$Y_{11} = \frac{A'}{B'} = -\frac{1}{2}$$

$$Y_{12} = -\frac{1}{B'} = -\frac{1}{2}$$

$$Y_{21} = -\frac{|T'|}{B'} = -\frac{1}{2}$$

$$Y_{22} = \frac{D'}{B'} = \frac{3}{2}[|T'| = A'D' - B'C' = 1].$$

10.7.3 ABCD or T-parameters in Terms of Other Parameters

(*i*) ABCD parameters in terms of Z-parameters.
The standard equations for ABCD and Z-parameters are

$$V_1 = AV_2 - BI_2 \qquad \text{(from 10.5)}$$
$$I_1 = CV_2 - DI_2 \qquad \text{(from 10.6)}$$

and

$$V_1 = Z_{11}I_1 + Z_{12}I_2 \qquad \text{(from 10.1)}$$
$$V_2 = Z_{21}I_1 + Z_{22}I_2. \qquad \text{(from 10.2)}$$

Now rearranging Equation (10.2) into the form of Equation (10.6),

$$I_1 = \frac{1}{Z_{21}}V_2 - \frac{Z_{22}}{Z_{21}}I_2. \qquad (10.27)$$

Putting the value of I_1 from Equation (10.27) into Equation (10.1),

$$V_1 = Z_{11}\left[\frac{V_2}{Z_{21}} - \frac{Z_{22}}{Z_{21}}I_2\right] + Z_{12}I_2$$

or

$$V_1 = \frac{Z_{11}}{Z_{21}}V_2 - \frac{|Z|}{Z_{21}}I_2. \qquad (10.28)\ [|Z| = Z_{11}Z_{22} - Z_{12}Z_{21}]$$

Comparing Equation (10.28) with Equation (10.5) and (10.27) with (10.6), we get

$$A = \frac{Z_{11}}{Z_{21}}, \quad B = \frac{|Z|}{Z_{21}}, \quad C = \frac{1}{Z_{21}}, \quad D = \frac{Z_{22}}{Z_{21}}.$$

(*ii*) ABCD parameters in terms of Y-parameters.
Similar to the above case, we get

$$A = -\frac{Y_{22}}{Y_{21}}, \quad B = -\frac{1}{Y_{21}}, \quad C = -\frac{|Y|}{Y_{21}}, \quad D = -\frac{Y_{11}}{Y_{21}}.$$

(*iii*) ABCD parameters in terms of $A'B'C'D'$ parameters. The standard equations for $A'B'C'D'$ parameters are

$$V_2 = A'V_1 - B'I_1 \qquad \text{(from 10.11)}$$
$$I_2 = C'V_1 - D'I_1. \qquad \text{(from 10.12)}$$

Equation (10.11) $\times$ D' $-$ Equation (10.12) $\times$ B', we get

$$V_1 = \frac{D\prime}{|T\prime|}V_2 - \frac{B\prime}{|T\prime|}I_2. \qquad (10.29)$$

Similarly we get,

$$I_1 = \frac{C'}{|T'|}V_2 - \frac{A'}{|T'|}I_2. \qquad (10.30)$$

Comparing Equation (10.29) with (10.5) and (10.30) with (10.6),

$$A = \frac{D'}{|T'|}, \quad B = \frac{B'}{|T'|}, \quad C = \frac{C'}{|T'|}, \quad D = \frac{A'}{|T'|}.$$

(*iv*) ABCD parameters in terms of h-parameters.
The h-parameter Equations (10.7) and (10.8) are

$$V_1 = h_{11}I_1 + h_{12}V_2$$
$$I_2 = h_{21}I_1 + h_{22}V_2.$$

Rearrange Equation (10.8),

$$I_1 = -\frac{h_{22}}{h_{21}}V_2 + \frac{I_2}{h_{21}}. \qquad (10.31)$$

Putting the value of I_1 from Equation (10.31) into Equation (10.7), we get

$$V_1 = h_{11}\left[-\frac{h_{22}}{h_{21}}V_2 + \frac{I_2}{h_{21}}\right] + h_{12}V_2$$
$$V_1 = -\frac{|h|}{h_{21}}V_2 + \frac{h_{11}}{h_{21}}I_2. \qquad (10.32)$$

Comparing Equation (10.32) with Equation (10.5) and (10.31) with Equation (10.6), we get

$$A = -\frac{|h|}{h_{21}}, \quad B = -\frac{h_{11}}{h_{21}}, \quad C = -\frac{h_{22}}{h_{21}} \text{ and } D = -\frac{1}{h_{21}}.$$

(v) ABCD parameters in terms of g-parameters.
Similar to the above case, we get

$$A = \frac{1}{g_{21}}, \quad B = \frac{g_{22}}{g_{21}}, \quad C = \frac{g_{11}}{g_{21}} \text{ and } D = \frac{|g|}{g_{21}}.$$

Example 10.10. *Calculate the value of the ABCD parameters. The other parameters value are as given below:*

(a) $g_{11} = 2, g_{12} = -1, g_{21} = 3, g_{22} = -2$
(b) $h_{11} = 1, h_{12} = -2, h_{21} = -3, h_{22} = 2$
(a) $Z_{11} = \frac{1}{3}, Z_{12} = \frac{2}{3}, Z_{21} = -\frac{1}{3}, Z_{22} = \frac{1}{6}$
(d) $A' = -1, B' = 2, C' = -2, D' = 3.$

Solution:

(a)
$$A = \frac{1}{g_{21}} = \frac{1}{3}, \quad B = \frac{g_{22}}{g_{21}} = -\frac{2}{3}$$

$$C = \frac{g_{11}}{g_{21}} = \frac{2}{3}$$

$$D = \frac{|g|}{g_{21}} = \frac{g_{11} \cdot g_{22} - g_{12} \cdot g_{21}}{g_{21}} = -\frac{1}{3}.$$

(b) ABCD parameters in terms of h-parameters

$$A = -\frac{|h|}{h_{21}} = -\frac{4}{3} \qquad [|h| = h_{11} \cdot h_{22} - h_{21} \cdot h_{12} = 1 \times 2 - 6 = -4]$$

$$B = \frac{-h_{11}}{h_{21}} = \frac{1}{3}$$

$$C = -\frac{h_{22}}{h_{21}} = \frac{2}{3}$$

$$D = -\frac{1}{h_{21}} = \frac{1}{3}$$

(c) ABCD parameters in terms of Z-parameters

$$A = \frac{Z_{11}}{Z_{21}} = -1$$

$$B = \frac{|Z|}{Z_{21}} = -\frac{5}{8} \times \frac{3}{1} = -\frac{5}{6}$$

$$C = \frac{1}{Z_{21}} = -3$$

$$D = \frac{Z_{22}}{Z_{21}} = -\frac{1}{2}$$

(d) ABCD parameters in terms of A'B'C'D' parameters

$$A = \frac{D'}{|T'|} = 3 \qquad\qquad [|T'| = A'D' - B'C' = 1]$$

$$B = \frac{B'}{|T'|} = 2$$

$$C = \frac{C'}{|T'|} = -2$$

$$D = \frac{A'}{|T'|} = -1$$

10.7.4 *h*-parameters in Terms of Other Parameters

(i) *h*-parameters in terms of Z-parameters.
The Z-parameters of Equations (10.1) and (10.2) are

$$V_1 = Z_{11}I_1 + Z_{12}I_2$$
$$V_2 = Z_{21}I_1 + Z_{22}I_2.$$

The *h*-parameters of Equations (10.7) and (10.8) are

$$V_1 = h_{11}I_1 + h_{12}V_2$$
$$I_2 = h_{21}I_1 + h_{22}V_2.$$

Rearranging Equation (10.2) into the form of Equation (10.8), we get

$$I_2 = -\frac{Z_{21}}{Z_{22}}I_1 + \frac{1}{Z_{22}}V_2. \qquad (10.33)$$

Putting the value of I_2 from Equation (10.33) into Equation (10.1),

$$V_1 = Z_{11}I_1 + Z_{12}\left[-\frac{Z_{21}}{Z_{22}}I_1 + \frac{1}{Z_{22}}V_2\right]$$

or

$$V_1 = \frac{|Z|}{Z_{22}}I_1 + \frac{Z_{12}}{Z_{22}}V_2. \qquad (10.34)$$

Comparing Equation (10.34) with Equation (10.7) and (10.33) with Equation (10.8), we get

$$h_{11} = \frac{|Z|}{Z_{22}}, \quad h_{12} = \frac{Z_{12}}{Z_{22}}, \quad h_{21} = -\frac{Z_{21}}{Z_{22}}, \quad h_{22} = \frac{1}{Z_{22}}.$$

(*ii*) *h*-parameters in terms of Y-parameters.
This is similar to the above case, and we get

$$h_{11} = \frac{1}{Y_{11}}, \quad h_{12} = -\frac{Y_{12}}{Y_{11}}, \quad h_{21} = \frac{Y_{21}}{Y_{11}}, \quad h_{22} = \frac{|Y|}{Y_{11}}.$$

(*iii*) *h*-parameters in terms of T-parameters or ABCD parameters.
The ABCD parameters of Equations (10.5) and (10.6) are

$$V_1 = AV_2 - BI_2$$
$$I_1 = CV_2 - DI_2.$$

Rearranging Equation (10.6) into the form of Equation (10.8), we get

$$I_2 = -\frac{1}{D}I_1 + \frac{C}{D}V_2. \tag{10.35}$$

Putting the value of I_2 from Equation (10.35) into Equation (10.5), we get

$$V_1 = AV_2 - B\left[-\frac{1}{D}I_1 + \frac{C}{D}V_2\right]$$

or

$$V_1 = \frac{B}{D}I_1 + \frac{|T|}{D}V_2. \tag{10.36}$$

Comparing Equation (10.36) with (10.7) and (10.35) with 10.8, we get

$$h_{11} = \frac{B}{D}, \quad h_{12} = \frac{|T|}{D}, \quad h_{21} = -\frac{1}{D}, \quad h_{22} = \frac{C}{D}.$$

(*iv*) *h*-parameters in terms of T′-parameters.
This is similar to the above case, and we get

$$h_{11} = \frac{B'}{A'}, \quad h_{12} = \frac{1}{A'}, \quad h_{21} = -\frac{|T'|}{A'} \text{ and } h_{22} = \frac{C'}{A'}.$$

(v) *h*-parameters in terms of g-parameters.
The *g*-parameters of Equations (10.9) and (10.10) are

$$I_1 = g_{11}V_1 + g_{12}I_2$$
$$V_2 = g_{21}V_1 + g_{22}I_2.$$

Equation (10.9) $\times g_{22}$ – Equation (10.10) $\times g_{12}$, we get

$$V_1 = \frac{g_{22}}{|g|}I_1 - \frac{g_{12}}{|g|}V_2. \tag{10.37}$$

Similarly, we get

$$I_2 = \frac{g_{21}}{|g|}I_1 - \frac{g_{11}}{|g|}V_2. \tag{10.38}$$

Comparing Equation (10.37) with Equation (10.7) and (10.38) with Equation (10.8), we get

$$h_{11} = \frac{g_{22}}{|g|}, \quad h_{12} = -\frac{g_{12}}{|g|}, \quad h_{21} = -\frac{g_{21}}{|g|}, \quad h_{22} = \frac{g_{11}}{|g|}.$$

10.8 CONDITION FOR RECIPROCITY

A two-port network is said to be reciprocal if the ratio of the excitation to the response is invariant to an interchange of the positions of the excitation and response in the network as shown in Figure 10.16.

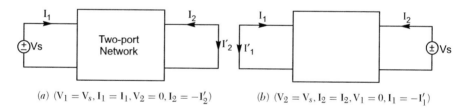

(a) $(V_1 = V_s, I_1 = I_1, V_2 = 0, I_2 = -I_2')$ (b) $(V_2 = V_s, I_2 = I_2, V_1 = 0, I_1 = -I_1')$

FIGURE 10.16

According to the condition of reciprocity,

$$\left[\frac{I_1}{V_1}\right]_{V_2=0} = \left[\frac{I_2}{V_2}\right]_{V_1=0}.$$

By the condition of Figure 10.16(a) and Figure 10.16(b),

$$\frac{I_1}{V_s} = \frac{I_2}{V_s} \text{ or } I_1 = I_2 \text{ or } -I_1' = -I_2' = \boxed{I_1' = I_2'}.$$

Networks containing resistors, inductors, and capacitors are generally reciprocal. Networks that have dependent sources are generally non-reciprocal.

(i) Condition of reciprocity for Z-parameters,

$$V_1 = Z_{11}I_1 + Z_{12}I_2$$
$$V_2 = Z_{21}I_1 + Z_{22}I_2.$$

By the condition of Figure 10.16(a) $V_1 = V_s$, $I_1 = I_1$, $V_2 = 0$, $I_2 = -I'_2$, we get

$$V_s = Z_{11}I_1 - Z_{12}I'_2$$
$$0 = Z_{21}I_1 - Z_{22}I'_2.$$

By the above two equations, we get

$$I'_2 = \frac{Z_{21}V_s}{|Z|}. \qquad (10.39)\ [|Z| = Z_{11}Z_{22} - Z_{21}Z_{12}]$$

As in Figure 10.16(b) $V_2 = V_s$, $I_2 = I_2$, $V_1 = 0$, $I_1 = -I'_1$, and we get

$$0 = -Z_{11}I'_1 + Z_{12}I_2$$
$$V_s = -Z_{21}I'_1 + Z_{22}I_2.$$

Hence,

$$I'_1 = \frac{Z_{12}}{|Z|}V_s, \qquad (10.40)$$

but for the reciprocal network $I'_1 = I'_2$, therefore,

$$\boxed{Z_{12} = Z_{21}} \qquad (10.41)$$

(ii) Condition of reciprocity for Y-parameters,

$$I_1 = Y_{11}V_1 + Y_{12}V_2$$
$$I_2 = Y_{21}V_1 + Y_{22}V_2.$$

By the condition of Figure 10.16(a) $V_1 = V_s$, $I_1 = I_1$, $V_2 = 0$, $I_2 = -I'_2$,

$$I'_2 = -Y_{21}V_s. \qquad (10.42)$$

By the condition of Figure 10.16(b) $V_2 = V_s$, $I_2 = I_2$, $V_1 = 0$, $I_1 = -I'_1$,

$$I'_1 = -Y_{12}V_s. \qquad (10.43)$$

For reciprocity $I'_1 = I'_2$, therefore,

$$Y_{12} = Y_{21}. \qquad (10.44)$$

(*iii*) Condition of reciprocity for T-parameters,

$$V_1 = AV_2 - BI_2$$
$$I_1 = CV_2 - DI_2.$$

By the condition of Figure 10.16(*a*) $V_1 = V_s$, $I_1 = I_1$, $V_2 = 0$, $I_2 = -I_2'$,

$$I_2' = \frac{V_s}{B}. \tag{10.45}$$

By the condition of Figure 10.16(*b*) $V_2 = V_s$, $I_2 = I_2$, $V_1 = 0$, $I_1 = -I_1'$,

$$I_1' = V_s \left(\frac{AD - BC}{B} \right). \tag{10.46}$$

For reciprocity $I_1' = I_2'$, therefore,

$$AD - BC = 1 \text{ or } |T| = 1. \tag{10.47}$$

(*iv*) Condition of reciprocity for $A'B'C'D'$ or T'-parameters and condition of reciprocity in the case of T'-parameters is similar, to the case of T-parameters,

$$A'D' - B'C' = 1 \text{ or } |T'| = 1.$$

(v) Condition of reciprocity for *h*-parameters,

$$V_1 = h_{11}I_1 + h_{12}V_2$$
$$I_2 = h_{21}I_1 + h_{22}V_2.$$

By the condition of Figure 10.16(*a*) $V_1 = V_s$, $I_1 = I_1$, $V_2 = 0$, $I_2 = -I_2'$,

$$I_2' = -V_s \frac{h_{21}}{h_{11}}. \tag{10.48}$$

By the condition of Figure 10.16(*b*) $V_2 = V_s$, $I_2 = I_2$, $V_1 = 0$, $I_1 = -I_1'$,

$$I_1' = V_s \frac{h_{12}}{h_{11}}. \tag{10.49}$$

From the definition of reciprocity, $I_1' = I_2'$, we get

$$h_{12} = -h_{21}. \tag{10.50}$$

(vi) Condition of reciprocity for *g*-parameters
In this case, the condition of reciprocity is similar to that of *h*-parameters,

$$g_{12} = -g_{21}. \tag{10.51}$$

Example 10.11. *Check whether the given network as shown in Figure 10.17 is reciprocal or not.*

Solution: The two loop equations are

$$V_1 = 2I_1 + 2I_2 + 2(I_1 + I_2)$$

or

$$V_1 = 4I_1 + 4I_2 \tag{1}$$
$$V_2 = 2I_2 + 2(I_1 + I_2)$$

or

$$V_2 = 2I_1 + 4I_2. \tag{2}$$

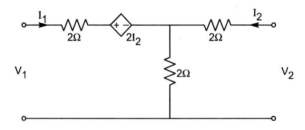

FIGURE 10.17

The values of the Z-parameters are

$$Z_{11} = 4, \quad Z_{12} = 4, \quad Z_{21} = 2, \quad Z_{22} = 4,$$

and we know the condition of reciprocity for the Z-parameters,
$Z_{12} = Z_{21}$, but in this case $Z_{12} \neq Z_{21}$, therefore, the given network is not a reciprocal network.

10.9 CONDITION FOR SYMMETRY

A two-port network is said to be symmetrical if the ports of the two-port network can be interchanged without changing the port voltages and currents as shown in Figure 10.18(a) and (b), and the condition of symmetry is

$$\frac{V_1}{I_1}\bigg|_{I_2=0} = \frac{V_2}{I_2}\bigg|_{I_1=0} \quad \text{or} \quad \frac{V_s}{I_1}\bigg|_{I_2=0} = \frac{V_s}{I_2}\bigg|_{I_1=0}.$$

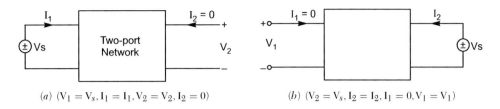

(a) ($V_1 = V_s, I_1 = I_1, V_2 = V_2, I_2 = 0$)   (b) ($V_2 = V_s, I_2 = I_2, I_1 = 0, V_1 = V_1$)

FIGURE 10.18

(*i*) Condition of symmetry for Z-parameters.
By the condition of Figure 10.18(*a*) $V_1 = V_s$, $I_1 = I_1$, $V_2 = V_2$, $I_2 = 0$,

$$V_s = Z_{11}I_1$$

$$\left.\frac{V_s}{I_1}\right|_{I_2=0} = Z_{11}. \tag{10.52}$$

By the condition of Figure 10.18(*b*) $V_2 = V_s$, $I_2 = I_2$, $I_1 = 0$, $V_1 = V_1$,

$$V_s = Z_{22}I_2$$

$$\left.\frac{V_s}{I_2}\right|_{I_1=0} = Z_{22}. \tag{10.53}$$

From the definition of symmetry $\left.\dfrac{V_s}{I_1}\right|_{I_2=0} = \left.\dfrac{V_s}{I_2}\right|_{I_1=0}$,

$$Z_{11} = Z_{22}. \tag{10.54}$$

(*ii*) Condition of symmetry for Y-parameters,

$$I_1 = Y_{11}V_1 + Y_{12}V_2$$
$$I_2 = Y_{21}V_1 + Y_{22}V_2.$$

By the condition of Figure 10.18(*a*) $V_1 = V_s$, $I_1 = I_1$, $I_2 = 0$, $V_2 = V_2$,

$$I_1 = Y_{11}V_s + Y_{12}V_2$$
$$O = Y_{21}V_s + Y_{22}V_2$$
$$I_1 = Y_{11}V_s + Y_{12}\left\{\frac{-Y_{21}}{Y_{22}}\right\}V_s$$
$$\frac{V_s}{I_1} = \frac{Y_{22}}{Y_{11} \cdot Y_{22} - Y_{12} \cdot Y_{21}}. \tag{10.55}$$

By the condition of Figure 10.18(b) $V_2 = V_s$, $I_2 = I_2$, $I_1 = 0$, $V_1 = V_1$,

$$O = Y_{11}V_1 + Y_{12}V_s$$

$$I_2 = Y_{21}V_1 + Y_{22}V_s$$

$$\frac{V_s}{I_2} = \frac{Y_{11}}{Y_{11} \cdot Y_{22} - Y_{12} \cdot Y_{21}} \qquad (10.56)$$

$$\frac{V_s}{I_1}\bigg|_{I_2=0} = \frac{V_s}{I_2}\bigg|_{I_1=0}.$$

$$Y_{11} = Y_{22} \qquad (10.57)$$

(*iii*) Condition of symmetry for T-parameters.
The T-parameter equations are

$$V_1 = AV_2 - BI_2$$

$$I_1 = CV_2 - DI_2.$$

By the condition of Figure 10.18(*a*) $V_1 = V_s$, $I_1 = I_1$, $I_2 = 0$, $V_2 = V_2$, we get

$$V_s = AV_2$$

$$I_1 = CV_2$$

then,

$$\frac{V_s}{I_1} = \frac{A}{C}. \qquad (10.58)$$

By the condition of Figure 10.18(*b*) $V_2 = V_s$, $I_2 = I_2$, $I_1 = 0$, $V_1 = V_1$,

$$V_1 = AV_s - BI_2$$

$$O = CV_s - DI_2 \quad \text{or} \quad \frac{V_s}{I_2} = \frac{D}{C}. \qquad (10.59)$$

The condition for symmetry is

$$\frac{V_s}{I_1}\bigg|_{I_2=0} = \frac{V_s}{I_2}\bigg|_{I_1=0}, \text{ and we get by Equations (10.57) and (10.58)}$$

$$A = D. \qquad (10.60)$$

(*iv*) Condition of symmetry for T'-parameters.
Condition of symmetry in the case of T'-parameters is similar to T-parameters,

$$A' = D'. \qquad (10.61)$$

(v) Condition of symmetry for h-parameters.

The h-parameter equations are

$$V_1 = h_{11}I_1 + h_{12}V_2$$
$$I_2 = h_{21}I_1 + h_{22}V_2.$$

By the condition of Figure 10.18(a) $V_1 = V_s$, $I_1 = I_1$, $I_2 = 0$, $V_2 = V_2$,

$$V_s = h_{11}I_1 + h_{12}V_2$$
$$0 = h_{21}I_1 + h_{22}V_2$$
$$\frac{V_s}{I_1} = \frac{h_{11} \cdot h_{22} - h_{12} \cdot h_{21}}{h_{22}}.$$

By the condition of Figure 10.18(b) $V_2 = V_s$, $I_2 = I_2$, $I_1 = 0$, $V_1 = V_1$,

$$V_1 = h_{12}V_s$$
$$I_2 = h_{22}V_s \text{ or } \frac{V_s}{I_2} = \frac{1}{h_{22}}.$$

The condition for symmetry is

$$\left. \frac{V_s}{I_1} \right|_{I_2=0} = \left. \frac{V_s}{I_2} \right|_{I_1=0}$$
$$\frac{h_{11} \cdot h_{22} - h_{12} \cdot h_{21}}{h_{22}} = \frac{1}{h_{22}}$$
$$h_{11} \cdot h_{22} - h_{12} \cdot h_{21} = 1 \text{ or } |h| = 1.$$

(vi) Condition of symmetry for g-parameters.

The case of g-parameters is similar to the case of h-parameters,

$$g_{11} \cdot g_{22} - g_{12} \cdot g_{21} = 1 \text{ or } |g| = 1$$

TABLE 10.2   Parameter Relationships for Symmetrical and Reciprocal Networks

Parameters	Condition of Symmetry	Condition of Reciprocity
[Z]	$Z_{11} = Z_{22}$	$Z_{12} = Z_{21}$
[Y]	$Y_{11} = Y_{22}$	$Y_{12} = Y_{21}$
[T] or [ABCD]	$A = D$	$AD - BC = 1$ or $|T| = 1$
[T′] or [A′B′C′D′]	$A' = D'$	$A'D' - B'C' = 1$ or $|T'| = 1$
[h]	$h_{11} \cdot h_{12} - h_{12} \cdot h_{21} = 1$	$h_{12} = -h_{21}$
[g]	$g_{11} \cdot h_{22} - h_{12} \cdot h_{21} = 1$	$g_{12} = -g_{21}$

or

$$\begin{vmatrix} g_{11} & g_{12} \\ g_{21} & g_{22} \end{vmatrix} = 1.$$

Example 10.12. *Check whether the given network as shown in Figure 10.18(c) is symmetrical.*

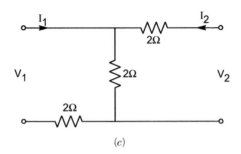

(c)

FIGURE 10.18

Solution: The two loop equations are

$$V_1 = 2(I_1 + I_2) + 2I_1$$

or

$$V_1 = 4I_1 + 2I_2 \tag{1}$$
$$V_2 = 2I_2 + 2(I_1 + I_2)$$

or

$$V_2 = 2I_1 + 4I_2. \tag{2}$$

We get the Z-parameter by Equations (1) and (2),

$$Z_{11} = 4, \quad Z_{12} = 2, \quad Z_{21} = 2, \quad Z_{22} = 4.$$

In this case $\boxed{Z_{11} = Z_{22} = 4}$ therefore the given network is symmetrical.

10.10 INTERCONNECTIONS OF TWO-PORT NETWORKS

Two-port networks may be interconnected in various configurations, such as series, parallel, cascade, series-parallel, and parallel-series connections. For each configuration a certain set of parameters may be more useful than others to describe the network.

10.10.1 Series Connection

Figure 10.19 shows a series connection of two-port networks N_a and N_b.

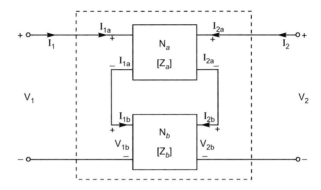

FIGURE 10.19   Series connection of two two-port networks.

For network N_a,

$$\begin{bmatrix} V_{1a} \\ V_{2a} \end{bmatrix} = \begin{bmatrix} Z_{11a} & Z_{12a} \\ Z_{21a} & Z_{22a} \end{bmatrix} \begin{bmatrix} I_{1a} \\ I_{2a} \end{bmatrix}$$

$$V_{1a} = Z_{11a}I_{1a} + Z_{12a}I_{2a} \tag{10.62}$$

$$V_{2a} = Z_{21a}I_{1a} + Z_{22a}I_{2a}. \tag{10.63}$$

For network N_b,

$$\begin{bmatrix} V_{1b} \\ V_{2b} \end{bmatrix} = \begin{bmatrix} Z_{11b} & Z_{12b} \\ Z_{21b} & Z_{22b} \end{bmatrix} \begin{bmatrix} I_{1b} \\ I_{2b} \end{bmatrix}$$

$$V_{1b} = Z_{11b}I_{1b} + Z_{12b}I_{2b} \tag{10.64}$$

$$V_{2b} = Z_{21b}I_{1b} + Z_{22b}I_{2b} \tag{10.65}$$

The condition for series connection is

$$I_{1a} = I_{1b} = I_1, \text{ and } I_2 = I_{2b} = I_2 \text{ (current same)}$$

$$V_1 = V_{1a} + V_{1b} \tag{10.66}$$

$$V_2 = V_{2a} + V_{2b}. \tag{10.67}$$

Putting the values of V_{1a} and V_{1b} from Equation (10.62) and Equation (10.64),

$$V_1 = Z_{11a}I_{1a} + Z_{12a}I_{2a} + Z_{11b}I_{1b} + Z_{12b}I_{2b}$$

$$= Z_{11a}I_1 + Z_{12a}I_2 + Z_{11b}I_1 + Z_{12b}I_2 \qquad [I_{1a} = I_{1b} = I_1, I_{2a} = I_{2b} = I_2]$$

$$V_1 = (Z_{11a} + Z_{11b})I_1 + (Z_{12a} + Z_{12b})I_2. \tag{10.68}$$

Putting the values of V_{2a} and V_{2b} from Equations (10.63) and (10.65) into Equation (10.67), we get

$$V_2 = (Z_{21a} + Z_{21b})I_1 + (Z_{22a} + Z_{22b})I_2. \qquad (10.69)$$

The Z-parameters of the series-connected combined network can be written as

$$V_1 = Z_{11}I_1 + Z_{12}I_2$$
$$V_2 = Z_{21}I_1 + Z_{22}I_2,$$

where

$$Z_{11} = Z_{11a} + Z_{11b}$$
$$Z_{12} = Z_{12a} + Z_{12b}$$
$$Z_{21} = Z_{21a} + Z_{21b}$$
$$Z_{22} = Z_{22a} + Z_{22b}$$

or in the matrix form,

$$[Z] = [Z_a] + [Z_b].$$

The overall Z-parameter matrix for series connected two-port networks is simply the sum of Z-parameter matrices of each individual two-port network connected in series.

10.10.2 Parallel Connection

Figure 10.20 shows a parallel connection of two two-port networks N_a and N_b. The resultant of two admittances connected in parallel is $Y_1 + Y_2$. So in parallel connection, the parameters are Y-parameters.

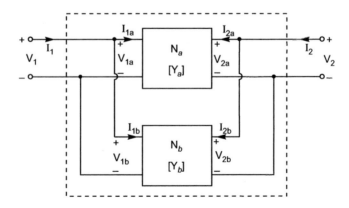

FIGURE 10.20   **Parallel connections for two two-port networks.**

For network N_a,

$$\begin{bmatrix} I_{1a} \\ I_{2a} \end{bmatrix} = \begin{bmatrix} Y_{11a} & Y_{12a} \\ Y_{21a} & Y_{22a} \end{bmatrix} \begin{bmatrix} V_{1a} \\ V_{2a} \end{bmatrix}$$

or

$$I_{1a} = Y_{11a}V_{1a} + Y_{12a}V_{2a} \tag{10.70}$$

$$I_{2a} = Y_{21a}V_{1a} + Y_{22a}V_{2a}. \tag{10.71}$$

For network N_b,

$$\begin{bmatrix} I_{1b} \\ I_{2b} \end{bmatrix} = \begin{bmatrix} Y_{11b} & Y_{12b} \\ Y_{21b} & Y_{22b} \end{bmatrix} \begin{bmatrix} V_{1b} \\ V_{2b} \end{bmatrix}$$

$$I_{1b} = Y_{11b}V_{1b} + Y_{12b}V_{2b} \tag{10.72}$$

$$I_{2b} = Y_{21b}V_{1b} + Y_{22b}V_{2b}. \tag{10.73}$$

Now the condition for parallel,

$$V_{1a} = V_{1b} = V_1, \quad V_{2a} = V_{2b} = V_2 \quad \text{[Same voltage]}$$

and

$$I_1 = I_{1a} + I_{1b} \tag{10.74}$$

$$I_2 = I_{2a} + I_{2b} \tag{10.75}$$

$$I_1 = Y_{11a}V_{1a} + Y_{12a}V_{2a} + Y_{11b}V_{1b} + Y_{12b}V_{2b}$$

$$= Y_{11a}V_1 + Y_{12a}V_2 + Y_{11b}V_1 + Y_{12b}V_2$$

$$I_1 = (Y_{11a} + Y_{11b})V_1 + (Y_{12a} + Y_{12b})V_2. \tag{10.76}$$

Similarly,

$$I_2 = (Y_{21a} + Y_{21b})V_1 + (Y_{22a} + Y_{22b})V_2. \tag{10.77}$$

The Y-parameters of the parallel connected combined network can be written as

$$I_1 = Y_{11}V_1 + Y_{12}V_2$$

$$I_2 = Y_{21}V_1 + Y_{22}V_2.$$

Comparing the above equation with Equation (10.76) and (10.77), we get

$$Y_{11} = Y_{11a} + Y_{11b}$$

$$Y_{12} = Y_{12a} + Y_{12b}$$

$$Y_{21} = Y_{21a} + Y_{21b}$$

$$Y_{22} = Y_{22a} + Y_{22b}$$

or in the matrix form

$$[Y] = [Y_a] + [Y_b].$$

The overall Y-parameter matrix for parallel connected two-port networks is simply the sum of Y-parameter matrices of each individual two-port network connected in parallel.

Example 10.13. *Two identical networks as shown in Figure 10.21 are connected in series. Determine the Z-parameters of the combined network.*

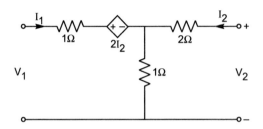

FIGURE 10.21

Solution: The two loop equations are

$$V_1 = I_1 + 2I_2 + 1 \cdot (I_1 + I_2)$$

or

$$V_1 = 2I_1 + 3I_2 \tag{1}$$
$$V_2 = 2I_2 + 1 \cdot (I_1 + I_2)$$

or

$$V_2 = I_1 + 3I_2. \tag{2}$$

The Z-parameters of the network are

$$Z_{11} = 2, \quad Z_{12} = 3, \quad Z_{21} = 1, \quad Z_{22} = 3.$$

For network N_a,

$$[Z_a] = \begin{bmatrix} 2 & 3 \\ 1 & 3 \end{bmatrix} = [Z_b]. \qquad \text{[given networks are identical]}$$

In series connection,

$$[Z] = [Z_a] + [Z_b] = 2[Z_a] = \begin{bmatrix} 4 & 6 \\ 2 & 6 \end{bmatrix}.$$

Example 10.14. *Two identical networks as shown in Figure 10.21 are connected in parallel. Determine the Y-parameters of the combined network.*

Solution: By the last question (Example 10.13),

$$V_1 = 2I_1 + 3I_2 \tag{1}$$
$$V_2 = I_1 + 3I_2. \tag{2}$$

By Equations (1) and (2),

$$I_2 = -\frac{1}{3}V_1 + \frac{2}{3}V_2 \tag{3}$$

and

$$I_1 = V_1 - V_2. \tag{4}$$

The Y-parameters of the network are

$$Y_{11} = 1, \quad Y_{12} = -1, \quad Y_{21} = -\frac{1}{3}, \quad Y_{22} = \frac{2}{3}.$$

For network N_a,

$$[Y_a] = \begin{bmatrix} 1 & -1 \\ -\frac{1}{3} & \frac{2}{3} \end{bmatrix} = [Y_b]. \qquad \text{[Networks are identical]}$$

In parallel connection,

$$[Y] = [Y_a] + [Y_b] = 2[Y_a]$$
$$[Y] = \begin{bmatrix} 2 & -2 \\ -\frac{2}{3} & \frac{4}{3} \end{bmatrix}.$$

10.10.3 Cascade Connection

Two two-port networks are said to be connected in cascade if the output port of the first network becomes the input port of the second network as shown in Figure 10.22(a). Two transmission lines are connected in this manner. So transmission is suitable for cascade connection.

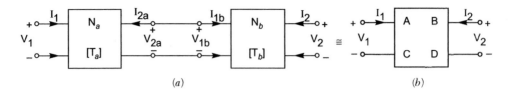

FIGURE 10.22   Cascade connection of two two-port networks.

If the T_a and T_b are T-parameters of the network N_a and N_b, respectively, then for network N_a,

$$\begin{bmatrix} V_1 \\ I_1 \end{bmatrix} = \begin{bmatrix} A_1 & B_1 \\ C_1 & D_1 \end{bmatrix} \begin{bmatrix} V_{2a} \\ -I_{2a} \end{bmatrix}$$

$$V_1 = A_1 V_{2a} - B_1 I_{2a} \qquad (10.78)$$

$$I_1 = C_1 V_{2a} - D_1 I_{2a}. \qquad (10.79)$$

For network N_b,

$$\begin{bmatrix} V_{1b} \\ I_{1b} \end{bmatrix} = \begin{bmatrix} A_2 & B_2 \\ C_2 & D_2 \end{bmatrix} \begin{bmatrix} V_2 \\ -I_2 \end{bmatrix}$$

$$V_{1b} = A_2 V_2 - B_2 I_2 \qquad (10.80)$$

$$I_{1b} = C_2 V_2 - D_2 I_2. \qquad (10.81)$$

From the figure,

$$I_{1b} = -I_{2a} \quad \text{and} \quad V_{2a} = V_{1b}.$$

Putting these values into Equations (10.78) and (10.79), we get

$$V_1 = A_1 V_{1b} + B_1 I_{1b} \qquad \text{from (10.78}a)$$

$$I_1 = C_1 V_{1b} + D_1 I_{1b}. \qquad \text{from (10.79}a)$$

Putting the values of V_{1b} and I_{1b} from Equations (10.80) and (10.81),

$$V_1 = A_1[A_2 V_2 - B_2 I_2] + B_1[C_2 V_2 - D_2 I_2]$$

$$V_1 = [A_1 A_2 + B_1 C_2]V_2 - [A_1 B_2 + B_1 D_2]I_2. \qquad (10.82)$$

Similarly,

$$I_1 = C_1[A_2 V_2 - B_2 I_2] + D_1[C_2 V_2 - D_2 I_2]$$

$$I_1 = [C_1 A_2 + D_1 C_2]V_2 - [C_1 B_2 + D_1 D_2]I_2. \qquad (10.83)$$

For the equivalent network,

$$\begin{bmatrix} V_1 \\ I_1 \end{bmatrix} = \begin{bmatrix} A & B \\ C & D \end{bmatrix} \begin{bmatrix} V_2 \\ -I_2 \end{bmatrix}$$

or

$$V_1 = AV_2 - BI_2$$
$$I_1 = CV_2 - DI_2.$$

Comparing Equations (10.82) and (10.83) with standard equations

$$A = A_1A_2 + B_1C_2, \quad B = A_1B_2 + B_1D_2,$$
$$C = C_1A_2 + D_1C_2, \quad D = C_1B_2 + D_1D_2$$

or in matrix form

$$\begin{bmatrix} A & B \\ C & D \end{bmatrix} = \begin{bmatrix} A_1A_2 + B_1C_2 & A_1B_2 + B_1D_2 \\ C_1A_2 + D_1C_2 & C_1B_2 + D_1D_2 \end{bmatrix} = \begin{bmatrix} A_1 & B_1 \\ C_1 & D_1 \end{bmatrix} \begin{bmatrix} A_2 & B_2 \\ C_2 & D_2 \end{bmatrix}.$$

Therefore,

$$\begin{bmatrix} A & B \\ C & D \end{bmatrix} = \begin{bmatrix} A_1 & B_1 \\ C_1 & D_1 \end{bmatrix} \begin{bmatrix} A_2 & B_2 \\ C_2 & D_2 \end{bmatrix} [T] = [T_a][T_b]. \qquad (10.84)$$

Thus, the overall ABCD matrix for the two cascaded networks is equal to the product of ABCD matrices of the individual networks taken in order. The result may be extended to n number of cascaded networks.

 Similarly, we can derive in terms of inverse transmission $[T']$ parameters,

$$[T'] = [T'_b] \cdot [T'_a] \qquad (10.85)$$

The overall T'-parameter matrix for the cascaded two-port network is simply the matrix product of the T'-parameter matrices for each individual two-port network in reverse order.

10.10.4 Series Parallel Connection

Two two-port networks are said to be connected in series-parallel. The input ports are connected in series and output ports are connected in parallel as shown in Figure 10.23.

For network N_a,

$$V_{1a} = h_{11a}I_{1a} + h_{12a}V_{2a}$$
$$I_{2a} = h_{21a}I_{1a} + h_{22a}V_{2a}.$$

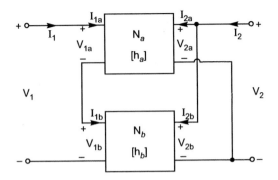

FIGURE 10.23 Series-parallel connection of two two-port networks.

For network N_b,

$$V_{1b} = h_{11b}I_{1b} + h_{12b}V_{2b}$$
$$I_{2b} = h_{21b}I_{1b} + h_{22b}V_{2b}.$$

But

$$V_1 = V_{1a} + V_{1b} \quad \text{and} \quad I_{1a} = I_{1b} = I_1 \text{ (at input port)}$$

and

$$V_2 = V_{2a} = V_{2b} \quad \text{and} \quad I_2 = I_{2a} + I_{2b} \text{ (at output port)}.$$

For the combined network

$$V_1 = h_{11}I_1 + h_{12}V_2$$
$$I_2 = h_{21}I_1 + h_{22}V_2$$
$$V_1 = (h_{11a} + h_{11b})I_1 + (h_{12a} + h_{12b})V_2 \tag{10.86}$$
$$I_2 = (h_{21a} + h_{21b})I_1 + (h_{22a} + h_{22b})V_2 \tag{10.87}$$
$$h_{11} = h_{11a} + h_{11b}, h_{12} = h_{12a} + h_{12b}, h_{21} = h_{21a} + h_{21b}, h_{22} = h_{22a} + h_{22b}$$

or in matrix form

$$[h] = [h_a] + [h_b].$$

The overall h-parameter matrix for series. Parallel connected two-port networks is simply the sum of h-parameter matrices of each individual two-port network connected in series parallel.

Example 10.15. *Two identical networks as shown in Figure 10.24 are connected in cascade. Determine the overall transmission parameters of the combined network.*

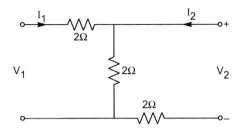

FIGURE 10.24

Solution: The two loop equations are

$$V_1 = 2I_1 + 2(I_1 + I_2)$$

or

$$V_1 = 4I_1 + 2I_2 \tag{1}$$

$$V_2 = 2(I_1 + I_2) + 2I_2$$

or

$$V_2 = 2I_1 + 4I_2$$

or

$$I_1 = \frac{V_2}{2} - 2I_2. \tag{2}$$

Putting the value of I_1 from Equation (2) into Equation (1),

$$V_1 = 4\left[\frac{V_2}{2} - 2I_2\right] + 2I_2$$

$$V_1 = 2V_2 + 6I_2. \tag{1a}$$

Comparing Equation (1a) and (2) with the standard equation of T-parameters, we get $A = 2, B = -6, C = \frac{1}{2}, D = 2$. These are the parameters for network N_a and the overall ABCD parameters are

$$\begin{bmatrix} A & B \\ C & D \end{bmatrix} = \begin{bmatrix} A_1 & B_1 \\ C_1 & D_1 \end{bmatrix}\begin{bmatrix} A_2 & B_2 \\ C_2 & D_2 \end{bmatrix} = \begin{bmatrix} 2 & -6 \\ \frac{1}{2} & 2 \end{bmatrix}\begin{bmatrix} 2 & -6 \\ \frac{1}{2} & 2 \end{bmatrix}$$

$$= \begin{bmatrix} 1 & -24 \\ 2 & 1 \end{bmatrix}.$$

Example 10.16. *Two identical networks as shown in Figure 10.24 are connected in series parallel. Determine the overall hybrid parameters of the combined network.*

Solution: The two loop equations are

$$V_1 = 4I_1 + 2I_2 \tag{1}$$
$$V_2 = 2I_1 + 4I_2. \tag{2}$$

The h-parameters of the network are shown in Figure 10.25,

$$I_2 = -\frac{1}{2}I_1 + \frac{1}{4}V_2. \tag{2a}$$

Comparing with $I_2 = h_{21}I_1 + h_{22}V_2$, we get

$$h_{21} = -\frac{1}{2}, \quad h_{22} = \frac{1}{4}$$

and

$$V_1 = 4I_1 + 2\left[-\frac{1}{2}I_1 + \frac{1}{4}V_2\right]$$
$$V_1 = 3I_1 + \frac{1}{2}V_2.$$

Comparing with $V_1 = h_{11}I_1 + h_{12}V_2$

$$h_{11} = 3, h_{12} = \frac{1}{2}$$

$$[h_a] = \begin{bmatrix} 3 & \frac{1}{2} \\ -\frac{1}{2} & \frac{1}{4} \end{bmatrix} = [h_b] \qquad \text{(because networks are identical)}$$

and for the series-parallel connection,

$$[h] = [h_a] + [h_b] = 2[h_a]$$
$$[h] = \begin{bmatrix} 6 & 1 \\ -1 & \frac{1}{2} \end{bmatrix}.$$

10.10.5 Parallel-series Connection

Figure 10.25 shows a parallel-series connection of two two-port networks. In this case input ports are connected in parallel while the output ports are connected in series.

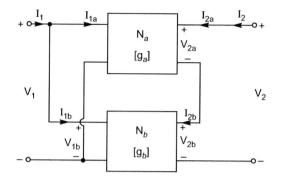

FIGURE 10.25   *Parallel-series connection of two two-port networks.*

The conditions at the input and output ports are

$$V_1 = V_{1a} = V_{1b} \quad \text{and} \quad I_1 = I_{1a} + I_{1b} \text{ (at input port)}$$
$$V_2 = V_{2a} + V_{2b} \quad \text{and} \quad I_2 = I_{2a} = I_{2b} \text{ (at output port).}$$

This is similar to the above case,

$$g_{11} = g_{11a} + g_{11b}, \quad g_{12} = g_{12a} + g_{12b},$$
$$g_{21} = g_{21a} + g_{21b}, \quad g_{22} = g_{22a} + g_{22b}$$

or in matrix form

$$[g] = [g_a] + [g_b].$$

The overall *g*-parameters matrix for parallel-series connected two-port networks is simply the sum of *g*-parameter matrices of each individual two-port network connected in parallel-series.

TABLE 10.3   Overall Parameter Matrix for Interconnections

Interconnection	Individual Parameter Matrix	Overall Parameter Matrix
Series-series	$[Z_a], [Z_b]$	$[Z] = [Z_a] + [Z_b]$
Parallel-Parallel	$[Y_a], [Y_b]$	$[Y] = [Y_a] + [Y_b]$
Cascade or Tandem	$[T_a], [T_b]$	$[T] = [T_a] \cdot [T_b]$
Series-parallel	$[h_a], [h_b]$	$[h] = [h_a] + [h_b]$
Parallel-series	$[g_a], [g_b]$	$[g] = [g_a] + [g_b]$

10.11 LATTICE NETWORKS

One of the common four-terminal two-port networks is the lattice, or bridge network, shown in Figure 10.26(a). Lattice networks are used in filter sections and are also used as attenuaters filter and attenuaters which we will discuss in Chapter 13. Lattice structures are sometimes used in preference to ladder structures in some special applications. Z_a and Z_d are called the series arms, Z_b and Z_c are called the diagonal arms. It can be observed that if Z_d is zero, the lattice structure becomes a π-section as shown in Figure 10.26(c). The lattice network is redrawn as a bridge network as shown in Figure 10.26(b).

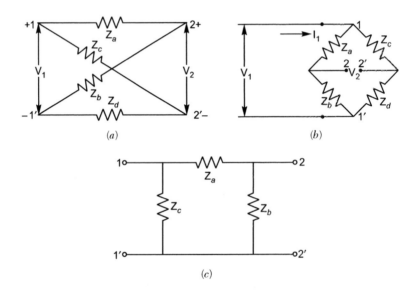

FIGURE 10.26

Z-parameters,

$$Z_{11} = \left.\frac{V_1}{I_1}\right|_{I_2=0}.$$

When

$$I_2 = 0, \quad V_1 = I_1 \frac{(Z_a + Z_b)(Z_d + Z_c)}{Z_a + Z_b + Z_c + Z_d}$$

Therefore

$$Z_{11} = \frac{(Z_a + Z_b)(Z_d + Z_c)}{Z_a + Z_b + Z_c + Z_d}.$$

If the network is symmetric, then $Z_a = Z_d, Z_b = Z_c$

Therefore

$$Z_{11} = \frac{Z_a + Z_b}{2}$$

$$Z_{21} = \frac{V_2}{I_1} \bigg|_{I_2=0}.$$

When $I_2 = 0$, V_2 is the voltage across $2\text{--}2'$

$$V_2 = V_1 \left[\frac{Z_b}{Z_a + Z_b} - \frac{Z_d}{Z_c + Z_d} \right].$$

Substituting the value of V_1 from Equation (10.87), we have

$$V_2 = \left[\frac{I_1(Z_a + Z_b)(Z_d + Z_c)}{Z_a + Z_b + Z_c + Z_d} \right] \left[\frac{Z_b(Z_c + Z_d) - Z_d(Z_a + Z_b)}{(Z_a + Z_b)(Z_c + Z_d)} \right]$$

$$\frac{V_2}{I_1} = \frac{Z_b(Z_c + Z_d) - Z_d(Z_a + Z_b)}{Z_a + Z_b + Z_c + Z_d} = \frac{Z_b Z_c - Z_a Z_d}{Z_a + Z_b + Z_c + Z_d}$$

Therefore

$$Z_{21} = \frac{Z_b Z_c - Z_a Z_d}{Z_a + Z_b + Z_c + Z_d}.$$

If the network is symmetric, $Z_a = Z_d$, $Z_b = Z_c$

$$Z_{21} = \frac{Z_b - Z_a}{2}.$$

When the input port is open, $I_1 = 0$,

$$Z_{12} = \frac{V_1}{I_2} \bigg|_{I_1=0}.$$

The network can be redrawn as shown in Figure 10.27.

$$V_1 = V_2 \left[\frac{Z_c}{Z_a + Z_c} - \frac{Z_d}{Z_b + Z_d} \right] \tag{10.88}$$

$$V_2 = I_2 \left[\frac{(Z_a + Z_c)(Z_d + Z_b)}{Z_a + Z_b + Z_c + Z_d} \right] \tag{10.89}$$

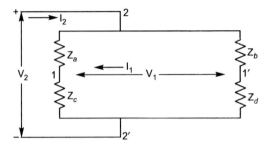

FIGURE 10.27

Substituting the value of V_2 into Equation (10.88), we get

$$V_1 = I_2 \left[\frac{Z_c(Z_b + Z_d) - Z_d(Z_a + Z_c)}{Z_a + Z_b + Z_c + Z_d} \right]$$

$$\frac{V_1}{I_2} = \frac{Z_c Z_b - Z_a Z_d}{Z_a + Z_b + Z_c + Z_d}.$$

If the network is symmetric, $Z_a = Z_d$, $Z_b = Z_c$,

$$\frac{V_1}{I_2} = \frac{Z_b^2 - Z_a^2}{2(Z_a + Z_b)}$$

Therefore

$$Z_{12} = \frac{Z_b - Z_a}{2}$$

$$Z_{22} = \left. \frac{V_2}{I_2} \right|_{I_2 = 0}.$$

From Equation (10.89), we have

$$\frac{V_2}{I_2} = \frac{(Z_a + Z_c)(Z_d + Z_b)}{Z_a + Z_d + Z_c + Z_d}.$$

If the network is symmetric, $Z_a = Z_d$, $Z_b = Z_c$,

$$Z_{22} = \frac{Z_a + Z_b}{2} = Z_{11}.$$

From the above equations, $Z_{11} = Z_{22} = \frac{Z_a + Z_b}{2}$
and

$$Z_{12} = Z_{21} = \frac{Z_b - Z_a}{2}$$

Therefore

$$Z_b = Z_{11} + Z_{12} \tag{10.90}$$

$$Z_a = Z_{11} - Z_{12}. \tag{10.91}$$

Example 10.17. *Obtain the lattice equivalent of the symmetrical T-network shown in Figure 10.28.*

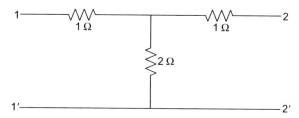

FIGURE 10.28

Solution: A two two-port network can be realized as a symmetric lattice if it is reciprocal and symmetric. The Z-parameters of the network are $Z_{11} = 3\,\Omega$, $Z_{12} = Z_{21} = 2\,\Omega$, $Z_{22} = 3\,\Omega$.

Since $Z_{11} = Z_{22}$, $Z_{12} = Z_{21}$, the given network is symmetrical and reciprocal.

Therefore the parameters of the lattice network are (from Equations (10.90) and (10.91))

$$Z_a = Z_{11} - Z_{12} = 1\,\Omega$$

$$Z_b = Z_{11} + Z_{12} = 5\,\Omega.$$

The lattice network is shown in Figure 10.28(a).

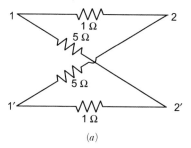

(a)

FIGURE 10.28

Example 10.18. *Obtain the lattice equivalent of the symmetric π-network shown in Figure 10.29.*

Solution: The Z-parameters of the given network are

$$Z_{11} = 6\,\Omega = Z_{22}, \quad Z_{12} = Z_{21} = 4\,\Omega.$$

Hence, the parameters of the lattice network are (from Equations (10.90) and (10.91))

$$Z_a = Z_{11} - Z_{12} = 2\,\Omega$$
$$Z_b = Z_{11} + Z_{12} = 10\,\Omega.$$

The lattice network is shown in Figure 10.29(a).

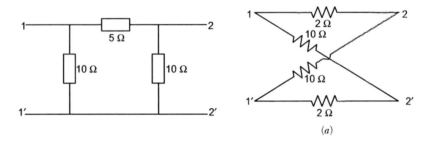

FIGURE 10.29

10.12   IMAGE PARAMETERS

The image impedance Z_{I1} and Z_{I2} of the two-port network shown in Figure 10.30 are two values of impedance such that, if port 1-1' of the network is terminated in Z_{I1}, the input impedance of port 2-2' is Z_{I2}, and if port 2-2' is terminated in Z_{I2}, the input impedance at port 1-1' is Z_{I1}.

Then, Z_{I1} and Z_{I2} are called image impedances of the two-port network shown in Figure 10.30. These parameters can be obtained in terms of two-port parameters.

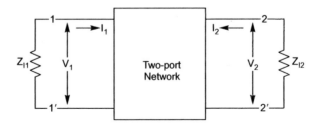

FIGURE 10.30

Defining equations for ABCD parameters are

$$V_1 = AV_2 - BI_2$$
$$I_1 = CV_2 - DI_2.$$

If the network is terminated in Z_{I2} at 2-2′ as shown in Figure 10.31,

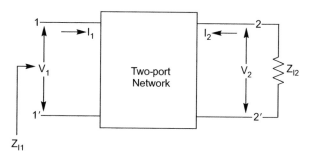

FIGURE 10.31

$$V_2 = -I_2 Z I_2 \text{ [direction of current is opposite]}$$

$$\frac{V_1}{I_1} = \frac{AV_2 - BI_2}{CV_2 - DI_2} = Z_{I1}$$

$$Z_{I1} = \frac{-AI_2 Z_{I2} - BI_2}{-CI_2 Z_{I2} - DI_2}$$

$$Z_{I1} = \frac{-AZ_{I2} - B}{-CZ_{I2} - D}$$

or

$$Z_{I1} = \frac{AZ_{I2} + B}{CZ_{I2} + D}. \tag{10.92}$$

Similarly, if the network is terminated in Z_{I1} at port 1-1′ as shown in Figure 10.32, then

$$V_1 = -I_1 Z_{I1}$$

$$\frac{V_2}{I_2} = Z_{I2}$$

Therefore

$$-Z_{I1} = \frac{V_1}{I_1} = \frac{AV_2 - BI_2}{CV_2 - DI_2}$$

$$-Z_{I1} = \frac{AI_2 Z_{I2} - BI_2}{CI_2 Z_{I2} - DI_2}$$

$$-Z_{I1} = \frac{AZ_{I2} - B}{CZ_{I2} - D},$$

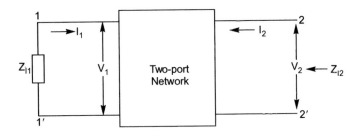

FIGURE 10.32

from which

$$Z_{I2} = \frac{DZ_{I1} + B}{CZ_{I1} + A}.$$ (10.93)

Substituting the value of Z_{I1} from Equation (10.92),

$$Z_{I2}\left[C\frac{(+AZ_{I2}+B)}{(CZ_{I2}+D)}+A\right] = D\left[\frac{+AZ_{I2}+B}{CZ_{I2}+D}\right]+B$$

$$2ACZ_{I2}^2 + Z_{I2}^2(AD+BC) = Z_{I2}(AD+BC)+2BD,$$

from which $Z_{I2} = \sqrt{\frac{BD}{AC}}$.

Similarly, we can find $Z_{I1} = \sqrt{\frac{AB}{CD}}$ from Equation (10.92).

If the network is symmetrical, then $A = D$.

Therefore

$$Z_{I1} = ZI2 = \sqrt{\frac{B}{C}}$$

If the network is symmetrical, the image impedances Z_{I1} and Z_{I2} are equal to each other; the image impedance is then called the *characteristic* impedance, or the *iterative* impedance (*i.e.*, if a symmetrical network is terminated in Z_L, its input impedance will also be Z_L, or its impedance transformation ratio is unity). Since a reciprocal symmetric network can be described by two independent parameters, the image parameters Z_{I1} and Z_{I2} are sufficient to characterize reciprocal symmetric networks. Z_{I1} and Z_{I2}, the two image parameters, do not completely define a network. A third parameter called an *image transfer constant ϕ* is also used to describe reciprocal networks. This parameter may be obtained from the voltage and current ratios.

If the image impedance Z_{I2} is connected across port 2-2′, then

$$V_1 = AV_2 - BI_2$$ (10.94)

$$V_2 = -I_2Z_{I2}$$ (10.95)

Therefore

$$V_1 = \left[A + \frac{B}{Z_{I2}}\right]V_2 \tag{10.96}$$

$$I_1 = CV_2 - DI_2 \tag{10.97}$$

$$I_1 = -[CZ_{I2} + D]I_2. \tag{10.98}$$

From Equation (10.96),

$$\frac{V_1}{V_2} = \left[A + \frac{B}{Z_{I2}}\right] = A + B\sqrt{\frac{AC}{BD}}$$

$$\frac{V_1}{V_2} = A + \sqrt{\frac{ABCD}{D}}. \tag{10.99}$$

From Equation (10.98),

$$\frac{-I_1}{I_2} = [CZ_{I2} + D] = D + C\sqrt{\frac{BD}{AC}}$$

$$\frac{-I_1}{I_2} = D + \sqrt{\frac{ABCD}{A}}. \tag{10.100}$$

Multiplying Equations (10.99) and (10.100) we have,

$$\frac{-V_1}{V_2} \times \frac{I_1}{I_2} = \left(\frac{AD + \sqrt{ABCD}}{D}\right)\left(\frac{AD + \sqrt{ABCD}}{A}\right)$$

$$\frac{-V_1}{V_2} \times \frac{I_1}{I_2} = (\sqrt{AD} + \sqrt{BC})^2$$

or

$$\sqrt{AD} + \sqrt{BC} = \sqrt{\frac{-V_1}{V_2} \times \frac{I_1}{I_2}}$$

$$\sqrt{AD} + \sqrt{AD - 1} = \sqrt{\frac{-V_1}{V_2} \times \frac{I_1}{I_2}}. \qquad (\because AD - BC = 1)$$

Let

$$\cosh\phi = \sqrt{AD}; \quad \sinh\phi = \sqrt{AD - 1}$$

$$\tanh\phi = \frac{\sqrt{AD - 1}}{\sqrt{AD}} = \sqrt{\frac{BC}{AD}}$$

$$\phi = \tanh^{-1}\sqrt{\frac{BC}{AD}}. \tag{10.101}$$

Also

$$e^{\phi} = \cosh \phi + \sinh \phi = \sqrt{-\frac{V_1 I_1}{V_2 I_2}} \qquad (10.102)$$

$$\phi = \log_e \sqrt{\left(-\frac{V_1 I_1}{V_2 I_2}\right)} = \frac{1}{2} \log_e \left(\frac{V_1}{V_2} \frac{I_1}{I_2}\right).$$

Since

$$V_1 = Z_{I1} I_1; \quad V_2 = -I_2 Z_{I2}$$

$$\phi = \frac{1}{2} \log_e \left[\frac{Z_{I1}}{Z_{I2}}\right] + \log \left[\frac{I_1}{I_2}\right].$$

For symmetrical reciprocal networks, $Z_{I1} = Z_{I2}$,

$$\phi = \log e \left[\frac{I_1}{I_2}\right] = \gamma, \qquad (10.103)$$

where γ is called the propagation constant and $\phi = \alpha + j\beta$.

Example 10.19. *Determine the image parameters of the T-network shown in Figure 10.33.*

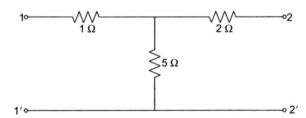

FIGURE 10.33

Solution: The ABCD parameters of the network are

$$A = \frac{6}{5}, \quad B = \frac{17}{5}, \quad C = \frac{1}{5}, \quad D = \frac{7}{5}.$$

Since the network is not symmetrical, ϕ, Z_{I1}, and Z_{I2} are to be evaluated to describe the network.

$$Z_{I1} = \sqrt{\frac{AB}{CD}} = \sqrt{\frac{\frac{6}{5} \times \frac{17}{5}}{\frac{1}{5} \times \frac{7}{5}}} = 3.817 \ \Omega$$

$$Z_{I2} = \sqrt{\frac{BC}{AC}} = \sqrt{\frac{\frac{17}{5} \times \frac{7}{5}}{\frac{6}{5} \times \frac{1}{5}}} = 4.453\,\Omega$$

$$\phi = \tanh^{-1}\sqrt{\frac{BC}{AD}} = \tanh^{-1}\sqrt{\frac{17}{42}}$$

$$\phi = 0.75$$

SOLVED PROBLEMS

Problem 10.1. *Determine the Z-, h-, and ABCD parameters of the network shown in Figure 10.34.*

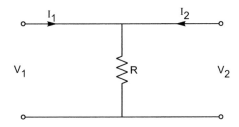

FIGURE 10.34

Solution:

$$V_1 = (I_1 + I_2)R = V_2$$

or

$$V_1 = V_2 = I_1R + I_2R. \tag{1}$$

Comparing Equation (1) with the standard equations of the Z-parameters We get

$$Z_{11} = R, \quad Z_{12} = R, \quad Z_{21} = R, \quad Z_{22} = R.$$

Now the standard equations for the *h*-parameters are

$$V_1 = h_{11}I_1 + h_{12}V_2 \tag{2}$$

$$I_2 = h_{21}I_1 + h_{22}V_2, \tag{3}$$

and Equation (1) can be written as

$$V_1 = V_2 \tag{4}$$

and

$$I_2 = -I_1 = \frac{V_2}{R}. \tag{5}$$

Comparing Equations (4), (5), with (2) and (3),

$$h_{11} = 0, \quad h_{12} = 1, \quad h_{21} = -1, \quad h_{22} = \frac{1}{R}.$$

The standard equations for the T-parameters are

$$V_1 = AV_2 - BI_2 \tag{6}$$
$$I_1 = CV_2 - DI_2. \tag{7}$$

Now rearranging Equation (5) in the form of Equation (7),

$$I_1 = \frac{V_2}{R} - I_2. \tag{8}$$

Comparing Equations (6) and (7) with (4) and (8),

$$A = 1, \quad B = 0, \quad C = \frac{1}{R}, \quad D = 1.$$

Problem 10.2. *Determine the Z-, h-, T- and Y- parameters of the network shown in Figure 10.35.*

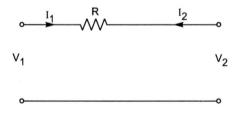

FIGURE 10.35

Solution: Applying KVL,

$$V_1 = I_1R + V_2 \tag{1}$$

and

$$I_1 = -I_2. \tag{2}$$

(*i*) Z-parameters: They do not exist, since currents I_1 and I_2 are not independent. They depend on each other. It is clear by Equation (2).

(*ii*) Now the standard equations for the *h*-parameters are

$$V_1 = h_{11}I_1 + h_{12}V_2 \tag{3}$$
$$I_2 = h_{21}I_1 + h_{22}V_2. \tag{4}$$

Now comparing Equations (1) and (2) with Equations (3) and (4),

$$h_{11} = R, \quad h_{12} = 1, \quad h_{21} = -1, \quad h_{22} = 0.$$

(*iii*) The standard equations for the T-parameters are

$$V_1 = AV_2 - BI_2 \tag{5}$$
$$I_1 = CV_2 - DI_2. \tag{6}$$

By Equations (1) and (2),

$$V_1 = V_2 - RI_2 \qquad (7) \quad [I_1 = -I_2]$$

Now comparing Equations (7) and (2) with (5) and (6),

$$A = 1, \quad B = R, \quad C = 0, \quad D = 1.$$

(*iv*) Now the standard equations for the Y-parameters are

$$I_1 = Y_{11}V_1 + Y_{12}V_2 \tag{8}$$
$$I_2 = Y_{21}V_1 + Y_{22}V_2. \tag{9}$$

Now rearranging Equation (1) into the form of Equation (8),

$$I_1 = \frac{V_1}{R} - \frac{V_2}{R} \tag{10}$$

$$-I_2 = \frac{V_1}{R} - \frac{V_2}{R} \quad \text{or} \quad I_2 = -\frac{V_1}{R} + \frac{V_2}{R}. \tag{11}$$

Now comparing Equations (10) and (11) with Equations (8) and (9),

$$Y_{11} = \frac{1}{R}, \quad Y_{12} = -\frac{1}{R}, \quad Y_{21} = -\frac{1}{R}, \quad Y_{22} = \frac{1}{R}.$$

Problem 10.3. *For the network shown in Figure 10.36 calculate:*
    (*i*) Z-   (*ii*) Y-   (*iii*) ABCD   (*iv*) h-   (*v*) g-parameters.

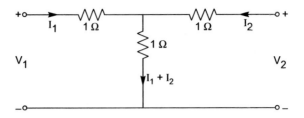

FIGURE 10.36

Solution: The loop equations for loop 1 and 2 are

$$V_1 = 1 \cdot I_1 + 1 \cdot (I_1 + I_2)$$
$$V_1 = 2I_1 + I_2 \tag{1}$$

and

$$V_2 = 1 \cdot I_2 + 1 \cdot (I_1 + I_2)$$
$$V_2 = I_1 + 2I_2. \tag{2}$$

(*i*) Z-parameters: The standard equations for the Z-parameters are

$$V_1 = Z_{11}I_1 + Z_{12}I_2 \tag{3}$$
$$V_2 = Z_{21}I_1 + Z_{22}I_2. \tag{4}$$

Compare Equation (1) with (3) and (2) with (4).
Now,

$$Z_{11} = 2, \quad Z_{12} = 1$$
$$Z_{21} = 1, \quad Z_{22} = 2.$$

(*ii*) Y-parameters: The standard equations for the Y-parameters are

$$I_1 = Y_{11}V_1 + Y_{12}V_2 \tag{5}$$
$$I_2 = Y_{21}V_1 + Y_{22}V_2. \tag{6}$$

Now by using Equations (1) and (2), we get

$$2V_2 - V_1 = 3I_2 \qquad [\text{Equation (2)} \times 2 - \text{Equation (1)}]$$

or

$$I_2 = -\frac{1}{3}V_1 + \frac{2}{3}V_2. \tag{7}$$

Similarly,

$$2V_1 - V_2 = 3I_1 \qquad \text{[Equation (1)} \times 2 - \text{Equation (2)]}$$

or

$$I_1 = \frac{2}{3}V_1 - \frac{1}{3}V_2. \tag{8}$$

Now compare Equation (5) with (8) and (4) with (7),

$$Y_{11} = \frac{2}{3}, \quad Y_{12} = -\frac{1}{3}, \quad Y_{21} = -\frac{1}{3}, \quad Y_{22} = \frac{2}{3}.$$

(*iii*) ABCD parameters: The standard equations for the ABCD parameters are

$$V_1 = AV_2 - BI_2 \tag{9}$$
$$I_1 = CV_2 - DI_2. \tag{10}$$

Now by Equation (7),

$$\frac{V_1}{3} = \frac{2}{3}V_2 - I_2$$

or

$$V_1 = 2V_2 - 3I_2. \tag{11}$$

Now compare Equation (9) with (11),

$$A = 2, \quad B = 3.$$

Now by Equation (8) and (11), put the value of V_1 into Equation (8),

$$I_1 = \frac{2}{3}(2V_2 - 3I_2) - \frac{1}{3}V_2$$
$$= \frac{4}{3}V_2 - 2I_2 - \frac{1}{3}V_2$$
$$I_1 = V_2 - 2I_2. \tag{12}$$

Now comparing Equation (10) with (12),

$$C = 1, \quad D = 2.$$

Hence, the ABCD parameters are $A = 2, B = 3, C = 1$ and $D = 2$.

(*iv*) *h*-parameters: The standard equations for the *h*-parameters are

$$V_1 = h_{11}I_1 + h_{12}V_2 \tag{13}$$
$$I_2 = h_{21}I_1 + h_{22}V_2. \tag{14}$$

Rearrange Equation (8),

$$\frac{2}{3}V_1 = I_1 + \frac{1}{3}V_2$$

or

$$V_1 = \frac{3}{2}I_1 + \frac{1}{2}V_2. \qquad (15)$$

Now compare Equation (13) and (15),

$$h_{11} = \frac{3}{2}, \quad h_{12} = \frac{1}{2}.$$

Now put the value of V_1 from Equation (15) into Equation (7),

$$I_2 = -\frac{1}{3}\left(\frac{3}{2}I_1 + \frac{1}{2}V_2\right) + \frac{2}{3}V_2$$

$$I_2 = -\frac{1}{2}I_1 - \frac{1}{6}V_2 + \frac{2}{3}V_2$$

$$I_2 = -\frac{1}{2}I_1 + \frac{1}{2}V_2. \qquad (16)$$

Now comparing Equation (14) with (16),

$$h_{21} = -\frac{1}{2}, \quad h_{22} = \frac{1}{2}.$$

Hence, the h-parameters are $h_{11} = \frac{3}{2}$, $h_{12} = \frac{1}{2}$, $h_{21} = -\frac{1}{2}$, and $h_{22} = \frac{1}{2}$.

(v) g-parameters: The standard equations for the "g" parameters are

$$I_1 = g_{11}V_1 + g_{12}I_2 \qquad (17)$$
$$V_2 = g_{21}V_1 + g_{22}I_2. \qquad (18)$$

Rearranging Equation (1),

$$2I_1 = V_1 - I_2$$
$$I_1 = \frac{1}{2}V_1 - \frac{1}{2}I_2. \qquad (19)$$

Comparing Equation (17) and (19),

$$g_{11} = \frac{1}{2}, \quad g_{12} = -\frac{1}{2}.$$

Putting the value of I_1 from Equation (19) into Equation (2),

$$V_2 = \frac{1}{2}V_1 - \frac{1}{2}I_2 + 2I_2$$

$$V_2 = \frac{1}{2}V_1 + \frac{3}{2}I_2. \tag{20}$$

Comparing Equation (20) with (18),

$$g_{21} = \frac{1}{2}, \quad g_{22} = \frac{3}{2}.$$

Hence, the g-parameters are $g_{11} = \frac{1}{2}$, $g_{12} = -\frac{1}{2}$, $g_{21} = \frac{1}{2}$, and $g_{22} = \frac{3}{2}$.

Problem 10.4. *For the π-network of Figure 10.37 obtain the Y-parameters and the h-parameters.*

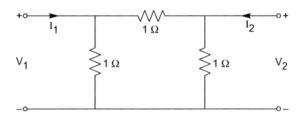

FIGURE 10.37

Solution: Applying mesh analysis,

$$V_1 = 1(I_1 - I_3) \tag{1}$$

$$V_2 = 1(I_2 + I_3) \tag{2}$$

$$1 \cdot I_3 + 1(I_2 + I_3) + (I_3 - I_1) = 0$$

$$3I_3 = I_1 - I_2 \implies I_3 = \left(\frac{I_1 - I_2}{3}\right). \tag{3}$$

Putting the value of I_3 from Equation (3) into Equation (1) and Equation (2),

$$V_1 = I_1 - \left(\frac{I_1 - I_2}{3}\right) = \frac{2}{3}I_1 + \frac{I_2}{3} \tag{4}$$

$$V_2 = I_2 + \left(\frac{I_1 - I_2}{3}\right) = \frac{1}{3}I_1 + \frac{2}{3}I_2. \tag{5}$$

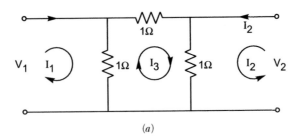

(a)

FIGURE 10.37

Now by Equations (4) and (5),

$$2V_2 - V_1 = I_2 \implies I_2 = -V_1 + 2V_2 \qquad (6)$$

and

$$2V_1 - V_2 = I_1. \qquad (7)$$

Comparing Equations (6) and (7) with the standard equations of the Y-parameters,

$$I_1 = Y_{11}V_1 + Y_{12}V_2$$
$$I_2 = Y_{21}V_1 + Y_{22}V_2,$$

we get,

$$Y_{11} = 2, \quad Y_{12} = -1, \quad Y_{21} = -1, \quad Y_{22} = 2.$$

Now for the h-parameters, the standard equations are

$$V_1 = h_{11}I_1 + h_{12}V_2$$
$$I_2 = h_{21}I_1 + h_{22}V_2.$$

Now by Equation (7),

$$V_1 = \frac{1}{2}I_1 + \frac{1}{2}V_2. \qquad (8)$$

Putting the value of V_1 from (8) to (6),

$$I_2 = -\left[\frac{1}{2}I_1 + \frac{1}{2}V_2\right] + 2V_2$$

$$I_2 = -\frac{1}{2}I_1 + \frac{3}{2}V_2. \qquad (9)$$

Now comparing Equations (8) and (9) with the standard equations of the h-parameters $h_{11} = \frac{1}{2}, h_{12} = \frac{1}{2}, h_{21} = -\frac{1}{2}$, and $h_{22} = \frac{3}{2}$.

Problem 10.5. *Find the hybrid parameters for the network of Figure 10.38.*

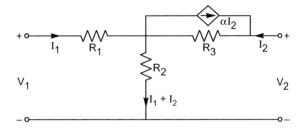

FIGURE 10.38

Solution: Applying KVL,

$$V_1 = I_1 R_1 + (I_1 + I_2)R_2$$
$$V_1 = I_1(R_1 + R_2) + I_2 R_2 \tag{1}$$
$$V_2 = (I_2 + \alpha I_1)R_3 + (I_1 + I_2)R_2$$
$$V_2 = I_1(\alpha R_3 + R_2) + I_2(R_2 + R_3). \tag{2}$$

Now the standard equations for the *h*-parameters are

$$V_1 = h_{11}I_1 + h_{12}V_2 \tag{3}$$
$$I_2 = h_{21}I_1 + h_{22}V_2. \tag{4}$$

Rearranging Equation (2),

$$I_2(R_2 + R_3) = V_2 - I_1(\alpha R_3 + R_2)$$
$$I_2 = -\left(\frac{\alpha R_3 + R_2}{R_2 + R_3}\right)I_1 + \left(\frac{1}{R_2 + R_3}\right)V_2. \tag{5}$$

Now comparing Equation (5) with (4),

$$h_{21} = -\left(\frac{\alpha R_3 + R_2}{R_2 + R_3}\right), \quad h_{22} = \frac{1}{R_2 + R_3}.$$

Now putting the value of I_2 into Equation (1),

$$V_1 = I_1(R_1 + R_2) + R_2\left[-\left(\frac{\alpha R_3 + R_2}{R_2 + R_3}\right)I_1 + \left(\frac{1}{R_2 + R_3}\right)V_2\right]$$
$$V_1 = I_1\left[R_1 + R_2 - \frac{R_2(\alpha R_3 + R_2)}{R_2 + R_3}\right] + \left(\frac{R_2}{R_2 + R_3}\right)V_2. \tag{6}$$

Comparing Equation (6) with Equation (3),

$$h_{11} = R_1 + R_2 - \frac{R_2(\alpha R_3 + R_2)}{R_2 + R_3}, \quad h_{12} = \left(\frac{R_2}{R_2 + R_3}\right).$$

Problem 10.6. *Two identical sections of the network shown in Figure 10.39 are cascaded. Calculate the transmission (ABCD) parameters of the resulting network.*

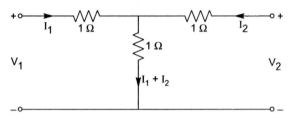

FIGURE 10.39

Solution: Applying KVL,

$$V_1 = I_1 \cdot 1 + (I_1 + I_2) \cdot 1$$
$$V_1 = 2I_1 + I_2 \tag{1}$$
$$V_2 = I_2 \cdot 1 + (I_1 + I_2) \cdot 1$$
$$V_2 = I_1 + 2I_2. \tag{2}$$

Now the standard equations for the T-parameters are

$$V_1 = AV_2 - BI_2 \tag{3}$$
$$I_1 = CV_2 - DI_2. \tag{4}$$

Rearranging Equation (2) into the form of Equation (4),

$$I_1 = V_2 - 2I_2. \tag{5}$$

Now comparing Equation (5) with (4),

$$C = 1, \quad D = 2.$$

Putting the value of I_1 from Equation (5) into Equation (1),

$$V_1 = 2(V_2 - 2I_2) + I_2$$
$$V_1 = 2V_2 - 3I_2. \tag{6}$$

Now comparing Equation (6) with (3),

$$A = 2, \quad B = 3.$$

Therefore, the overall T-parameters of the cascaded network are given by

$$[T] = \begin{bmatrix} A & B \\ C & D \end{bmatrix} \begin{bmatrix} A & B \\ C & D \end{bmatrix}$$

$$\begin{bmatrix} 2 & 3 \\ 1 & 2 \end{bmatrix} \begin{bmatrix} 2 & 3 \\ 1 & 2 \end{bmatrix} = \begin{bmatrix} 4+3 & 6+6 \\ 2+2 & 3+4 \end{bmatrix} = \begin{bmatrix} 7 & 12 \\ 4 & 7 \end{bmatrix}.$$

Problem 10.7. *For the circuit of Figure 10.40 find the Z- and h-parameters.*

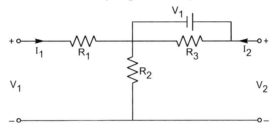

FIGURE 10.40

Solution: The KVL equations for the two loops are

$$V_1 = I_1 R_1 + (I_1 + I_2)R_2$$

or

$$V_1 = I_1(R_1 + R_2) + I_2 R_2 \qquad (1)$$
$$V_2 = V_1 + (I_1 + I_2)R_2$$

or

$$V_2 = I_1(R_1 + R_2) + I_2 R_2 + (I_1 + I_2)R_2$$
$$V_2 = I_1(R_1 + 2R_2) + I_2(2R_2). \qquad (2)$$

The standard equations for the Z-parameters are

$$V_1 = Z_{11}I_1 + Z_{12}I_2$$
$$V_2 = Z_{21}I_1 + Z_{22}I_2.$$

Now compare Equations (1) and (2) with the Z-parameter standard equations,

$$Z_{11} = (R_1 + R_2), \quad Z_{12} = R_2, \quad Z_{21} = (R_1 + 2R_2), \quad Z_{22} = 2R_2$$

The standard equations for the *h*-parameters are

$$V_1 = h_{11}I_1 + h_{12}V_2$$
$$I_2 = h_{21}I_1 + h_{22}V_2.$$

Rearranging Equation (2),

$$I_2 = -\frac{(R_1 + 2R_2)}{2R_2}I_1 + \frac{V_2}{2R_2}. \qquad (3)$$

Putting the value of I_2 from Equation (3) into Equation (1),

$$V_1 = I_1(R_1 + R_2) + R_2\left[-\left(\frac{R_1 + 2R_2}{2R_2}\right)I_1 + \frac{V_2}{2R_2}\right]$$

$$V_1 = I_1 = \left(R_1 + R_2 - \frac{R_1}{2} - R_2\right) + \frac{V_2}{2} = \frac{R_1}{2}I_1 + \frac{V_2}{2}. \qquad (4)$$

Comparing Equation (3) and (4) with the h-parameter standard equations we get,

$$h_{11} = \frac{R_1}{2}, \quad h_{12} = \frac{1}{2}, \quad h_{21} = -\frac{(R_1 + 2R_2)}{2R_2}, \quad \text{and} \quad h_{22} = \frac{1}{2R_2}.$$

Problem 10.8. *Figure 10.41 shows a network model for a common base transistor. Find the Z-parameters.*

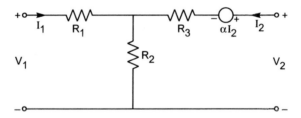

FIGURE 10.41

Solution: Applying KVL,

$$V_1 = I_1R_1 + (I_1 + I_2)R_2$$
$$V_1 = I_1(R_1 + R_2) + I_2R_2 \qquad (1)$$
$$V_2 = \alpha I_1 + I_2R_3 + (I_1 + I_2)R_2$$
$$V_2 = I_1(\alpha + R_2) + I_2(R_2 + R_3). \qquad (2)$$

Comparing Equations (1) and (2) with the standard equations of the Z-parameters,

$$V_1 = Z_{11}I_1 + Z_{12}I_2$$
$$V_2 = Z_{21}I_1 + Z_{22}I_2,$$

we get,

$$Z_{11} = R_1 + R_2, \quad Z_{12} = R_2$$
$$Z_{21} = (\alpha + R_2), \quad Z_{22} = (R_2 + R_3).$$

Problem 10.9. *For the network shown in Figure 10.42, find the*
(a) impedance and admittance parameters,
(b) transmission and inverse transmission parameters,
(c) hybrid and inverse hybrid parameters.

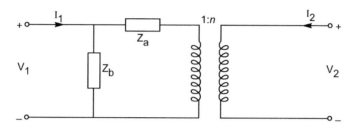

FIGURE 10.42

Solution: (*a*) Let the V_2' and I_2' be the input voltage and input current of the transformer, respectively, as shown in Figure 10.42(*a*), then the loop equations become

$$V_1 = (I_1 - I_2')Z_b \tag{1}$$

and

$$\frac{V_2'}{V_2} = \frac{1}{n} \text{ [transformation ratio]} = -\frac{I_2}{I_2'}$$
$$V_2 = nV_2' \quad \text{and} \quad V_2' = V_1 - I_2'Z_a \tag{2}$$

$$I_2 = -\frac{I_2'}{n} \implies I_2' = -nI_2. \tag{3}$$

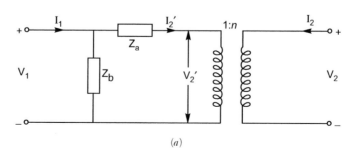

(*a*)

FIGURE 10.42

Putting the value of I_2' into Equation (1),

$$V_1 = (I_1 + I_2 n)Z_b = I_1 Z_b + n Z_b I_2. \tag{4}$$

By Equation (2),

$$
\begin{aligned}
V_2 &= n(V_1 - I_2' Z_a) = nV_1 - nX - nI_2 Z_a \\
&= n(I_1 Z_b + n Z_b I_2) + n2 Z_a I_2 \\
V_2 &= I_1(nZ_b) + I_2(n^2 Z_b + n^2 Z_a).
\end{aligned} \tag{5}
$$

The standard equations for the Z-parameters are

$$V_1 = Z_{11} I_1 + Z_{12} I_2 \tag{6}$$
$$V_2 = Z_{21} I_1 + Z_{22} I_2. \tag{7}$$

Now comparing Equation (4) with Equation (6) and (5) with Equation (7), we then get

$$Z_{11} = Z_b, \quad Z_{12} = nZ_b, \quad Z_{21} = nZ_b$$
$$Z_{22} = n^2(Z_a + Z_b),$$

and

$$[Y] = [Z]^{-1} = \begin{bmatrix} Z_b & nZ_b \\ nZ_b & n^2(Z_a + Z_b) \end{bmatrix}^{-1}$$

$$[Y] = \begin{bmatrix} \dfrac{Z_a + Z_b}{Z_a \cdot Z_b} & -\dfrac{1}{nZ_a} \\ -\dfrac{1}{nZ_a} & \dfrac{1}{n^2 Z_a} \end{bmatrix}.$$

(*b*) The standard equations for the transmission parameters are

$$V_1 = AV_2 - BI_2 \tag{8}$$
$$I_1 = CV_2 - DI_2. \tag{9}$$

Now rearrange Equation (5),

$$nZ_b I_1 = V_2 - I_2 n^2(Z_a + Z_b)$$

or

$$I_1 = \left(\frac{1}{nZ_b}\right) V_2 - \frac{n^2(Z_a + Z_b)}{nZ_b} I_2. \tag{10}$$

Compare Equation (10) with Equation (9),

$$C = \frac{1}{nZ_b}; \quad D = \frac{n(Z_a + Z_b)}{Z_b},$$

and put the value of I_1 from Equation (10) into Equation (4),

$$V_1 = Z_b \left[\left(\frac{1}{nZ_b} \right) V_2 - \frac{n(Z_a + Z_b)}{Z_b} I_2 \right] + nZ_b I_2$$

$$V_1 = \left(\frac{1}{n} \right) V_2 - I_2 \left[n(Z_a + Z_b) - nZ_b \right]$$

$$V_1 = \frac{V_2}{n} - nZ_a I_2. \tag{11}$$

Compare Equation (11) with (8),

$$A = \frac{1}{n}, \quad B = nZ_a$$

$$[T] = \begin{bmatrix} \dfrac{1}{n} & nZ_a \\ \dfrac{1}{nZ_b} & \dfrac{n(Z_a + Z_b)}{Z_b} \end{bmatrix}$$

and

$$A' = \frac{D}{\Delta T} = n\frac{(Z_a + Z_b)}{Z_b}$$

$$B' = \frac{B}{\Delta T} = nZ_a$$

$$C' = \frac{C}{\Delta T} = \frac{1}{nZ_b}$$

$$D' = \frac{A}{\Delta T} = \frac{1}{n}.$$

(*c*) The standard equations for the *h*-parameters are

$$V_1 = h_{11}I_1 + h_{12}V_2 \tag{12}$$
$$I_2 = h_{21}I_1 + h_{22}V_2. \tag{13}$$

Rearranging Equation (5),

$$V_2 - nZ_b I_1 = I_2 n^2 (Z_a + Z_b)$$

$$I_2 = \frac{V_2}{n^2(Z_a + Z_b)} - \frac{nZ_b I_1}{n^2(Z_a + Z_b)}. \tag{14}$$

Comparing Equation (14) with (13),

$$h_{21} = -\frac{Z_b}{n(Z_a + Z_b)}, \quad h_{22} = \frac{1}{n^2(Z_a + Z_b)}.$$

Now put the value of I_2 from Equation (14) into Equation (4),

$$V_1 = I_1 Z_b + n Z_b \left[\frac{V_2}{n^2(Z_a + Z_b)} - \frac{Z_b I_1}{n(Z_a + Z_b)} \right]$$

$$V_1 = I_1 \left[Z_b - \frac{Z_b^2}{(Z_a + Z_b)} \right] + \frac{n Z_b V_2}{n^2(Z_a + Z_b)}$$

$$V_1 = \left(\frac{Z_a \cdot Z_b}{Z_a + Z_b} \right) I_1 + \frac{Z_b V_2}{n(Z_a + Z_b)}. \tag{15}$$

Now comparing Equation (15) with (12),

$$h_{11} = \frac{Z_a \cdot Z_b}{Z_a + Z_b}, \quad h_{12} = \frac{Z_b}{n(Z_a + Z_b)}$$

$$[h] = \begin{bmatrix} \dfrac{Z_a \cdot Z_b}{Z_a + Z_b} & \dfrac{Z_b}{n(Z_a + Z_b)} \\ -\dfrac{Z_b}{n(Z_a + Z_b)} & \dfrac{1}{n^2(Z_a + Z_b)} \end{bmatrix}.$$

The inverse $[h]$ parameters $= [g] = [h]^{-1}$

$$[g] = \begin{bmatrix} \frac{1}{Z_b} & -n \\ n & n^2 Z_a \end{bmatrix}$$

Problem 10.10. *Starting from the definition of the parameters h_{11} and h_{21}, establish their relation to the Z- and Y-parameters. Specifically, show that $h_{11} = \frac{1}{Y_{11}}$ and $h_{22} = \frac{1}{Z_{22}}$.*

Solution: The standard equations for the h-parameters are

$$V_1 = h_{11} I_1 + h_{12} V_2 \tag{1}$$
$$I_2 = h_{21} I_1 + h_{22} V_2. \tag{2}$$

The standard equations for the Z-parameters are

$$V_1 = Z_{11} I_1 + Z_{12} I_2 \tag{3}$$
$$V_2 = Z_{21} I_1 + Z_{22} I_2. \tag{4}$$

The standard equations for the Y-parameters are

$$I_1 = Y_{11}V_1 + Y_{12}V_2 \tag{5}$$

$$I_2 = Y_{21}V_1 + Y_{22}V_2. \tag{6}$$

By Equation (4),

$$I_2 = \frac{V_2 - Z_{21}I_1}{Z_{22}}$$

$$I_2 = -\frac{Z_{21}}{Z_{22}}I_1 + \frac{1}{Z_{22}}V_2. \tag{7}$$

Compare Equation (2) and (7),

$$\boxed{h_{22} = \frac{1}{Z_{22}}} \text{ proved.}$$

Similarly, rearrange Equation (5),

$$Y_{11}V_1 = I_1 - Y_{12}V_2$$

$$V_1 = \frac{1}{Y_{11}}I_1 - \frac{Y_{12}}{Y_{11}}V_2. \tag{8}$$

Compare Equation (1) and (8) $\boxed{h_{11} = \frac{1}{Y_{11}}}$ **proved**.

Problem 10.11. *Determine the Z- and Y-parameters of the given network shown in Figure 10.43.*

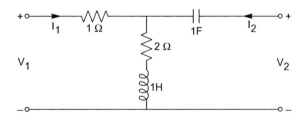

FIGURE 10.43

Solution: The *s*-domain representation of Figure 10.43 is shown in Figure 10.43(*a*). Apply KVL and find the loops equations.

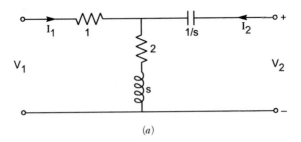

(a)

FIGURE 10.43

or

$$V_1 = I_1 + (2+s) \times (I_1 + I_2)$$
$$V_1 = I_1(3+s) + (2+s)I_2 \tag{1}$$
$$V_2 = \frac{I_2}{s} + (2+s)(I_1 + I_2)$$

or

$$V_2 = (2+s)I_1 + I_2 \left(\frac{1}{s} + 2 + s \right). \tag{2}$$

Now comparing Equations (1) and (2) with the standard equations of the Z-parameters,

$$Z_{11} = (3+s), \quad Z_{12} = (2+s), \quad Z_{21} = (2+s), \quad Z_{22} \left(\frac{1}{s} + 2 + s \right).$$

Equation $(1) \times \left(\frac{1}{s} + 2 + s \right) -$ equation$(2) \times (2+s)$, we get

$$I_1 = \frac{(s^2 + 2s + 1)}{(s^2 + 3s + 3)} V_1 - \frac{(2+s)s}{(s^2 + 3s + 3)} V_2. \tag{3}$$

Similarly, equation $(1) \times (2+s) -$ equation$(2) \times (3+s)$, we get

$$I_2 = -\frac{(2+s)s}{(s^2 + 3s + 3)} V_1 + \frac{(3+s)s}{(s^2 + 3s + 3)} V_2. \tag{4}$$

Now comparing Equations (3) and (4) with the standard equations of the Y-parameters,

$$Y_{11} = \frac{(s^2 + 2s + 1)}{(s^2 + 3s + 3)}, \quad Y_{12} = -\frac{(2+s)s}{(s^2 + 3s + 3)}, \quad Y_{21} = -\frac{(2+s)s}{(s^2 + 3s + 3)}$$

and

$$Y_{22} = \frac{(3+s)s}{(s^2 + 2s + 3)}.$$

Problem 10.12. *Find the short-circuit parameters of the circuit shown in Figure 10.44.*

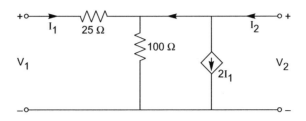

FIGURE 10.44

Solution: The Y-parameters are known as short-circuit parameters.
   Applying KVL,

$$V_1 = 25I_1 + 100(I_2 - I_1)$$

or

$$V_1 = -75I_1 + 100I_2 \tag{1}$$
$$V_2 = 100(I_2 - I_1)$$
$$V_2 = -100I_1 + 100I_2. \tag{2}$$

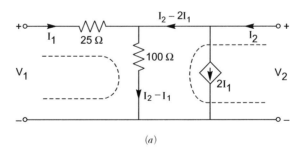

(*a*)

FIGURE 10.44

Equation (1) – Equation (2),

$$V_1 - V_2 = 25I_1 \implies I_1 = \frac{1}{25}V_1 - \frac{1}{25}V_2. \tag{3}$$

Similarly,

$$I_2 = \frac{4}{100}V_1 - \frac{3}{100}V_2. \tag{4}$$

The standard equations of the Y-parameters are

$$I_1 = Y_{11}V_1 + Y_{12}V_2 \tag{5}$$
$$I_2 = Y_{21}V_1 + Y_{22}V_2. \tag{6}$$

Compare Equation (3) with (5), and (4) with (6),

$$Y_{11} = \frac{1}{25}\,\mho, \quad Y_{12} = -\frac{1}{25}\,\mho, \quad Y_{21} = \frac{1}{25}\,\mho, Y_{22} = -\frac{3}{100}\,\mho.$$

Problem 10.13. *Determine transmission parameters of the T-network shown in Figure 10.45 considering three sections as shown. Assume they are connected in a cascade manner.*

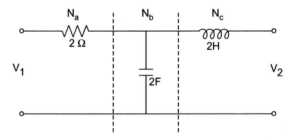

FIGURE 10.45

Solution: From Figure 10.45 we can say that the three different two-port networks are connected in cascade. So the overall transmission parameters of the network shown in the Figure are given as

$$T = [T_a][T_b][T_c],$$

where T_a, T_b, and T_c are the T-parameter matrices of the networks N_a, N_b and N_c respectively, for network N_a:

$$V_1 = 2I_1 + V_2 \quad \text{and} \quad I_1 = -I_2. \tag{1}$$

Now,

$$V_1 = V_2 - 2I_2. \tag{2}$$

Compare with the standard [T] parameters,

$$[A = 1, \ B = 2, \ C = 0, \ D = 1],$$

$$[T_a] = \begin{bmatrix} 1 & 2 \\ 0 & 1 \end{bmatrix}.$$

For network N_b,

$$V_1 = V_2 = (I_1 + I_2) \times \frac{1}{2s}$$

$$V_1 = V_2 \tag{3}$$

and

$$I_1 = 2sV_2 - I_2. \tag{4}$$

Compare with standard [T] parameters $[A = 1, B = 0, C = 2s, D = 1]$,

$$[T_b] = \begin{bmatrix} 1 & 0 \\ 2s & 1 \end{bmatrix}.$$

for N_c, similarly to N_a,

$$[T_c] = \begin{bmatrix} 1 & 2s \\ 0 & 1 \end{bmatrix}.$$

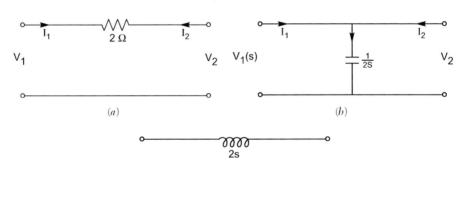

FIGURE 10.45

Therefore, the overall transmission parameters of the given network are

$$[T] = [T_a][T_b][T_c] = \begin{bmatrix} 1 & 2 \\ 0 & 1 \end{bmatrix}\begin{bmatrix} 1 & 0 \\ 2s & 1 \end{bmatrix}\begin{bmatrix} 1 & 2s \\ 0 & 1 \end{bmatrix}$$

$$= \begin{bmatrix} 1+4s & 2 \\ 2s & 1 \end{bmatrix}\begin{bmatrix} 1 & 2s \\ 0 & 1 \end{bmatrix} = \begin{bmatrix} 4s+1 & 2(4s^2+s+1) \\ 2s & 4s^2+1 \end{bmatrix}.$$

Problem 10.14. *Find the hybrid parameters of the network shown in Figure 10.46.*

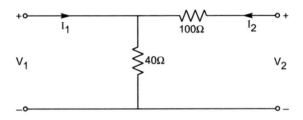

FIGURE 10.46

Solution: Applying KVL,

$$V_1 = 40(I_1 + I_2) \tag{1}$$

$$V_2 = 100I_2 + 40(I_1 + I_2)$$

$$V_2 = 40I_1 + 140I_2. \tag{2}$$

Now the standard equations for the h-parameters are

$$V_1 = h_{11}I_1 + h_{12}V_2 \tag{3}$$

$$I_2 = h_{21}I_1 + h_{22}V_2. \tag{4}$$

Now rearranging Equation (2) into the form of Equation (4),

$$I_2 = -\frac{2}{7}I_1 + \frac{1}{140}V_2. \tag{5}$$

Putting the value of I_2 from Equation (5) into (1),

$$V_1 = 40I_1 + 40\left[-\frac{2}{7}I_1 + \frac{1}{140}V_2\right]$$

$$= 40\left(1 - \frac{2}{7}\right)I_1 + \frac{2}{7}V_2$$

$$V_1 = \frac{200}{7}I_1 + \frac{2}{7}V_2. \tag{6}$$

Comparing Equation (3) with (6), and (4) with (5),

$$h_{11} = \frac{200}{7}\ \Omega, \quad h_{12} = \frac{2}{7}, \quad h_{21} = -\frac{2}{7}, \quad \text{and} \quad h_{22} = \frac{1}{140}\ \mho.$$

Problem 10.15. *Determine the transmission parameters of the network shown in Figure 10.47 using the concept of inter-connection of two two-port networks* N_1 *and* N_2 *in cascade.*

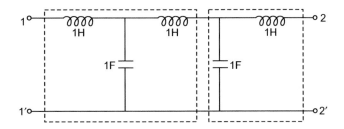

FIGURE 10.47

Solution: For network N_1, as shown in Figure 10.47(a),
applying KVL,

$$V_1 = sI_1 + (I_1 + I_2) \times \frac{1}{s}$$

$$V_1 = I_1 \left(s + \frac{1}{s} \right) + \frac{I_2}{s}$$

$$V_1 = I_1 \left(\frac{s^2 + 1}{s} \right) + \frac{I_2}{s} \tag{1}$$

$$V_2 = sI_2 + (I_1 + I_2) \times \frac{1}{s}$$

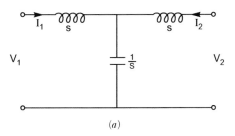

(a)

FIGURE 10.47

or

$$V_2 = \frac{I_1}{s} + \left(\frac{s^2 + 1}{s}\right)I_2. \tag{2}$$

The standard equations of the [T] parameters are

$$V_1 = AV_2 - BI_2 \tag{3}$$
$$I_1 = CV_2 - DI_2. \tag{4}$$

Rearrange Equation (2) into the form of Equation (4),

$$V_2 - \left(\frac{s^2 + 1}{s}\right)I_2 = \frac{I_1}{s}$$

$$I_1 = sV_2 - (s^2 + 1)I_2. \tag{5}$$

Putting the value of I_1 from Equation (5) into Equation (1),

$$V_1 = \left(\frac{s^2 + 1}{s}\right)[sV_2 - (s^2 + 1)I_2] + \frac{I_2}{s}$$

$$V_1 = (s^2 + 1)V_2 - \left[\frac{(s^2 + 1)^2}{s} - \frac{1}{s}\right]I_2$$

$$V_1 = (s^2 + 1)V_2 - (2s + s^3)I_2. \tag{6}$$

Comparing Equation (6) with (3) and (5) with (4),

$$A = (s^2 + 1), \quad B = s^3 + 2s$$
$$C = s, \quad D = s^2 + 1,$$

for network N_2.

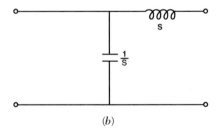

(b)

FIGURE 10.47

Similarly, we can calculate the [T] parameter for N_2,

$$A = 1, \quad B = s, \quad C = s, \quad D = s^2 + 1.$$

Now, the overall transmission parameters of the network shown in the figure are given by

$$[T] = [T_{N_1}][T_{N_2}]$$

$$= \begin{bmatrix} 1 + s^2 & s^3 + 2s \\ s & s^2 + 1 \end{bmatrix} \begin{bmatrix} 1 & s \\ s & s^2 + 1 \end{bmatrix}$$

$$= \begin{bmatrix} s^4 + 3s^2 + 1 & s^5 + 4s^3 + 3s \\ s^3 + 2s & s^4 + 3s^2 + 1 \end{bmatrix}.$$

Problem 10.16. *The network shown in Figure 10.48 contains a voltage controlled current source for the given values of the parameters. Determine the Z- and Y-parameters.*

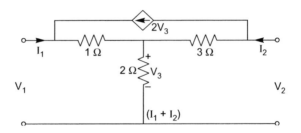

FIGURE 10.48

Solution: Applying KVL,

$$V_1 = (I_1 + 2V_3) \times 1 + 2(I_1 + I_2) \tag{1}$$

and

$$V_2 = (I_2 - 2V_3) \times 3 + 2(I_1 + I_2) \tag{2}$$

and

$$V_3 = 2(I_1 + I_2).$$

Putting the values of V_3 into Equations (1) and (2),

$$V_1 = I_1 + 4I_1 + 4I_2 + 2I_1 + 2I_2$$

or

$$V_1 = 7I_1 + 6I_2 \tag{3}$$

$$V_2 = (I_2 - 4I_1 - 4I_2) \times 3 + 2I_1 + 2I_2$$

$$V_2 = -12I_1 - 9I_2 + 2I_1 + 2I_2$$

or

$$V_2 = -10I_1 - 7I_2. \tag{4}$$

Comparing (3) and (4) with the standard Z-parameters equation,

$$Z_{11} = 7, \quad Z_{12} = 6, \quad Z_{21} = -10, \quad Z_{22} = -7.$$

By using (3) and (4), we get

$$I_1 = -\frac{7}{11}V_1 - \frac{6}{11}V_2 \tag{5}$$

$$I_2 = \frac{10}{11}V_1 + \frac{7}{11}V_2. \tag{6}$$

Now comparing Equations (5) and (6) with the standard equation of the Y-parameters we get,

$$Y_{11} = -\frac{7}{11}\,\mho, \quad Y_{12} = -\frac{6}{11}\,\mho, \quad Y_{21} = \frac{10}{11}\,\mho, \quad Y_{22} = \frac{7}{11}\,\mho.$$

Problem 10.17. *Find the Z- and ABCD parameters of the given network shown in Figure 10.49.*

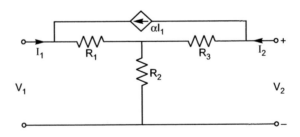

FIGURE 10.49

Solution: Applying KVL equations,

$$V_1 = (I_1 + \alpha I_1)R_1 + (I_1 + I_2)R_2$$

or

$$V_1 = I_1(R_1 + \alpha R_1 + R_2) + I_2 R_2 \tag{1}$$

$$V_2 = (I_2 - \alpha I_1)R_3 + (I_1 + I_2)R_2$$

or

$$V_2 = I_1(R_2 - \alpha R_3) + I_2(R_2 + R_3). \tag{2}$$

Now comparing Equations (1) and (2) with the standard equations of the Z-parameters,

$$Z_{11} = (R_1 + \alpha R_1 + R_2), \quad Z_{12} = R_2,$$
$$Z_{21} = (R_2 - \alpha R_3), \qquad Z_{22} = (R_2 + R_3).$$

Now the standard equations for the ABCD parameters are

$$V_1 = AV_2 - BI_2 \tag{3}$$
$$I_1 = CV_2 - DI_2. \tag{4}$$

Rearranging Equation (2) into the form of Equation (4),

$$I_1 = \frac{V_2}{(R_2 - \alpha R_3)} - \frac{(R_2 + R_3)}{(R_2 - \alpha R_3)}I_2. \tag{5}$$

Putting the value of I_1 from Equation (5) into Equation (1),

$$V_1 = \left(\frac{R_1 + \alpha R_1 + R_2}{R_2 - \alpha R_3}\right)V_2 - \frac{(1+\alpha)(R_1 R_2 + R_2 R_3 + R_3 R_1)}{(R_2 - \alpha R_3)}I_2. \tag{6}$$

Now comparing Equations (5) and (6) with Equations (4) and (3),

$$A = \frac{1}{R_2 - \alpha R_3}, \quad B\left(\frac{R_2 + R_3}{R_2 - \alpha R_3}\right), \quad C\left(\frac{R_1 + \alpha R_1 + R_2}{R_2 - \alpha R_3}\right)$$

and

$$D = \frac{(1+\alpha)(R_1 R_2 + R_2 R_3 + R_3 R_1)}{(R_2 - \alpha R_3)}.$$

Problem 10.18. *Find the Z- and h-parameters of the given network shown in Figure 10.50.*

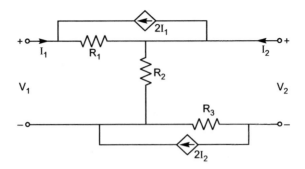

FIGURE 10.50

Solution: Applying KVL,

$$V_1 = 3I_1R_1 + (I_1 + I_2)R_2$$

or

$$V_1 = I_1(3R_1 + R_2) + I_2R_2 \qquad (1)$$
$$V_2 = (I_1 + I_2)R_2 + 3I_2R_3 \implies V_2 = I_1R_2 + I_2(R_2 + 3R_3) \qquad (2)$$

Comparing Equations (1) and (2) with the standard equations of the Z-parameters,

$$Z_{11} = 3R_1 + R_2, \quad Z_{12} = R_2, \quad Z_{21} = R_2, Z_{22} = R_2 + 3R_3.$$

Now the standard equations for the h-parameters are

$$V_1 = h_{11}I_1 + h_{12}V_2 \qquad (3)$$
$$I_2 = h_{21}I_1 + h_{22}V_2. \qquad (4)$$

Rearrange Equation (2) into the form of Equation (4),

$$I_2 = -\left(\frac{R_2}{R_2 + 3R_3}\right)I_1 + \frac{V_2}{(R_2 + 3R_3)}. \qquad (5)$$

Putting the value of I_2 from Equation (5) into Equation (1),

$$V_1 = I_1(3R_1 + R_2) + R_2\left[-\left(\frac{R_2}{R_2 + 3R_3}\right)I_1 + \frac{V_2}{(R_2 + 3R_3)}\right].$$

After solving we get

$$V_1 = \frac{3(R_1R_2 + R_2R_3 + 3R_3R_1)}{(R_2 + 3R_3)}I_1 + \left(\frac{R_2}{R_2 + 3R_3}\right)V_2. \qquad (6)$$

Now comparing Equations (6) and (5) with Equations (3) and (4), we get

$$h_{11} = \frac{3(R_1R_2 + R_2R_3 + 3R_3R_1)}{(R_2 + 3R_3)},$$

$$h_{12} = \left(\frac{R_2}{R_2 + 3R_3}\right), \quad h_{21} = -\left(\frac{R_2}{R_2 + 3R_3}\right)$$

and

$$h_{22} = \frac{1}{(R_2 + 3R_3)}.$$

Problem 10.19. *Find the Z- and h-parameters of the given network shown in Figure 10.51.*

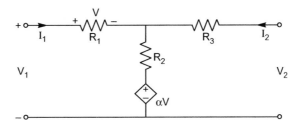

FIGURE 10.51

Solution: Applying KVL,

$$V_1 = I_1R_1 + (I_1 + I_2)R_2 + \alpha V$$

where $V = I_1R_1$.

Now,

$$V_1 = I_1(R_1 + R_2 + \alpha R_1) + I_2R_2 \tag{1}$$
$$V_2 = I_2R_3 + (I_1 + I_2)R_2 + \alpha I_1R_1$$
$$V_2 = I_1(R_2 + \alpha R_1) + I_2(R_2 + R_3). \tag{2}$$

Now comparing Equations (1) and (2) with the standard Z-parameter equations,

$$Z_{11} = (R_1 + R_2 + \alpha R_1), \quad Z_{12} = R_2, \quad Z_{21} = (R_2 + \alpha R_1) \quad \text{and}$$
$$Z_{22} = (R_2 + R_3).$$

Now the standard equations for the h-parameters are

$$V_1 = h_{11}I_1 + h_{12}V_2 \tag{3}$$
$$I_2 = h_{21}I_1 + h_{22}V_2. \tag{4}$$

Rearranging Equation (2) into the form of Equation (4),

$$I_2 = -\frac{(R_2 + \alpha R_1)}{(R_2 + R_3)}I_1 + \frac{V_2}{(R_2 + R_3)}. \tag{5}$$

Putting I_2 from Equation (5) into Equation (1),

$$V_1 = I_1(R_1 + R_2 + \alpha R_1) + R_2\left[-\left(\frac{R_2 + \alpha R_1}{R_2 + R_3}\right)I_1 + \frac{V_2}{R_2 + R_3}\right]$$
$$V_1 = \left[\frac{R_1R_2 + R_2R_3 + R_3R_1(1+\alpha)}{R_2 + R_3}\right]I_1 + \left(\frac{R_2}{R_2 + R_3}\right)V_2. \tag{6}$$

Comparing Equations (6) and (5) with Equations (3) and (4),

$$h_{11} = \frac{R_1R_2 + R_2R_3 + R_3R_1(1+\alpha)}{R_2 + R_3}, \quad h_{12} = \left(\frac{R_2}{R_2 + R_3}\right),$$
$$h_{21} = -\left(\frac{R_2 + \alpha R_1}{R_2 + R_3}\right), \quad h_{22} = \left(\frac{1}{R_2 + R_3}\right).$$

Problem 10.20. *Determine the Z- and ABCD parameters of the given network shown in Figure 10.52.*

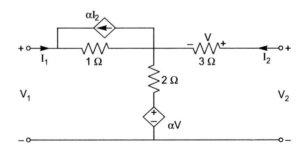

FIGURE 10.52

Solution: Two loops equations are:

$$V_1 = (I_1 + \alpha I_2) \times 1 + 2(I_1 + I_2) + \alpha V \quad [V = 3I_2]$$

$$V_1 = I_1(3) + I_2(\alpha + 2 + 3\alpha) \implies V_1 = 3I_1 + (4\alpha + 2)I_2 \qquad (1)$$

$$V_2 = 3I_2 + 2(I_1 + I_2) + \alpha \times 3I_2$$

$$V_2 = 2I_1 + I_2(5 + 3\alpha). \qquad (2)$$

Now comparing Equations (1) and (2) with the standard Z-parameters equations,

$$Z_{11} = 3, \quad Z_{12} = (4\alpha + 2), \quad Z_{21} = 2, \quad Z_{22} = (5 + 3\alpha).$$

Now the standard equations for the ABCD parameters are

$$V_1 = AV_2 - BI_2 \qquad (3)$$

$$I_1 = CV_2 - DI_2. \qquad (4)$$

Now rearrange Equation (2) into the form of Equation (4),

$$I_1 = \frac{V_2}{2} - \left(\frac{5 + 3\alpha}{2}\right) I_2. \qquad (5)$$

Now putting the value of I_1 from Equation (5) into Equation (1),

$$V_1 = 3\left[\frac{V_2}{2} - \left(\frac{5 + 3\alpha}{2}\right) I_2\right] + (4\alpha + 2)I_2$$

or

$$V_1 = \frac{3}{2}V_2 - \left(\frac{11 + \alpha}{2}\right) I_2. \qquad (6)$$

Comparing Equations (6) and (5) with Equations (3) and (4),

$$A = \frac{3}{2}, \quad B = \left(\frac{11 + \alpha}{2}\right), \quad C = \frac{1}{2}, \quad D = \left(\frac{5 + 3\alpha}{2}\right).$$

Problem 10.21. *(a) Derive expressions for the Y-parameters in terms of* A, B, C, *and* D *parameters of a two-port network.*

(b) Two networks have general ABCD parameters as follows:

Parameter	Network I	Network II
A	1.50	$\frac{5}{3}$
B	11 Ω	4 Ω
C	0.25 ℧	1 ℧
D	2.5 Ω	3.0

If the two networks are connected with their inputs and outputs in parallel, obtain the admittance matrix of the resulting network.

Solution: For network I,

$$Y_{11} = \frac{D}{B} = \frac{2.5}{11}\, \mho \qquad \begin{bmatrix} |T| = AD - BC \\ = 1.5 \times 2.5 - 11 \times 0.25 \\ = 1 \end{bmatrix}$$

$$Y_{12} = -\frac{|T|}{B} = -\frac{1}{11}\, \mho$$

$$Y_{21} = -\frac{1}{B} = -\frac{1}{11}\, \mho$$

$$Y_{22} = \frac{A}{B} = \frac{1.50}{11} = \frac{3}{22}\, \mho.$$

For network II,

$$Y_{11} = \frac{D}{B} = \frac{3}{4}\, \mho \qquad \left[|T| = AD - BC = \frac{5}{3} \times 3 - 4 = 1 \right]$$

$$Y_{12} = -\frac{|T|}{B} = -\frac{1}{4}\, \mho$$

$$Y_{21} = -\frac{1}{B} = -\frac{1}{4}\, \Omega$$

$$Y_{22} = \frac{A}{B} = \frac{5}{3 \times 4} = \frac{5}{12}\, \mho.$$

In the parallel-parallel connection,

$$[Y] = [Y_a] + [Y_b]$$

$$[Y] = \begin{bmatrix} \dfrac{5}{22} & -\dfrac{1}{11} \\ -\dfrac{1}{11} & \dfrac{3}{22} \end{bmatrix} + \begin{bmatrix} \dfrac{3}{4} & -\dfrac{1}{4} \\ -\dfrac{1}{4} & \dfrac{5}{12} \end{bmatrix}$$

$$= \begin{bmatrix} \dfrac{43}{44} & -\dfrac{15}{44} \\ -\dfrac{15}{44} & \dfrac{73}{132} \end{bmatrix}.$$

Problem 10.22. *Find the Y-parameters for the network shown in Figure 10.53.*

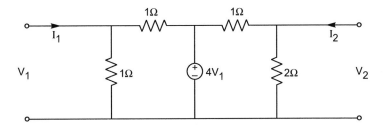

FIGURE 10.53

Solution: By applying mesh analysis,

$$V_1 = I_1 + I_3 \tag{1}$$

$$-4V_1 + I_3 + (I_3 + I_1) = 0$$

$$4V_1 = 2I_3 + I_1 = 4(I_1 + I_3)$$

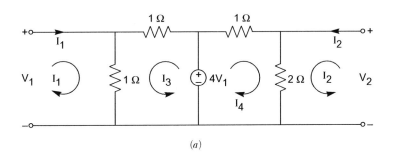

(a)

FIGURE 10.53

or

$$I_3 = -1.5I_1 \tag{2}$$

$$V_2 = 2(I_2 + I_4) \tag{3}$$

$$4V_1 = I_4 + 2(I_2 + I_4)$$

$$4(I_1 + I_3) = 2I_2 + 3I_4$$

$$4I_1 + 4 \times -1.5I_1 = 2I_2 + 3I_4$$

$$-2(I_1 + I_2) = 3I_4$$

$$I_4 = -\frac{2}{3}(I_1 + I_2). \tag{4}$$

Putting the value of I_3 into Equation (1) and I_4 into Equation (3), we get

$$V_1 = -\frac{1}{2}I_1 \tag{5}$$

$$V_2 = 2I_2 - \frac{4}{3}(I_1 + I_2)$$

$$V_2 = -\frac{4}{3}I_1 + \frac{2}{3}I_2. \tag{6}$$

By Equation (5) and (6),

$$I_1 = -2V_1 + 0V_2 \tag{7}$$
$$I_2 = -8V_1 + 3V_2. \tag{8}$$

Comparing Equations (7) and (8) with the standard equations of the Y-parameters, we get

$$Y_{11} = -2, \quad Y_{12} = 0, \quad Y_{21} = -8, \quad Y_{22} = 3.$$

Problem 10.23. *Determine the Z-, ABCD, and h-parameters of the given network shown in Figure 10.54.*

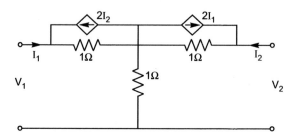

FIGURE 10.54

Solution: The two loop equations are

$$V_1 = 1 \cdot (I_1 + 2I_2) + 1 \cdot (I_1 + I_2)$$

or

$$V_1 = 2I_1 + 3I_2 \tag{1}$$
$$V_2 = 1(2I_1 + I_2) + 1(I_1 + I_2)$$
$$V_2 = 3I_1 + 2I_2. \tag{2}$$

By Equations (1) and (2) the Z-parameters are

$$Z_{11} = 2, \quad Z_{12} = 3, \quad Z_{21} = 3, \quad Z_{22} = 2.$$

Now the standard equations for the ABCD parameters are

$$V_1 = AV_2 - BI_2$$
$$I_1 = CV_2 - DI_2.$$

Rearranging Equation (2),

$$I_1 = \frac{V_2}{3} - \frac{2}{3}I_2. \tag{3}$$

Putting the value of I_1 from Equation (3) into Equation (1),

$$V_1 = 2\left[\frac{V_2}{3} - \frac{2}{3}I_2\right] + 3I_2$$

$$V_1 = \frac{2}{3}V_2 + \frac{5}{3}I_2 \tag{3}$$

$$A = \frac{2}{3}, \quad B = -\frac{5}{3}, \quad C = \frac{1}{3}, \quad \text{and} \quad D = \frac{2}{3}.$$

The standard equations for the h-parameters are

$$V_1 = h_{11}I_1 + h_{12}V_2$$
$$I_2 = h_{21}I_1 + h_{22}V_2.$$

Rearranging Equation (2),

$$I_2 = -\frac{3}{2}I_1 + \frac{V_2}{2} \tag{4}$$

$$h_{21} = -\frac{3}{2} \quad \text{and} \quad h_{22} = \frac{1}{2}.$$

Putting the value of I_2 from Equation (4) into Equation (1),

$$V_1 = 2I_1 + 3\left[-\frac{3}{2}I_1 + \frac{V_2}{2}\right]$$

$$V_1 = -\frac{5}{2}I_1 + \frac{3}{2}V_2 \tag{5}$$

$$h_{11} = -\frac{5}{2}, \quad h_{12} = \frac{3}{2}.$$

Problem 10.24. *Find the open circuit transfer impedance $\frac{V_2}{I_1}$ and the open circuit voltage ratio $\frac{V_2}{V_1}$ for the ladder network shown Figure 10.55.*

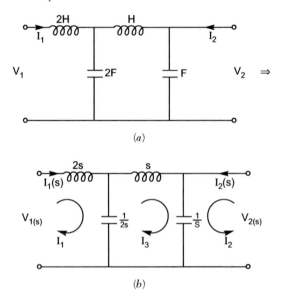

FIGURE 10.55

Solution: First convert the s-domain representation as shown in Figure 10.55(b). The loop equations are

$$V_1 = 2sI_1 + (I_1 - I_3)\frac{1}{2s} \tag{1}$$

$$I_3s + (I_3 + I_2) \times \frac{1}{s} + (I_3 - I_1) \times \frac{1}{2s} = 0 \implies I_3 = \frac{I_1 - 2I_2}{(2s^2 + 3)}$$

$$V_2 = \frac{1}{s}(I_2 + I_3) \quad \text{or} \quad V_2 = \frac{1}{s}\left[I_2 + \frac{I_1 - 2I_2}{2s^2 + 3}\right]. \tag{2}$$

Given, $I_2 = 0$, putting the value of I_2 into Equations (1) and (2), now by Equations (1) and (2),

$$V_1 = 2sI_1 + \left[I_1 - \frac{I_1 - 2I_2}{2s^2 + 3}\right]\frac{1}{2s}$$

$$V_1 = 2sI_1 + \left[I_1 - \frac{I_1}{2s^2 + 3}\right] \times \frac{1}{2s} \implies V_1 = \left[\frac{4s^4 + 7s^2 + 1}{s(2s^2 + 3)}\right]I_1$$

$$V_2 = \frac{I_1}{s(2s^2 + 3)} \implies \frac{V_2}{I_1} = \frac{1}{s(2s^2 + 3)}.$$

Now,

$$\frac{V_2}{V_1} = \frac{I_1}{s(2s^2+3)} \times \frac{s(2s^2+3)}{(4s^4+7s^2+1)I_1}$$

$$\frac{V_2}{V_1} = \frac{1}{4s^4+7s^2+1}.$$

Problem 10.25. *Determine the image parameters of the T-network shown in Figure 10.56.*

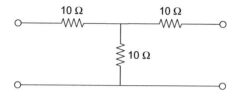

FIGURE 10.56

Solution: By KVL,

$$V_1 = 10I_1 + 10(I_1 + I_2)$$
$$V_1 = 20I_1 + 10I_2 \tag{1}$$
$$V_2 = 10I_2 + 10(I_1 + I_2)$$
$$V_2 = 10I_1 + 20I_2. \tag{2}$$

The standard T-parameters equations are

$$V_1 = AV_2 - BI_2$$
$$I_1 = CV_2 - DI_2.$$

By Equation (2),

$$I_1 = \frac{1}{10}V_2 - 2I_2.$$

By comparison, $C = \frac{1}{10}, D = 2,$

$$V_1 = 20\left(\frac{1}{10}V_2 - 2I_2\right) + 10I_2$$
$$V_1 = 2V_2 - 30I_2$$

$$A = 2, \quad B = 30$$

$$A = D = 2 \quad [\text{network is symmetrical}]$$

$$Z_{i_1} = Z_{i_2} = \sqrt{\frac{B}{C}} = \sqrt{\frac{30}{\frac{1}{10}}} = 17.32$$

$$\theta = \tanh^{-1}\sqrt{\frac{BC}{AD}} = \tanh^{-1}\sqrt{\frac{30 \times \frac{1}{10}}{2 \times 2}} = \tanh^{-1}\sqrt{\frac{3}{4}}.$$

Problem 10.26. *Determine the Y-parameters for the network shown in Figure 10.57.*

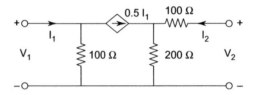

FIGURE 10.57

Solution: Applying KVL,

$$V_1 = 100(I_1 - 0.5I_1)$$

$$V_1 = 50I_1 \tag{1}$$

$$V_2 = 100I_2 + 200(I_2 + 0.5I_1)$$

$$V_2 = 100I_1 + 300I_2. \tag{2}$$

From Equations (1) and (2), we get

$$I_1 = \frac{1}{50}V_1$$

$$I_2 = -\frac{1}{150}V_1 + \frac{V_2}{300}$$

$$Y_{11} = \frac{1}{50}\ \mho, \quad Y_{12} = 0, \quad Y_{21} = -\frac{1}{150}\ \mho, \quad Y_{22} = \frac{1}{300}\ \mho.$$

QUESTIONS FOR DISCUSSION

1. What is a port?
2. Write the standard Z-parameter equations.

3. Write the standard Y-parameter equations.
4. Write the standard h-parameter equations.
5. Write the standard ABCD parameter equations.
6. Why are ABCD parameters known as transmission parameters?
7. Write the standard g-parameter equations.
8. What is the unit of Z_{11}, Y_{11}, A, and h_{12}?
9. Why are Z-parameters known as open circuit parameters?
10. Why are Y-parameters known as short circuit parameters?
11. What is the relation between [Z] and [Y]?
12. Why are h-parameters known as hybrid parameters?
13. Write the standard inverse hybrid parameter equations.

OBJECTIVE QUESTIONS

1. When the terminal voltage and currents of a two-port network are indicated in a figure as if the two-port is reciprocal, then
   (a) $\dfrac{Z_{12}}{Y_{12}} = Z_{12}^2 - Z_{11} \cdot Z_{22}$
   (b) $Z_{12} = \dfrac{1}{Y_{22}}$
   (c) $h_{12} = -h_{21}$
   (d) $AD - BC = 0$.

2. If a two-port network is passive, then we have, with the usual notation, the following relationship
   (a) $h_{12} = h_{21}$ (b) $h_{12} = -h_{21}$
   (c) $h_{11} = h_{22}$ (d) $h_{11}h_{22} - h_{12}h_{21} = 1$.

3. A two-port device is defined by the following pair of equations $i_1 = 2V_1 + V_2$ and $i_2 = V_1 + V_2$. Its impedance parameters $(Z_{11}, Z_{12}, Z_{21}, Z_{22})$ are given by
   (a) $(2, 1, 1, 1)$ (b) $(1, -1, -1, 2)$
   (c) $(1, 1, 1, 2)$ (d) $(2, -1, -1, 1)$.

4. With the usual notations, if a two-port resistive network satisfies the condition: $A = D = \frac{3}{2}$ $B = \frac{4}{3}C$. The Z_{11} of the network is
   (a) $\dfrac{5}{3}$ (b) $\dfrac{4}{3}$ (c) $\dfrac{2}{3}$ (d) $\dfrac{1}{3}$.

5. In a two-port network, the condition for reciprocity in terms of the h-parameters is
   (a) $h_{12} = h_{21}$ (b) $h_{11} = h_{22}$
   (c) $h_{11} = -h_{22}$ (d) $h_{12} = -h_{21}$.

6. The h-parameters of the network shown in Figure 10.58 will be

$(a) \begin{bmatrix} 1 & 1 \\ 1 & 1 \end{bmatrix}$ $(b) \begin{bmatrix} 0 & 1 \\ -1 & 1 \end{bmatrix}$ $(c) \begin{bmatrix} -1 & 1 \\ 1 & -1 \end{bmatrix}$ $(d) \begin{bmatrix} 0 & -1 \\ 2 & \frac{1}{2} \end{bmatrix}$.

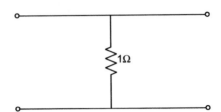

FIGURE 10.58

7. The Y_{21} parameter of the 3 Ω network shown in Figure 10.59 will be

$(a) \frac{1}{6} \mho$ $(b) -\frac{1}{6} \mho$ $(c) \frac{1}{3} \mho$ $(d) -\frac{1}{3} \mho$.

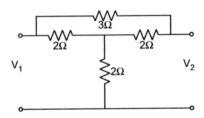

FIGURE 10.59

8. A T-network is shown in Figure 10.60. Its Y_{sc} matrix will be (units in Siemens)

$(a) \begin{bmatrix} \dfrac{10}{100} & \dfrac{5}{200} \\ \dfrac{5}{200} & \dfrac{10}{200} \end{bmatrix}$ $(b) \begin{bmatrix} \dfrac{10}{200} & \dfrac{-5}{200} \\ \dfrac{-5}{200} & \dfrac{10}{200} \end{bmatrix}$

$(c) \begin{bmatrix} \dfrac{15}{200} & \dfrac{-5}{200} \\ \dfrac{-5}{200} & \dfrac{15}{200} \end{bmatrix}$ $(d) \begin{bmatrix} \dfrac{15}{200} & \dfrac{5}{200} \\ \dfrac{5}{200} & \dfrac{15}{200} \end{bmatrix}$.

9. For a two-port network to be reciprocal, it is necessary that
    $(a) Z_{11} = Z_{22}$ and $Y_{21} = Y_{12}$ $(b) Z_{11} = Z_{22}$ and $AD - BC = 0$
    $(c) h_{21} = -h_{12}$ and $AD - BC = 0$ $(d) Y_{21} = Y_{12}$ and $h_{21} = -h_{12}$.

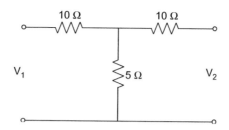

FIGURE 10.60

10. Two two-port networks with transmission parameters A_1, B_1, C_1, D_1, and A_2, B_2, C_2, D_2, respectively, are cascaded. The transmission parameter matrix of the cascaded network will be

(a) $\begin{bmatrix} A_1 & B_1 \\ C_1 & D_1 \end{bmatrix} + \begin{bmatrix} A_2 & B_2 \\ C_2 & D_2 \end{bmatrix}$ (b) $\begin{bmatrix} A_1 & B_1 \\ C_1 & D_1 \end{bmatrix} \begin{bmatrix} A_2 & B_2 \\ C_2 & D_2 \end{bmatrix}$

(c) $\begin{bmatrix} A_1 A_2 & B_1 B_2 \\ C_1 C_2 & D_1 D_2 \end{bmatrix}$ (d) $\begin{bmatrix} (A_1 A_2 + C_1 C_2) & (A_1 A_2 - B_1 D_2) \\ (C_1 A_2 - D_1 C_2) & (C_1 C_2 + D_1 D_2) \end{bmatrix}$.

11. A two-port network is reciprocal if and only if
   (a) $Z_{11} = Z_{22}$ (b) $BC - AD = -1$
   (c) $Y_{12} = -Y_{21}$ (d) $h_{12} = h_{21}$.

12. The parameter Y_{12} of the two-port network as shown in Figure 10.61 is
   (a) Y_a (b) $-Y_a$ (c) Y_b (d) $-Y_b$.

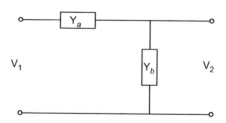

FIGURE 10.61

13. A two-port network is symmetrical if
   (a) $Z_{11} Z_{22} - Z_{12} Z_{21} = 1$ (b) $AD - BC = 1$
   (c) $h_{11} h_{22} - h_{12} h_{21} = 1$ (d) $Y_{11} Y_{22} - Y_{12} Y_{21} = 1$.

14. The circuit shown in Figure 10.62 is equivalent to a load of
   (a) $\frac{4}{3} \Omega$ (b) $\frac{8}{3} \Omega$ (c) 4Ω (d) 2Ω.

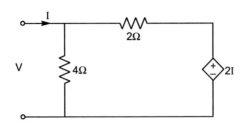

FIGURE 10.62

15. The short circuit admittance matrix of the network shown in Figure 10.63 is

$(a) \begin{bmatrix} \frac{1}{Z} & -\frac{1}{Z} \\ -\frac{1}{Z} & \frac{1}{Z} \end{bmatrix}$ $(b) \begin{bmatrix} \frac{1}{Z} & -\frac{1}{Z} \\ \frac{1}{Z} & \frac{1}{Z} \end{bmatrix}$

$(c) \begin{bmatrix} \frac{1}{Z} & \frac{1}{Z} \\ -1 & 1 \end{bmatrix}$ $(d) \begin{bmatrix} \frac{1}{Z} & 1 \\ 1 & \frac{1}{Z} \end{bmatrix}.$

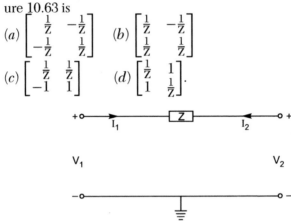

FIGURE 10.63

16. The two-port network α and β having ABCD parameters as

$$A_\alpha = D\alpha = 4 \quad \text{and} \quad A_\beta = D_\beta = 3$$
$$B_\alpha = 5, C_\alpha = 3 \quad \text{and} \quad B_\beta = 4 \quad \text{and} \quad C_\beta = 2$$

are connected in cascade in the order of $\alpha; \beta$. The equivalent A parameter of the combination is
(a) 17 (b) 22 (c) 24 (d) 31.

17. When a number of two-port networks are connected in cascade, the individual
(a) Z_{oc} matrices are added (b) Y_{sc} matrices are added
(c) Chain matrices are multiplied (d) H-matrices are multiplied.

18. A two-port network is defined by the relations

$$I_1 = 2V_1 + V_2, \quad I_2 = 2V_1 + 3V_2, \quad \text{then } Z_{12} \text{ is}$$

(a) $-2\,\Omega$ (b) $-1\,\Omega$ (c) $-\frac{1}{2}\,\Omega$ (d) $-\frac{1}{4}\,\Omega.$

UNSOLVED PROBLEMS

1. Find the Z- and h-parameters for the given network shown in Figure 10.64

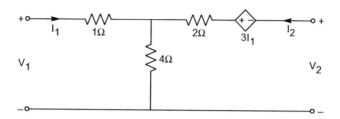

FIGURE 10.64

2. Find the Z- and h-parameters for the given network shown in Figure 10.65.

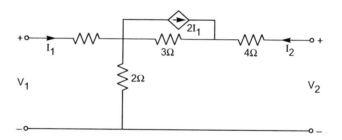

FIGURE 10.65

3. Prove that $Z_{11} = (\frac{s^2+2s+1}{s})$ and $Y_{11} = (\frac{3s}{3s^2+5s+3})$ for the given network shown in Figure 10.66.

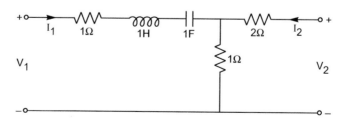

FIGURE 10.66

4. Two similar networks as shown in Figure 10.67 are connected in cascade. Calculate the overall transmission parameters for the combined network.

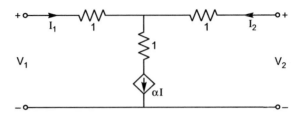

FIGURE 10.67

5. Two similar networks are connected in series as shown in Figure 10.68. Determine the equivalent Z-parameters of the combined network.

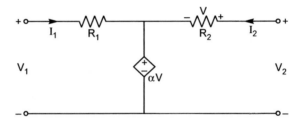

FIGURE 10.68

6. Determine the Z and ABCD parameters of the given network shown in Figure 10.69.

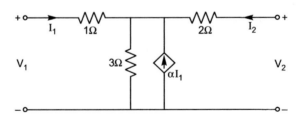

FIGURE 10.69

7. Two similar networks are connected in parallel as shown in Figure 10.70. Determine the equivalent Y-parameters of the combined network.

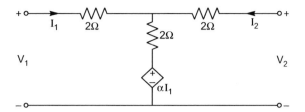

FIGURE 10.70

8. Determine the transmission parameters of the network shown in Figure 10.71. Using the concept of interconnection of two two-port networks N_1 and N_2 in cascade.

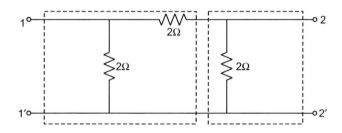

FIGURE 10.71

9. Determine the transmission parameters of the network shown in Figure 10.72, using the concept of interconnection of two-port networks N_1 and N_2 in cascade.

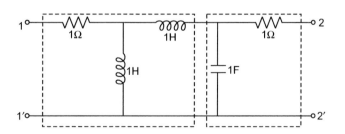

FIGURE 10.72

10. For the network shown in Figure 10.73 which contains a controlled source, find the Y- and Z-parameters.

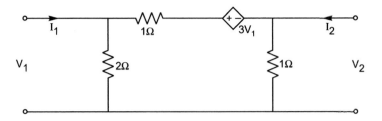

FIGURE 10.73

11. Show that the h-parameters will not exist for a two-port network when $Z_{22} = 0$.
12. Find the Z- and inverse transmission parameters for the given network shown in Figure 10.74.

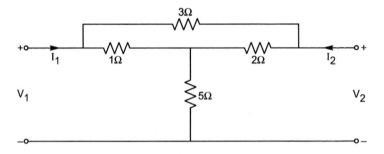

FIGURE 10.74

13. Determine the ABCD and g-parameters of the given network shown in Figure 10.75.

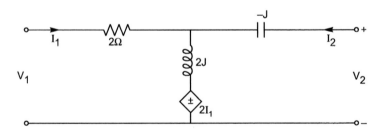

FIGURE 10.75

14. Determine the Z- and Y-parameters of the network shown in Figure 10.76.

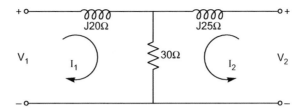

FIGURE 10.76

15. 15. Find the Z-parameters for the given network shown in Figure 10.77.

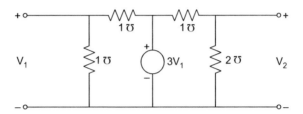

FIGURE 10.77

16. For the two-port network shown in Figure 10.78, calculate the value of Y_{12} and Z_{22}.

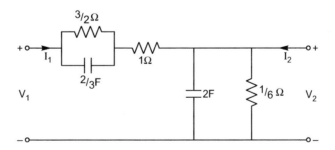

FIGURE 10.78

17. Determine ABCD parameters, if the values of the other parameters are as given below:
    (a) $Z_{11} = 1, Z_{12} = -2, Z_{21} = 2, Z_{22} = 1$
    (b) $g_{11} = 1, g_{12} = -2, g_{21} = 2, g_{22} = 1$.
18. Check whether the given network is symmetrical

19. Check whether the given network shown in Figure 10.79 is reciprocal.

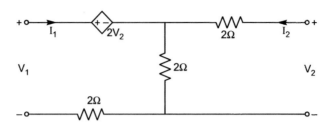

FIGURE 10.79

20. Discuss the condition of symmetry and reciprocity for h-parameters.
21. Determine the Z-and ABCD parameters of the given network shown in Figure 10.80.

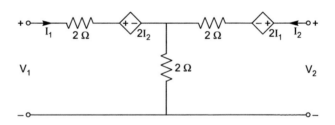

FIGURE 10.80

22. Two networks are connected in series as shown in Figure 10.81(a) and (b). Determine the Y-parameters of the combined network.

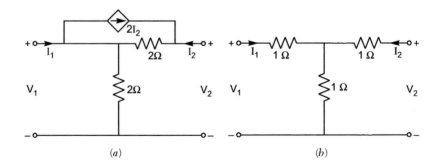

FIGURE 10.81

23. Two identical networks shown in Figure 10.81(b) are connected in series parallel. Calculate the equivalent h-parameters of the combined network.

24 Determine the value of Z_{12}, A and h_{22} for given network shown in Figure 10.82.

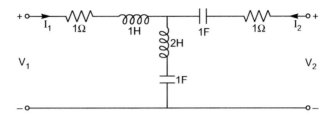

FIGURE 10.82

25. Two identical networks shown in Figure 10.81(b) are connected in parallel-series. Calculate the equivalent g-parameters of the combined network.

26. Determine the h and g-parameters of the given network shown in Figure 10.83.

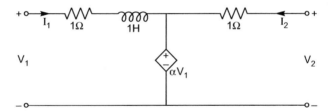

FIGURE 10.83

27. Verify symmetry and reciprocity conditions for the given network shown in Figure 10.84.

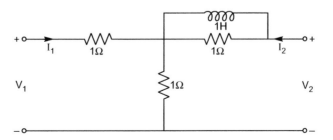

FIGURE 10.84

28. Two transmission lines are connected in cascade, whose ABCD parameters are

$$\begin{bmatrix} A_1 & B_1 \\ C_1 & D_1 \end{bmatrix} = \begin{bmatrix} 1 & 10\angle 30° \\ 0 & 1 \end{bmatrix}; \begin{bmatrix} A_2 & B_2 \\ C_2 & D_2 \end{bmatrix} = \begin{bmatrix} 1 & 0 \\ 0.25\angle 30° & 1 \end{bmatrix},$$

respectively. Find the resultant ABCD parameters.

29. A $200 \, \text{Km}, 3 - \phi, 50 \, \text{Hz}$ transmission line has the following data $A = D = 0.938\angle 1.2$, $B = 131.2\angle 72.3°\Omega/\text{phase}$, $C = 0.001\angle 90°$ Siemens/phase. The sending end voltage is 230 KV. Determine:
(a) The receiving end voltage when the load is disconnected.
(b) The line charging current.
(c) The maximum power that can be transmitted at a receiving end voltage of 220 KV and the corresponding load reactive power required at the receiving end.

30. For the two-port network shown in Figure 10.85, determine the admittance matrix.

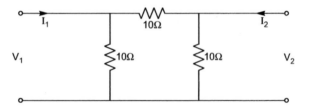

FIGURE 10.85

31. Define the terms of driving-point impedance, transfer impedance, and transfer function of a network. The network shown in Figure 10.86 is driven by a current source and is terminated by a resistor at port 2. For this terminated two-port network, calculate
(i) transfer function $g_{21}(s)$, $a_{21}(a)$, $Z_{21}(s)$ and $Y_{21}(s)$
(ii) driving-point impedance $Z_{11}(s)$.

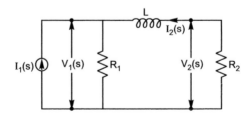

FIGURE 10.86

11 NETWORK SYNTHESIS

11.1 INTRODUCTION

The network, the excitation, and the response. If any two of the three quantities are given, the third may be found for linear networks. If the network and the excitation are given and the response is to be determined, the problem is defined as analysis. When the excitation and the response are given, and it is required to determine a network, the problem is defined as synthesis. If the network and the response are given to find the excitation there is generally no accepted name and it is not common.

Synthesis is the process of finding a network corresponding to a given driving-point impedance or admittance. In this chapter, we will consider some aspects of passive network synthesis, mainly driving-point immittance functions (impedance or admittance functions),

$$F(s) = \frac{R(s)}{E(s)} \tag{11.1}$$

where $F(s)$ is the network function, which is the ratio of response $R(s)$ to the excitation $E(s)$.

11.2 ANALYSIS AND DESIGN

- Analysis of a system is the investigation of properties and the behavior (response) of an existing system.
- Design of a system is the choice and arrangement of system components to perform a specific task.
- Design by analysis is accomplished by modifying the characteristics of an existing system.
- Design by synthesis is accomplished by defining the form of the system directly from its specifications.

11.3   ELEMENTS OF THE REALIZABILITY THEORY

By Equation (11.1),

$$F(s) = \frac{R(s)}{E(s)}.$$

The first step in a synthesis procedure is to determine whether $F(s)$ can be realized as a physical passive network. $F(s)$ should be stable and satisfy the following conditions:

(*i*) $F(s)$ cannot have poles in the right half of the s-plane; all poles should lie in the left half of the s-plane.

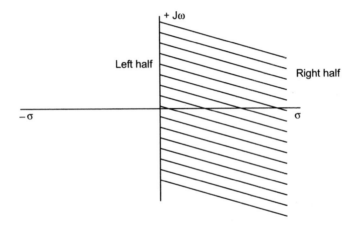

FIGURE 11.1

(*ii*) $F(s)$ should have simple poles, not multiples on the imaginary $J\omega$-axis.
(*iii*) The degree of the numerator of $F(s)$ cannot exceed the degree of the denominator by more than unity.

11.4   HURWITZ POLYNOMIAL

In the last chapter we discussed Routh-Hurwitz criteria for stable or unstable systems. An immitance function is realizable if it is a Hurwitz polynomial $P(s)$ and satisfies the following conditions:

(*i*) $P(s)$ is real when s is real.
(*ii*) The roots of $F(s)$ have real parts which are zero or negative.

Example 11.1. *Draw the poles and zeros of* $F(s) = \frac{(s+1)}{s(s-2)}$ *on the s-plane and check whether the system is stable or unstable.*

Solution:

$$F(s) = \frac{(s+1)}{s(s-2)}$$

Zeros:

$$s = -1$$

Poles:

$$s = 0, 2$$

The pole-zero configuration is shown in Figure 11.2.

One pole lies in the right half of the s-plane, therefore, this is not a stable system.

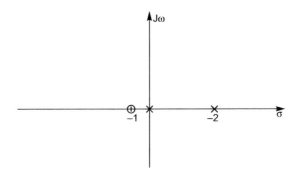

FIGURE 11.2

Properties of a Hurwitz polynomial:

A Hurwitz polynomial P (s) has the following properties:

(i) If the polynomial $P(s)$ can be written as

$$P(s) = a_n s^n + a_{n-1}s^{n-1} + a_{n-2}s^{n-2} + \cdots + a_1 s + a_0,$$

then all the coefficients a_i must be positive and no terms of s should be missing unless the polynomial is even or odd.

(ii) If $M(s)$ and $N(s)$ are the even and odd parts of the hurwitz polynomial $P(s)$, so that $P(s) = M(s) + N(s)$, then $M(s)$ and $N(s)$ both have roots on the $J\omega$-axis only.

(iii) If $P(s)$ is either even or odd, all its roots lie on the $J\omega$-axis (including the origin).

(*iv*) The continued fraction expansion of the ratio $\Psi(s) = \left(\dfrac{M(s)}{N(s)}\right)^{\pm 1}$ of a

Hurwitz polynomial yields all positive quotient terms as,

$$\Psi(s) = \frac{M(s)}{N(s)} \text{ or } \frac{N(s)}{M(s)}. \tag{11.2}$$

In the case of $\dfrac{M(s)}{N(s)}$, the M (s) degree is higher than N(s).

In the case of $\dfrac{N(s)}{M(s)}$, the N (s) degree is higher than M(s)

$$\Psi(s) = \frac{M(s)}{N(s)} \text{ or } \frac{N(s)}{M(s)}$$

$$= q_1 s + \cfrac{1}{q_2 s + \cfrac{1}{q_3 s + \cfrac{1}{\cdots + \cfrac{1}{q_n s}}}}.$$

If the polynomial $P(s) = M(s) + N(s)$ is Hurwitz, and the quotients $q_1, q_2, \ldots, q_n$ must be positive.

(*v*) If $P(s)$ is a Hurwitz polynomial then $P_1(s) = P(s)$. $W(s)$ is also a Hurwitz polynomial if $W(s)$ is a Hurwitz polynomial. It means the product of the two Hurwitz polynomials is a Hurwitz polynomial.

(*vi*) If the polynomial is either even or odd, it is not possible to obtain the continued fraction expansion. In such cases, the polynomial $P(s)$ is Hurwitz if the ratio of $P(s)$ and its derivative $P'(s)$ gives a continued fraction expansion,

$$\Psi(s) = \frac{P(s)}{P'(s)}, \tag{11.3}$$

where $P(s)$ is either an even or odd polynomial.

11.5 PROCEDURE FOR OBTAINING THE CONTINUED FRACTION EXPANSION

To obtain the continued fraction expansion,

$$\Psi(s) = \frac{M(s)}{N(s)},$$

where the degree of the $M(s)$ polynomial is higher than the $N(s)$ polynomial, then we obtain a single quotient and a remainder

$$\Psi(s) = q_1 s + \frac{R_1(s)}{N(s)} \qquad\qquad (11.4)$$

or

$$N(s)) \; M(s) \; (q_1 s$$
$$\frac{M_1(s)}{R_1(s)}. \qquad\qquad [R_1(s) = M(s) - M_1(s)]$$

Now the degree of the remainder $R_1(s)$ is one lower than the degree of the $N(s)$. Therefore, invert the second term and divide

$$\Psi(s) = q_1 s + \frac{1}{\frac{N(s)}{R_1(s)}}$$

$$= q_1 s + \frac{1}{q_2 s + \frac{R_2(s)}{R_1(s)}}. \qquad\qquad (11.5)$$

Now the degree of the remainder $R_2(s)$ is one lower than the degree of the $R_1(s)$. Therefore, once again invert and divide, and we get

$$\Psi(s) = q_1 s + \frac{1}{q_2 s + \frac{1}{q_3 s + \frac{R_3(s)}{R_2(s)}}}.$$

The number of quotients $(q_1, q_2, \ldots, q_n)$ depends on the degree of the polynomial. If the degree of the polynomial is 4 then there are four quotients and all quotients should be positive for the Hurwitz polynomial.

Example 11.2. *Check whether or not the given polynomial* $P(s)$ *is Hurwitz.*

$$P(s) = 2s^4 + s^3 + 7s^2 + 3s + 2$$

Solution: Since all coefficients of $P(s)$ are positive, $P(s)$ is real for s real. Now by the continued fraction,

$$M(s) = 2s^4 + 7s^2 + 2$$
$$N(s) = s^3 + 3s.$$

Continued fraction

$$\Psi(s) = \frac{M(s)}{N(s)} = \frac{2s^4 + 7s^2 + 2}{s^3 + 3s}$$

$$
\begin{array}{r}
s^3 + 3s\overline{)\,2s^4 + 7s^2 + 2\,}(2s \\
\underline{2s^4 + 6s^2} \\
s^2 + 2\overline{)\,s^3 + 3s\,}(s \\
\underline{s^3 + 2s} \\
s\overline{)\,s^2 + 2\,}(s \\
\underline{s^2} \\
2\overline{)\,s\,}(\tfrac{1}{2}s \\
\underline{s} \\
\overline{\text{X.}}
\end{array}
$$

The continued fraction expansion of $\Psi(s)$ is

$$\Psi(s) = \frac{M(s)}{N(s)} = 2s + \cfrac{1}{s + \cfrac{1}{s + \cfrac{1}{\cfrac{1}{2}s}}}$$

$$q_1 = 2, q_2 = 1, q_3 = 1, q_4 = \frac{1}{2}.$$

Since all the coefficients are positive, $P(s)$ is a Hurwitz polynomial.

Example 11.3. *Check whether or not the given polynomial $P(s)$ is a Hurwitz.*

$$P(s) = 2s^4 + s^3 + 2s^2 + 4s + 1$$

Solution: Since all coefficients of $P(s)$ are positive, $P(s)$ is real for s real.
    Now by continued fraction,

$$M(s) = 2s^4 + 2s^2 + 1$$
$$N(s) = s^3 + 4s$$
$$\Psi(s) = \frac{M(s)}{N(s)} = \frac{2s^4 + 2s^2 + 1}{s^3 + 4s}$$

$$s^3 + 4s) \; 2s^4 + 2s^2 + 1 \; (2s$$
$$\underline{2s^4 + 8s^2}$$

$$-6s^2 + 1) \; s^3 + 4s \; (-\frac{1}{6}s$$
$$\underline{s^3 - \frac{1}{6}s}$$

$$\frac{25}{6}s) \; -6s^2 + 1 \; (-\frac{36}{25}s$$
$$\underline{\pm 6s^2}$$

$$1) \; \frac{25}{6}s \; (\frac{25}{6}s$$
$$\underline{\frac{25}{6}s}$$

$$\text{X.}$$

$$\Psi(s) = \frac{2s^4 + 2s^2 + 1}{s^3 + 4s} = 2s + \cfrac{1}{-\frac{1}{6}s + \cfrac{1}{-\frac{36}{25}s + \cfrac{1}{\frac{25}{6}s}}}$$

$$q_1 = 2, \quad q_2 = -\frac{1}{6}, \quad q_3 = -\frac{36}{25}, \quad q_4 = \frac{25}{6}.$$

The degree of the polynomial is 4, hence, there are four quotients and two are negative (q_2 and q_3). Therefore, P(s) is not a Hurwitz polynomial.

Example 11.4. *Check whether or not the given polynomial* P(s) *is a Hurwitz.*

$$P(s) = 2s^5 + 3s^4 + 6s^3 + 5s^2 + 3s + 4$$

Solution: Since all coefficients are positive, P(s) is real for s real.

Now by continued fraction,

$$N(s) = 2s^5 + 6s^3 + 3s$$
$$M(s) = 3s^4 + 5s^2 + 4$$
$$\Psi(s) = \frac{N(s)}{M(s)} = \frac{2s^5 + 6s^3 + 3s}{3s^4 + 5s^2 + 4} \quad \text{[degree of N(s) is higher]}$$

$$3s^4 + 5s^2 + 4 \Big) \; 2s^5 + 6s^2 + 3s \; \Big(\frac{2}{3}s$$

$$\underline{2s^5 + \frac{10}{3}s3}$$

$$\frac{8}{3}s^3 + 3s \Big) \; 3s^4 + 5s^2 + 4 \; \Big(\frac{9}{8}s$$

$$\underline{3s^4 + \frac{27}{8}s^2}$$

$$\frac{13}{8}s^2 + 4 \Big) \; \frac{8}{3}s^3 + 3s \; \Big(\frac{64}{39}s$$

$$\underline{\frac{8}{3}s^3 + \frac{254}{39}s}$$

$$-\frac{137}{39}s \Big) \; \frac{13}{8}s^2 + 4 \; \Big(-\frac{13 \times 39}{137 \times 8}s$$

$$\underline{\frac{13}{8}s^2}$$

$$4 \Big) -\frac{137}{39}s \; \Big(4 \times -\frac{137}{39}s$$

$$\underline{-\frac{137}{39}s}$$

$$X.$$

$$\Psi(s) = \frac{2s^5 + 6s^3 + 3s}{3s^4 + 5s^2 + 4} = \frac{2}{3}s + \cfrac{1}{\frac{9}{8}s + \cfrac{1}{\frac{64}{39}s + \cfrac{1}{-\frac{507}{1096}s + \cfrac{1}{-\frac{548}{39}s}}}}$$

$$q1 = \frac{2}{3}, \quad q2 = \frac{9}{8}, \quad q3 = \frac{64}{39}, \quad q4 = -\frac{507}{1096}, \quad q5 = -\frac{548}{39}.$$

The degree of polynomial P(s) is 5. Hence, there are five quotients and two are negative (q_4 and q_5). Therefore, P(s) is not a Hurwitz polynomial.

Example 11.5. *Check whether the given polynomial P(s) is a Hurwitz or not.*

$$P(s) = s^4 + 2s^2 + 1$$

Solution: Polynomial P(s) has only even terms and all coefficients are positive.

Now by continued fraction,

$$\Psi(s) = \frac{P(s)}{P'(s)} = \frac{s^4 + 2s^2 + 1}{4s^3 + 4s}$$

$$4s^3 + 4s) \ s^4 + 2s^2 + 1 \ (\frac{1}{4}s$$
$$\underline{s^4 + s^2}$$
$$s^2 + 1) \ 4s^3 + 4s \ (4s$$
$$\underline{4s^3 + 4s}$$
$$\underline{\text{X}.}$$

The degree of polynomial P(s) is 4. There should be four quotients but the fraction vanishes and we get only two quotients,

$$q_1 = \frac{1}{4}, q_2 = 4.$$

The other two quotients we get from Q $(s) = s^2 + 1$

$$\Psi'(s) = \frac{Q(s)}{Q'(s)} = \frac{s^2 + 1}{2s}$$

$$2s) \ s^2 + 1 \ (\frac{1}{2}s$$
$$\underline{s^2}$$
$$1) \ 2s \ (2s$$
$$\underline{2s}$$
$$\underline{\text{X}}$$

$$q_3 = \frac{1}{2} \quad \text{and} \quad q_4 = 2.$$

Thus, all the quotients are positive. Therefore, P(s) is a Hurwitz polynomial.

Example 11.6. *Check whether or not the given polynomial P(s) is a Hurwitz.*

$$P(s) = s^5 + 2s^3 + 4s$$

Solution: The polynomial P(s) has only odd terms and all coefficients are positive.

Now by continued fraction,

$$\Psi(s) = \frac{P(s)}{P'(s)} = \frac{s^5 + 2s^3 + 4s}{5s^4 + 6s^2 + 4}$$

$$5s^4 + 6s^2 + 4) \; s^5 + 2s^3 + 4s \; (\frac{1}{5}s$$

$$s^5 + \frac{6}{5}s^3 + \frac{4}{5}s$$

$$\overline{}$$

$$\frac{4}{5}s^3 + \frac{16}{5}s) \; 5s^4 + 6s^2 + 4 \; (\frac{25}{4}s$$

$$5s^4 + 20s^2$$

$$\overline{}$$

$$-14s^2 + 4) \; \frac{4}{5}s^3 + \frac{16}{5}s \; (-\frac{4}{70}s$$

$$\frac{4}{5}s^3 - \frac{16}{70}s$$

$$\overline{}$$

$$\frac{24}{7}s) \; -14s^2 + 4 \; (-\frac{7 \times 14}{24}s$$

$$-14s2$$

$$\overline{}$$

$$4) \; \frac{24}{7}s \; (\frac{24 \times 4}{7}s$$

$$\frac{24}{7}s$$

$$\overline{}$$

$$\text{X.}$$

The quotients are.

$$q_1 = \frac{1}{5}, \quad q_2 = \frac{25}{4}, \quad q_3 = -\frac{4}{70}, \quad q_4 = -\frac{98}{24}, \quad q_5 = \frac{96}{7}.$$

Two quotients are negative, therefore, P(s) is not a Hurwitz polynomial.

Example 11.7. *Find the range of 1a' so that* P(s) *is a Hurwitz polynomial.*

$$P(s) = 2s^4 + s^3 + as^2 + 3s + 2$$

Solution: By continued fraction,

$$\mathbf{M}(s) = 2s^4 + as^2 + 2$$

$$\mathbf{N}(s) = s^3 + 3s$$

$$\Psi(s) = \frac{\mathbf{M}(s)}{\mathbf{N}(s)} = \frac{2s^4 + as^2 + 2}{s^3 + 3s}$$

$$s^3 + 3s)\ 2s^4 + as^2 + 2\ (2s$$
$$\underline{2s^4 + 6s^2}$$

$$(a-6)s^2 + 2)\ s^3 + 3s\ (\frac{1}{(a-6)}s$$
$$\underline{s^3 + \left(\frac{2}{a-6}\right)s}$$

$$\left(\frac{3a-20}{a-6}\right)s)(a-6)s^2 + 2(\frac{(a-6)^2}{3a-2a}s$$
$$\underline{(a-6)s^2}$$

$$2)\left(\frac{3a-20}{a-6}\right)s(\frac{1}{2}\left(\frac{3a-20}{a-6}\right)s$$
$$\underline{\left(\frac{3a-20}{a-6}\right)s}$$

$$\mathrm{X}$$

$$q_1 = 2,$$
$$q_2 = \frac{1}{a-6}$$
$$q_3 = \frac{(a-6)^2}{3a-20}$$
$$q_4 = \frac{1}{2}\left(\frac{3a-20}{a-6}\right).$$

For a Hurwitz polynomial, all quotients $(q_1, q_2, q_3$ and $q_4)$ should be positive

$$a - 6 > 0 \quad \text{or} \quad a > 6$$
$$3a - 20 > 0$$
$$a > \frac{20}{3}.$$

Thus, the range of a is $a > \frac{20}{3}$.

11.6   POSITIVE REAL FUNCTIONS

Positive real functions are important because they represent physically realizable passive driving-point immittances (impedance or admittance). A function $F(s) = \dfrac{N(s)}{D(s)}$ is a positive real function (p.r.f.) if the following conditions are satisfied:

(*i*)   F(*s*) is real for s real. F(*s*) is purely real.   [$s = \sigma + J\Omega$]

(*ii*) D(*s*) is a Hurwitz polynomial.

(*iii*) F(*s*) may have poles on the Jω-axis. Poles should be simple, not multiple, and their residues should be real and positive.

(*iv*) The real part of F(*s*) is greater than or equal to zero for $\sigma \geq 0$,

$$R_e F(s) \geq 0 \text{ for } \sigma \geq 0$$

or

$$R_e F(J\omega) \geq 0 \text{ for all } \omega. \qquad (11.6)$$

A simplification of condition (*iv*) is possible.

Let

$$F(s) = \frac{N(s)}{D(s)} = \frac{M_1(s) + N_1(s)}{M_2(s) + N_2(s)}, \qquad (11.7)$$

where $M_i(s)$ are even functions and $N_i(s)$ are odd functions,

$$F(s) = \frac{M_1 + N_1}{M_2 + N_2} \times \frac{M_2 - N_2}{M_2 - N_2} \qquad \text{[rationalizing]}$$

$$F(s) = \frac{M_1 M_2 - N_1 N_2}{M_2^2 - N_2^2} + \frac{N_1 M_2 - M_1 N_2}{M_2^2 - N_2}.$$

We see that the products $M_1 M_2$ and $N_1 N_2$ are even functions, while $N_1 M_2$ and $M_1 N_2$ are odd functions,

$$E_V[F(s)]|_{s=J\omega} = \left.\frac{M_1 M_2 - N_1 N_2}{M_2^2 - N_2^2}\right|_{s=J\omega} = R_e[F(J\Omega)]$$

and

$$R_e[F(J\omega)] \geq 0 \text{ for all } \omega$$

$$[M_2(J\omega)]^2 - [N_2(J\omega)]^2 = M_2\omega^2 + N_2(\omega)^2 \geq 0.$$

Now $M_1(J\omega)M_2(J\omega) - N_1(J\omega)N_2(J\omega) \geq 0$ for all ω.

Let

$$A(\omega^2) = M_1(J\omega).M_2(J\omega) - N_1(J\omega).N_2(J\omega) \geq 0. \qquad (11.8)$$

11.6.1 Properties of Positive Real Functions (P.R.F.)

(i) The coefficients of the numerator and denominator polynomials in F $(s) = \frac{N(s)}{D(s)}$ are real and positive as a consequence.

    (a) Y(s) is real when s is real.

    (b) Complex poles and zeros of Y(s) occur in conjugate pairs.

    (c) The scale factor H $= \frac{a_0}{b_0}$ is real and positive.

(ii) The poles and zeros of F(s) have either negative or zero real parts.

(iii) Poles of F(s) on the imaginary axis must be simple and their residues must be real and positive.

(iv) If F(s) is positive real then $\frac{1}{F(s)}$ is also positive real. This property implies that if a driving-point impedance Z(s) is p.r., then its reciprocal $\frac{1}{Z(s)} = $ Y(s), the driving-point admittance, is also p.r.

(v) The sum of the two p.r.f. [$F_1(s)$, $F_2(s)$] must be a p.r.f. but it is not necessary that their difference be also a p.r.f.

(vi) A p.r.f. always satisfies the condition [R_eF(Jω)] ≥ 0 for all ω.

(vii) The degree of the numerator and denominator polynomials in F(s) differ at most by 1. Thus, the number of finite poles and finite zeros of F(s) differ at most by 1.

($viii$) The terms of lowest degree in the numerator and denominator polynomials of F(s) differ in degree at most by 1. So F(s) has neither multiple poles nor zeros at the origin.

11.6.2 Procedure for Testing for Positive Real Functions

1. Inspection Test. It is required that:

    (a) All polynomial coefficients should be real and positive.

    (b) Higher degrees of numerator and denominator polynomials differ at most by 1.

    (c) Numerator and denominator terms of lowest degree differ at most by 1.

    (d) Imaginary axis poles and zeros should be simple.

    (e) There should be no missing terms in numerator and denominator polynomials unless all even or all odd terms are missing.

2. Test for necessary and sufficient conditions:

    (a) If F $(s) = \frac{N(s)}{D(s)}$, then N (s) + D(s) must be a Hurwitz. This requires that:

        (i) the continued fraction expansion of the Hurwitz test give only real and positive quotients.

        (ii) the continued fraction not end prematurely.

(b) In order that $R_e F(J\omega) \geq 0$ for all ω, it is necessary and sufficient that $A(\omega^2) = M_1 M_2 - N_1 N_2|_{s=J\omega}$ have no real positive roots of odd multiplicity.

Example 11.8. *Check whether the function* $F(s) = \frac{2s^2 + s + 1}{s^2 + s + 2}$ *is p.r. or not.*

Solution:

(a) Inspection test: All coefficients are real and positive, and higher and lower degree difference between the numerator and denominator is zero. There are no missing terms and all poles and zeros are simple. Therefore, $F(s)$ may be a positive real function.

(b) Test for necessary and sufficient conditions:

$$M_1 = 2s^2 + 1, \quad N_1 = s, \quad M_2 = s^2 + 2, \quad N_2 = s$$

$$M_1 M_2 - N_1 N_2|_{s=J\omega}$$

$$[(2s^2 + 1)(s^2 + 2) - s^2]_{s=J\omega}$$

$$[2s^4 + 5s^2 + 2 - s^2]_{s=J\omega}$$

$$[2s^4 + 4s^2 + 2]_{s=J\omega} = 2\omega^4 - 4\omega^2 + 2 = 2(\omega^2 - 1)^2.$$

This function is always positive or zero, therefore, the function is a positive real function.

Example 11.9. *Check whether or not the function is p.r.*

$$F(s) = \frac{s + 2}{s + 1}$$

Solution:

(a) Inspection test: All coefficients are positive, no term is missing. The higher degree and lower degree difference is zero. Therefore, $F(s)$ may be a positive real function:

(b) Necessary and sufficient conditions:

$$F(s) = \frac{s + 2}{s + 1}$$

$$F(J\omega) = \frac{2 + J\omega}{1 + J\omega} \times \frac{1 - J\omega}{1 - J\omega}$$

$$F(J\omega) = \left(\frac{2 + \omega^2}{1 + \omega^2}\right) - J\left(\frac{\omega}{1 + \omega^2}\right)$$

$$R_e F(J\omega) = \left(\frac{2 + \omega^2}{1 + \omega^2}\right) \geq 0 \text{ for all } \omega.$$

Therefore, $F(s)$ is a p.r.f.

NOTE We can also solve by $M_1M_2 - N_1N_2|_{s=J\omega}$,

where

$$M_1 = 2, \quad M_2 = 1, \quad N_1 = s, \quad N_2 = s$$
$$A(\omega^2) = M_1M_2 - N_1N_2|_{s=J\omega} = 2 - s^2|_{s=J\omega}$$
$$= 2 + \omega^2 \geq 0 \text{ for all } \omega.$$

Example 11.10. *Justify* $F(s) = s + \sqrt{s^2 + 1}$ *is p.r.f.*

Solution:
$$F(J\omega) = J\omega + \sqrt{1 - \omega^2}$$
$$R_e F(J\omega) = \sqrt{1 - \omega^2}$$

$R_e F(J\omega) \geq 0$ for all ω for p.r.f.
Therefore, $F(s)$ is not a p.r.f.

11.7 SYNTHESIS OF ONE-PORT NETWORKS WITH TWO ELEMENTS (R, L, C)

In the last chapter, we studied two-port networks. In this chapter we will solve the synthesis problems for one-port networks only. Table 11.1 shows various relationships of passive elements (R, L, and C).

TABLE 11.1 V-I relationship of R, L, and C

	R	L	C
(a) Symbol	—/\/\/—	——	—⊣⊢—
(b) Relationship between voltage $v(t)$ and current $i(t)$	$v(t) = Ri(t)$	$v(t) = L\dfrac{di(t)}{dt}$	$v(t) = \dfrac{1}{C}\int i(t)dt$
(c) Transform equation initial conditions set equal to zero	$V(s) = RI(s)$	$V(s) = LsI(s)$	$V(s) = I\dfrac{1}{Cs}(s)$
(d) Impedance, Z(s)	$\dfrac{V(s)}{I(s)} = R$	$\dfrac{V(s)}{I(s)} = Ls$	$\dfrac{V(s)}{I(s)} = \dfrac{1}{Cs}$
(e) Admittance, Y(s)	$G = \dfrac{1}{R}$	$\dfrac{1}{Ls}$	Cs

Figure 11.3 shows a series connection of n one terminal-pair networks with each network identified in terms of its driving-point impedance. The total impedance of the combined networks is the sum of the impedances of the individual networks:

$$Z(s) = Z_1(s) + Z_2(s) + \cdots + Z_n(s). \tag{11.9}$$

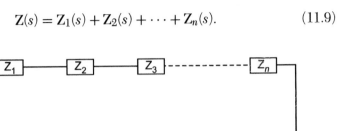

FIGURE 11.3

For the parallel connection of networks shown in Figure 11.4, the total admittance is the sum of the admittances of the individual networks

$$Y(s) = Y_1(s) + Y_2(s) + Y_3(s) + \cdots + Y_n(s). \tag{11.10}$$

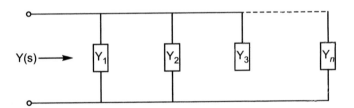

FIGURE 11.4

For a network made up of sub-networks as shown in Figure 11.5, find the total impedance or admittance.

$$Z(s) = Z_1 + \frac{1}{Y_2 + Y_3}$$

$$= Ls + \frac{1}{Cs + \frac{1}{R}} = Ls + \frac{R}{1 + RCs} = \frac{Ls(1 + RCs) + R}{1 + RCs}$$

$$Z(s) = \frac{RLCs^2 + Ls + R}{1 + RCs} = \frac{s^2 + s + 1}{s + 1} \qquad [R = 1\Omega, L = 1H, C = 1F]$$

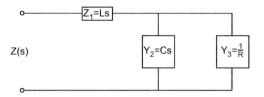

FIGURE 11.5

A very important form of series-parallel is the ladder network as shown in Figure 11.6.

$$Z(s) = Z_1(s) + \cfrac{1}{Y_2(s) + \cfrac{1}{Z_3(s) + \cfrac{1}{Y_4(s) + \cfrac{1}{Z_5(s) + \cfrac{1}{Y_6(s) + \cdots}}}}}$$

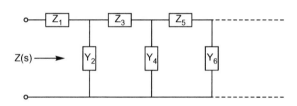

FIGURE 11.6

For an element ladder made up of inductors in the series, and capacitors in the parallel or shunt arms:

$$Z(s) = L_1s + \cfrac{1}{C_2s + \cfrac{1}{L_3s + \cfrac{1}{C_4s}}}.$$

If $L_1 = L_3 = 1H$ and $C_2 = C_4 = 1$ F, then,

$$Z(s) = s + \cfrac{1}{s + \cfrac{1}{s + \cfrac{1}{s}}} = \frac{s^4 + 3s^2 + 1}{s^3 + 2s}.$$

The networks to be synthesized are either LC, RL, or RC.

Networks with the properties of LC, RL and RC functions are different. First, we will discuss the properties of a particular type for one-port network, and then we will synthesize it.

There are a number of methods for synthesizing a one-port network. We only consider the four basic forms as follows:

11.7.1 Cauer-I form: Function Should be Impedance Function $Z(s)$

Cauer-I form obtained by cont-inued fraction.

Expansion of $Z(s)$ in the descending order general representation of the Cauer-I form is shown in Figure 11.7.

$$Z(s) = Z_1 + \cfrac{1}{Y_2 + \cfrac{1}{Z_2 + \cfrac{1}{Y_4 + \cdots}}}$$

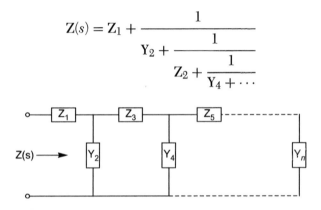

FIGURE 11.7   General representation of the Cauer-I form.

$Z(s) = \frac{N(s)}{D(s)}$ if the degree of $N(s)$ is lower than the degree of $D(s)$. Then starting the continued fraction as $Z(s) = \frac{1}{\frac{D(s)}{N(s)}}$.

In this case the value of Z_1 will be zero therefore the value of the starting element will be zero and this form will be represented as shown in Figure 11.8.

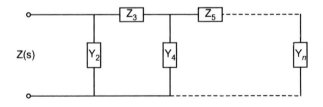

FIGURE 11.8

Example 11.11. *Synthesize the Cauer-I form of the given impedance function:*

$$Z(s) = \frac{(s^2 + 1)(s^2 + 4)}{s(s^2 + 2)} = \frac{s^4 + 5s^2 + 4}{s^3 + 2s}.$$

Solution:

$$s^3 + 2s)\overline{s^4 + 5s^2 + 4}(s \leftarrow Z_1(s)$$
$$\underline{s^4 + 2s^2}$$

$$3s^2 + 4)\overline{s^3 + 2s}(\frac{1}{3}s \leftarrow Y_2(s)$$
$$s^3 + \frac{4}{3}s$$

$$\frac{2}{3}s)\overline{3s^2 + 4}(\frac{9}{2}s \leftarrow Z_3(s)$$
$$3s^2$$

$$4)\frac{2}{3}s(\frac{2}{12}s \leftarrow Y_4(s)$$
$$\frac{2}{3}s$$

$$\overline{X}$$

$$Z(s) = s + \cfrac{1}{\underset{Z_1(s)}{\cfrac{1}{3}s + \cfrac{1}{\underset{Y_2(s)}{\cfrac{9}{2}s + \cfrac{1}{\underset{Z_3(s)}{\cfrac{1}{6}s}}}}}}$$
$$Y_4(s)$$

The first element is an inductor of value 1 H $[Z_1(s) = sL = s]$, therefore, L = 1H.

The second element is a capacitor of value $\frac{1}{3}$F $[Y_2(s) = sC, = \frac{1}{3}s$ after comparing C = $\frac{1}{3}$.

The third element is an inductor of value $\frac{9}{2}$H $[Z_3(s) = sL = \frac{9}{2}s, L = \frac{9}{2}$H after comparing.

The fourth element is a capacitor of value $\frac{1}{6}$F $[Y_4(s) = sC = \frac{1}{6}s]C = \frac{1}{6}$F. The cauer-I form is shown in Figure 11.9.

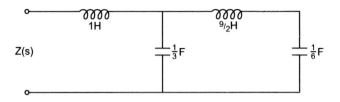

FIGURE 11.9

Example 11.12. *Synthesize the Cauer-I form of the given admittance function*

$$Y(s) = \frac{(s^2 + 1)(s^2 + 5)}{s(s^2 + 3)}.$$

Solution: The function should be $Z(s)$

$$Z(s) = \frac{s^3 + 3s}{s^4 + 6s^2 + 5} = 0 + \frac{1}{\frac{s^4 + 6s^2 + 5}{s^3 + 3s}}$$

[Numerator degree is lower than denominator]

$$
\begin{array}{r}
s^3 + 3s)s^4 + 6s^2 + 5(s \\
\underline{s^4 + 3s^2} \\
3s^2 + 5)s^3 + 3s(\frac{1}{3}s \\
\underline{s^3 + \frac{5}{3}s} \\
\frac{4}{3}s)3s^2 + 5(\frac{9}{4}s \\
\underline{3s^2} \\
5)\frac{4}{3}s(\frac{4}{15}s \\
\underline{\frac{4}{3}s} \\
X
\end{array}
$$

$$Z(s) = o + \cfrac{1}{s + \cfrac{1}{\frac{1}{3}s + \cfrac{1}{\frac{9}{4}s + \cfrac{1}{\frac{4}{15}s}}}}$$

$Z_1(s)$ ↓ s ↓

$Y_2(s)$ ↓ $\frac{1}{3}s$

$Z_3(s)$ ↓ $\frac{9}{4}s$

$Y_4(s)$ ↓ $\frac{4}{15}s$

$Z_5(s)$ ↓

The first element is zero, the second element is a capacitor of value 1 F, the third element is an inductor $\frac{1}{3}$H, the fourth element is a capacitor of $\frac{9}{4}$ F, and the fifth element is an inductor of $\frac{4}{15}$ H as shown in Figure 11.10.

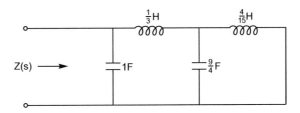

FIGURE 11.10

11.7.2 Cauer-II Form

The function should be impedance function Z(s). The Cauer-II form obtained by continued fraction expansion of Z(s) in the ascending order, general representation of the Cauer-II form is shown in Figure 11.11.

$$Z(s) = Z_1 + \cfrac{1}{Y_2 + \cfrac{1}{Z_3 + \cfrac{1}{Y_4 + \cdots}}}$$

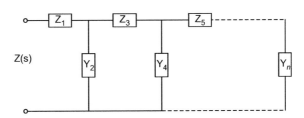

FIGURE 11.11   General representation of the Cauer-II form.

$Z(s) = \frac{N(s)}{D(s)}$ if degree of N(s) is lower than D(s), then start the continued fraction as $Z(s) = \frac{1}{\frac{D(s)}{N(s)}}$.

In this case the value of Z_1 will be zero therefore the value of the starting element will be zero and this form will be represented as shown in Figure 11.12. But this case is not applicable everywhere. If the pole at the origin exists and the numerator degree is lower than the denominator, then the form will become as shown in Figure 11.12.

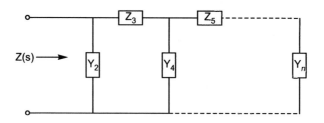

FIGURE 11.12

Example 11.13. *Synthesize the Cauer-II form of the given impedance function,*

$$Z(s) = \frac{\left(s^2 + 1\right)\left(s^2 + 4\right)}{s\left(s^2 + 2\right)}.$$

Solution:

$$Z(s) = \frac{s^4 + 5s^2 + 4}{s^3 + 2s} = \frac{4 + 5s^2 + s^4}{2s + s^3}$$

$$2s + s^3)\,4 + 5s^2 + s^4\,(\frac{2}{s}$$
$$\underline{4 + 2s^2}$$
$$\qquad 3s^2 + s^4)\,2s + s^3\,(\frac{2}{3s}$$
$$\qquad \underline{2s + \frac{2}{3}s^3}$$
$$\qquad\qquad \frac{1}{3}s^3)\,3s^2 + s^4\,(\frac{9}{s}$$
$$\qquad\qquad \underline{3s^2}$$
$$\qquad\qquad\qquad s^4)\,\frac{1}{3}s^3\,(\frac{1}{3s}$$
$$\qquad\qquad\qquad \underline{\frac{1}{3}s^3}$$
$$\qquad\qquad\qquad\qquad \mathrm{X}$$

$$Z(s) = \frac{2}{s} + \cfrac{1}{\cfrac{2}{3s} + \cfrac{1}{\cfrac{9}{s} + \cfrac{1}{\cfrac{1}{3s}}}}$$

For the Cauer-II form,

$$Z(s) = \boxed{\frac{1}{C_1 s}} + \cfrac{1}{\boxed{\frac{1}{L_2 s}} + \cfrac{1}{\boxed{\frac{1}{C_3 s}} + \cfrac{1}{\boxed{\frac{1}{L_4 s}}}}}$$

$$\begin{bmatrix} Z &=& sL &=& \dfrac{1}{sC} \\[2mm] Y &=& sC &=& \dfrac{1}{sL} \end{bmatrix}$$

Z_1 Y_2 Z_3 Y_4

$$C_1 = \frac{1}{2}\text{F}, L_2 = \frac{3}{2}\text{H}, C_3 = \frac{1}{9}\text{F}, L_4 = 3\text{H}$$

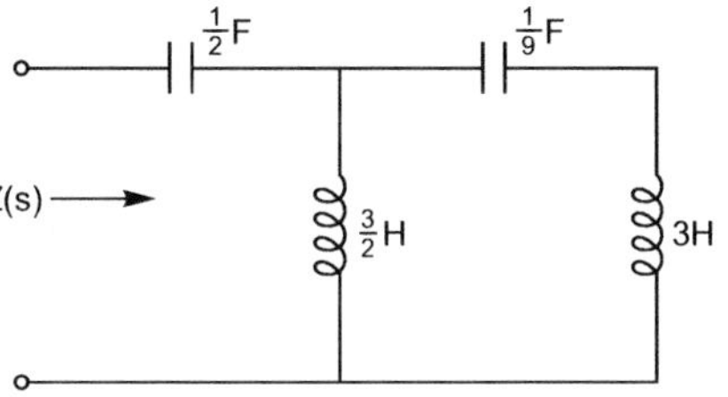

(a)

FIGURE 11.12   Cauer-II form.

Example 11.14. *Synthesize the Cauer-II form of the given admittance function,*

$$Y(s) = \frac{\left(s^2 + 1\right)\left(s^2 + 5\right)}{s\left(s^2 + 3\right)}.$$

Solution: In the Cauer-II form, the function should be $Z(s)$ and apply the continued partial fraction method in ascending order,

$$\mathbf{Z}(s) = \frac{s^3 + 3s}{s^4 + 6s^2 + 5} = \frac{3s + s^3}{5 + 6s^2 + s^4} = \frac{1}{\frac{5 + 6s^2 + s^4}{3s + s^3}}$$

$$3s + s^3 \overline{)\; 5 + 6s^2 + s^4} \left(\frac{5}{3s} \leftarrow Y_2\right.$$
$$\underline{\quad 5 + \frac{5}{3}s^2 \quad}$$
$$\frac{13}{3}s^2 + s^4 \overline{)\; 3s + s^3} \left(\frac{9}{13s} \leftarrow Z_3\right.$$
$$\underline{\quad 3s + \frac{9}{13}s^3 \quad}$$

$$\frac{4}{13}s^3)\frac{13}{3}s^2 + s^4(\frac{169}{12s} \leftarrow Y_4$$

$$\frac{13}{3}s^2$$

$$s^4)\frac{4}{13}s^3(\frac{4}{13s} \leftarrow Z_5$$

$$\frac{4}{13}s^3$$

$$\overline{X}$$

$$\begin{bmatrix} Z &=& sL &=& \dfrac{1}{sC} \\[2mm] Y &=& sC &=& \dfrac{1}{sL} \end{bmatrix}$$

$$Z(s) = \boxed{0} + \cfrac{1}{\boxed{\dfrac{5}{3s}} + \cfrac{1}{\boxed{13s} + \cfrac{1}{\boxed{\dfrac{169}{12s}} + \cfrac{1}{\boxed{\dfrac{4}{13s}}}}}}$$

$Z_1 \quad Y_2 \quad Z_3 \quad Y_4 \quad Z_5$.

The first element is zero because the degree of the numerator is lower than the denominator. The Cauer-II form is shown in Figure 11.13.

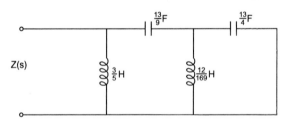

FIGURE 11.13 Cauer-II form.

11.7.3 Foster-I Form

The function should be impedance function $Z(s)$. The Foster-I form obtained by the partial fraction method general representation of the Foster-I form is shown in Figure 11.14.

All residues should be positive. If any residue is negative, apply partial fraction on $\frac{Z(s)}{s}$.

Example 11.15. *Synthesize the Foster-I form of the given impedance function,*

$$Z(s) = \frac{(s^2 + 1)(s^2 + 4)}{s(s^2 + 2)}.$$

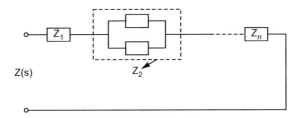

FIGURE 11.14   General representation of the Foster-I form.

Solution: For the Foster-I form the function should be impedance function $Z(s)$. Applying the partial fraction method, then the condition for a partial fraction is the numerator degree should be lower than the denominator.

$$Z(s) = \frac{s^4 + 5s^2 + 4}{s^3 + 2s}$$

$$s^3 + 2s) \; s^4 + 5s^2 + 4 \; (s$$
$$\underline{s^4 + 2s^2}$$
$$3s^2 + 4$$

Now

$$Z(s) = s + \frac{3s^2 + 4}{s(s^2 + 2)},$$

$\frac{3s^2+4}{s(s^2+2)}$. Now the degree of the numerator is lower than the denominator

$$\frac{3s^2 + 4}{s(s^2 + 2)} = \frac{A}{s} + \frac{Bs + C}{s^2 + 2}$$

$$= \frac{A(s^2 + 2) + (Bs + C)(s)}{s(s^2 + 2)}$$

$$3s^2 + 4 = s^2(A + B) + sC + 2A$$

$$A + B = 3, \quad 2A = 4 \quad \text{or} \quad A = 2 \quad \text{and} \quad C = 0$$

$$B = 1$$

$$Z(s) = s + \frac{2}{s} + \frac{s}{s^2 + 2} \Rightarrow Z(s) = \boxed{s} + \boxed{\frac{1}{\frac{1}{2}s}} + \boxed{\frac{1}{s + \frac{2}{s}}}$$
$$\underset{Z_1}{\swarrow} \qquad \underset{Z_2}{\downarrow} \quad \underset{Z_3}{\downarrow}$$

$$Z_3 = \frac{1}{s + \frac{2}{s}} = \frac{1}{Y_a + Y_b}.$$

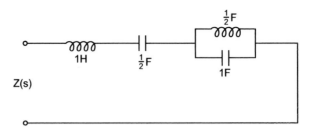

FIGURE 11.15   Foster-I form.

Example 11.16. *Synthesize the Foster-I form of the given admittance function,*

$$Y(s) = \frac{(s^2 + 1)(s^2 + 5)}{s(s^2 + 3)}.$$

Solution: For the Foster-I form, the function should be the impedance function $Z(s)$. Applying the partial fraction method,

$$Z(s) = \frac{s(s^2 + 3)}{(s^2 + 1)(s^2 + 5)} = \frac{As + B}{s^2 + 1} + \frac{Cs + D}{s^2 + 5}$$

$$s^3 + 3s = (As + B)(s^2 + 5) + (Cs + D)(s^2 + 1)$$

$$s^3 + 3s = s^3(A + C) + s^2(B + D) + s(5A + C) + 5B + D.$$

After comparing,

$$A + C = 1, B + D = 0, 5A + C = 3, 5B + D = 0$$

$$B = D = 0$$

$$A = \frac{1}{2}, \quad C = \frac{1}{2}$$

$$Z(s) = \frac{s}{2s^2 + 2} + \frac{s}{2s^2 + 10} \Rightarrow Z(s) = \underbrace{\left[\cfrac{1}{2s + \cfrac{2}{s}}\right]}_{Z_1} + \underbrace{\left[\cfrac{1}{2s + \cfrac{10}{s}}\right]}_{Z_2}$$

$$Z_1 = \frac{1}{Y_a + Y_b} \quad \text{and} \quad Z_2 = \frac{1}{Y_c + Y_d}.$$

Y_a and Y_b are in parallel because in parallel the equivalent admittance is the sum of individual admittance. Similarly, Y_c and Y_d are in parallel.

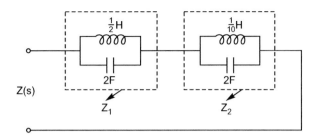

FIGURE 11.16 **Foster-I form.**

11.7.4 Foster-II Form

The function should be the admittance function $Y(s)$, the Foster-II form obtained by the partial fraction method. The general representation of the Foster-II form is shown in Figure 11.17.

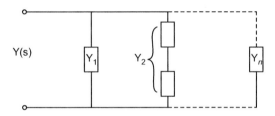

FIGURE 11.17 **General representation of the Foster-II form.**

All residues should be positive. If any residue is negative apply partial fraction on $\frac{Y(s)}{s}$.

Example 11.17. *Synthesize the Foster-II form of the given impedance function,*

$$Z(s) = \frac{(s^2 + 1)(s^2 + 4)}{s(s^2 + 2)}.$$

Solution: For the Foster-II form, the function should be the admittance function $Y(s)$. Apply the partial fraction method. In this case the numerator degree is lower than the denominator,

$$Y(s) = \frac{s(s^2 + 2)}{(s^2 + 1)(s^2 + 4)} = \frac{As + B}{s^2 + 1} + \frac{Cs + D}{s^2 + 4}$$

$$s^3 + 2s = (As + B)(s^2 + 4) + (Cs + D)(s^2 + 1)$$

$$= s^3(A + C) + s^2(B + D) + s(4A + C) + 4B + D.$$

After comparing we get,

$$A = \frac{1}{3}, B = D = 0, C = \frac{2}{3}$$

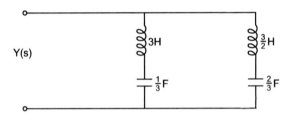

$$Y(s) = \frac{\frac{1}{3}s}{s^2+1} + \frac{\frac{2}{3}s}{s^2+1} = \frac{1}{\left[3s + \frac{s}{3}\right]} + \frac{1}{\left[\frac{3}{2}s + \frac{3}{2s}\right]}$$

$$Y_1 = \frac{1}{Z_a + Z_b} \text{ and } Y_2 = \frac{1}{Z_c + Z_d}.$$

Z_a and Z_b are in series, because in a series the equivalent impedance is the sum of the individual impedance.

Y(s)

3H $\frac{1}{3}$F $\frac{3}{2}$H $\frac{2}{3}$F

FIGURE 11.18 **Foster-II form.**

Example 11.18. *Synthesize the Foster-II form of the given admittance function,*

$$Y(s) = \frac{(s^2+1)(s^2+5)}{s(s^2+3)}.$$

Solution: For the Foster-II form, the immittance function should be $Y(s)$. Apply the partial fraction method. In this case the numerator degree is higher but for a partial fraction, the numerator degree should be lower.

$$Y(s) = \frac{(s^2+1)(s^2+5)}{s(s^2+3)} = \frac{s^4 + 6s^2 + 5}{s^3 + 3s}$$

$$s^3 + 3s)s^4 + 6s^2 + 5(s$$
$$\frac{s^4 + 3s^2}{3s^2 + 5}$$

$$Y(s) = s + \frac{3s^2 + 5}{s(s^2+3)}$$

Now by the partial fraction,

$$\frac{3s^2 + 5}{s(s^2 + 3)} = \frac{A}{s} + \frac{Bs + C}{s^2 + 3}$$

$$3s^2 + 5 = s^2(A + B) + Cs + 3A.$$

After a coefficient we get,

$$A = \frac{5}{3}, \quad B = \frac{4}{3}, \quad C = 0$$

$$Y(s) = s + \frac{\frac{5}{3}}{s} + \frac{\frac{4}{3}s}{s^2 + 3}$$

$$Y(s) = \boxed{s} + \boxed{\frac{1}{\frac{3}{5}s}} + \boxed{\frac{1}{\frac{3}{4}s + \frac{9}{4s}}} \left[Y_3 = \frac{1}{Z_a + Z_b} = \frac{1}{\frac{3}{4}s + \frac{9}{4s}} \right].$$

$$\underset{Y_1}{\downarrow} \qquad \underset{Y_2}{\downarrow} \qquad \underset{Y_3}{\downarrow}$$

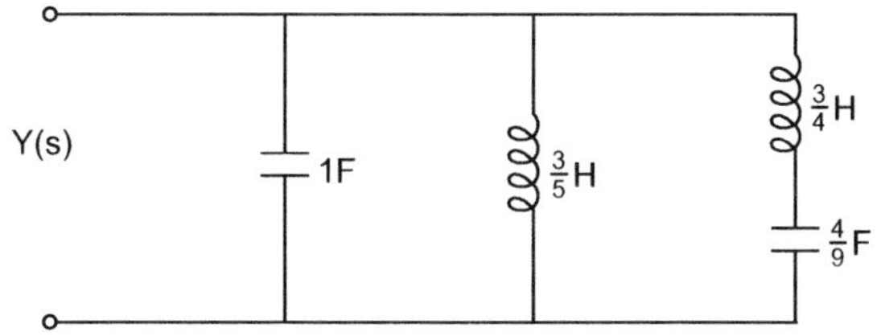

FIGURE 11.19   Foster-II form.

11.8   LC IMMITTANCE FUNCTION OR LC NETWORK SYNTHESIS

Consider an impedance function $Z(s)$ which is positive real and has the property that $R_e Z(J\omega) \geq 0$ for all ω if $Z(s)$ and is written in the form

$$Z(s) = \frac{M_1 + N_1}{M_2 + N_2},$$

where M_1, M_2 are even parts of the numerator and denominator and N_1, N_2 are odd parts.

The average power dissipated by the one-port passive network is

$$\text{Average power} = \frac{1}{2} R_e[Z(J\omega)] \cdot I_2,$$

where I is the input current.

For a pure LC network, it is known that the power dissipated is zero. Therefore,

$$\frac{1}{2} R_e[Z(J\omega)] \cdot I_2 = 0$$

or

$$R_e[Z(J\omega)] = 0.$$

Since we know that

$$R_e[Z(J\omega)] = \text{Even parts } [Z(J\omega)] \text{ [condition for p.r.f.]}$$
$$= \frac{M_1(J\omega)M_2(J\omega) - N_1(J\omega)N_2(J\omega)}{M_2^2(J\omega) - N_2^2(J\omega)}$$

or

$$M_1(J\omega) \cdot M_2(J\omega) - N_1(J\omega) \cdot N_2(J\omega) = 0. \qquad (11.11)$$

For existence of the equation either of the following cases must hold:

(*i*) $M_1 = 0$ and $N_2 = 0$
(*ii*) $M_2 = 0$ and $N_1 = 0$.

In case (I), $Z(s)$ is

$$Z(s) = \frac{N_1(s)}{M_2(s)}.$$

In case (II),

$$Z(s) = \frac{M_1(s)}{N_2(s)}.$$

Consider the example of an LC immittance function given by

$$F(s) = \frac{a_4 s^4 + a_2 s^2 + a_0}{b_5 s^5 + b_3 s^3 + b_1 s} \qquad (11.12)$$

$$F(s) = \frac{K(s^2 + \omega_1^2)(s^2 + \omega_3^2)}{s(s^2 + \omega_2^2)(s^2 + \omega_4^2)}. \qquad (11.13)$$

We see from the above equations, the following properties of LC immittance functions:

1. $Z_{L-C}(s)$ and Y_{L-C} are the ratio of even to odd or odd to even polynomials.
2. Since both $M_i(s)$ and $N_J(s)$ are Hurwitz, they have only imaginary roots or we can say that the roots lie on the imaginary axis (including origin).
3. The poles and zeros interlace (or alternate) on the $J\omega$-axis.
4. The highest and lowest powers of the numerator and denominator must differ by unity.

The function $F(s) = \frac{(s^2+1)(s^2+9)}{s(s^2+4)}$, whose pole zero diagram shown in Figure 11.20, is an LC immittance function or the elements of $F(s)$ is inductor and capacitor (L and C).

All roots lie on the imaginary axis therefore $F(s)$ is an LC immittance function.

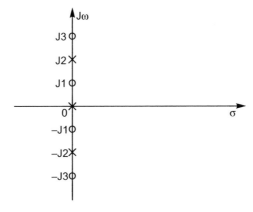

FIGURE 11.20   Pole-zero diagram for the general immittance function.

11.9   RC IMPEDANCE OR RL ADMITTANCE FUNCTION

The RC impedance or RL admittance function following properties:

1. The poles and zeros lie on the negative real axis of the s-plane (including origin).

2. The poles and zeros should interlace (or alternate) along the negative real axis.
3. The residues of the poles of Z_{RC} or Y_{RL} must be real and positive.
4. The pole should be near the origin or at the origin for the RC impedance function.

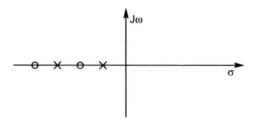

FIGURE 11.21   Pole-zero diagram for general RC impedance function.

In the pole-zero diagram the pole is near the origin for the impedance function, therefore, this is a RC network (passive elements are R and C).

Example 11.19. *Check whether the given function Z(s) is an RC network,*

$$Z(s) = \frac{(s+1)(s+3)}{s(s+2)}.$$

Solution:

$$\text{Zeros at } s = -1, -3$$
$$\text{Poles at } s = 0, -2$$

The pole-zero configuration is shown in Figure 11.22.

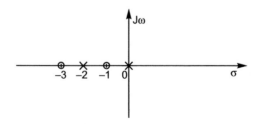

FIGURE 11.22

The pole is at the origin, therefore, this function $Z(s)$ is an RC impedance function.

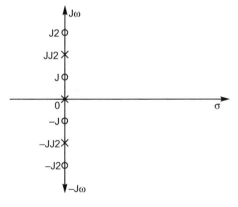
NOTE ▶ An RC impedance, Z_{R-C} can also be realized as an RL admittance (Y_{R-L}). All the properties as in Section 11.9 are the same. It is therefore important to specify whether a function is to be realized as an RC impedance or an RL admittance.

Example 11.20. *Check whether the given function $Z(s)$ is an LC or RC network,*

$$Z(s) = \frac{(s^2 + 1)(s^2 + 4)}{s(s^2 + 2)}.$$

Solution:

$$\text{Zeros at } s = +J, -J, +2J, -2J$$
$$\text{Poles at } s = 0, \sqrt{2J}, -\sqrt{2J}.$$

The pole-zero configuration is shown in Figure 11.23. All the roots lie on the imaginary axis. Therefore, function $Z(s)$ is an L-C impedance function.

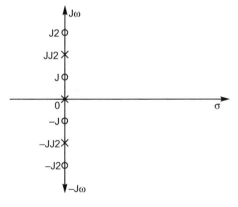

FIGURE 11.23

11.10 RL IMPEDANCE OR RC ADMITTANCE FUNCTION

The RL impedance or RC admittance function has the following properties:

1.   The poles and zeros lie on the negative real axis of the complex s-plane (including origin).

2. The poles and zeros interlace (or alternate) along the negative real axis.
3. The residues of the poles of Z_{R-L} or Y_{R-C} must be real and positive. If negative, the residues of $\frac{Z_{R-L}}{s}$ or $\frac{Y_{R-C}}{s}$ must be real and positive.
4. The zero should be near the origin or at origin for an RL impedance function.

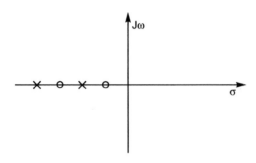

FIGURE 11.24   **Pole-zero diagram for the general RL impedance function.**

Example 11.21. *Check whether the given function Z(s) is an LC, RC, or RL network,*

$$Z(s) = \frac{s(s+3)}{(s+2)(s+4)}.$$

Solution:

$$\text{Zeros at } s = 0, -3$$
$$\text{Poles at } s = -2, -4$$

The pole-zero configuration is shown in Figure 11.25. All the roots lie in the left half of the s-plane and zero is at the origin. Therefore, $Z(s)$ is an RL network.

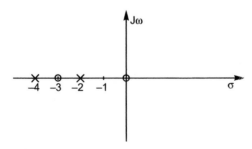

FIGURE 11.25

SOLVED PROBLEMS

Problem 11.1. *Check whether the polynomial* P(s) *is a Hurwitz,*

$$P(s) = 3s^4 + s^3 + 10s^2 + 3s + 3.$$

Solution: All the coefficients of the polynomial are real and positive. It may be a Hurwitz polynomial.

Now by continued fraction,

$$M(s) = 3s^4 + 10s^2 + 3$$
$$N(s) = s^3 + 3s$$
$$\Psi(s) = \frac{M(s)}{N(s)} = \frac{3s^4 + 10s^2 + 3}{s^3 + 3s}$$

$$s^3 + 3s)3s^4 + 10s^2 + 3(3s$$
$$\underline{3s^4 + 9s^2}$$
$$s^2 + 3)s^3 + 3s(s$$
$$\underline{s^3 + 3s}$$
$$X$$

$$q_1 = 3, \quad q_2 = 1.$$

Now by

$$Q(s) = s^2 + 3$$
$$Q'(s) = 2s$$
$$\Psi'(s) = \frac{Q(s)}{Q'(s)} = \frac{s^2 + 3}{2s + 3}$$

$$2s) s^2 + 3 \left(\frac{1}{2}s\right)$$
$$\underline{s^2}$$
$$3) 2s \left(\frac{2}{3}s\right)$$
$$\underline{2s}$$
$$X$$

$$q_3 = \frac{1}{2}, \quad q_4 = \frac{2}{3}.$$

All quotients are positive, therefore, P(s) is a Hurwitz polynomial.

Problem 11.2. *Determine the range of* α *so that* P(s) *is a Hurwitz,*

$$P(s) = s^3 + \alpha s^2 + 2s + 3.$$

Solution: All coefficients a_i must be positive, therefore, $\alpha > 0$. Now by continued fraction,

$$N(s) = s^3 + 2s$$
$$M(s) = \alpha s^2 + 3$$
$$\Psi(s) = \frac{N(s)}{M(s)} = \frac{s^3 + 2s}{\alpha s^2 + 3}$$

$$\alpha s^2 + 3) \; s^3 + 2s \; (\frac{1}{\alpha}s$$
$$\underline{s^3 + \frac{3}{\alpha}s}$$
$$(\frac{2\alpha - 3}{\alpha})s \;) \; \alpha s^2 + 3(\; (\frac{\alpha^2}{2\alpha - 3})s$$
$$\underline{\alpha s^2}$$
$$3 \;)(\frac{2\alpha - 3}{\alpha})s(\; (\frac{2\alpha - 3}{3\alpha})s$$
$$\underline{(\frac{2\alpha - 3}{\alpha})s}$$
$$\mathrm{X.}$$

Now the quotients are

$$q_1 = \frac{1}{\alpha}, \quad q_2 = \frac{\alpha^2}{2\alpha - 3}, \quad q_3 = \left(\frac{2\alpha - 3}{3\alpha}\right)$$

$$\alpha > 0$$

$$2\alpha - 3 > 0 \quad \text{or} \quad \alpha > \frac{3}{2}.$$

Thus, the range of α is $\alpha > 1.5$.

Problem 11.3. *Test whether the following polynomial is a Hurwitz,*

$$P(s) = s^3 + 2s^2 + s + 2.$$

Solution: Even and odd parts of $P(s)$ are

$$M(s) = 2s^2 + 2$$
$$N(s) = s^3 + s.$$

Now by the continued fraction $\Psi(s) = \frac{N(s)}{M(s)} = \frac{s^3+s}{2s^2+2}$

$$2s^2 + 2\,)s^3 + s\,(\,\frac{1}{2}s$$
$$\underline{s^3 + s}$$
$$\text{X.}$$

We see that the division has been terminated prematurely. This is a special case,

$$M(s) = 2s^2 + 2$$
$$M'(s) = 4s$$
$$\Psi'(s) = \frac{M(s)}{M\prime(s)} = \frac{2s^2 + 2}{4s}$$

$$4s)\,2s^2 + 2\,(\frac{1}{2}s$$
$$\underline{2s^2}$$
$$2)\,4s\,(2s$$
$$\underline{4s}$$
$$\text{X.}$$

The highest order in P(s) is 3. Therefore, there should be three quotients and all quotients should be positive for a Hurwitz polynomial $\alpha_1 = \frac{1}{2}, \alpha_2 = \frac{1}{2}, \alpha_3 = 2$. All quotients are positive. Hence, P(s) is a Hurwitz polynomial.

Problem 11.4. *Test whether the following polynomial is a Hurwitz,*

$$P(s) = s^5 + 12s^4 + 45s^3 + 44s + 48.$$

Solution: All the coefficients α_i are real and positive, hence, P(s) may be a Hurwitz polynomial. Now by continued fraction,

$$M(s) = 12s^4 + 60s^2 + 48$$
$$N(s) = s^5 + 45s^3 + 44s$$
$$\Psi(s) = \frac{N(s)}{M(s)} = \frac{s^5 + 45s^3 + 44s}{12s^4 + 60s^2 + 48}$$
$$12s^4 + 60s^2 + 48)s^5 + 45s^3 + 44s\,(\frac{1}{12}s$$
$$\underline{s^5 + 5s^3 + 4s}$$
$$40s^3 + 40s)12s^4 + 60s^2 + 48\,(\frac{12}{40}s$$

$$\dfrac{12s^4 + 12s^2}{}$$

$$48s^2 + 48)40s^3 + 40s \ (\dfrac{40}{48}s$$
$$\underline{40s^3 + 40s}$$
$$\text{X.}$$

$$q_1 = \dfrac{1}{12}, \quad q_2 = \dfrac{12}{40}, \quad q_3 = \dfrac{40}{48}.$$

Because of the cancellation of factor $48s^2 + 48$, the expansion ends prematurely

$$Q(s) = 48s^2 + 48$$
$$Q'(s) = 96\,s$$
$$\Psi'(s) = \dfrac{Q(s)}{Q'(s)} = \dfrac{48s^2 + 48}{96s}$$

$$96s)48s^2 + 48(\dfrac{1}{2}\,s$$
$$\underline{48s^2}$$
$$48)96s(2s$$
$$\underline{96s}$$
$$\text{X}$$

$$q_4 = \dfrac{1}{2}, \quad q_5 = 2.$$

All the quotients are positive, therefore, $P(s)$ is a Hurwitz polynomial.

Problem 11.5. *Test whether the given function* $F(s)$ *represents a p.r.f.* $F(s) = (\frac{s+3}{s^2+5s+1})$.

Solution: Inspection test: No term is missing in the numerator and denominator and all the coefficients are positive. The difference between the higher order and lower order is 1 and 0, respectively, hence $F(s)$ may be a positive real function.

Necessary and sufficient conditions:

$$M_1 = 3, N_1 = s$$
$$M_2 = s^2 + 1, N_2 = 5s$$
$$A(\omega^2) = M_1 M_2 - N_1 N_2|_{s=J\omega}$$
$$A(\omega^2) = 3(s^2 + 1) - 5s^2|_{s=J\omega}$$

$$A(\omega^2) = 3 - 2s^2|_{s=J\omega}$$
$$A(\omega^2) = 3 + 2\omega^2$$
$$A(\omega^2) \geq 0 \text{ for all } \omega.$$

Therefore, $F(s)$ is a p.r.f.

Problem 11.6. *Realize the following* RC *driving-point impedance function in the* (i) *Foster-I form and the* (ii) *Cauer-II form,*

$$Z(s) = \frac{s^2 + 4s + 3}{s^2 + 2s}.$$

Solution: Foster-I form: Using the partial fraction method,

$$Z(s) = \frac{s^2 + 4s + 3}{s^2 + 2s} = 1 + \frac{2s + 3}{s(s + 2)}.$$

NOTE ▶ In the partial fraction method, the order of the numerator should be lower than the denominator,

$$A(s) = \frac{2s + 3}{s(s + 2)} = \frac{A}{s} + \frac{B}{s + 2}$$

$$A = \left[\frac{2s + 3}{s + 2}\right]_{s=0} = \frac{3}{2}$$

$$B = [(s + 2)A(s)]_{s=-2} = \left[\frac{2s + 3}{s}\right]_{s=-2} = \frac{1}{2}.$$

Now,

$$Z(s) = 1 + \frac{\frac{3}{2}}{s} + \frac{\frac{1}{2}}{(s + 2)} = Z_1 + Z_2 + Z_3.$$

Therefore, the synthesized network is as shown in Figure 11.26.

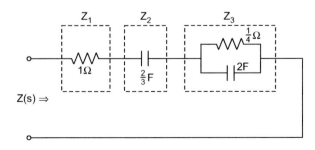

FIGURE 11.26 Foster-I form.

II. Cauer-II form: $Z(s) = \frac{3+4s+s^2}{2s+s^2}$ [lower order term first]
Now by continued fraction,

$$2s + s^2)3 + 4s + s^2(\frac{3}{2s}$$
$$3 + \frac{3}{2}s$$
$$\overline{\hspace{2cm}}$$
$$\frac{5}{2}s + s^2)2s + s^2(\frac{4}{5}$$
$$2s + \frac{4}{5}s^2$$
$$\overline{\hspace{2cm}}$$
$$\frac{1}{5}s^2)\frac{5}{2}s + s^2(\frac{25}{2s}$$
$$\frac{5}{2}s$$
$$\overline{\hspace{2cm}}$$
$$s^2)\,\frac{1}{5}s^2(\frac{1}{5}$$
$$\frac{1}{5}s^2$$
$$\overline{\hspace{2cm}}$$
$$X.$$

Therefore, the synthesized network is as shown in Figure 11.27.

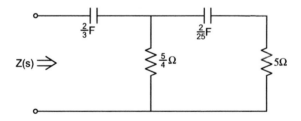

$Z(s) \Rightarrow$

FIGURE 11.27

Problem 11.7. *Check whether the following function is a Hurwitz or not,*

$$P(s) = (s^2 + 3s + 1)(s^3 + 2s^2 + s + 1).$$

Solution:

$$P(s) = P_1(s)P_2(s)$$

where,

$$P_1(s) = s^2 + 3s + 1$$

and

$$P_2(s) = s^3 + 2s^2 + s + 1.$$

We know the product of two Hurwitz polynomials is a Hurwitz.

For $P_1(s)$,

$$M(s) = s^2 + 1$$
$$N(s) = 3s$$

$$3s)\ s^2 + 1\ (\frac{1}{3}s$$
$$\underline{s^2}$$
$$\qquad 1)\ 3s\ (3s$$
$$\qquad \underline{3s}$$
$$\qquad X.$$

$q_1 = \dfrac{1}{3}$ and $q_2 = 3$, therefore $P_1(s)$ is a Hurwitz polynomial.

For $P_2(s)$,

$$M(s) = 2s^2 + 1$$
$$N(s) = s^3 + s$$
$$\Psi(s) = \frac{N(s)}{M(s)} = \frac{s^3 + s}{2s^2 + 1}$$

$$2s^2 + 1)s^3 + s(\frac{1}{2}s$$
$$\underline{s^3 + \frac{1}{2}s}$$
$$\qquad \frac{1}{2}s)2s^2 + 1(4s$$
$$\qquad \underline{2s^2}$$
$$\qquad\qquad 1)\ \frac{1}{2}s\ (\frac{1}{2}s$$
$$\qquad\qquad \underline{\frac{1}{2}s}$$
$$\qquad\qquad X$$

$$q_1 = \frac{1}{2}, \quad q_2 = 4, \quad q_3 = \frac{1}{2}.$$

Therefore, $P_2(s)$ is also Hurwitz polynomial $P(s) = P_1(s)P_2(s)$. Thus, $P(s)$ is also a Hurwitz polynomial.

We can solve this question by another method. First, multiply $P_1(s)$ and $P_2(s)$ and then apply the continued fraction expansion on $P(s)$.

Problem 11.8. *Determine the range of β such that the polynomial $P(s) = s^4 + s^3 + 4s^2 + \beta s + 2$ is Hurwitz.*

Solution:

$$M(s) = s^4 + 4s^2 + 2$$
$$N(s) = s^3 + \beta s.$$

Now by the continued fraction,

$$s^3 + \beta s)s^4 + 4s^2 + 2(s$$
$$\underline{s^4 + \beta s^2}$$
$$(4 - \beta)s^2 + 3)s^3 + \beta s(\dfrac{s}{4 - \beta}$$
$$\underline{s^3 + \dfrac{3}{4 - \beta}s}$$
$$(\beta - \dfrac{3}{4 - \beta})s)(4 - \beta)s^2 + 3(\,(\dfrac{4 - \beta}{4\beta - \beta^2 - 3})s.$$
$$\underline{(4 - \beta)s^2}$$
$$3$$

All quotients must be positive, *i.e.*,
(*i*) $4 - \beta > 0$ or $\beta < 4$
(*ii*) $4\beta - \beta^2 - 3 > 0$ or $\beta^2 - 4\beta + 3 < 0$ or $(\beta - 1)(\beta - 3) < 0$ *i.e.*, $1 < \beta < 3$.
Therefore, by condition (*i*) and (*ii*) the required range of β is $1 < \beta < 3$.

Problem 11.9. *Determine the relationship between α and β for the given function. $P(s)$ is a Hurwitz,*

$$P(s) = 2s^3 + \alpha s^2 + 3s + \beta.$$

Solution:

$$M(s) = \alpha s^2 + \beta$$
$$N(s) = 2s^3 + 3s$$
$$\Psi(s) = \dfrac{N(s)}{M(s)} \quad \text{[degree of N}(s)\text{ is higher than M}(s)]$$
$$= \dfrac{2s^3 + 3s}{\alpha s^2 + \beta}$$

$$\alpha s^2 + \beta)2s^3 + 3s(\dfrac{2}{\alpha}s$$

$$\dfrac{2s^3 + \dfrac{2\beta}{\alpha}s}{\left(\dfrac{3\alpha - 2\beta}{\alpha}\right)s\ )\alpha s^2 + \beta(\ \left(\dfrac{\alpha^2}{3\alpha - 2\beta}\right)s}$$

$$q_1 = \dfrac{2}{\alpha}$$

$$q_2 = \dfrac{\alpha^2}{3\alpha - 2\beta}$$

$$\dfrac{\alpha s^2}{\beta)\left(\dfrac{3\alpha - 2\beta}{\alpha}\right)s(\left(\dfrac{3\alpha - 2\beta}{\alpha\beta}\right)s}$$

$$\dfrac{\left(\dfrac{3\alpha - 2\beta}{\alpha}\right)s}{\mathrm{X}}$$

$$q_3 = \dfrac{3\alpha - 2\beta}{\alpha\beta}$$

$$3\alpha - 2\beta > 0$$

$$3\alpha > 2\beta$$

or

$$2\beta < 3\alpha$$

$$\beta < 1.5\alpha.$$

Problem 11.10. *Which of the following functions is an* RL *driving-point impedance? Why? Synthesize the realizable impedance in Foster's-I form [F_1].*

$$F_1(s) = \dfrac{(s+1)(s+8)}{(s+2)(s+4)}, \quad F_2(s) = \dfrac{(s+2)(s+4)}{(s+3)(s+5)}$$

Solution: For $F_1(s)$:

$$F_1(s) = \dfrac{(s+1)(s+8)}{(s+2)(s+4)}$$

$$\text{zeros} : s = -1, -8$$

$$\text{poles} : s = -2, -4$$

$F_1(s)$ is not a driving-point impedance, since poles and zeros are not interlacing at a negative real axis as shown in Figure 11.28.

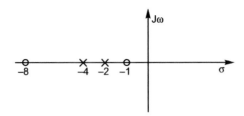

FIGURE 11.28

For $F_2(s)$:

$$F_2(s) = \frac{(s+2)(s+4)}{(s+3)(s+5)}$$

$$\text{zeros at } s = -2, -4$$

$$\text{poles at } s = -3, -5$$

It is clear by Figure 11.29 that it is an RL driving-point impedance, since the poles and zeros are interlacing at a negative real axis and zero is nearest the origin.

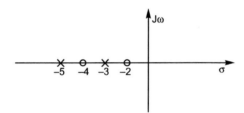

FIGURE 11.29

Now

$$F_2(s) = \frac{(s+2)(s+4)}{(s+3)(s+5)} = \frac{s^2+6s+8}{s^2+8s+15} = 1 - \frac{2s+7}{(s+3)(s+5)}.$$

There is a negative sign, hence this is a special case. Now we apply

$$\frac{F_2(s)}{s} = \frac{(s+2)(s+4)}{s(s+3)(s+5)}.$$

Using the partial fraction expansion,

$$\frac{F_2(s)}{s} = \frac{A}{s} + \frac{B}{(s+3)} + \frac{C}{(s+5)}$$

$$A = \left[\frac{(s+2)(s+4)}{(s+3)(s+5)} \right]_{s=0} = \frac{8}{15}$$

$$B = \left[\frac{(s+2)(s+4)}{s(s+5)}\right]_{s=-3} = \frac{1}{6}$$

$$C = \left[\frac{(s+2)(s+4)}{s(s+3)}\right]_{s=-5} = \frac{3}{10}.$$

Therefore,

$$F_2(s) = \frac{8}{15} + \frac{\frac{1}{6}s}{s+3} + \frac{\frac{3}{10}s}{(s+5)}$$

and the synthesized network is shown in Figure 11.30.

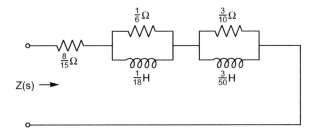

FIGURE 11.30

Problem 11.11. *An impedance function has the pole-zero pattern as shown in Figure 11.31. If $Z(-2) = 3$, synthesize the impedance in the Foster-II form* (F_{II}).

Solution: From the pole-zero diagram,

$$Z(s) = k\frac{(s+1)(s+5)}{s(s+3)}$$

$$Z(-2) = \frac{k(-1)(3)}{(-2)(1)} = \frac{3}{2}k = 3 \text{ (given)}$$

$$k = 2 \quad \text{therefore,}$$

$$Z(s) = \frac{2(s+1)(s+5)}{s(s+3)}.$$

Since the pole is near the origin, the given function is an RC impedance function. But for the Foster-II form, the function should be $Y(s)$,

$$Y(s) = \frac{s(s+3)}{2(s+1)(s+5)}$$

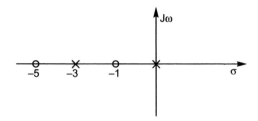

FIGURE 11.31

or

$$\frac{Y(s)}{s} = \frac{(s+3)}{2(s+1)(s+5)} = \frac{A}{(s+1)} + \frac{B}{(s+5)}$$

$$A = \left[\frac{(s+3)}{2(s+5)}\right]_{s=-1} = \frac{1}{4}$$

$$B = \left[\frac{(s+3)}{2(s+1)}\right]_{s=-5} = \frac{1}{4}.$$

Therefore,

$$Y(s) = \frac{\frac{1}{4}s}{(s+1)} + \frac{\frac{1}{4}s}{(s+5)}.$$

The synthesized network is as shown in Figure 11.32.

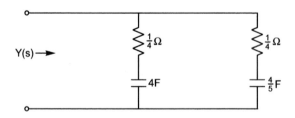

FIGURE 11.32

Problem 11.12. *An* LC *impedance function for a one-port network is given by*

$$Z(s) = \frac{2(s^2+1)(s^2+9)}{s(s^2+4)}.$$

Synthesize the network in
(i) Foster's type-I form
(ii) Cauer's type-II form

Solution: Foster I-form: In the Foster-I form, the function should be $Z(s)$. Apply the partial fraction method. For the partial fraction, it is a necessary and sufficient condition that the degree of the numerator should be lower than the denominator.

$$Z(s) = \frac{2(s^2+1)(s^2+9)}{s(s^2+4)} = \frac{2s^4 + 20s^2 + 18}{s^3 + 4s}$$

$$s^3 + 4s)\overline{2s^4 + 20s^2 + 18}(2s$$
$$\underline{2s^4 + 8s^2}$$
$$12s^2 + 18$$

$$Z(s) = 2s + \frac{12s^2 + 18}{s(s^2+4)}$$

Using partial fraction expansion,

$$\frac{12s^2 + 18}{s(s^2+4)} = \frac{A}{s} + \frac{Bs + C}{s^2 + 4}.$$

After solving,

$$A = \frac{9}{2}, \quad C = 0 \quad \text{and} \quad B = \frac{15}{2}$$

$$Z(s) = 2s\frac{9}{2s} + \frac{15s}{2(s^2+4)} = 2s + \frac{1}{\frac{2}{9}s} + \frac{1}{\left(\frac{2}{15}s + \frac{8}{15}s\right)}$$

$$Z(s) = Z_1 + Z_2 + \frac{1}{Y_3 + Y_4}.$$

The Foster-I form is shown in Figure 11.33.

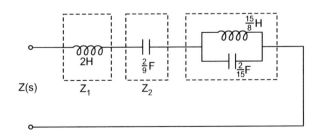

$Z(s)$ $\qquad$ Z_1 $\qquad$ Z_2

FIGURE 11.33 **Foster-I form.**

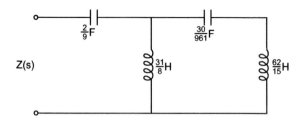

FIGURE 11.34

Cauer-II form: The function should be $Z(s)$. Apply the continued partial fraction method in ascending order,

$$Z(s) = \frac{2s^4 + 20s^2 + 18}{s^3 + 4s} = \frac{18 + 20s^2 + 2s^4}{4s + s^3}.$$

The continued fraction expansion is

$$4s + s^3)18 + 20s^2 + 2s^4(\frac{18}{4s}$$

$$\underline{18 + \frac{9}{2}s^2}$$

$$\frac{31}{2}s^2 + 2s^4)4s + s^3(\frac{8}{31s}$$

$$\underline{4s + \frac{16}{31}s^3}$$

$$\frac{15}{31}s^3)\frac{31}{2}s^2 + 2s^4(\frac{(31)^2}{30s}$$

$$\underline{\frac{31}{2}s^2}$$

$$2s^4)\frac{15}{31}s^3(\frac{15}{62s}$$

$$\underline{\frac{15}{31}s^3}$$

$$\text{X.}$$

The final synthesized network is shown in Figure 11.34.

Problem 11.13. *Check the positive realness of the following functions:*

$$(i)\ \frac{s^2 + s + 6}{s^2 + s + 1} \qquad (ii)\ \frac{s^2 + 6s + 5}{s^2 + 9s + 14}.$$

Solution:

(*i*)
$$F(s) = \frac{s^2 + s + 6}{s^2 + s + 1}$$

Inspection test: All the coefficients are positive; there are no missing terms, and the higher and lower degree difference in the numerator and denominator is zero. Therefore, the function $F(s) = \frac{s^2+s+6}{s^2+s+1}$ may be a positive real function.

The necessary and sufficient conditions:

$$M_1 = s^2 + 6; N_1 = s$$
$$M_2 = s^2 + 1; N_2 = s$$
$$A(\omega^2) = M_1 M_2 - N_1 N_2|_{s=J\omega} = (s^2 + 6)(s^2 + 1) - s^2|_{s=J\omega}$$
$$= s^4 + 7s^2 + 6 - s^2|_{s=J\omega}$$
$$A(\omega^2) = \omega^4 - 6\omega^2 + 6 \qquad (1)$$

and by Equation (1) $A(\omega^2) \geq 0$ for all ω, therefore, $F(s)$ is a positive real function.

(*ii*)
$$F(s) = \frac{s^2 + 6s + 5}{s^2 + 9s + 14} = \frac{N(s)}{D(s)}$$

(*a*) Inspection test: All the coefficients are positive, there are no missing terms, and the higher and lower degree difference in the numerator and denominator is zero. Therefore the function

$F(s)$ may be positive real function.

(*b*) Hurwitz test:

$$M(s) = s^2 + 14$$
$$N(s) = 9s$$
$$\Psi(s) = \frac{M(s)}{N(s)} = \frac{s^2 + 14}{9s}$$

$$9s)s^2 + 14 \,(\frac{1}{9}\,s$$
$$\underline{s^2}$$
$$14)9s\,(\frac{9}{14}\,s$$

$$\frac{9s}{\text{X.}}$$

$$q_1 = \frac{1}{9}, q_2 = \frac{9}{14}$$

q_1 and q_2 are positive therefore $D(s)$ is a Hurwitz polynomial

(c)
$$A(\omega^2) = M_1 M_2 - N_1 N_2|_{s=J\omega}$$
$$M_1 = s^2 + 5, N_1 = 6s, M_2 = s^2 + 14, N_2 = 9s$$
$$A(\omega^2) = (s^2 + 5)(s^2 + 14) - 54s^2|_{s=J\omega}$$
$$= s^4 + 19s^2 + 70 - 54s^2|_{s=J\omega}$$
$$= \omega^4 + 35\omega^2 + 70,$$

therefore, $A(\omega^2) > 0$ for all ω.
Thus $F(s)$ is a p.r.f.

Problem 11.14. *Synthesize*

$$Z(s) = \frac{(s+1)(s+3)}{s(s+2)} \quad in\ Cauer's\text{-}I\ form.$$

Solution:

$$Z(s) = \frac{s^2 + 4s + 3}{s^2 + 2s}$$

Now by continued fraction expansion,

$$s^2 + 2s)s^2 + 4s + 3(1$$
$$\underline{s^2 + 2s}$$
$$2s + 3)s^2 + 2s(\frac{s}{2}$$
$$\underline{s^2 + \frac{3}{2}s}$$
$$\frac{1}{2}s)2s + 3(4$$
$$\underline{2s}$$
$$3)\frac{1}{2}s\ (\frac{1}{6}s$$

$$\frac{\frac{1}{2}s}{\text{X.}}$$

$$Z(s) = 1 + \cfrac{1}{\underset{\underset{Y_1}{\overset{\downarrow}{}}}{\cfrac{s}{2}} + \cfrac{1}{\underset{\underset{Y_2}{\overset{\downarrow}{}}}{4 + \cfrac{1}{\cfrac{1}{6}s}}}}$$

$$\underset{Z_1}{\overset{\downarrow}{}} \quad \underset{Z_2}{\overset{\downarrow}{}}$$

$$\left[\text{we know } Z = \frac{1}{Y}\right]$$

Therefore, the synthesized network is as shown in Figure 11.35.

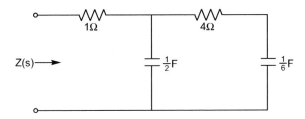

FIGURE 11.35   Cauer-I form.

Problem 11.15. *Synthesize*

$$Z(s) = \frac{(s+5)}{(s+1)(s+6)} \text{ in Foster's II form.}$$

Solution: For the Foster-II form, the function should be $Y(s)$

$$Y(s) = \frac{(s+1)(s+6)}{(s+5)}.$$

Now we know the partial fraction numerator order should be lower than the denominator,

$$\frac{Y(s)}{s} = \frac{(s+1)(s+6)}{s(s+5)} = \frac{s^2 + 7s + 6}{s^2 + 5s}$$

$$s^2 + 5s \overline{)s^2 + 7s + 6}\,(1$$
$$\underline{s^2 + 5s}$$
$$2s + 6$$

$$\frac{Y(s)}{s} = 1 + \frac{2s + 6}{s(s + 5)} = 1 + \frac{A}{s} + \frac{B}{s + 5}$$

$$\frac{2s + 6}{s(s + 5)} = \frac{A}{s} + \frac{B}{(s + 5)}.$$

$A = \frac{6}{5}, B = \frac{4}{5}$ by using partial fraction,

$$\frac{Y(s)}{s} = 1 + \frac{\frac{6}{5}}{s} + \frac{\frac{4}{5}}{s + 5}$$

$$Y(s) = s + \frac{6}{5} + \frac{\frac{4}{5}s}{s + 5} = Y_1 + Y_2 + Y_3.$$

Three admittances are connected and they are in parallel, as shown in Figure 11.36.

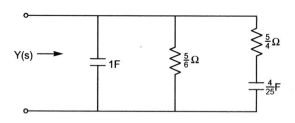

FIGURE 11.36   Foster-II form.

Problem 11.16. *Realize the following RC driving-point impedance function in*

  (*i*) *Foster-I form*   (*ii*) *Cauer-I form:*

$$Z(s) = \frac{s^2 + 6s + 8}{s^2 + 4s + 3}$$

Solution: (*i*) Foster-I form:

$$Z(s) = \frac{s^2 + 6s + 8}{s^2 + 4s + 3} = 1 + \frac{2s + 5}{(s + 1)(s + 3)}$$

and

$$\frac{2s + 5}{(s + 1)(s + 3)} = \frac{A}{s + 1} + \frac{B}{(s + 3)}$$

$$A = \left[\frac{2s + 5}{s + 3}\right]_{s=-1} = \frac{3}{2}$$

$$B = \left[\frac{2s + 5}{s + 1}\right]_{s=-3} = \frac{1}{2}.$$

Now,

$$\frac{2s+5}{(s+1)(s+3)} = \frac{\frac{3}{2}}{(s+1)} + \frac{\frac{1}{2}}{(s+3)}.$$

Therefore,

$$Z(s) = 1 + \frac{\frac{3}{2}}{(s+1)} + \frac{\frac{1}{2}}{(s+3)} = Z_1 + Z_2 + Z_3.$$

The Foster-I form is shown in Figure 11.37.

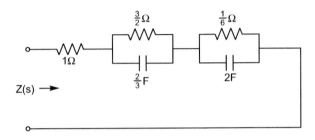

FIGURE 11.37

(ii) Cauer-I form:

$$s^2 + 4s + 3)s^2 + 6s + 8(1$$
$$\underline{s^2 + 4s + 3}$$
$$2s + 5)s^2 + 4s + 3(\frac{1}{2}s$$
$$\underline{s^2 + \frac{5}{2}s}$$
$$\frac{3}{2}s)2s + 5(\frac{4}{3}$$
$$\underline{2s + 4}$$
$$1)\frac{3}{2}s + 3(\frac{3}{2}\,s$$
$$\underline{\frac{3}{2}\,s}$$
$$3)1(\frac{1}{3}$$
$$\underline{1}$$
$$X.$$

Now by continued fraction,

$$Z(s) = 1 + \cfrac{1}{\cfrac{1}{2}s + \cfrac{1}{\cfrac{4}{3} + \cfrac{1}{\cfrac{3}{2}s + \cfrac{1}{\cfrac{1}{3}}}}}$$

Therefore, the Cauer-I form is shown in Figure 11.38.

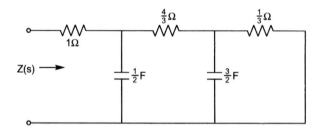

FIGURE 11.38

Problem 11.17. *Find the driving-point impedance* Z(s) *for the network shown in Figure 11.39.*

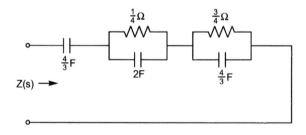

FIGURE 11.39

Solution:

$$Z(s) = Z_1 + Z_2 + Z_3$$

$$Z_1 = \frac{1}{sC} = \frac{1}{\frac{4}{3}s} = \frac{3}{4s}$$

$$Z_2 = \frac{\frac{1}{4} \times \frac{1}{2s}}{\frac{1}{4} + \frac{1}{2s}} = \frac{\frac{1}{2}}{(s+2)}$$

$$Z_3 = \frac{\frac{3}{4} \times \frac{3}{4s}}{\frac{3}{4} + \frac{3}{4s}} = \frac{\frac{3}{4}}{s+1}$$

$$Z(s) = \frac{3}{4s} + \frac{1}{2s+4} + \frac{3}{4s+4} = \frac{6s+12+4s}{4s(2s+4)} + \frac{3}{4s+4}$$

$$Z(s) = \frac{8s^2 + 17s + 6}{4s(s+1)(s+2)}$$

Problem 11.18. *Find the first and second Cauer forms of the LC networks realizing the impedance function*

$$Z(s) = \frac{s^4 + 10s^2 + 9}{s^3 + 4s}.$$

Solution: Cauer-I form:

By continued fraction expansion,

$$s^3 + 4s)\overline{s^4 + 10s^2 + 9}(s \leftarrow Z$$
$$\underline{s^4 + 4s^2}$$
$$6s^2 + 9)\overline{s^3 + 4s}(\frac{s}{6} \leftarrow Y$$
$$\underline{s^3 + \frac{2}{3}s}$$
$$\frac{5}{2}s)\overline{6s^2 + 9}(\frac{12}{5}s \leftarrow Z$$
$$\underline{6s^2}$$
$$9)\frac{5}{2}s(\frac{5}{18}s \leftarrow Y$$
$$\underline{\frac{5}{2}s}$$
$$X.$$

The Cauer-I form is shown in Figure 11.40.

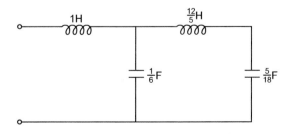

FIGURE 11.40   Cauer I-form.

Cauer-IInd form: the second form is obtained by forming continued fraction expansion with the polynomials arranged in reverse order,

$$4s + s^3)9 + 10s^2 + 9s^4(\frac{9}{4s} \leftarrow Z$$

$$\underline{9 + \frac{9}{4}s^2}$$

$$\frac{31}{4}s^2 + s^4)4s + s^3(\frac{16}{31s} \leftarrow Y$$

$$\underline{4s + \frac{16}{31}s^3}$$

$$\frac{15}{31}s^3)\frac{31}{4}s^2 + s^4(\frac{(31)^2}{60s} \leftarrow Z$$

$$\underline{\frac{31}{4}s^2}$$

$$s^4)\frac{15}{31}s^3(\frac{15}{31s} \leftarrow Y$$

$$\underline{\frac{15}{31}s^3}$$

$$X.$$

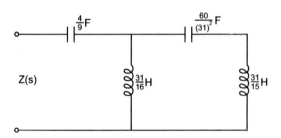

FIGURE 11.41   Cauer-II form.

Problem 11.19. *An impedance is given by*

$$Z(s) = \frac{8(s^2 + 1)(s^2 + 3)}{s(s^2 + 2)(s^2 + 4)}.$$

Realize the network in (i) Foster-I form, (ii) Cauer-II form.

Solution:

$$Z(s) = \frac{8(s^2 + 1)(s^2 + 3)}{s(s^2 + 2)(s^2 + 4)}$$

Zeros are at $s = +J, -J, +\sqrt{3}J, -\sqrt{3}J$.

Poles are at $s = 0, \sqrt{2}J, -\sqrt{2}J, 2J, -2J$.

All the roots lie on the imaginary axis, therefore $Z(s)$ is a LC network. For the Foster-I form, the function should be $Z(s)$. Applying the partial fraction method,

$$Z(s) = \frac{8(s^2 + 1)(s^2 + 3)}{s(s^2 + 2)(s^2 + 4)} = \frac{A}{s} + \frac{Bs + C}{s^2 + 2} + \frac{Ds + E}{s^2 + 4}$$

$$8(s^4 + 4s^2 + 3) = s^4(A + B + D) + s^3(C + E) + s^2(6A + 4B + 2D) + s(4C + 2E) + 8A$$

we get $A = 3, B = 2, D = 3, C = E = 0$

$$Z(s) = \frac{3}{s} + \frac{2s}{s^2 + 2} + \frac{3s}{s^2 + 4}$$

$$Z(s) = \frac{1}{\frac{1}{3}s} + \frac{1}{\frac{s}{2} + \frac{1}{s}} + \frac{1}{\frac{s}{3} + \frac{4}{3s}}$$

$$Z(s) = Z_1 + Z_2 + Z_3.$$

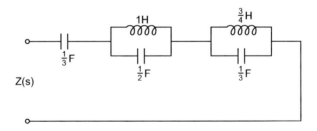

FIGURE 11.42   Foster-I form.

Cauer-II form: The function should be $Z(s)$ and applying continued fraction in ascending order,

$$Z(s) = \frac{8(s^2 + 1)(s^2 + 3)}{s(s^2 + 2)(s^2 + 4)} = \frac{8(s^4 + 4s^2 + 3)}{s(s^4 + 6s^2 + 8)}$$

$$= \frac{8s^4 + 32s^2 + 24}{s^5 + 6s^3 + 8s} = \frac{24 + 32s^2 + 8s^4}{8s + 6s^3 + s^5}$$

$$8s + 6s^3 + s^5)24 + 32s^2 + 8s^4(\frac{3}{s}$$
$$\underline{24 + 18s^2 + 3s^4}$$
$$14s^2 + 5s^4)8s + 6s^3 + s^5(\frac{4}{7s}$$
$$8s + \frac{20}{7}s^3$$
$$\overline{\frac{22}{7}s^3 + s^5)14s^2 + 5s^4(\frac{49}{11s}}$$
$$14s^2 + \frac{49}{11}s^4$$
$$\overline{\frac{6}{11}s4)\frac{22}{7}s^3 + s^5(\frac{121}{21s}}$$
$$\frac{22}{7}s^3$$
$$\overline{s^5)\frac{6}{11}s^4(\frac{6}{11s}}$$
$$\frac{6}{11}s^4$$
$$\overline{\text{X.}}$$

The Cauer-II form is shown in the figure.

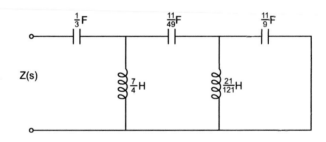

FIGURE 11.43

Problem 11.20. *Realize the following RC driving-point impedance function in (i) foster-I form.*

(ii) Cauer-II form $Z(s) = \dfrac{s^2 + 4s + 3}{s^2 + 2s}.$

Solution: Foster-I form:

$$Z(s) = \frac{(s+1)(s+3)}{s(s+2)} = 1 + \frac{A}{s} + \frac{B}{s+2}.$$

Using the partial fraction method,

$$Z(s) = 1 + \frac{3}{2s} + \frac{1}{2(s+1)}.$$

Therefore, the synthesized network is as shown in Figure 11.44(a).

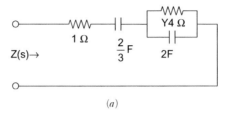

(a)

FIGURE 11.44

Cauer-II form:

$$Z(s) = \frac{3 + 4s + s^2}{2s + s^2}$$

By continued fraction expansion,

$$2s + s^2)\overline{3 + 4s + s^2}(\frac{3}{2s} \leftrightarrow Z_1$$

$$\underline{3 + \frac{3}{2}s}$$

$$\frac{5}{2}s + s^2)\overline{2s + s^2}(\frac{4}{5} \leftrightarrow Y_2$$

$$\underline{2s + \frac{4}{5}s^2}$$

$$\frac{1}{5}s^2)\overline{\frac{5}{2}s + s^2}(\frac{25}{25} \leftrightarrow Z_3$$

$$\underline{\frac{5}{2}s}$$

$$s^2)\overline{\frac{1}{5}s^2}(\frac{1}{5} \leftrightarrow Y_4$$

$$\underline{\frac{1}{5}s^2}$$

$$X.$$

Therefore, the synthesized network is as shown in Figure 11.44(b).

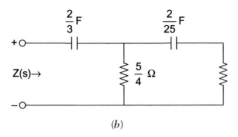

(b)

FIGURE 11.44

Problem 11.21. *Realize the following functions in first and second Foster form,*

$$(i)\ Z(s) = \frac{(s^2 + 1)(s^2 + 16)}{s(s^2 + 9)} \qquad (ii)\ Z(s) = \frac{(s + 2)(s + 5)}{(s + 1)(s + 4)}.$$

Solution: (i) Foster-I form:

$$Z(s) = \frac{(s^2 + 1)(s^2 + 16)}{s(s^2 + 9)} = s + \frac{8s^2 + 16}{s(s^2 + 9)}$$

$$= s + \frac{A}{s} + \frac{Bs}{s^2 + 9}$$

$$Z(s) = s + \frac{\frac{16}{9}}{s} + \frac{\frac{56}{9s}}{s^2 + 9}.$$

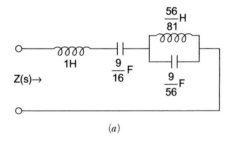

(a)

FIGURE 11.45

Foster-II form:

$$Y(s) = \frac{s(s^2 + 9)}{(s^2 + 1)(s^2 + 16)} = \frac{As}{s^2 + 1} + \frac{Bs}{s^2 + 16}$$

$$= \frac{\frac{8}{15}s}{s^2 + 1} + \frac{\frac{7}{15}s}{s^2 + 16}.$$

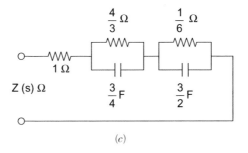

(b)

FIGURE 11.45

(*ii*) Foster-I form:

$$Z(s) = \frac{(s+2)(s+5)}{(s+1)(s+4)} = 1 + \frac{2s+6}{(s+1)(s+4)}$$

$$Z(s) = 1 + \frac{A}{s+1} + \frac{B}{s+4} = 1 + \frac{\frac{4}{3}}{s+1} + \frac{\frac{2}{3}}{s+4}.$$

Therefore, the Foster-I form of the given $Z(s)$ is shown in Figure 11.45(c).

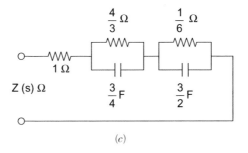

(c)

FIGURE 11.45

Foster-II form:

$$\frac{Y(s)}{s} = \frac{(s+1)(s+4)}{s(s+2)(s+5)} = \frac{A}{s} + \frac{B}{s+2} + \frac{C}{s+5}$$

$$Y(s) = \frac{2}{5} + \frac{\frac{1}{3}s}{s+2} + \frac{\frac{4}{15}s}{s+5}.$$

Therefore, the Foster-II form is shown in Figure 11.45(d).

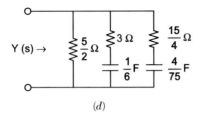

(d)

FIGURE 11.45

Problem 11.22. *Test the following polynomial for its Hurwitz character,*

$$P(s) = s^8 + 3s^7 + 10s^6 + 24s^5 + 35s^4 + 57s^3 + 50s^2 + 36s + 24.$$

Solution: $P(s) = s^8 + 3s^7 + 10s^6 + 24s^5 + 35s^4 + 57s^3 + 50s^2 + 36s + 24$

$$M(s) = s^8 + 10s^6 + 35s^4 + 50s^2 + 24$$
$$N(s) = 3s^7 + 24s^5 + 57s^3 + 36s$$
$$\Psi(s) = \frac{M(s)}{N(s)} = \frac{s^8 + 10s^6 + 35s^4 + 50s^2 + 24}{3s^7 + 24s^5 + 57s^3 + 36s}$$

$$3s^7 + 24s^5 + 57s^3 + 36s \overline{)\, s^8 + 10s^6 + 35s^4 + 50s^2 + 24} \,(\frac{s}{3}$$
$$\underline{s^8 + 8s^6 + 19s^4 + 12s^2}$$

$$2s^6 + 16s^4 + 38s^2 + 24 \overline{)\, 3s^7 + 24s^5 + 57s^3 + 36s} \,(\frac{3}{2}s$$
$$\underline{3s^7 + 24s^5 + 57s^3 + 36s}$$
$$\overline{ \text{X.}}$$

The continued fraction expansion terminated prematurely,

$$P_1(s) = 3s^7 + 24s^5 + 57s^3 + 36s$$
$$= 3s(s^6 + 8s^4 + 19s^2 + 12) = W(s) \cdot P(s)$$
$$P'(s) = 6s^5 + 32s^3 + 38s$$
$$\Psi'(s) = \frac{P(s)}{P'(s)}$$

$$6s^5 + 32s^3 + 38s \overline{)\, s^6 + 8s^4 + 19s^2 + 12} \,(\frac{1}{6}s$$
$$\underline{s^6 + \frac{16}{3}s^4 + \frac{19}{3}s^2}$$

$$\frac{8}{3}s^4 + \frac{38}{3}s^2 + 12 \overline{)\, 6s^5 + 32s^3 + 38s} \left(\frac{9}{4}s\right.$$

$$\underline{6s^5 + \frac{57}{2}s^3 + 27s}$$

$$\frac{7}{2}s^3 + 11s \overline{)\, \frac{8}{3}s^4 + \frac{38}{3}s^2 + 12} \left(\frac{16}{21}s\right.$$

$$\underline{\frac{8}{3}s^4 + \frac{176}{21}s^2}$$

$$\frac{30}{7}s^2 + 12 \overline{)\, \frac{7}{2}s^3 + 11s} \left(\frac{49}{60}s\right.$$

$$\underline{\frac{7}{2}s^3 + \frac{49}{5}s}$$

$$\frac{6}{5}s \overline{)\, \frac{30}{7}s^2 + 12} \left(\frac{25}{7}s\right.$$

$$\underline{\frac{30}{7}s^2}$$

$$12 \overline{)\, \frac{6}{5}s} \left(\frac{6}{60}s\right.$$

$$\underline{\frac{6}{5}s}$$

$$\text{X.}$$

Since all quotient terms are positive, the polynomial P(s) is a Hurwitz and hence the given polynomial is a Hurwitz.

QUESTIONS FOR DISCUSSION

1. Define a Hurwitz polynomial.
2. Check whether the following function is stable $F(s) = \frac{(s+2)}{s(s+3)}$.
3. If any quotients in a Hurwitz polynomial are negative, is it a Hurwitz polynomial?
4. Check if the given function is a Hurwitz, $F(s) = s^2 + 1$.
5. What is the range of a for which the given function is a Hurwitz $P(s) = s^3 + 2s^2 + as + 1$?
6. What is the value of Z(s), as shown in Figure 11.46?
7. What is the value of Y(s), as shown in Figure 11.47?

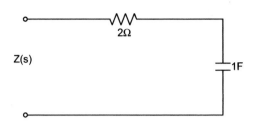

FIGURE 11.46

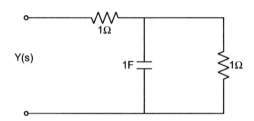

FIGURE 11.47

OBJECTIVE QUESTIONS

1. The driving-point impedance of the circuit shown in the Figure 11.48 is

    (a) $\dfrac{s^2 + Rs}{1 + Rs}$

    (b) $\dfrac{Rs^2 + s + R}{Rs + 1}$

    (c) $\dfrac{s^2(R + 1) + 3}{1 + Rs}$

    (d) $\dfrac{1}{(1 + Rs)^2}$.

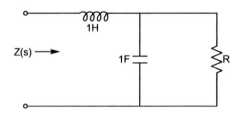

FIGURE 11.48

2. The Foster-I form is shown in Figure 11.49. $Z(s)$ is

    (a) $\dfrac{(s + 2)(s + 5)}{(s + 1)(s + 4)}$

    (b) $\dfrac{(s + 1)(s + 4)}{(s + 2)(s + 5)}$

    (c) $\dfrac{(s + 2)(s + 5)}{(s + 1)(s + 3)}$

    (d) $\dfrac{(s + 1)(s + 3)}{(s + 2)(s + 5)}$.

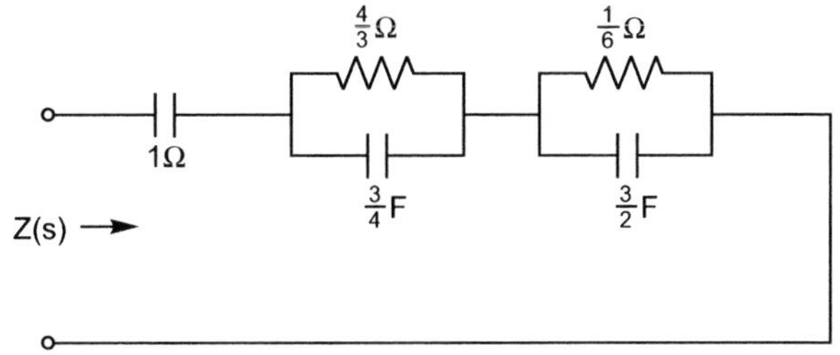

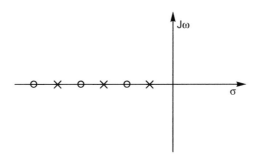

FIGURE 11.49

3. $M(s) + N(s)$ is a Hurwitz polynomial, where $M(s)$ is the even and $N(s)$ is the odd part. All the zeros of $M(s) - N(s)$ are located:
   (a) in the left half of the s-plane.   (b) in the right half of the s-plane.
   (c) in the entire σ-plane.   (d) on the Jω-axis only.

4. The pole-zero configuration of an impedance function is shown in Figure 11.50. The network is:
   (a) RL realizable   (b) RC realizable
   (c) RL realizable   (d) RLC realizable.

FIGURE 11.50

5. An LC-driving point impedance function is
   (a) $\dfrac{s^3 + s^2 + s + 1}{s^2 + 2s + 5}$ (b) $\dfrac{s^4 + 2s^2 + 1}{s^2 + 5}$
   (c) $\dfrac{s^4 + 1}{s^3 + 2}$ (d) $\dfrac{2(s^2 + 1)}{s}$.

6. $F(s) = \dfrac{(s+1)(s+3)}{s(s+2)}$ represents an:
   (a) RC impedance   (b) RC admittance
   (c) RC impedance and an RL admittance   (d) RL admittance.

7. A Hurwitz polynomial has
   (a) only zeros in the left-half of the s-plane
   (b) only poles in the left half of the s-plane
   (c) zeros anywhere in the s-plane
   (d) poles on the Jω axis.

UNSOLVED PROBLEMS

1. Explain the difference in the philosophy between the Foster and Cauer forms of synthesis of a given driving-point impedance.

$$\text{Realize the network function, } Y(s) = \frac{(s + 2)(s + 4)}{(s + 1)(s + 3)},$$

   as a cauer network.

2. State clearly the conditions to be fulfilled for a function to be positive real. Test the following polynomial for the Hurwitz property:

$$2s^6 + s^5 + 13s^4 + 6s^3 + 56s^2 + 25s + 25.$$

3. (a) State the properties of driving-point impedance of an
   (1) LC network and   (2) RC network.
   (b) Realize the network function, $Z(s) = \frac{s(s^2+4)}{2(s^2+1)(s^2+9)}$
   as first and second forms of cauer networks.

4. Test the following polynomial for the Hurwitz property,

$$F(s) = s^4 + s^3 + 5s^2 + 3s + 4.$$

5. For the network shown in Figure 11.51 find the driving-point impedance $Z(s)$. Locate the poles and zeros of this function.

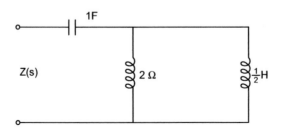

FIGURE 11.51

6. Point out the difference in the philosophy between the Foster and Cauer forms of synthesis of a given driving-point impedance.

7. Realize the following impedance function using both first and second forms of a Foster and cauer canonical network,

$$Z(s) = \frac{(s+1)(s+4)}{s(s+2)}.$$

8. Determine the relationship between α and β for the given Hurwitz, polynomial $P(s)$

$$P(s) = 2s^3 + \alpha s^2 + \beta s + 2.$$

9. Show that the function $P(s) = \frac{3s^2+5}{s(s^2+1)}$ is not a positive real function.

10. Test whether $G(s) = \frac{(s^2+6s+5)}{(s^2+9s+14)}$ is a positive real function.

11. Test whether the following polynomials are Hurwitz or not.
   (a) $F(s) = s^4 + 7s^3 + 6s^2 + 21s + 8$
   (b) $F(s) = s^3 + 4s^2 + 5s + 2$
   (c) $F(s) = s^7 + 2s^6 + 2s^5 + s^4 + 4s63 + 8s^2 + 8s + 4$
   (d) $F(s) = s^4 + 4s^3 + 3s + 2$
   (e) $F(s) = s^5 + 7s^4 + 5s^3 + s^2 + s$
   (f) $F(s) = s^6 + s^5 + 3s^4 - 2s^3 + 4s + 3$
   (g) $F(s) = s^4 + s^3 + 5s^2 + 3s + 4$
   (h) $F(s) = s^5 + s^4 + 6s^3 + 4s^2 + 8s + 3.$

Chapter 12

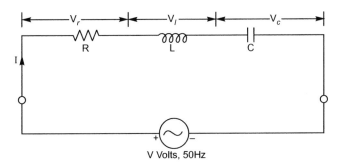

RESONANCE AND SELECTIVITY

12.1 INTRODUCTION

A two-terminal network, in general, offers a complex impedance consisting of resistive and reactive components. If a sinusoidal voltage is applied to such a network, the current is then out of phase with the applied voltage. Under special circumstances, however, the impedance offered by the network is purely resistive. The phenomenon is called *resonance* and the frequency of the applied signal at that resonance is called the *frequency of resonance* or resonant frequency. The nature of resonance depends upon whether the inductance and the capacitance are in series or in parallel. Accordingly, we classify the resonant circuits into the following two categories:

(*i*) Series resonant circuit, and,
(*ii*) Parallel resonant circuit.

12.2 SERIES RESONANCE

Figure 12.1 shows a series RLC circuit. A sinusoidal voltage V sends a current I through the circuit. The circuit is said to be resonant when the resultant reactance is zero, (*i.e.*, the circuit is purely resistive or we can say the imaginary part should be equal to zero).

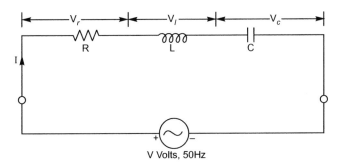

FIGURE 12.1   Series RLC circuit connected to a voltage source V.

635

The impedance Z of the circuit is given by

$$Z = R + j\omega L + \frac{1}{j\omega C}, \tag{12.1}$$

where

> Z and R in ohms
> L is in Henrys
> C is in Farads

and ω is the angular frequency of the applied voltage in radians/sec.
The current,

$$I = \frac{V}{Z} = \frac{V}{R + j\left(\omega L - \frac{1}{\omega C}\right)}. \tag{12.2}$$

At resonance, the imaginary part of Z should be equal to zero

$$\omega L - \frac{1}{\omega C} = 0 \Rightarrow \omega L = \frac{1}{\omega \cdot C}, \tag{12.3}$$

where ω_0 is the frequency of resonance in radians/seconds or the resonant frequency. From Equation (12.3), we get

$$\omega_r = \frac{1}{\sqrt{LC}} \tag{12.4}$$

or

$$f_r = \frac{1}{2\pi\sqrt{LC}}, \tag{12.5}$$

where f_r is the frequency of resonance in Hertz.
At resonance, putting

$$\omega L = \frac{1}{\omega C} \text{ into Equation (12.2), we get } I = \frac{V}{R}.$$

The condition is called *series resonance*.
Series resonance at any desired frequency f may be obtained by varying either L or C or both. For fixed values of L and C series resonance may be achieved by varying the frequency of the applied signal. This type of circuit is known as an acceptor circuit because the value of the current at resonance is maximum.

Example 12.1. *A series RLC circuit consists of a resistance 100 Ω, an induc-tance of 0.1 Henry, and a capacitance of 0.01 μF. Calculate the frequency of resonance in Hertz and in radians/seconds.*

Solution: The frequency of resonance is

$$\omega_r = \frac{1}{\sqrt{LC}} = \frac{1}{\sqrt{0.1 \times 0.001 \times 10^{-6}}}$$
$$= 10^5 \text{ radians/sec.}$$

$$f_r = \frac{\omega_r}{2\pi} = \frac{10^5}{2\pi} = 15{,}920 \text{ Hertz.}$$

Example 12.2. *Two impedances $Z_1 = a + jb$ and $Z_2 = c - jd$ are con-nected in series. Find the condition for resonance.*

Solution: Two impedances are connected in series

$$Z_{eq} = Z_1 + Z_2 = a + jb + c - jd = a + c + j(b - d).$$

At resonance the circuit will be purely resistive, therefore, the imaginary part of Z_{eq} should be equal to zero,

$$b - d = 0 \Rightarrow \boxed{b = d}.$$

This is the condition for resonance.

Example 12.3. *Find X_L for the condition of resonance in the given circuit:*

Solution:

$$Z_1 = 2 + jX_L$$
$$Z_2 = 4 - 2j$$

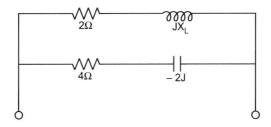

FIGURE 12.2

Z_1 and Z_2 are connected in parallel

$$Z_{eq} = \frac{Z_1 \cdot Z_2}{Z_1 + Z_2} = \frac{(2 + jX_L)(4 - 2j)}{6 + j(X_L - 2)}$$

$$= \frac{8 - 4j + 4jX_L + 2X_L}{6 + j(X_L - 2)}$$

$$= \frac{(8 + 2X_L) + j(4X_L - 4)}{[6 + j(X_L - 2)]} \times \frac{6 - j(X_L - 2)}{6 - j(X_L - 2)}$$

$$= \frac{6(8 + 2X_L) + 4(X_L - 1)(X_L - 2) + j[-(8 + 2X_L)(X_L - 2) - 6(4X_L - 4)]}{36 + (X_L - 2)^2}.$$

Now the imaginary part of Z_{eq} should be equal to zero at resonance.

$$-(8 + 2X_L)(X_L - 2) + 6(4X_L - 4) = 0$$

or

$$(4 + X_L)(X_L - 2) - 6(2X_L - 2) = 0$$

$$X_L^2 + 2X_L - 8 - 12X_L + 12 = 0$$

$$X_L^2 - 10X_L + 4 = 0$$

$$X_L = \frac{+10 \pm \sqrt{100 - 16}}{2} = \frac{10 \pm \sqrt{84}}{2}$$

$$X_{L_1} = 9.58, \quad X_{L_2} = 0.417.$$

There are two values for X_L.

Example 12.4. *Find the condition for resonance in the given circuit:*

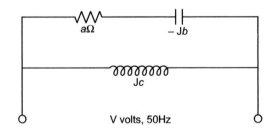

FIGURE 12.3

Solution:

$$Z_1 = a - jb$$
$$Z_2 = jc$$

$$Z_{eq} = \frac{Z_1 \cdot Z_2}{Z_1 + Z_2}$$

$$= \frac{(a - jb)jc}{a - jb + jc}$$

$$= \frac{ajc + bc}{a + j(c - b)} \quad [j^2 = -1].$$

Now multiply by conjugate $[a - j(c - b)]$,

$$Z_{eq} = \frac{bc + jac}{a + j(c - b)} \times \frac{[a - j(c - b)]}{[a - j(c - b)]}$$

$$Z_{eq} = \frac{abc + ja^2c - j(c - b)bc + ac(c - b)}{a^2 + (c - b)^2}$$

or

$$Z_{eq} = \frac{abc + ac(c - b) + j(a^2c - bc^2 + b^2c)}{a^2 + (c - b)^2}.$$

The imaginary part should be equal to zero,

$$a^2c - bc^2 + b^2c = 0$$

$$a^2 - bc + b^2 = 0$$

$$a^2 = bc - b^2$$

$$\boxed{a = \sqrt{b(c - b)}}.$$

This is the condition for resonance.

Example 12.5. *A series* RLC *circuit resonates at* 10^4 Hz. *The value of inductance is 0.02 Henry. Calculate the value of the capacitances.*

Solution: Let the capacitance be C Farads.
Then

$$10^4 = \frac{1}{2\pi \sqrt{0.02\,C}}.$$

Hence,

$$C = \frac{1}{\left(2\pi \times 10^4\right)^2 \times 0.02}$$

$$= 1.268 \times 10^{-8} \text{ Farad.}$$

12.3   PHASOR DIAGRAM OF SERIES RLC CIRCUIT

The nature of the phasor diagram of a series RLC circuit depends on the frequency f of the applied signal in relation to the frequency of resonance f_0. Three different cases may be considered:

$(i) f = f_r,$ $(ii) f < f_r,$ and $(iii) f > f_r.$

With $f = f_0$, the reactance X_L of inductor L equals the reactance $X_c (= \frac{1}{\omega C})$ of capacitor C.

The circuit is then purely resistive and the voltage V across the circuit is in phase with the current I. The phasor diagram for the series RLC circuit of Figure 12.4 for the series resonance condition is shown in Figure 12.4(a).

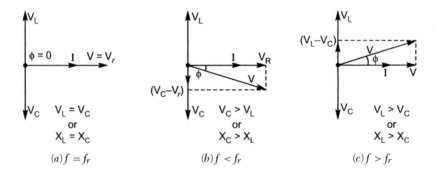

(a) $f = f_r$ (b) $f < f_r$ (c) $f > f_r$

FIGURE 12.4   Phasor diagram of the series RLC circuit.

With $f < f_r$, the negative reactance X_c of capacitor C exceeds the positive reactance X_L of inductor L. Hence, the overall reactance is capacitive. Current I leads the applied voltage V. Voltage V_L leads the current by 90° while the voltage V_c lags behind current I by 90°. The phasor diagram is shown in Figure 12.4(b).

With $f > f_r$, reactance X_L exceeds the reactance X_c. The overall reactance is inductive. Current I lags behind the applied voltage V. The phasor diagram is shown in Figure 12.4(c).

Example 12.6. *A series RLC circuit consists of a resistance* R = 10 Ω, *inductance* L = 0.2 H, *and capacitance* C = 0.2 μF. *Calculate the frequency of resonance. A10 volts sinusoidal voltage at the frequency of resonance is applied across the circuit. Draw the phasor diagram showing the value of each phasor. Also calculate the values of (i) the current and the (ii) voltage across* R, C, *and* L *and draw the phasor diagram when* 10 *volt* 850 *Hz voltage is applied to the circuit.*

Solution: The frequency of resonance,

$$f_r = \frac{1}{2\pi\sqrt{LC}}$$

$$= \frac{1}{2\pi\sqrt{0.2 \times 0.2 \times 10^{-6}}}$$

$$= 796 \text{ Hz}.$$

At

$$f = f_r$$

$$I = \frac{V}{R} = \frac{10 \text{ volts}}{10 \text{ ohms}} = 1 \text{ amp.}$$

$$V_r = I \times R = 1 \times 10 = 10 \text{ volts}$$

$$V_L = I \times \omega L = 1 \times 2\pi \times 796 \times 0.2 = 10^3 \text{volts}$$

$$V_C = V_L = 10^3 \text{volts}.$$

Hence, the phasor diagram is as shown in Figure 12.5.

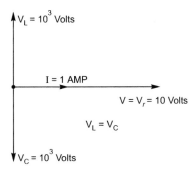

FIGURE 12.5

At

$$f = 850 \text{ Hz}$$

$$\omega L = 2\pi \times 850 \times 0.2 = 1068 \ \Omega$$

$$\frac{1}{\omega C} = \frac{1}{2\pi \times 850 \times 0.2 \times 10^{-6}} = 936.5 \ \Omega$$

$$I = \frac{V}{R + j\left(\omega L - \frac{1}{\omega C}\right)}$$

$$= \frac{10}{10 + j(1068 - 936.5)} = \frac{10}{10 + j131}.$$

Magnitude

$$I = \frac{10}{\sqrt{10^2 + (131)^2}}$$
$$= 76.15 \times 10^{-3} \text{amp}.$$

Let θ be the phase angle of current I relative to voltage V. Then

$$\theta = -\tan^{-1} \frac{131}{10} = -85.6°$$
$$V_r = IR = 76.15 \times 10^{-3} \times 10$$
$$= 76.15 \times 10^{-2} \text{ volt}.$$
$$V_C = \frac{I}{\omega C} = 76.15 \times 10^{-3} \times 936.5$$
$$= 71.32 \text{ volts}.$$
$$V_L = I \times \omega L$$
$$= 76.15 \times 10^{-3} \times 1068 = 81.32 \text{ volts}.$$
$$V_L - V_C = 81.32 - 71.32 = 10 \text{ volts}.$$

The phase diagram is shown in Figure 12.6.

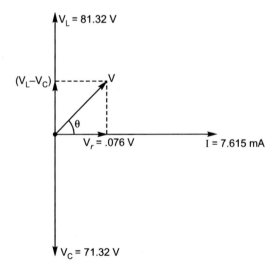

FIGURE 12.6

12.4   REACTANCE AND IMPEDANCE CURVES OF A SERIES RLC CIRCUIT

The reactance of inductor $L = X_L = \omega L$ and its impedance $Z_i = 0 + j\omega L = 0 + jX_L$. Thus, the impedance of an inductor is always imaginary, leading the resistor R by 90°. Curve (a) in Figure 12.7 shows the variation of $X_L(= \Omega L)$ with the frequency. This is a straight line. The reactance of capacitor C is

$$X_c = \frac{1}{\omega C} = \frac{1}{2\pi f_0}$$

and its impedance, $Z_c = 0 + \frac{1}{j\omega C} = 0 - j\frac{1}{\omega C} = 0 - jX_0$.

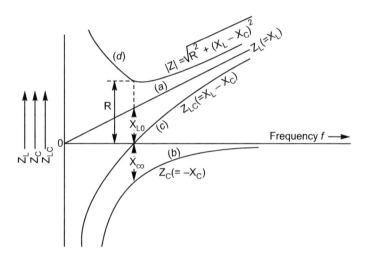

FIGURE 12.7   Vibration of impedances of inductor L, capacitor C, L, and C jointly and total series circuit.

Thus, the impedance of a capacitor is always imaginary and lags behind the resistor by 90°. Curve (b) in Figure 12.7 shows the variation of $Z_c(= -\frac{1}{\omega C})$ with frequency. This is a hyperbolic curve.

The impedance of L and C in series is given by

$$Z_{LC} = 0 + j\omega L + \frac{1}{j\omega C}$$
$$= 0 + j(X_L - X_C).$$

The variation of $(X_L$ and $X_C)$ is shown by curve C in Figure 12.7. This curve has both positive and negative values depending on whether X_L is greater than or less than X_C. Also this curve crosses the X-axis at frequency f_r which is the frequency of series resonance. At this frequency $X_C = X_L$. At a frequency greater than f_r, $X_L < X_C$, i.e., $(X_L - X_C)$, is positive and the circuit is inductive in nature. On the other hand, at frequencies less than f_r, $X_L < X_c$, i.e., $(X_L - X_C)$ is negative and the circuit is capacitive in nature.

The impedance of the entire RLC circuit is given by,

$$Z = R + \left(\omega L - \frac{1}{\omega C}\right) \tag{12.6}$$

$$= R + j(X_L - X_c). \tag{12.7}$$

Hence,

$$|Z| = \sqrt{R^2 + (X_L - X_C)^2}. \tag{12.8}$$

Curve (d) in Figure 12.7 shows the nature of the variation of Z with frequency. At $f = f_r$, $Z = R$ since $X_L = X_C$. This is the minimum value of Z. Therefore, the value of the current is maximum.

12.5 VIBRATION OF CURRENT AND VOLTAGES WITH FREQUENCY IN A SERIES RLC CIRCUIT

In a series RLC circuit, impedance Z is given by

$$Z = R + j\left(\omega L - \frac{1}{\omega C}\right).$$

Hence, current I in the series circuit is given by

$$I = \frac{V}{Z} = \frac{V}{R + j\left(\omega L - \frac{1}{\omega C}\right)}.$$

Magnitude

$$|I| = \frac{V}{\sqrt{R^2 + \left(\omega L - \frac{1}{\omega C}\right)^2}}. \tag{12.9}$$

At resonance, $\omega L = \frac{1}{\omega C}$ and $Z = R$ and this is the minimum value of Z. Hence, from Equation (12.8), at resonance,

$$I = \frac{V}{R}.$$

Hence, the current is maximum at resonance, therefore this type of circuit is known as an acceptor circuit.

At frequencies either below or above the resonance frequency, Z is greater than R so that I is smaller than $\frac{V}{R}$. The curve marked I in Figure 12.8 shows the nature of variation of current I with frequency.

The voltage across the capacitor C is given by

$$V_c = I \frac{1}{j\omega C} = \frac{1}{j\omega C} \cdot \frac{V}{R + j\left(\omega L - \frac{1}{\omega C}\right)}. \qquad (12.10)$$

Hence, magnitude

$$V_c = \frac{V}{\omega C \sqrt{R^2 + \left(\omega L - \frac{1}{\omega C}\right)^2}}. \qquad (12.11)$$

The curve marked V_c in Figure 12.8 gives the variation of voltage V_c with frequency.

The frequency f_c at which V_c is maximum may be obtained by equating $\frac{dV_c^2}{d\omega}$ to zero. This results in

$$f_c = \frac{1}{\omega C} \sqrt{\frac{1}{LC} - \frac{R^2}{2L^2}}. \qquad (12.12)$$

Obviously $f_c < f_r$.

The voltage across the inductor L is given by

$$V_L = Ij\omega L = \frac{V \cdot j\omega L}{R + j\left[\omega L - \left(\frac{1}{\omega C}\right)\right]}. \qquad (12.13)$$

The magnitude

$$|V_L| = \frac{V \cdot \omega L}{\sqrt{R^2 + \left(\omega L - \frac{1}{\omega C}\right)^2}}. \qquad (12.14)$$

The curve marked V_L in Figure 12.8 gives the variation of V_L with frequency. The frequency f_L at which V_L is maximum may be obtained by equating $\frac{dV_L^2}{d\omega}$ to zero. Thus, we get

$$f_L = \frac{1}{2\pi \sqrt{LC - \frac{C^2 R^2}{2}}}. \qquad (12.15)$$

Obviously, $f_L > f_r$.

It may be seen from Equation (12.15) that $f_L > f_r$.

Voltages V_c and V_l have equal values and opposite phases at the frequency of resonance f_r as shown by their point of intersection P.

If R is extremely small, both f_L and f_C tend to equal f_r.

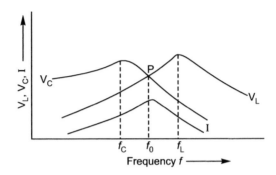

FIGURE 12.8 Variation of I_c, V_c, and V_l with frequency in a series *RLC* circuit.

Example 12.7. *A series RLC circuit consists of resistance* R $= 20\,\Omega$, *inductance* L $= 0.01$H, *and capacitance* C $= 0.04\,\mu$F. *Calculate the frequency of resonance. If a* 10 *volts voltage of frequency equal to the frequency of resonance is applied to this circuit, calculate the values of voltages* V_c *and* V_L *across* C *and* L, *respectively. Find the frequencies at which these voltages* V_c *and* V_L *are maximum.*

Solution:

$$f_r = \frac{1}{2\pi\sqrt{LC}}$$

$$= \frac{1}{2\pi\sqrt{0.01 \times 0.04 \times 10^{-6}}}$$

$$= 7960 \text{ Hz.}$$

At

$$f = f_0,$$

$$I = \frac{V}{R} = \frac{10 \text{ volts}}{20 \text{ ohms}} = 0.5 \text{ amp.}$$

Hence,

$$V_L = I \cdot \omega_0 L$$

$$= 0.5 \times 2\pi \times 7960 \times 0.01 = 250 \text{ volts.}$$

$$V_0 = I \cdot \frac{1}{\omega_0 C}$$

$$= \frac{0.5}{2\pi \times 7960 \times 0.04 \times 10^{-6}} = 250 \text{ volts.}$$

The frequency at which V_c is maximum is given by

$$f_c = \frac{1}{2\pi} \sqrt{\frac{1}{LC} - \frac{R^2}{2L^2}}$$

$$= \frac{1}{2\pi} \sqrt{\frac{1}{0.01 \times 0.04 \times 10^{-6}} - \frac{(20)^2}{2(0.01)^2}}$$

$$= 7955 \text{ Hz.}$$

The frequency at which V_l is maximum is given by

$$f_L = \frac{1}{2\pi \sqrt{LC - \frac{C^2 R^2}{2}}}$$

$$= \frac{1}{2\pi \sqrt{\left(0.01 \times 0.04 \times 10^{-6}\right) - \frac{\left(0.04 \times 10^{-6} \times 20\right)^2}{2}}}$$

$$= 7960 \text{ Hz.}$$

12.6   SELECTIVITY AND BANDWIDTH

A series RLC circuit gives an unequal response to voltages of different frequencies. At the frequency of resonance the impedance is minimum and the current is maximum. As the frequency of the applied voltage is either reduced or increased from this resonance frequency, the impedance increases and the current falls. Figure 12.9 shows the variation of current I with frequency. Thus, a series RLC circuit possesses frequency selectivity.

Figure 12.9 shows current *versus* frequency curves of a series RLC circuit for a small value of R. The frequencies f_1 and f_2 at which current I falls to $\frac{1}{\sqrt{2}}$ (or 0.707) of its maximum value $I_0 (= \frac{V}{R})$ are called *half-power frequencies*. The bandwidth $(f_2 - f_1)$ is called the *half-power bandwidth* or simply the *bandwidth* of the circuit.

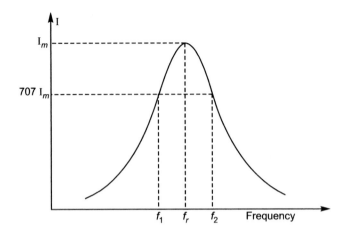

FIGURE 12.9 Current versus frequency curve of a series RLC circuit.

Selectivity of a resonant circuit is defined as the ratio of resonant frequency f_r to the half power bandwidth, thus selectivity.

$$= \frac{\text{Resonance frequency}}{3 - \text{dB Bandwidth}} = \frac{f_r}{(f_2 - f_1)}. \tag{12.16}$$

The current in the series RLC circuit is given by,

$$I = \frac{V}{R + j\left(\omega L - \frac{1}{\omega C}\right)}.$$

Let ω_2 be such a frequency that

$$\omega_2 L - \frac{1}{\omega_2 C} = R. \tag{12.17}$$

Then at frequency ω_2,

$$I_2 = \frac{V}{R + jR}.$$

Magnitude,

$$I_2 = \frac{V}{\sqrt{2}R} = \frac{I_0}{\sqrt{2}}.$$

Thus, ω_2 radians/sec. (or f_2 Hertz) gives the upper half-power frequency. Similarly, let ω_1 be such a frequency that,

$$\omega_1 L - \frac{1}{\omega_1 C} = -R.$$

Then the current at frequency $\omega 1$ is given by,

$$I_2 = \frac{V}{R - jR}.$$

Magnitude,

$$I_1 = \frac{V}{\sqrt{2}R} = \frac{I_0}{\sqrt{2}}.$$

Thus, ω_1 radians/sec. (or f_1 Hertz) forms the lower half-power frequency.

Frequencies f_1 and f_2 are also called *half-power frequencies* because the power dissipation in the circuit at these frequencies is half of the power dissipation at the resonant frequency f_r. This may be seen as below:

$$P_0 = \text{Power dissipation at } f_r = I_0^2 R$$

$$P_1 = \text{Power dissipation at } f_1 = I_1^2 R = \frac{I_0^2 R}{2} = \frac{P_0}{2}$$

$$P_2 = \text{Power dissipation at } f_2 = I_2^2 R = \frac{I_0^2 R}{2} = \frac{P_0}{2}.$$

12.7 QUALITY FACTOR OR Q-FACTOR

(*a*) **Q of a Coil.** Every inductor possesses a small resistance in addition to its inductance. The lower the value of this resistance R, the better the quality of the coil. The quality factor or the *Q-factor* of an inductor at the operating frequency ω is defined as the ratio of reactance of the coil to its resistance.

Thus for a coil,

$$Q = \frac{\omega L}{R} = \frac{X_L}{R}, \tag{12.18}$$

where L is the effective inductance of the coil in Henrys and R is the effective resistance of the coil in Ohms.

Obviously, Q is a dimensionless ratio.

The Q-factor may be defined as

$$Q = 2\pi \frac{\text{Maximum energy stored per cycle}}{\text{Energy dissipated per cycle}}. \qquad (12.19)$$

Thus, consider a sinusoidal voltage V of frequency ω radians/seconds applied to an inductor L of effective internal resistance R as shown in Figure 12.10(a). Let the resulting peak current through the coil be I_m. Then the maximum energy stored in the inductor $= \frac{1}{2}L.I_{m^2}$.

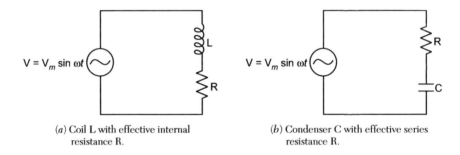

(a) Coil L with effective internal
resistance R.

(b) Condenser C with effective series
resistance R.

FIGURE 12.10 **RL and RC circuits connected to a sinusoidal voltage sources.**

The average power dissipated in the inductor per cycle

$$= \left(\frac{I_m}{\sqrt{2}}\right)^2 R.$$

Hence, the energy dissipated in the inductor per cycle

$$= \text{ Power} \times \text{Periodic time for one cycle}$$

$$= \left(\frac{I_m}{\sqrt{2}}\right)^2 R.T$$

$$= \frac{I_m^2}{2} \cdot R \cdot \frac{1}{f}.$$

Hence,

$$Q = 2\pi \cdot \frac{\frac{1}{2}LI_m^2}{\frac{I_m^2}{2} \cdot \frac{R}{f}}$$

$$= \frac{2\pi f L}{R} = \frac{\omega L}{R} = \frac{X_L}{R}.$$

(*b*) **Q of a capacitor.** Figure 12.10(*b*) shows a capacitor C with small series resistance R associated within. The Q-factor or the quality factor of a capacitor at the operating frequency is defined as the ratio of the reactance of the capacitor to its series resistance.

Thus,

$$Q = \frac{1}{\omega CR}. \tag{12.20}$$

In this case also, the Q is a dimensionless quantity. Equation (12.19) giving the alternative definition of Q also holds good in this case. Thus, for the circuit of Figure 12.10(*b*), on application of a sinusoidal voltage of value V volts and frequency ω, the maximum energy stored in the capacitor

$$= \frac{1}{2} C V_{max}^2,$$

where V_{max} is the maximum value of voltage across the capacitance C.

But if

$$R < \omega C$$

then

$$V_{max} = \frac{I_m}{\omega C},$$

where I_m is the maximum value of current through C and R.

Hence, the maximum energy stored in capacitor C is

$$\frac{1}{2} \cdot \frac{I_m^2}{\omega^2 C}.$$

Energy dissipated per cycle

$$= \frac{I_m^2 R}{2f}.$$

Hence,

$$Q = 2\pi \frac{\frac{1}{2} \cdot \frac{I_m^2}{\omega^2 C}}{\frac{I_m^2 R}{2f}} = \frac{2\pi \cdot f}{\omega^2 CR} = \frac{1}{\omega CR} = \frac{X_C}{R}.$$

Often a lossy capacitor is represented by a capacitance C with a high resistance R_p in shunt as shown in Figure 12.11.

Then for the capacitor of Figure 12.11, the maximum energy stored in the capacitor

$$= \frac{1}{2}C \cdot V_{max}^2,$$

where V_{max} is the maximum value of the applied voltage. The average power dissipated in resistance R_p

$$= \frac{\left(\frac{V_m}{\sqrt{2}}\right)^2}{R_p} = \frac{V_m^2}{2R_p}.$$

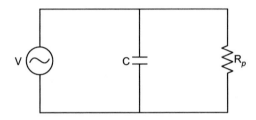

FIGURE 12.11   Alternative method of representing a lossy capacitor.

Energy dissipated per cycle

$$= \frac{V_m^2}{2R_p} \cdot \frac{1}{f}. \tag{12.21}$$

Hence,

$$Q = 2\pi \cdot \frac{\frac{1}{2}CV_m^2}{\frac{V_m^2}{2R_pf}} = \omega CR_p. \tag{12.22}$$

12.8   SERIES RESONANCE AND SELECTIVITY OF SERIES RLC CIRCUIT WITH FREQUENCY VARIABLE

The impedance of a series RLC circuit is given by

$$Z = R + j\left(\omega L - \frac{1}{\omega C}\right) \tag{12.23}$$

where ω is any frequency.

For ω close to the resonant frequency ω_0, Equation (12.23) does not yield very accurate results.

It is desirable in such a case to use a special relation valid only for frequencies close to the resonant frequency. Thus, Equation (12.26) may be put as,

$$
\begin{aligned}
Z &= R\left[1 + j\left(\frac{\omega L}{R} - \frac{1}{\omega CR}\right)\right] \\
&= R\left[1 + j\left(\frac{\omega_r L}{R} \cdot \frac{\omega}{\omega_r} - \frac{1}{\omega_r CR} \cdot \frac{\omega_r}{\omega}\right)\right].
\end{aligned}
\tag{12.24}
$$

At resonance,

$$
\omega_r L = \frac{1}{\omega_r C}.
$$

Hence,

$$
\frac{\omega_r L}{R} = \frac{1}{\omega_r CR} = Q_r,
\tag{12.25}
$$

where Q_r is the value of Q at resonance.

Hence, Equation (12.24) may be put as,

$$
Z = R\left[1 + jQ_r\left(\frac{\omega}{\omega_r} - \frac{\omega_r}{\omega}\right)\right].
\tag{12.26}
$$

The *fractional frequency variation* is given by

$$
\delta = \frac{\omega - \omega_r}{\omega_r}.
\tag{12.27}
$$

Hence,

$$
\frac{\omega}{\omega_0} = 1 + \delta.
$$

Hence, Equation (12.26) may be put as,

$$
\begin{aligned}
Z &= R\left[1 + jQ_0\left(1 + \delta - \frac{1}{1 + \delta}\right)\right] \\
&= R\left[1 + jQ_0\delta\left(\frac{2 + \delta}{1 + \delta}\right)\right].
\end{aligned}
\tag{12.28}
$$

If ω is close to ω_0, then $\delta < 1$.

Hence,

$$2 + \delta \approx 2$$

and

$$1 + \delta \approx 1.$$

Hence, Equation (12.28) reduces to,

$$Z = R[1 + j2Q_0\delta]. \tag{12.29}$$

Hence, the current

$$I = \frac{V}{Z}$$

$$= \frac{V}{R[1 + 2jQ_0\delta]}. \tag{12.30}$$

Figure 12.12 shows the variation of current I with the frequency as obtained from Equation (12.29).

At resonance,

$$\delta = 0.$$ Hence, from Equation (12.29),

$$Z = R \text{ and } I = \frac{V}{R} = I_0.$$

At half-power frequencies f_1 and f_2, the magnitude of the reactive component of impedance equals the resistance, $i.e.,$

$$\left(\omega L - \frac{1}{\omega C}\right) = \pm R.$$

At these half-power frequencies f_1 and f_2, the current

$$I = \frac{I_0}{2} = 0.707 I_0.$$

From Equation (12.29), at half-power frequencies,

$$2Q_r\delta = \pm 1. \tag{12.31}$$

In Equation (12.31), the positive sign applies for the upper half-power frequency f_2, while the negative sign applies for the lower half-power frequency f_1.

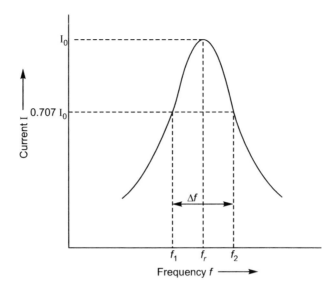

FIGURE 12.12 Variation of current with frequency in a series RLC current.

For upper half-power frequency f_2 from Equation (12.31),

$$2Q_r \left(\frac{f_2 - f_r}{f_r} \right) = +1.$$

Hence,

$$\frac{f_2 - f_r}{f_r} = \frac{1}{2Q_r}. \tag{12.32}$$

Similarly,

$$\frac{f_r - f_1}{f_r} = \frac{1}{2Q_r}. \tag{12.33}$$

From Equations (12.32) and (12.33) we see that $(f_2 - f_r) = (f_r - f_1)$, *i.e.*, the half-power frequencies f_2 and f_1 are symmetrically disposed with respect to the resonant frequency f_r. Also the response curve, *i.e.*, current *versus* frequency curve, is symmetrical about the resonant frequency f_r.

But Equations (12.32) and (12.33) are true only when $\delta \ll 1$. This condition is satisfied only when the circuit is highly selective, *i.e.*, when Q_r is large. For the low Q circuit, *i.e.*, when R is large, the half-power frequencies f_1 and f_2 are away from the resonant frequency f_r. The impedance Z is then given by

Equation (12.28) instead of Equation (12.29). Half-power frequencies must then be calculated from Equation (12.28), *i.e.*, from the following equation,

$$Q_r \delta \left(\frac{2 + \delta}{1 - \delta} \right) = \pm 1. \tag{12.34}$$

Thus, for the low selectivity circuit, for example, $(f_2 - f_0) \neq (f_0 - f_1)$, the response curve is not symmetrical about the resonant frequency.

Referring back to Equations (12.32) and (12.33) for the high Q circuit, by addition of the equations, we get

$$\frac{f_r - f_1}{f_r} = \frac{1}{Q_r}. \tag{12.35}$$

Let the half-power bandwidth $(f_2 - f_1)$ be indicated by Δf. Then selectivity

$$= \frac{f_0}{\Delta f} = Q_r. \tag{12.36}$$

From Equation (12.36), we find for a given value of f, the bandwidth Δf varies inversely with Q_r. The larger the value of Q_r, the lesser the bandwidth and more selective is the circuit.

At a lower half-power frequency $\omega 1$,

$$\omega_1 L - \frac{1}{\omega_1 C} = -R. \tag{12.37}$$

Similarly, at an upper half-power frequency ω_2,

$$\omega_2 L - \frac{1}{\omega_2 C} = +R. \tag{12.38}$$

Adding Equations (12.37) and (12.38), we get

$$(\omega_1 + \omega_2)L - \left(\frac{1}{\omega_1} + \frac{1}{\omega_2} \right) \frac{1}{C} = 0$$

or

$$(\omega_1 + \omega_2)L = \frac{1}{C} \left(\frac{\omega_1 + \omega_2}{\omega_1 \omega_2} \right).$$

Hence,

$$\omega_1 \omega_2 = \frac{1}{LC}. \tag{12.39}$$

But

$$\omega_0^2 = \frac{1}{LC}.$$

Hence,

$$\omega_1\omega_2 = \omega_0^2 = \frac{1}{LC} \Rightarrow \omega_0 = \sqrt{\omega_1 \cdot \omega_2} \qquad (12.39\,a)$$

and we can also write

$$f_0 = \sqrt{f_1 \cdot f_2}.$$

12.9   SELECTIVITY OR SERIES RLC CIRCUIT WITH A CAPACITANCE VARIABLE

In certain cases, we feed to a series RLC circuit a fixed frequency signal and it is required to vary the capacitance C to obtain the condition of resonance. In such a case, as C varies, the response current in the circuit varies as shown in Figure 12.13. The maximum response is reached when the value of C is such that the circuit resonates at the frequency of the applied signal.

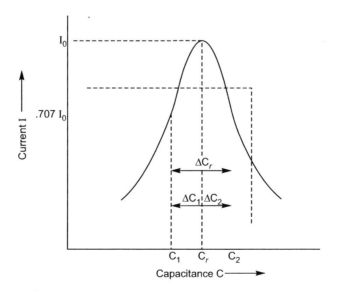

FIGURE 12.13   Response of a series RLC circuit with a C variable.

Let C_0 be the value of C at resonance and let C_1 and C_2 be the values of capacitance C at which the magnitude of reactance $(\omega L - \frac{1}{\omega C})$ equals the resistance R.

Then

$$\omega L - \frac{1}{\omega C_1} = -R \qquad\qquad (12.40)$$

and

$$\omega L - \frac{1}{\omega C_2} = R. \qquad\qquad (12.41)$$

At resonance

$$\omega L = \frac{1}{\omega C_0}. \qquad\qquad (12.42)$$

From Equations (12.40) and (12.41), we get

$$\frac{1}{\omega C_1} - \frac{1}{\omega C_2} = 2R. \qquad\qquad (12.43)$$

However, if the circuit is highly selective, then

$$\Delta C_1 < C_r \quad \text{and} \quad \Delta C_2 < C_r.$$

Hence,

$$C_2 = C_r + \Delta C_2 \approx C_r$$

and

$$C_1 = C_r - \Delta C_1 \approx C_r.$$

For a highly selective circuit, ΔC_1 and ΔC_2 are small and almost equal. Then

$$C_1 + C_2 \approx 2C_r \qquad\qquad (12.44)$$

and

$$C_1 \cdot C2 \approx C_r^2. \qquad\qquad (12.45)$$

Substituting the value of $C_1 C_2$ from Equation (12.45) into Equation (12.43), we get

$$\frac{C_2 - C_1}{\omega C_r^2} = 2R. \qquad\qquad (12.46)$$

Combining Equations (12.46) and (12.42), we get

or
$$\frac{C_2 - C_1}{C_r} \cdot \omega L = 2R$$

or
$$\frac{C_2 - C_1}{C_r} = \frac{2R}{\omega L} = \frac{1}{Q_0}. \tag{12.47}$$

$(C_2 - C_1)$ gives the total variation in C in moving from one half-power condition to the other. The quantity $\frac{C_r}{C_1 - C_2}$ thus gives the selectivity of the series-tuned circuit with C variables and this equals $\frac{Q_r}{2}$, as may be seen from Equation (12.47).

Figure 12.14 gives curves showing the variation of current I, voltage V_l across inductor L, and voltage V_c across capacitor C as C is varied. Voltage $V_l(= \omega L \cdot I)$ varies in the same manner as current I and becomes maximum at resonance, *i.e.*, when $C = C_0$. At resonance $= V_l = V_c$, *i.e.*,

$$\omega L = \frac{1}{\omega C_0}.$$

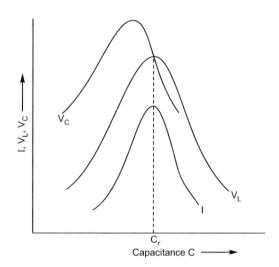

FIGURE 12.14   Variation of current and voltage V_l and V_c with a variation of C.

12.10 SELECTIVITY OF SERIES RLC CIRCUIT WITH AN L VARIABLE

In certain cases, we feed to a series RLC circuit, a fixed frequency signal, and it is required to vary the inductance L to obtain the condition of resonance.

In such a case, as L is varied, the current varies as shown in Figure 12.15. The maximum response is obtained when the value of L is such that the circuit resonates at the frequency of the applied signal.

Let L_r be the value of L at resonance and let L_1 and L_2 be the values of L at which the magnitude of reactance $\left(\omega L - \frac{1}{\omega C}\right)$ equals the resistance R. Then

$$\omega L_1 - \frac{1}{\omega C} = -R \tag{12.48}$$

and

$$\omega L_2 - \frac{1}{\omega C} = R. \tag{12.49}$$

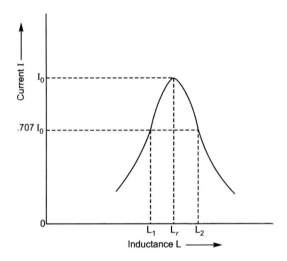

FIGURE 12.15 Response of a series RLC circuit with an L variable.

At resonance,

$$\omega L_r = \frac{1}{\omega C}. \tag{12.50}$$

Subtracting Equation (12.48) from Equation (12.49), we get

$$\omega L_2 - \omega L_1 = 2R$$

or

$$\frac{L_2 - L_1}{L_0} = \frac{2R}{\omega L_r} = \frac{2}{Q_r}. \tag{12.51}$$

$(L_2 - L_1)$ gives the total variation in L in moving from one-half power condition to the other. The quantity $\frac{L}{L_2-L_1}$ thus gives the selectivity of the series RLC circuit with the L variable and this equals $\frac{Q_r}{2}$, as may be seen from Equation (12.51).

Figure 12.16 gives curves showing a variation of current I, voltage V_l across inductor L, and voltage V_c across capacitor C as L is varied. Voltage V_c, being equal to $\frac{1}{\omega C}$, varies in the same manner as current I and becomes maximum at resonance, *i.e.*, when $L = L_r$.

At resonance,

$$V_l = V_c \quad i.e., \quad \omega L_0 = \frac{1}{\omega C}.$$

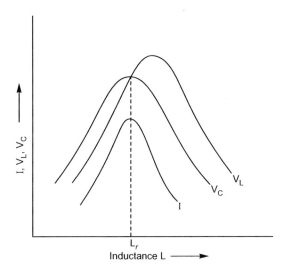

FIGURE 12.16 Variation of current I and voltages V_c and V_l with a variation of L.

12.11 PARALLEL RESONANCE

A parallel resonant circuit consists of an inductor L in parallel with a capacitor C as shown in Figure 12.17. R is a small resistance associated with the coil. The capacitor C is assumed to be lossless. The tuned circuit is driven by a voltage source V. Such a parallel-tuned circuit is commonly used in tuned amplifiers, oscillators, etc.

Analysis of a parallel-tuned circuit may be done more conveniently in terms of admittances instead of impedances. Thus, admittance of the

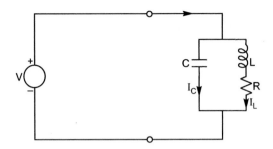

FIGURE 12.17 **Parallel resonant circuit.**

inductive branch is given by

$$Y_1 = \frac{1}{R + j\omega L}$$

$$= \frac{R - j\omega L}{R^2 + \omega^2 L^2}. \tag{12.52}$$

Admittance of capacitor C is given by

$$Y_2 = j\omega C. \tag{12.53}$$

Total admittance

$$Y_{eq} = Y_1 + Y_2$$

$$= \frac{R}{R^2 + \omega^2 L^2} - j\left[\frac{\omega L}{R^2 + \omega^2 L^2} - \omega C\right]. \tag{12.54}$$

At resonance, the imaginary part should be equal to zero,

$$\frac{\omega_r L}{R^2 + \omega^2 L^2} - \omega_r C = 0$$

or

$$R^2 + \omega_r^2 L^2 = \frac{L}{C}. \tag{12.55}$$

Hence,

$$\omega_r = \sqrt{\frac{1}{LC} - \frac{R^2}{L^2}} \tag{12.56}$$

or

$$f_r = \frac{1}{2\pi}\sqrt{\frac{1}{LC} - \frac{R^2}{L^2}}. \tag{12.57}$$

If coil resistance (R) is very small ($R \cong 0$), then $f_r = \frac{1}{2\pi} \sqrt{\frac{1}{LC}}$.

Equation (12.57) may also be put as

$$f_r = \frac{1}{2\pi} \sqrt{\frac{1}{LC}} \cdot \sqrt{1 - \frac{CR^2}{L}}. \qquad (12.58)$$

Considering the three elements, L, C, and R in series, the Q_r is given by $\frac{\omega_{rs}L}{R}$ or $\frac{1}{\omega_{rs}CR}$, where ω_{rs} is the series resonant frequency in radians/sec. Then

$$Q_{rs}^2 = \frac{\omega_{rs}L}{R} \cdot \frac{1}{\omega_{rs}CR} = \frac{L}{CR^2}. \qquad (12.59)$$

Substituting Q_{rs} for $\frac{1}{R^2C}$, Equation (12.59) yields

$$f_0 = \frac{1}{2\pi} \sqrt{\frac{1}{LC}} \cdot \sqrt{1 - \frac{1}{Q_{rs}^2}}. \qquad (12.60)$$

But the frequency of series resonance is given by

$$f_{rs} = \frac{1}{2\pi} \cdot \sqrt{\frac{1}{LC}}. \qquad (12.61)$$

Hence,

$$f_r = f_{rs} \sqrt{1 - \frac{1}{Q_{rs}^2}}. \qquad (12.62)$$

From Equation (12.62) we find that the frequency of parallel resonance f_r differs from the frequency of series resonance f_{rs}. However, if Q_{rs} exceeds 10, then the factor $\sqrt{1 - \frac{1}{Q_0^2}} \approx 1$ and $f_r = f_{rs}$.

12.12 IMPEDANCE OF PARALLEL-TUNED CIRCUIT

(*a*) **At resonance.** From Equation (12.54), the admittance of the parallel-tuned circuit at resonance is given by

$$Y_a = \frac{R}{R^2 + \omega_0^2 L^2}. \qquad (12.63) \quad \text{[Real part only]}$$

But from Equation (12.59), $R^2 + \omega_r^2 L^2 = \frac{L}{C}$.

Hence,

$$Y_a = \frac{R}{\frac{L}{C}} = \frac{1}{\frac{L}{CR}} = \frac{1}{Z_a}. \tag{12.64}$$

Hence, impedance at resonance is given by

$$Z_a = R_a = \frac{L}{CR}. \tag{12.65}$$

This is a pure resistance and is often called the *dynamic resistance* of the parallel-tuned circuit.

From Equation (12.65) we see that the lower the resistance R of the coil, the higher the value of the resistance R_a at resonance. Therefore, current flowing through the circuit is very small, hence, parallel resonance circuits are known as rejector circuits.

The current down from the supply source at resonance is called the *make up current* and is equal to $\frac{V}{\frac{L}{CR}} = \frac{VCR}{L}$. The current in the capacitor branch or inductor branch is called the *forced oscillatory current* and is equal to $V\omega_r C$.

Hence,

$$\frac{\text{Oscillatory current}}{\text{Make up current for supply}} = \frac{V\omega_r C}{\frac{VCR}{L}}$$

$$= \frac{L\omega_r}{R} = Q_r. \tag{12.66}$$

Thus, the parallel-tuned circuit at resonance exhibits current magnification of $Q_r (= \frac{\omega_r L}{R})$ while the series resonant circuit exhibits voltage magnification of the same value Q_r.

The dynamic resistance R_a of the parallel-tuned circuit may also be expressed in terms of Q_r. Thus,

$$R_a = \frac{L}{CR} = \frac{L\omega_r}{R} \cdot \frac{1}{\omega_r L}$$

$$= \frac{1}{\omega_r C} = Q_r. \tag{12.67}$$

Also,

$$R_a = \frac{L}{CR} = L\omega_r \frac{1}{\omega_r CR} = \omega_r L Q_r. \tag{12.68}$$

(*b*) **Impedance at a frequency close to resonant frequency.** The impedance of the parallel-tuned circuit of Figure 12.17 at any frequency ω is given by

$$Z = \frac{(R + j\omega L)(\frac{1}{j\omega C})}{R + j(\omega L - \frac{1}{\omega C})} \tag{12.69}$$

$$= \frac{R(1 + j\frac{\omega L}{R})(\frac{1}{j\omega C})}{R[1 + j\frac{\omega L}{R}(1 - \frac{1}{\omega^2 LC})]}$$

$$= \frac{\frac{L}{CR} + \frac{1}{j\omega C}}{1 + j\frac{\omega L}{R}(1 - \frac{1}{\omega^2 LC})}. \tag{12.70}$$

Equation (12.70) gives the general expression for impedance of a parallel circuit at any frequency ω. In most cases, however, we are concerned with frequencies close to the resonant frequency of the parallel-tuned circuit. For such frequencies, let δ give the fractional *frequency deviation or the fractional detuning* defined as below

$$\delta = \frac{\omega - \omega_r}{\omega_r}. \tag{12.71}$$

Then

$$\frac{\omega}{\omega_r} = 1 + \delta. \tag{12.72}$$

Hence,

$$\frac{\omega L}{R} = \frac{\omega_0 L}{R \cdot \frac{\omega}{\omega_0}} = Q_r(1 + \delta). \tag{12.73}$$

For $Q_r > 10$,

$$\omega_r \approx \frac{1}{\sqrt{LC}}.$$

Hence,

$$\frac{1}{\omega^2 LC} = \frac{\omega_r^2}{\omega^2}. \tag{12.74}$$

Hence, Equation (12.70) yields

$$Z = \frac{\frac{L}{CR}(1 - j\frac{R}{\omega L})}{1 + jQ_r(1 + \delta)(1 - \frac{\omega_r^2}{\omega^2})} \tag{12.75}$$

$$= \frac{L}{CR} \cdot \frac{[1 - j\frac{1}{Q_r(1+\delta)}]}{1 + jQ_r(1 + \delta)[1 - j\frac{1}{(1+\delta)^2}]}$$

$$= \frac{L}{CR} \cdot \frac{1 - j\frac{1}{Q_r(1+\delta)}}{1 + jQ_r\left[(1+\delta) - \frac{1}{(1+\delta)}\right]}$$

$$= \frac{L}{CR} \cdot \frac{1 - j\frac{1}{Q_r(1+\delta)}}{1 + jQ_r\delta\left(\frac{2+\delta}{1+\delta}\right)}. \tag{12.76}$$

At parallel resonance, $\omega = \omega_r$ and $\delta = 0$.

Hence, at resonance, the impedance of the parallel circuit is

$$Z_0 = \frac{L}{CR}\left(1 - j\frac{1}{Q_0}\right). \tag{12.77}$$

For $Q_r > 10$, $\frac{1}{Q_r}$ is small in comparison with unity and may, therefore, be neglected.

Hence,

$$Z_0 \approx \frac{L}{CR}. \tag{12.78}$$

This impedance at resonance is a pure resistance and may be denoted R_t.

Equation (12.76) gives the impedance of the parallel circuit for any value of δ. However, if the signal frequency is close to resonance, *i.e.*, $\delta < 1$, then neglecting δ in comparison with unity, Equation (12.76) may be written as

$$Z = \frac{L}{RC} \cdot \frac{1 - j\frac{1}{Q_r}}{1 + f2\delta Q_r}. \tag{12.79}$$

Further, if $Q_r > 10$, $\frac{1}{Q_r} < 1$, Then Equation (12.79) reduces to

$$Z = \frac{L}{CR} \cdot \frac{1}{1 + j2\delta Q_r} \tag{12.80}$$

$$= \frac{R_t}{1 + j2\delta Q_r} \tag{12.81}$$

or

$$\frac{Z}{R_t} = \frac{1}{1 + j2\delta Q_r}. \tag{12.82}$$

Figure 12.18 gives the nature of variation of relative impedance $\frac{Z}{R_t}$ of the parallel-tuned circuit with frequency f. The impedance is maximum at resonance and falls off suply for frequencies either below or above this. The frequencies f_1 and f_2 at which the impedance falls to $\frac{R_t}{\sqrt{2}}$ (or 0.707 R_t) constitute the half-power frequencies.

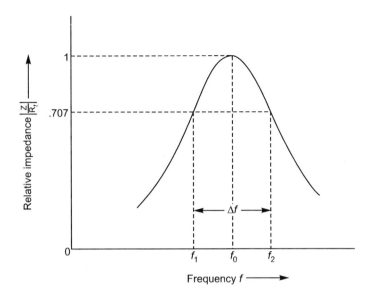

FIGURE 12.18 **Variation of relative impedance of a parallel-tuned circuit with frequency.**

12.13 SELECTIVITY AND BANDWIDTH OF A PARALLEL-TUNED CIRCUIT

Figure 12.19(a) shows a voltage source V of width impedance R_g driving the parallel-tuned circuit. Figure 12.19(b) shows an equivalent circuit in which the voltage generator has been replaced by the equivalent current generator by using the source transformation method.

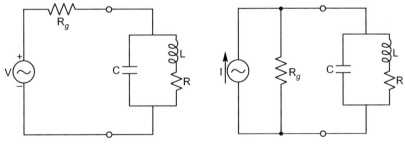

(a) Parallel-tuned circuit driven by a voltage source.

(b) Equivalent circuit using the equivalent current generator.

FIGURE 12.19 **A parallel-tuned circuit driven from a voltage source and its equivalent circuit.**

We proceed to find the selectivity and bandwidth of the parallel circuit neglecting the source resistance R_g. Under this condition, a constant current I flows through the parallel circuit and the output voltage across the parallel circuit is given by

$$V_0 = IZ.$$

Current I being constant, the output voltage V_0 varies exactly in the same fashion as the impedance Z of the parallel circuit. The response of the parallel circuit, *i.e.*, the output voltage V_0, is maximum at resonance and falls off on either side in accordance with the curve of Figure 12.18.

At the half-power frequencies f_1 and f_2, the impedance Z falls to a value $\frac{Z_r}{\sqrt{2}}$. Hence, from Equation (12.82)

$$|1 + j2\delta Q_r| = \sqrt{2}$$

or

$$2\delta Q_r = \pm 1$$

or

$$\delta = \pm \frac{1}{2Q_r}. \tag{12.83}$$

At upper half-power frequency f_2,

$$\frac{f_2 - f_r}{f_r} = \frac{1}{2Q_r}.$$

At lower half-power frequency f_1,

$$\frac{f_1 - f_r}{f_r} = \frac{1}{2Q_r}.$$

Hence,

$$(f_2 - f_1) = \frac{f_r}{Q_r} \tag{12.84}$$

or

$$Q_r = \frac{f_r}{(f_2 - f_1)} = \frac{f_r}{\Delta f} = \frac{f_r}{\text{Bandwidth}}. \tag{12.85}$$

This Q_r is a measure of the selectivity of the parallel resonant circuit.

The higher the value of Q_r the more selective the circuit. Q_r may also be put as

$$Q_r = \frac{\omega_r L}{R} = \frac{1}{\omega_r CR} = \frac{1}{R}\sqrt{\frac{L}{C}}. \qquad (12.86)$$

QUESTIONS FOR DISCUSSION

1. What is resonance ?
2. What is the difference between series and parallel resonance ?
3. What do you mean by current resonance ?
4. What is resonant frequency ?
5. What is the value of the power factor at resonance ?
6. What is the phase difference between voltage and current at resonance ?
7. What is bandwidth ?
8. What is the relation between time period and frequency ?
9. What is quality factor ?

OBJECTIVE QUESTIONS

1. Quality factor depends on
   (a) R   (b) L   (c) C   (d) R, L & C.
2. In a series RLC circuit, the effect of frequency is to produce
   (a) maximum voltage across the capacitor at a frequency of $\omega_C =$
   $$\sqrt{\frac{1}{LC} - \frac{R^2}{2L^2}}$$
   (b) maximum voltage across inductance at a frequency of $\omega_L =$
   $$\frac{1}{\sqrt{LC - \frac{C^2 R^2}{2}}}$$
   (c) both (a) and (b) are true
   (d) maximum voltage at resonance angular frequency.
3. The quality of a coil Q-factor is defined as
   (a) $Q = \dfrac{\omega_L}{R}$

   (b) $Q = \dfrac{1}{\omega_{CR}}$

   (c) $Q = 2\pi\,\dfrac{\text{Maximum energy stored per cycle}}{\text{Energy dissipated per cycle}}$

   (d) all of these.

4. A highly selective circuit has high
   (a) Q   (b) capacitance to produce resonance at a fixed frequency
   (c) R   (d) either (a) or (b).
5. Source impedance of a highly selective series circuit is
   (a) very low   (b) high   (c) very high   (d) none of these.
6. A coil at 200 KHz frequency has inductive reactance $X_L = 1000\,\Omega$ and self-resistance $R = 10\,\Omega$. Its Q factor is
   (a) 10   (b) 20   (c) 100   (d) 200.

UNSOLVED PROBLEMS

1. Two impedances $Z_1 = a + Jb$ and $Z_2 = c - Jd$ are connected in (a) series and (b) parallel. Determine the condition of resonance in each case.
2. A coil has impedance $(2 + 2J)$ connected in series with the capacitor $10\,\mu F$. Determine the frequency at which resonance occurs and also determine the Q-factor.
3. A series RLC circuit consists of resistance $R = 10\,\Omega$, $L = 0.05$ H, and $C = 0.02\,\mu F$. Calculate the resonant frequency. Also determine the frequencies at which the voltage across the capacitor and inductor are maximum (individually).
4. Prove for series RLC circuit the value of the current is maximum.
5. Prove for series resonant circuit $Bw = \frac{R}{2\pi L}$ where Bw = Bandwidth.
6. Two impedances $Z_1 = 2 + Jb$ and $Z_2 = C + 3J$ are connected in parallel. Determine the condition of resonance.

13

FILTERS

13.1 INTRODUCTION

An electric wave filter, or simply a filter, is an electric network which passes or allows unattenuated transmission of an electric signal within a certain frequency range, and stops transmission of electric signals outside this range.

Filters may be classified according to their uses:

(a) *Low-pass filters:* These filters reject all frequencies above a specified value.

(b) *High-pass filters:* These filters reject all frequencies below a specified value.

(c) *Band-pass filters:* A band-pass filter passes or allows transmission of a band of frequencies and rejects all frequencies beyond this band.

(d) *Band-stop filters:* A band-stop or *elimination filter* rejects transmission of a limited band of frequencies but allows transmission of all other frequencies.

An *ideal filter* offers *zero attenuation* in the pass band, and *infinite attenuation* in the stop band.

13.2 PARAMETERS OF A FILTER

The following parameters characterize a typical filter:

(i) *Characteristic impedance* Z_0: This should be chosen so that the filter may fit into a given line or between two pieces of equipment.

(ii) *Pass band:* Filter should have very low (ideally zero) attenuation in the pass band and high attenuation in the stop band or attenuation band.

(iii) *Stop band:* A band in which ideal filters have to stop frequencies.

(iv) *Cut-off frequency:* This frequency demarcates or separates the pass band and the stop band.

13.3   DECIBEL AND NEPER

The attenuation of a filter can be expressed in decibels (dB) or Nepers.

13.3.1   Bel

A *Bel* is fundamentally a unit of power ratio at the two-ports of a network. Let P_i be the power input to a network at the sending end or input end and P_0 be the output power of the network at the receiving end, then the output power/input power ratio in Bels is given by

$$\text{Power transfer ratio or power gain in Bels} = \log_{10} \frac{P_0}{P_i}. \qquad (13.1)$$

The power transfer ratio in Bels comes out to be negative ($P_i > P_0$).

$$\text{Hence, power loss or attenuation in Bels} = \log_{10} \frac{P_i}{P_0}. \qquad (13.2)$$

A Bel is a large unit. A more convenient unit is a *decibel* (abbreviated as dB) which is one-tenth of a Bel.

Thus, the power gain in

$$d\text{B} = 10 \log_{10} \left(\frac{P_0}{P_i} \right) \qquad (13.3)$$

and the power attenuation in

$$d\text{B} = 10 \log_{10} \left(\frac{P_i}{P_0} \right). \qquad (13.4)$$

The power P_i and P_0 are associated with equal impedance, then the power ratio may be expressed as the square of the voltage ratio or current ratio $[P = \frac{V^2}{R} = I^2 R]$.

Now the power gain in

$$d\text{B} = 10 \log_{10} \left| \frac{V_0}{V_i} \right|^2 \qquad (13.5)$$

$$= 20 \log_{10} \left| \frac{V_0}{V_i} \right|. \qquad (13.6)$$

Similarly for current, the power gain in

$$d\text{B} = 10 \log_{10} \left| \frac{I_0}{I_i} \right|^2 = 20 \log_{10} \left| \frac{I_0}{I_i} \right| \qquad (13.7)$$

and the power attenuation in

$$dB = 10 \log_{10} \left| \frac{V_i}{V_0} \right|^2 = 20 \log_{10} \left| \frac{V_i}{V_0} \right|. \tag{13.8}$$

Power attenuation in

$$dB = 10 \log_{10} \left| \frac{I_i}{I_0} \right|^2 = 20 \log_{10} \left| \frac{I_i}{I_0} \right|. \tag{13.9}$$

13.3.2 Neper

A Neper is fundamentally a unit of current ratio or voltage ratio.

$$\text{Attenuation in Nepers} = \log_e \left| \frac{I_i}{I_0} \right| = \log_e \left| \frac{V_i}{V_0} \right|. \tag{13.10}$$

13.4 PROPAGATION CONSTANT

Figure 13.1 shows a two-port network terminated in Z_0. For this termination, the ratio $\frac{V_1}{V_2}$ is the same as $\frac{I_1}{I_2}$.

It is convenient to express this ratio in exponential form,

$$\frac{V_1}{V_2} = \frac{I_1}{I_2} = e^\gamma = e^{\alpha + j\beta}. \tag{13.11}$$

γ is a known as a propagation constant ($\gamma = \alpha + j\beta$). α, the real part of the propagation constant, is known as an attenuation constant, and β, the imaginary part, is known as a *phase shift constant* or *phase factor* of the

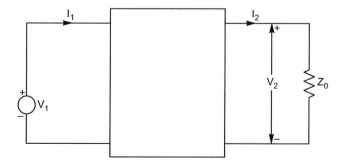

FIGURE 13.1 A two-port network terminated in characteristic impedance.

network. Both α and β are the functions of frequency. The transmission parameters (ABCD) can be expressed in terms of γ,

$$V_1 = AV_2 + BI_2 = AV_2 + \frac{BV_2}{Z_0} = \left(A + \frac{B}{Z_0}\right)V_2 \tag{13.12}$$

$$\left[\text{where } I_2 = \frac{V_2}{Z_0}\right]$$

$$I_1 = CV_2 + DI_2 = (CZ_0 + A)I_2. \tag{13.13}$$

Since

$$AD - BC = 1 \Rightarrow A^2 - BC = 1(A = D) \tag{13.14}$$

and

$$Z_0 = \sqrt{\frac{B}{C}}, \quad \text{we get} \tag{13.15}$$

$$\frac{V_1}{V_2} = A + \frac{B}{Z_0} = A + B\sqrt{\frac{C}{B}} = A + \sqrt{BC} = A + \sqrt{A^2 - 1} \tag{13.16}$$

$$\frac{I_1}{I_2} = C\sqrt{\frac{B}{C}} + A = A + \sqrt{BC} = A + \sqrt{A^2 - 1}. \tag{13.17}$$

By the equation, it is clear

$$\frac{V_1}{V_2} = \frac{I_1}{I_2} = A + \sqrt{A^2 - 1} = e^\gamma \tag{13.18}$$

$$A + \sqrt{A^2 - 1} = e^\gamma$$

$$\cosh \gamma = \frac{e^\gamma + e^{-\gamma}}{2} = \frac{1}{2}\left[A + \sqrt{A^2 - 1} + \frac{1}{A + \sqrt{A^2 - 1}}\right]$$

$$\cosh \gamma = \frac{1}{2}\left[\frac{(A + \sqrt{A^2 - 1})^2 + 1}{A + \sqrt{A^2 - 1}}\right]$$

$$\cosh \gamma = \frac{1}{2}\left[\frac{A^2 + A^2 - 1 + 2A\sqrt{A^2 - 1} + 1}{A + \sqrt{A^2 - 1}}\right] = A \tag{13.19}$$

$$\sinh \gamma = (\cosh^2 \gamma - 1)^{1/2} = \sqrt{A^2 - 1} = \sqrt{BC} \tag{13.20}$$

since $Z_0 = \sqrt{\dfrac{B}{C}} = $ by Equation (13.15).

Using Equations (13.20) and (13.15), we get

$$B = Z_0 \sinh \gamma \qquad (13.21)$$

$$C = \frac{\sinh \gamma}{Z_0}. \qquad (13.22)$$

Therefore, the general circuit parameters of two-port networks are in terms of propagation constants:

$$A = D = \cosh \gamma$$
$$B = Z_0 \sinh \gamma$$
$$C = \frac{\sinh \gamma}{Z_0}.$$

13.5 CLASSIFICATION OF FILTERS

There are four common types of filters classified according to the nature of their work.

13.5.1 Low-pass Filters

These filters reject all frequencies above cut-off frequency (f_L). The attenuation characteristic of an ideal Low-Pass (LP) filter is shown in Figure 13.2(b).

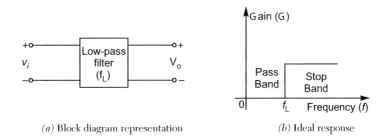

<table>
<tr><td>(a) Block diagram representation</td><td>(b) Ideal response</td></tr>
</table>

FIGURE 13.2 **Low-pass filter.**

Thus, the range of the low-pass filter $= 0$ to f_L and the stop-band frequency range is above f_L.

13.5.2 High-pass Filters

These filters reject all frequencies below the cut-off frequency f_H. The attenuation characteristic of an ideal High-Pass (HP) filter is shown in Figure 13.3(b).

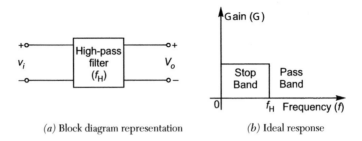

(a) Block diagram representation   (b) Ideal response

FIGURE 13.3   High-pass filter.

Thus, the range of a high-pass filter $= f_H$ to ∞ and the stop-band frequency range $= 0$ to f_H.

13.5.3   Band-pass Filters

These filters allow transmission of frequencies between two designated cut-off frequencies and reject all other frequencies as shown in Figure 13.4.

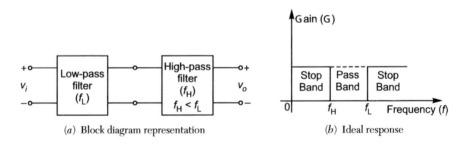

(a) Block diagram representation   (b) Ideal response

FIGURE 13.4   Band-pass filter.

Thus, the range of the band-pass filter $= f_H$ to $f_L (f_H < f_L)$ and the stop-band frequency range $= 0$ to f_H and f_L to ∞.

A band-pass filter is generally obtained by a series or cascade connection of a low-pass filter and high-pass filter as shown in Figure 13.4(a).

13.5.4   Band-stop or band-elimination Filters

A band-stop or band-elimination filter rejects transmission of a limited band of frequencies but allows transmission of all other frequencies as shown in Figure 13.5(b).

A band-stop filter is obtained by a parallel connection of a low-pass filter and a high-pass filter as shown in Figure 13.5(a) where the cut-off frequency

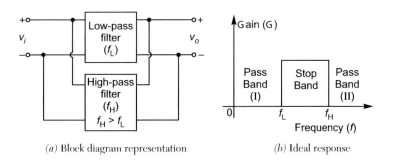

(a) Block diagram representation   (b) Ideal response

FIGURE 13.5   Band-stop or band-elimination filters.

TABLE 13.1   Frequency Range of Different Types of Filters

Types of Filter	Pass Band	Attenuation Band
Low-pass	$0 \rightarrow f_L$	$f_L \rightarrow \infty$
High-pass	$f_H \rightarrow \infty$	$0 \rightarrow f_H$
Band-pass ($f_L > f_H$)	$f_H \rightarrow f_L$	$0 \rightarrow f_H, f_L \rightarrow \infty$
Band-stop ($f_H > f_L$)	$0 \rightarrow f_L, f_H \rightarrow \infty$	$f_L \rightarrow f_H$

f_H of the high-pass filter is greater than the cut-off frequency f_L of low-pass filter.

The first band-pass frequencies will pass through the low-pass filter and the second pass band frequencies will pass through the high-pass filter.

Thus, the range of the stop band filter = 0 to f_L and f_H to ∞ and the stop-band frequency range = f_L to f_H.

13.6 BASIC FILTER NETWORKS

In general, a *filter network* consists of one or more sections of symmetrical T-networks and their half-sections as shown in Figure 13.6.

Z_1 is the total impedance in the T-section and may be considered to be built up of two unsymmetrical half-sections as shown in Figure 13.6.

Figure 13.7 shows a symmetrical π-section filter. Here Z_1 is also the total impedance in the series arm, while Z_2 is the total impedance of the shunt arm π-section and may be considered to be built up of two unsymmetrical half-sections, as shown in Figure 13.7(b).

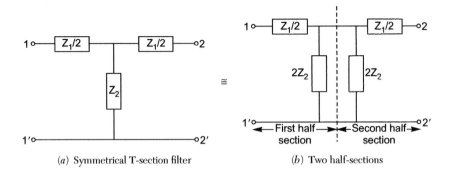

(a) Symmetrical T-section filter

(b) Two half-sections

FIGURE 13.6 **Symmetrical T-section filter.**

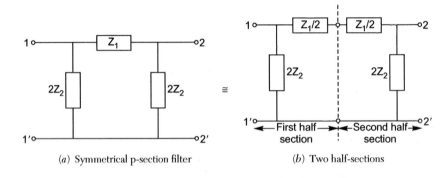

(a) Symmetrical p-section filter

(b) Two half-sections

FIGURE 13.7 **Symmetrical π-section filter.**

The transfer constant θ of a two-port network in terms of ABCD parameters is given by the relation

$$\theta = \cosh^{-1} \sqrt{AD}. \tag{13.23}$$

For a symmetrical network such as the T-section of Figure 13.6(a), the transfer constant θ becomes the propagation constant γ, and is given by

$$\theta = \gamma = \alpha + j\beta = \cosh^{-1} \sqrt{AD}. \tag{13.24}$$

But for the T-section of Figure 13.6(a)

$$A = \frac{Z_{11}}{Z_{21}} = 1 + \frac{Z_1}{2Z_2}$$

$$D = \frac{Z_{22}}{Z_{21}} = 1 + \frac{Z_1}{2Z_2} = A \tag{13.25}$$

$$\gamma = \cosh^{-1} A = \cosh^{-1}\left(1 + \frac{Z_1}{2Z_2}\right) \tag{13.26}$$

or

$$1 + \frac{Z_1}{2Z_2} = \cosh \gamma = 1 + 2\sinh^2 \frac{\gamma}{2}.$$

Hence,

$$\sinh \frac{\gamma}{2} = \sqrt{\frac{Z_1}{4Z_2}}.$$

or

$$\frac{\gamma}{2} = \sinh^{-1} \sqrt{\frac{Z_1}{4Z_2}}. \qquad (13.27)$$

Hence,

$$\theta = \gamma = 2\sinh^{-1} \sqrt{\frac{Z_1}{4Z_2}} \qquad (13.28) \quad (\gamma = \alpha + j\beta)$$

If γ or θ is purely imaginary, $\alpha = 0$ and $\gamma = j\beta$, then the attenuation offered by the filter is zero, and the transmission is said to be free or unattenuated.

If Z_1 and Z_2 are reactances and may, under ideal conditions, be assumed to be pure, then the ratio $\frac{Z_1}{4Z_2}$ will [$Z_1 = R + jX$] be real, and either positive or negative, depending whether Z_1 and Z_2 are of the same or different kinds.

Equation (13.28) may be written as

$$\sinh \left(\frac{\alpha}{2} + j\frac{\beta}{2} \right) = \sqrt{\frac{Z_1}{4Z_2}}$$

or

$$\sinh \frac{\alpha}{2} \cos \frac{\beta}{2} + j \cosh \frac{\alpha}{2} \sin \frac{\beta}{2} = \sqrt{\frac{Z_1}{4Z_2}}. \qquad (13.29)$$

The following two cases are considered.

(a) $\frac{Z_1}{Z_2} > 0$ or $\frac{Z_1}{4Z_2}$ is positive and real.

If impedances Z_1 and Z_2 have the same sign, then the ratio of Z_1 and Z_2 is real and positive, therefore,

$$\sinh \frac{\alpha}{2} \cos \frac{\beta}{2} = \sqrt{\frac{Z_1}{4Z_2}} \tag{13.30}$$

and

$$\cosh \frac{\alpha}{2} \sin \frac{\beta}{2} = 0 \tag{13.31}$$

or

$$\sin \frac{\beta}{2} = 0 \Rightarrow \frac{\beta}{2} = \pm n\pi$$
$$\beta = \pm 2n\pi, \tag{13.32}$$

where $n = 0, 1, 2, \ldots, n$ and also $\cos \frac{\beta}{2} = 1$ by the above condition. Hence, from Equation (13.30)

$$\sinh \frac{\alpha}{2} = \sqrt{\frac{Z_1}{4Z_2}} \tag{13.33}$$

or

$$\alpha = 2 \sinh^{-1} \sqrt{\frac{Z_1}{4Z_2}}. \tag{13.34}$$

Therefore, if $\frac{Z_1}{4Z_2}$ is positive, α is real which means attenuation takes place. Accordingly, the frequency range over which $\frac{Z_1}{4Z_2} > 0$ constitutes the attenuation band or stop-band.

(b) $\dfrac{Z_1}{Z_2} < 0$ or $\dfrac{Z_1}{4Z_2}$ is imaginary.

In this case by Equation (13.29)

$$\cosh \frac{\alpha}{2} \sin \frac{\beta}{2} = \sqrt{\frac{Z_1}{4Z_2}} \tag{13.35}$$

and

$$\sinh \frac{\alpha}{2} \cos \frac{\beta}{2} = 0, \tag{13.36}$$

then either

$$\sinh \frac{\alpha}{2} = 0 \tag{13.37}$$

or

$$\cos \frac{\beta}{2} = 0. \tag{13.38}$$

First taking $\sinh \frac{\alpha}{2} = 0, \alpha = 0$ attenuation is zero. The frequency range over which $\frac{Z_1}{4Z_2} < 0$ and $\alpha = 0$ constitutes the band-pass.

Since

$$\sinh \frac{\alpha}{2} = 0, \quad \cosh \frac{\alpha}{2} = 1.$$

By Equation (13.35),

$$\sin \frac{\beta}{2} = \sqrt{\frac{Z_1}{4Z_2}}$$

or

$$\beta = 2 \sin^{-1} \sqrt{\frac{Z_1}{4Z_2}}. \tag{13.39}$$

Within the pass band Equation (13.39) gives the value of phase constant β as a function of reactances Z_1 and Z_2, *i.e.*, as a function of frequency when $\frac{Z_1}{4Z_2}$ is negative and $|\frac{Z_1}{4Z_2}| > 1$ yields $\sin \frac{\beta}{2} > 1$. This is absurd. This signifies that for free transmission, $-1 < \frac{Z_1}{4Z_2} < 0$, thus for free transmission through a filter, $\frac{Z_1}{4Z_2}$ must be negative and numerically less than unity,

$$-1 < \frac{Z_1}{4Z_2} < 0. \tag{13.40}$$

Now

$$\cos \frac{\beta}{2} = 0, \text{ hence } \sin \frac{\beta}{2} = \pm 1$$

$$\frac{\beta}{2} = (2n - 1)\frac{\pi}{2} \text{ where } n = 1, 2, \ldots, n.$$

Now by Equation (13.35)

$$\cosh \frac{\alpha}{2} = \sqrt{\left|\frac{Z_1}{4Z_2}\right|}$$

or

$$\alpha = 2\cosh^{-1}\sqrt{\left|\frac{Z_1}{4Z_2}\right|}. \tag{13.41}$$

In this case, both Z_1 and Z_2 are purely imaginary.

13.7 CUT-OFF FREQUENCIES

The frequencies at which conditions change from transmission to attenuation are known as *cut-off frequencies*.

Equation (13.40) defines these cut-off frequencies. Thus, these cut-off frequencies are given by the equations

$$\frac{Z_1}{4Z_2} = 0 \tag{13.42}$$

or

$$Z_1 = 0$$

and

$$\frac{Z_1}{4Z_2} = -1 \quad \text{or} \quad Z_1 + 4Z_2 = 0. \tag{13.43}$$

The characteristic impedance of a filter in pass and stop-bands:
The characteristic impedance is given by

$$Z_0 = \sqrt{\frac{B}{C}},$$

where $B = \frac{Z_{11} \cdot Z_{22} - Z_{12} \cdot Z_{21}}{Z_{21}}$ and

$$C = \frac{1}{Z_{21}}.$$

Hence,

$$Z_0 = \sqrt{Z_{11} \cdot Z_{22} - Z_{12} \cdot Z_{21}}. \tag{13.44}$$

For the T-section of Figure 13.6(a)

$$Z_{11} = Z_{22} = Z_2 + \frac{Z_1}{2}$$

and

$$Z_{12} = Z_{21} = Z_2.$$

Hence,

$$Z_{0T} = \sqrt{\left(Z_2 + \frac{Z_1}{2}\right)^2 - Z_2^2}$$

$$Z_{0T} = \sqrt{Z_1 Z_2 + \frac{Z_1^2}{4}} = \sqrt{Z_1 Z_2 \left(1 + \frac{Z_1}{4Z_2}\right)}. \tag{13.45}$$

Let the filter network be constituted by pure reactances only

$$Z_1 = jX_1, \quad Z_2 = jX_2. \qquad \text{[Z_1 and Z_2 nature is opposite]}$$

Then,

$$Z_{0T} = \sqrt{-X_1 X_2 \left(1 + \frac{X_1}{4X_2}\right)}. \tag{13.46}$$

13.7.1 In the case of the pass-band

We have noted that in the pass-band, Z_1 and Z_2 (and hence X_1 and X_2) are of opposite sign, and further $-1 < \frac{X_1}{4X_2} < 0$. Hence, in the pass-band, both the quantities $-X_1 X_2$ and $(1 + \frac{X_1}{4X_2})$ are positive. Hence, by Equation (13.46), the entire quantity becomes positive. Therefore, Z_{0T} is real (*i.e.*, resistive), and let it be indicated by R_0.

Accordingly, when such a network is terminated in a resistive load equal to R_0, then the input impedance of the network is also R_0. The network then transmits power to the resistive load without attenuation.

13.7.2 In the stop-band with $\frac{X_1}{4X_2} > 0$

X_1 and X_2 are of the same sign and further $\frac{X_1}{4X_2} > 0$. Using substitution of this condition in Equation (13.46) we get a negative real quantity. Hence, the characteristic impedance Z_0 is a pure reactance:

The characteristic impedance of π-network:

For π-section as shown in Figure 13.7(*a*),

$$Z_{11} = Z_{22} = \frac{(2Z_2)(Z_1 + 2Z_2)}{Z_1 + 4Z_2} = \frac{\frac{Z_1}{2} + Z_2}{1 + \frac{Z_1}{4Z_2}} \tag{13.47}$$

$$Z_{12} = Z_{21} = \frac{Z_2}{1 - \frac{Z_1}{4Z_2}} \tag{13.48}$$

$$Z_{0\pi} = \sqrt{Z_{11}Z_{22} - Z_{12}Z_{21}} = \sqrt{\left(\frac{Z_2 + \frac{Z_1}{2}}{1 + \frac{Z_1}{4Z_2}}\right)^2 - \left(\frac{Z_2}{1 + \frac{Z_1}{4Z_2}}\right)^2}$$

$$= \frac{1}{1 + \frac{Z_1}{4Z_2}}\sqrt{\left(Z_2 + \frac{Z_1}{2}\right)^2 - Z_2^2}$$

$$= \frac{\sqrt{\frac{Z_1^2}{4} + Z_1 Z_2}}{1 + \frac{Z_1}{4Z_2}}$$

$$= \sqrt{\frac{Z_1 Z_2}{1 + \frac{Z_1}{4Z_2}}} \tag{13.49}$$

$$= \frac{Z_1 Z_2}{\sqrt{Z_1 Z_2 \left(1 + \frac{Z_1}{4Z_2}\right)}}. \tag{13.50}$$

Hence,

$$Z_{0\pi} = \frac{Z_1 Z_2}{Z_{0T}}. \tag{13.51}$$

Assuming Z_1 and Z_2 to be pure reactances and substituting these into Equation (13.49), we find that in the case of the π-network and also in the pass-band, the characteristic impedance is a pure resistance while in the stop-band, the characteristic impedance is purely reactive.

13.8   CONSTANT-K FILTERS

In a constant-k filter, the series impedance Z_1 and the shunt impedance Z_2 are related by the relation,

$$Z_1 Z_2 = k^2 \tag{13.52}$$

where k is a constant independent of the frequency. In almost all cases, the product $Z_1 Z_2$ is real and has the dimensions of resistance squared.
Then

$$Z_1 Z_2 = k^2 = R_0^2.$$

R_0 is known as the "design impedance" of the filter section.
The constant-k for the π-sections are also referred to as the prototype filters.

13.8.1 Constant-k low-pass filters

Figure 13.8 shows T- and π-sections of a constant-k low-pass filter.

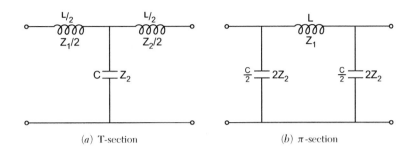

(a) T-section   (b) π-section

FIGURE 13.8 **T- and π-sections of constant-k low-pass filter.**

In both T- and π-sections,

$$\text{total series impedance} = Z_1 = j\omega L.$$
$$\text{total shunt impedance} = Z_2 = \frac{1}{j\omega C}.$$

Hence,

$$Z_1 Z_2 = \frac{L}{C}, \tag{13.53}$$

therefore, design resistance

$$R_0 = \sqrt{Z_1 Z_2} = \sqrt{\frac{L}{C}} \tag{13.54}$$

and

$$\frac{Z_1}{4Z_2} = -\frac{\omega^2 LC}{4}. \tag{13.55}$$

The pass-band is given by

$$-1 < \frac{Z_1}{4Z_2} < 0.$$

Hence, the two cut-off frequencies are given by the relations

$$\frac{Z_1}{4Z_2} = 0 \quad \text{and} \quad \frac{Z_1}{4Z_2} = -1.$$

If $\frac{Z_1}{4Z_2} = 0$ yields $Z_1 = j\omega L = 0$, *i.e.*, $\omega = 0$ or $f = 0$, this is the first cut-off frequency

if

$$\frac{Z_1}{4Z_2} = -1 \quad \text{yields} \quad -\frac{\omega^2 LC}{4} = -1$$

or

$$\omega = \frac{2}{\sqrt{LC}}$$

$$f_L = \frac{1}{\pi\sqrt{LC}}. \tag{13.56}$$

This is the second cut-off frequency.

Thus, the pass band of this constant-k low-pass filter extends from 0 to f_L.

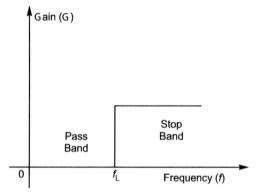

Characteristic impedance:

By Equation (13.45),

$$Z_{0T} = \sqrt{Z_1 Z_2 \left(1 + \frac{Z_1}{4Z_2}\right)}.$$

Putting the values of $Z_1 Z_2$ and $(1 + \frac{Z_1}{4Z_2})$ by Equation (13.53) and (13.55) we get,

$$Z_{0T} = \sqrt{\frac{L}{C}\left(1 - \frac{\omega^2 LC}{4}\right)} = \sqrt{\frac{L}{C}}\sqrt{1 - \left(\frac{f}{f_C}\right)^2} \tag{13.57}$$

$$Z_{0T} = R_0\sqrt{1 - \left(\frac{f}{f_C}\right)^2}. \tag{13.58}$$

In the pass-band, $f < f_C$ so that Z_{0T} is real.

In the attenuation band $f < f_C$, so that Z_{0T} is imaginary. Similarly, the characteristic impedance of a π-section filter by Equation (13.51) is

$$Z_{0\pi} = \frac{Z_1 Z_2}{Z_{0T}} = \frac{\frac{L}{C}}{\sqrt{\frac{L}{C}}\sqrt{1 - \left(\frac{f}{f_C}\right)^2}} = \frac{\sqrt{\frac{L}{C}}}{\sqrt{1 - \left(\frac{f}{f_C}\right)^2}} \qquad (13.59)$$

$$Z_{0\pi} = \frac{R_0}{\sqrt{1 - \left(\frac{f}{f_C}\right)^2}}. \qquad (13.60)$$

In the pass band $f < f_C$ so that $Z_{0\pi}$ is a real and positive quantity. In the attenuation band, $f < f_C$ so that $Z_{0\pi}$ becomes imaginary. The figure shows the curves showing the variation of $Z_{0\pi}$ and Z_{0T} with frequency in the pass band for a constant k low-pass filter.

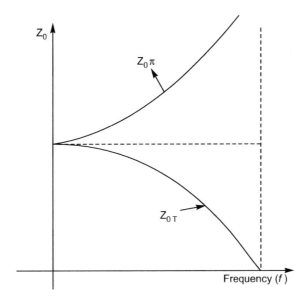

FIGURE 13.9

13.8.2 Design of Constant-k Low-pass Filters

Often it is required to design a constant-k filter for a given cut-off frequency and for feeding a given load resistance. So far as the cut-off frequency is concerned, we may obtain any cut-off frequency by suitable choice of elements forming the filter. A perfect impedance match between the filter and the load may be achieved at only one frequency. In practice the load resistance is made equal to Z_0 of the filter at zero frequency in the case of the low-pass filter.

$$R_0 = \sqrt{\frac{L}{C}}$$

and the cut-off frequency

$$f_C = \frac{1}{\pi\sqrt{LC}}.$$

By the above two equations, we get

$$L = \frac{R_0}{\pi f_C} \tag{13.61}$$

$$C = \frac{1}{\pi f_C R_0}. \tag{13.62}$$

Equations (13.61) and (13.62) give the values of elements L and C in terms of cut-off frequency f_C and design resistance R_0.

Example 13.1. *For the given T-section low-pass filter, determine the cut-off frequency and nominal characteristic impedance R_0.*

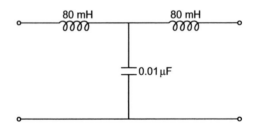

FIGURE 13.10

Solution: Total series inductance

$$L = 160 \text{ mH}$$
$$C = 0.01 \text{ } \mu\text{F}.$$

Cut-off frequency

$$f_C = \frac{1}{\pi\sqrt{LC}}$$

$$= \frac{1}{3.14\sqrt{160 \times 10^{-3} \times 10^{-8}}}$$

$$= 7.962 \times 10^3 \text{ Hz} = 7.962 \text{ KHz}$$

$$R_0 = \sqrt{\frac{L}{C}} = \sqrt{\frac{160 \times 10^{-3}}{10^{-8}}} = \sqrt{16 \times 10^6} = 4 \times 10^3$$

$$R_0 = 4 \text{ K}\Omega.$$

Example 13.2. *Design the constant-k low-pass T- and π-section filters having cut-off frequency = 3000 Hz and nominal characteristic impedance* $R_0 = 600 \ \Omega.$

Solution: By Equations (13.61) and (13.62),

$$L = \frac{R_0}{\pi f_C} = \frac{600}{\pi \times 3000} \text{H}$$

$$L = 63.68 \text{ mH}$$

$$C = \frac{1}{\pi R_0 f_C} = \frac{1}{\pi \times 600 \times 3000} \text{F}$$

$$= 0.1753 \ \mu\text{F}.$$

Hence, the required T- and π-section filters are as shown in Figure 13.11.

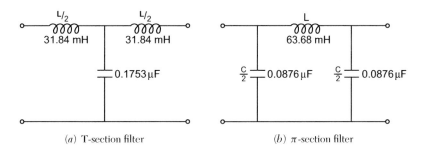

(a) T-section filter   (b) π-section filter

FIGURE 13.11

13.8.3 Constant-k high-pass filters

Figure 13.12 shows T- and π-sections of a constant-k high-pass filter.

In both T- and π-sections, total Series impedance $(Z_1) = \frac{1}{j\omega C}$ and total shunt impedance $(Z_2) = j\omega L$.

Now,

$$Z_1 Z_2 = \frac{L}{C}. \tag{13.63}$$

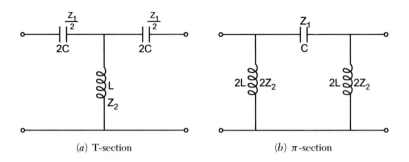

(a) T-section (b) π-section

FIGURE 13.12   Constant-k high-pass filter.

Thus, in the case of a high-pass filter as well, design resistance

$$R_0 = \sqrt{Z_1 Z_2} = \sqrt{\frac{L}{C}}. \qquad (13.64)$$

Cut-off frequency: The pass band is given by,

$$-1 < \frac{Z_1}{4Z_2} < 0.$$

Hence, the two cut-off frequencies are given by the relations,

$$\frac{Z_1}{4Z_2} = 0 \quad \text{yields} \quad Z_1 = 0 \Rightarrow \frac{1}{j\omega C} = 0, \text{ i.e., } \omega = \infty$$

and

$$\frac{Z_1}{4Z_2} = -1 \text{ yields} \; - \frac{1}{\omega C^2 4LC} = -1, \quad \text{where } \omega C \text{ is the cut-off frequency}$$

or

$$\omega_C = \frac{1}{2\sqrt{LC}} \Rightarrow f_H = \frac{1}{4\pi \sqrt{LC}}. \qquad (13.65)$$

Equation (13.65) gives the lower cut-off frequency, thus the pass band of the constant-k high-pass filter is from is f_H to ∞.

The characteristic impedance Z_0 and the characteristic impedance of the π-section filter in accordance with Equation (13.45) is given by

$$Z_{0T} = \sqrt{Z_1 Z_2 \left(1 + \frac{Z_1}{4Z_2}\right)}$$

$$Z_1 Z_2 = \frac{L}{C} \quad \text{and} \quad \frac{Z_1}{4Z_2} = -\frac{1}{4\omega^2 LC}, \quad \text{we get}$$

$$Z_{0T} = \sqrt{\frac{L}{C}\left(1 - \frac{1}{4\omega^2 LC}\right)} \tag{13.66}$$

$$= \sqrt{\frac{L}{C}}\sqrt{1 - \left(\frac{f_C}{f}\right)^2} \tag{13.67}$$

$$= R_0 \sqrt{1 - \left(\frac{f_C}{f}\right)^2}. \tag{13.68}$$

In the pass band, $f_C < f$ so that Z_{0T} is real.

In the attenuation band $f_C < f$ so that Z_{0T} becomes imaginary. The characteristic impedance of an π-section filter in accordance with Equation (13.51) is given by

$$Z_{0\pi} = \frac{Z_1 Z_2}{Z_{0T}} = \frac{\frac{L}{C}}{\sqrt{\frac{L}{C}}\sqrt{1 - \left(\frac{f_C}{f}\right)^2}} \tag{13.69}$$

$$= \frac{R_0}{\sqrt{1 - \left(\frac{f_C}{f}\right)^2}}. \tag{13.70}$$

In the pass band, $f_C < f$ so that $Z_{0\pi}$ is real and positive.

In the attenuation band, $f < f_C$ so that $Z_{0\pi}$ becomes imaginary. From Equations (13.68) and (13.70), we get

$$Z_{0T} Z_{0\pi} = R_0^2. \tag{13.71}$$

Figure 13.13 shows the curves of Z_{0T} and $Z_{0\pi}$ with frequency in the pass band for a constant-k high-pass filter.

13.8.4 Design of Constant-k High-pass Filters

$$R_0 = \sqrt{\frac{L}{C}}$$

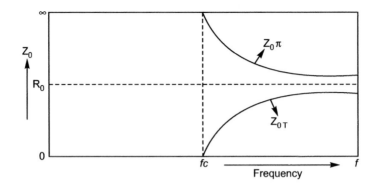

FIGURE 13.13   Variation of Z_0 with frequency for a T- and π-sections constant-*k* high-pass filter.

and

$$f_C = f_H = \frac{1}{4\pi \sqrt{LC}}.$$

By the above Equations (13.64) and (13.65), we get

$$L = \frac{R_0}{4\pi f_C} = \frac{R_0}{4\pi f_H} \qquad (13.72)$$

and

$$C = \frac{1}{4\pi f_C R_0} = \frac{1}{4\pi R_0 f_H}. \qquad (13.73)$$

Example 13.3. *Design a constant-k high-pass T- and π-sections filters having cut-off frequency of 3 KHz and nominal characteristic impedance* $R_0 = 600\ \Omega$.

Solution:

$$L = \frac{R_0}{4\pi f_C} = \frac{600}{4\pi \times 3000}$$

$$L = 15.9\ \text{mH}$$

$$C = \frac{1}{4\pi f_C R_0} = \frac{1}{4\pi \times 3000 \times 600}\ \text{F}$$

$$C = 0.0442\ \mu\text{F}.$$

Hence, the required T- and π-section filters are as shown in Figure 13.14.

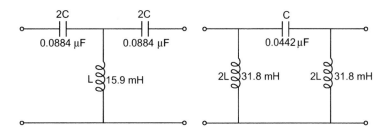

FIGURE 13.14   Constant-k high-pass filter for example 13.3.

Example 13.4. *Design a T-section constant-k high-pass filter having a cut-off frequency of 10 KHz and nominal characteristic resistance $R_0 = 600\,\Omega$. Also find (i) its characteristic impedance and phase constant at 25 KHz and (ii) attenuation of 5 KHz.*

Solution: (i) By Equations (13.72) and (13.73),

$$L = \frac{R_0}{4\pi f_C} = \frac{600}{4\pi \times 10^4} = 4.78 \text{ mH}$$

$$C = \frac{1}{4\pi R_0 f_C} = \frac{1}{4\pi \times 600 \times 10^4}F = 0.0133\,\mu F.$$

The T-section constant-k high-pass filter is shown in Figure 13.15.

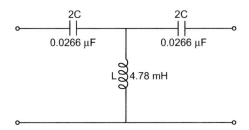

FIGURE 13.15

By Equation (13.68),

$$Z_{0T} = R_0\sqrt{1 - \left(\frac{f_C}{f}\right)^2} = 600\sqrt{1 - \left(\frac{10}{25}\right)^2} = 545\,\Omega$$

and

$$\beta = 2\sin^{-1}\left(\frac{f_C}{f}\right) = 2\sin^{-1}\left(\frac{10}{25}\right) = 47.2°.$$

(*ii*) Attenuation constant

$$\alpha = 2\cosh^{-1}\left(\frac{f_C}{f}\right) \text{ Nepers}$$

$$= 2\cosh^{-1}\left(\frac{10}{2}\right) = 2.6 \text{ Nepers.}$$

13.8.5 Band-pass Filter

There are two common methods of obtaining the band-pass characteristics:

(*i*) Conventional LC filter.
(*ii*) Use of two coupled parallel-tuned circuits.

The first type of band-pass filter is considered here. It may in its simplest form be obtained by connecting a high-pass filter and low-pass filter in cascade or tandem. The low-pass filter has a cut-off frequency higher than that of the high-pass filter. The overlapping region of the pass bands of the two constituent filters constitutes the pass band of the band-pass filter. Such a combination filter is not economical. Hence, in practice, low-pass filter action and high-pass filter actions are combined into a single filter action. Figure 13.16 shows a simple band-pass filter.

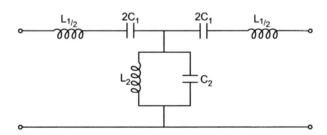

FIGURE 13.16   Band-pass filter.

It is a conventional T-section filter in which the series impedance Z_1 is a series resonant circuit (L_1 and C_1 in series) while the shunt impedance Z_2 is a parallel-tuned circuit (L_2 and C_2 in parallel). components $L_1, C_1, L_2,$ and C_2 are so chosen that the series resonance frequency of the series arm equals the parallel resonance frequency of the shunt arms.

For the series arm,

$$\omega_C{}^2 = \frac{1}{L_1 C_1} \tag{13.74}$$

and for the shunt arm,

$$\omega_C{}^2 = \frac{1}{L_2 C_2}. \tag{13.75}$$

Therefore,

$$L_1 C_1 = L_2 C_2 = \frac{1}{\omega_C{}^2}. \tag{13.76}$$

Nominal characteristic resistance R_0:

$$Z_1 = j\omega L_1 + \frac{1}{j\omega C_1} = j\left(\frac{\omega^2 L_1 C_1 - 1}{\omega C_1}\right) \tag{13.77}$$

$$Z_2 = \frac{j\omega L_2\left(\frac{1}{j\omega C_2}\right)}{j\omega L_2 + \frac{1}{j\omega C_2}} = j\frac{\omega L_2}{\left(1 - \omega^2 L_2 C_2\right)}. \tag{13.78}$$

Hence,

$$Z_1 Z_2 = \frac{L_2(\omega^2 L_1 C_1 - 1)}{C_1(\omega^2 L_2 C_2 - 1)}. \tag{13.79}$$

But by Equation (13.76),

$$L_1 C_1 = L_2 C_2.$$

Hence, Equation (13.79) reduces to

$$Z_1 Z_2 = \frac{L_2}{C_1} = \frac{L_1}{C_2} = R_0^2, \tag{13.80}$$

where R_0 is the nominal characteristic resistance (design resistance) of the band-pass filter.

Cut-off frequencies: Since $Z_1 Z_2$ is independent of frequency, the filter is a constant-k filter. The pass band of the filter is accordingly given by

$$-1 < \frac{Z_1}{4Z_2} < 0.$$

The cut-off frequency is given by

$$-1 = \frac{Z_1}{4Z_2} \quad \text{or} \quad Z_1 = -4Z_2$$

or

$$Z_1^2 = -4Z_1Z_2 = -4R_0^2. \tag{13.81}$$

Hence,

$$Z_1 = \pm j2R_0. \tag{13.82}$$

Equation (13.82) has two cut-off frequencies. At cut-off frequency $f_1, Z_1 = -j2R_0$, at cut-off frequency f_2,

$$Z_1 = j2R_0.$$

Accordingly the impedance Z_1 of the series arm at frequency f_1 is negative of the impedance Z_1 at frequency f_2.

Thus,

$$\left(\omega_1 L_1 - \frac{1}{\omega_1 C_1}\right) = -\left(\omega_2 L_1 - \frac{1}{\omega_2 C_1}\right)$$

or

$$1 - \omega_1^2 L_1 C_1 = \frac{\omega_1}{\omega_2}(\omega_2^2 L_1 C_1 - 1). \tag{13.83}$$

But

$$L_1 C_1 = \frac{1}{\omega C^2}.$$

Hence, Equation (13.83) reduces,

$$1 - \left(\frac{f_1}{f_C}\right)^2 = \frac{f_1}{f_2}\left[\left(\frac{f_2}{f_C}\right)^2 - 1\right]$$

$$f_C^2 - f_1^2 = \frac{f_1}{f_2}(f_2^2 - f_C^2)$$

or

$$f_C^2(f_1 + f_2) = f_1 f_2(f_1 + f_2)$$

$$f_C^2 = f_1 f_2 \Rightarrow f_C = \sqrt{f_1 f_2}. \tag{13.84}$$

Thus, the f_C resonant frequency of the shunt arm or series arm is the geometric mean of the cut-off frequencies.

Figure 13.17 shows the nature of variation of characteristics impedance Z_0 with frequency for a constant-k band-pass filter.

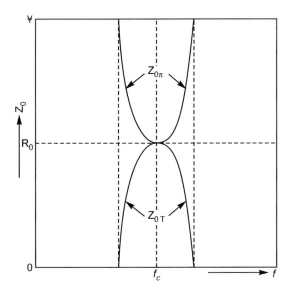

FIGURE 13.17 Variation of characteristics impedance in a constant-*k* band-pass filter.

13.8.6 Design of Constant-k Band-pass Filter

The design of a band-pass filter consists of finding the values of its circuit elements for the desired values of the nominal characteristic impedance R_0 and cut-off frequencies f_1 and f_2.

At a lower cut-off frequency $f_1, Z_1 = -2jR_0$

or

$$\frac{1}{\omega_1 C_1} - \omega L_1 = 2R_0$$

or

$$1 - \frac{f_1^2}{f_C^2} = 4\pi f_1 R_0 C_1$$

or

$$1 - \frac{f_1^2}{f_1 f_2} = 4\pi R_0 f_1 C_1$$

or

$$f_2 - f_1 = 4\pi R_0 f_1 f_2 C_1.$$

Hence,

$$C_1 = \frac{f_2 - f_1}{4\pi\, R_0 f_1 f_2} = \frac{1}{4\pi\, R_0}\left[\frac{1}{f_1} - \frac{1}{f_2}\right]. \tag{13.85}$$

By Equation (13.76),

$$L_1 = \frac{1}{\omega_C^2 C_1} = \frac{1}{4\pi^2 f_C^2 C_1} = \frac{1}{4\pi^2 f_1 f_2 C_1}. \tag{13.86}$$

Substituting the value of C_1 from Equation (13.85), we get

$$L_1 = \frac{R_0}{\pi\,(f_2 - f_1)}. \tag{13.87}$$

From Equation (13.80),

$$L_2 = C_1 R_0^2.$$

Substituting the value of C_1 from Equation (13.85), we get

$$L_2 = \frac{R_0^2 (f_2 - f_1)}{4\pi\, R_0 f_1 f_2} = \frac{R_0}{4\pi}\left[\frac{1}{f_1} - \frac{1}{f_2}\right] \tag{13.88}$$

and

$$C_2 = \frac{L_1}{R_0^2} = \frac{1}{\pi\, R_0 (f_2 - f_1)}. \tag{13.89}$$

Example 13.5. *Design a constant-k (prototype) band-pass filter section having cut-off frequencies of 2 KHz and 5 KHz and a nominal characteristic impedance of* 1KΩ.

Solution:

$$L_1 = \frac{R_0}{\pi\,(f_2 - f_1)} = \frac{1000}{3.14 \times 3000} = 106 \text{ mH}$$

$$C_1 = \frac{(f_2 - f_1)}{4\pi\, R_0 f_1 f_2} = \frac{3000}{4\pi \times 1000 \times 2000 \times 5000} = 0.0239\,\mu\text{F}$$

$$C_2 = \frac{L_1}{R_0^2} = \frac{1}{\pi\, R_0 (f_2 - f_1)} = \frac{106 \times 10^{-3}}{10^6} = 0.106\,\mu\text{F}$$

$$L_2 = \frac{R_0 (f_2 - f_1)}{4\pi f_1 f_2} = \frac{1000(5000 - 2000)}{4\pi \times 5000 \times 2000} = 0.0238 \text{ H} = 23.8 \text{ mH}.$$

Hence, the band-pass filter is as shown in Figure 13.18.

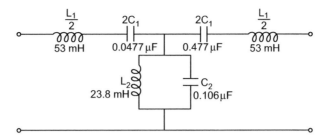

FIGURE 13.18   Band-pass filter.

13.8.7   Band-stop Filter or Band-elimination Filter

In certain devices it is necessary to eliminate or disallow a band of frequencies, say from f_1 and f_2, and to allow transmission of frequencies on the lower and higher sides as shown in Figure 13.5.

A band-stop filter may in its simplest form be obtained by connecting in parallel a low-pass filter with high-pass filter such that the cut-off frequency f_1 (or f_H) of the high-pass filter is higher than the cut-off frequency f_2 (or f_L) of the low-pass filter. The overlapping attenuation bands of the two filters then constitutes the stop band of the band-stop filter. Such a combination is uneconomical. Hence, in practice, the low-pass filter action and the high-pass filter action are combined into a single filter section. Figure 13.19 shows a constant-k band-stop filter.

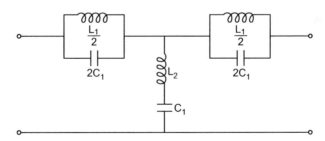

FIGURE 13.19   Band-stop filter.

It is a conventional T-section filter in which the series arm impedance Z_1 is a parallel-tuned circuit (L_1 and C_1 in parallel) while the shunt arm impedance Z_2 is a series-tuned circuit (L_2 and C_2 in series). The components $L_1, C_1, L_2,$ and C_2 are so chosen that the shunt resonance frequency of the series arm equals the series resonance frequency of the shunt arm.

Then for the series arm,

$$\omega_C{}^2 = \frac{1}{L_1 C_1} \qquad (13.90)$$

and for the shunt arm,

$$\omega_C{}^2 = \frac{1}{L_2 C_2}. \qquad (13.91)$$

Hence,

$$L_1 C_1 = L_2 C_2 = \frac{1}{\omega_C{}^2}. \qquad (13.92)$$

Nominal characteristic impedance R_0:
By Figure 13.19,

$$Z_1 = 2 \frac{\frac{j\omega L_1}{2}\left(\frac{1}{j\omega^2 C_1}\right)}{\frac{j\omega L_1}{2} + \frac{1}{j\omega^2 C_1}} = j\frac{\omega L_1}{1 - \omega^2 L_1 C_1} \qquad (13.93)$$

$$Z_2 = j\omega L_2 + \frac{1}{j\omega C_2} = j\left(\frac{\omega^2 L_2 C_2 - 1}{\omega C_2}\right). \qquad (13.94)$$

Hence,

$$Z_1 Z_2 = \frac{L_1}{C_2}\left(\frac{\omega^2 L_2 C_2 - 1}{\omega^2 L_1 C_1 - 1}\right).$$

But

$$L_1 C_1 = L_2 C_2.$$

Therefore, we get

$$Z_1 Z_2 = \frac{L_1}{C_2} = \frac{L_2}{C_1} = R_0^2, \qquad (13.95)$$

where R_0 is the design resistance or the nominal characteristic impedance of the band-stop filter.

Cut-off frequencies:

Since $Z_1 Z_2$ is independent of the frequency, the filter is a constant-k filter. The pass band of the filter is accordingly given by

$$-1 < \frac{Z_1}{4Z_2} < 0.$$

The cut-off frequency is given by

$$\frac{Z_1}{4Z_2} = -1$$

or

$$Z_1 = -4Z_2$$

or

$$Z_1^2 = -4Z_1Z_2 = -4R_0^2.$$

Hence,

$$Z_1 = \pm 2jR_0. \tag{13.96}$$

Equation (13.96) defines the two cut-off frequencies.

A cut-off frequency f_1 when Z_1 equals $+j2R_0$ and another cut-off frequency f_2 when Z_1 equals $-j2R_0$. Accordingly, the impedance Z_1 of the series arm as frequency f_1 is negative of the impedance Z_1 at frequency f_2.

$$\frac{\omega_1 L_1}{1 - \omega_1^2 L_1 C_1} = -\left(\frac{\omega_2 L_1}{1 - \omega_2^2 L_1 C_1}\right)$$

or

$$1 - \omega_1^2 L_1 C_1 = \frac{\omega_1}{\omega_2}(\omega_2^2 L_1 C_1 - 1). \tag{13.97}$$

But

$$L_1 C_1 = \frac{1}{\omega_C^2}.$$

Therefore,

$$1 - \frac{\omega_1^2}{\omega_C^2} = \frac{\omega_1}{\omega_2}\left[\left(\frac{\omega_2}{\omega_C}\right)^2 - 1\right]$$

or

$$1 - \left(\frac{f_1}{f_C}\right)^2 = \frac{f_1}{f_2}\left[\left(\frac{f_2}{f_C}\right)^2 - 1\right]$$

$$f_C^2 - f_1^2 = f_1 f_2 - \frac{f_1}{f_2}f_C^2$$

or

$$f_C = \sqrt{f_1 f_2}. \tag{13.98}$$

13.8.8 Design of Constant-k Band-stop Filters

For a given value of design resistance R_0 and for the desired cut-off frequencies f_1 and f_2, it is possible to determine the values of the circuit components of the constant-k band-stop filter.

Thus, at a lower cut-off frequency f_1,

$$Z_1 = j2R_0$$

or

$$j\frac{\omega_1 L_1}{1 - \omega_1^2 L_1 C_1} = j2R_0$$

or

$$1 - \omega_1^2 L_1 C_1 = \frac{\omega_1 L_1}{2R_0}$$

or

$$1 - \frac{f_1^2}{f_C^2} = \frac{\pi f_1 L_1}{R_0}$$

or

$$1 - \frac{f_1^2}{f_1 f_2} = \frac{\pi f_1 L_1}{R_0}$$

or

$$f_2 - f_1 = \frac{\pi f_1 f_2 L_1}{R_0}$$

$$L_1 = \frac{R_0(f_2 - f_1)}{\pi f_1 f_2}. \tag{13.99}$$

By Equations (13.95) and (13.92),

$$C_2 = \frac{L_1}{R_0^2} = \frac{(f_2 - f_1)}{R_0 \pi f_1 f_2}$$

$$C_1 = \frac{1}{\omega C^2 L_1} = \frac{1}{4\pi^2 f_C^2} \times \frac{\pi f_1 f_2}{R_0(f_2 - f_1)} \qquad [f_C^2 = f_1 f_2]$$

$$C_1 = \frac{1}{4\pi R_0(f_2 - f_1)} \tag{13.100}$$

and

$$L_2 = C_1 R_0^2 = \frac{R_0}{4\pi (f_2 - f_1)}. \tag{13.101}$$

Example 13.6. *Determine the value of the parameters for the stop-band filter, having cut-off frequencies of 2 KHz and 5 KHz and a nominal characteristic impedance of 1 K Ω.*

Solution:

$$f_1 = 2 \text{ KHz} = 2000 \text{ Hz}$$

$$f_2 = 5 \text{ KHz} = 5000 \text{ Hz}$$

$$R_0 = 1 \text{ K}\Omega = 1000 \text{ }\Omega$$

$$L_1 = \frac{R_0(f_2 - f_1)}{\pi f_1 f_2} = \frac{1000 \times 3000}{3.14 \times 2000 \times 5000} = 95.5 \text{ mH}$$

$$C_2 = \frac{f_2 - f_1}{R_0 \pi f_1 f_2} = \frac{L_1}{R_0^2} = \frac{95.5 \times 10^{-3}}{10^6} = 95.5 \text{ nF}$$

$$C_1 = \frac{1}{4\pi R_0(f_2 - f_1)} = \frac{1}{4\pi \times 10^3 \times 3 \times 10^3} = 26.5 \text{ nF}$$

$$L_2 = C_1 R_0^2 = 26.5 \times 10^{-9} \times 10^6 = 26.5 \text{ mH}.$$

The design of the band-stop filter is shown in Figure 13.20.

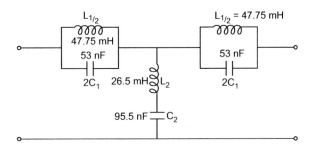

FIGURE 13.20   Band-stop filter.

13.9   m-DERIVED FILTERS

We studied constant-k filters in the last section. Constant-k filters have two main drawbacks:

(*i*) In the attenuation band, the attenuation does not increase rapidly with frequency beyond the cut-off frequency and

(*ii*) In the pass band, the characteristic impedance Z does not remain constant but varies widely from the nominal value R_0.

Thus, by connecting three identical constant-*k* low-pass filter sections, attenuation in the attenuation band beyond the cut-off frequency gets tripled as shown in Figure 13.21. Therefore, use of several identical constant-*k* sections improves the attenuation characteristic in the attenuation band.

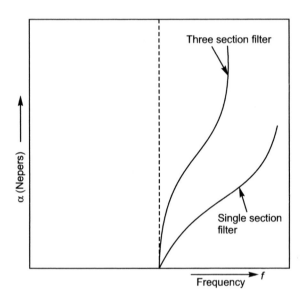

FIGURE 13.21   Attenuation characteristic for one and three section constant-*k* low-pass filters.

13.9.1 *m*-derived T-section

Figure 13.22 shows a general constant-*k* section of a T-type and *m*-derived T-section.

The series arm of the *m*-derived section has been obtained by multiplying the series arm element of the constant-*k* section by a constant *m*. The shunt arm element Z_2 is also modified in such a way that the modifying filter has the same characteristic impedance.

For the constant-*k* T-section, the characteristic impedance is given by

$$Z_{0T} = \sqrt{\frac{Z_1^2}{4} + Z_1 Z_2}.$$

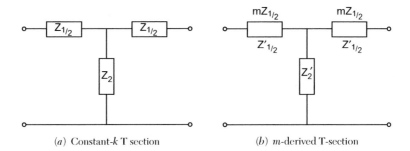

(a) Constant-k T section   (b) m-derived T-section

FIGURE 13.22   T-section filter.

For the m-derived T-section, the characteristic impedance is given by

$$Z'_{0T} = \sqrt{\frac{m^2 Z_1^2}{4} + m Z_1 Z'_2}$$ (13.102)

but for the same characteristic Z_{0T} and Z'_{0T} should be the same.
Hence,

$$\frac{Z_1^2}{4} + Z_1 Z_2 = \frac{m^2 Z_1^2}{4} + m Z_1 Z'_2$$

or

$$Z'_2 = \frac{Z_2}{m} + \left(\frac{1-m^2}{4m}\right) Z_1.$$ (13.103)

It shows that the shunt arm Z'_2 consists of two impedances in series as shown in Figure 13.23. m must have a value between 0 to 1 (*i.e.*, $0 < m < 1$) if m exceeds unity. The factor $(\frac{1-m^2}{4m})$ in the expression of Z'_2 will be negative.

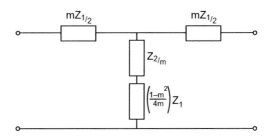

FIGURE 13.23 m-derived T-network.

13.9.2 *m*-derived π-section Filter

Figure 13.24 shows a general constant-*k* section of an π-type and an *m*-derived π-section.

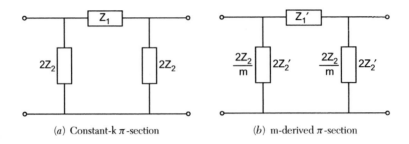

(a) Constant-k π-section   (b) m-derived π-section

FIGURE 13.24   *m*-derived π-section and constant-k π-section.

The shunt elements of the *m*-derived section are obtained by multiplying the shunt arm elements of the constant-*k* section by the factor $\frac{1}{m}$. The section, the arm element Z'_1 of the *m*-derived π-section, must also be modified such that the modified section has the same characteristic impedance.

For the constant-*k* π section, the characteristic impedance is given by

$$Z_0\pi = \sqrt{\frac{Z_1 Z_2}{1 + \frac{Z_1}{4Z_2}}}.$$

For the *m*-derived π-section, the characteristic impedance is given by

$$Z'_{o\pi} = \sqrt{\frac{Z'_1 \times \frac{Z_2}{m}}{1 + \frac{Z'_1}{\frac{4Z_2}{m}}}}. \tag{13.104}$$

For the same response, $Z_{0\pi} = Z'_{0\pi}$

$$\frac{Z_1 Z_2}{1 + \frac{Z_1}{4Z_2}} = \frac{Z'_1 \frac{Z_2}{m}}{1 + \frac{Z'_1}{\frac{4Z_2}{m}}}$$

or

$$Z_{1m}\left(1 + \frac{mZ'_1}{4Z_2}\right) = Z'_1\left(1 + \frac{Z_1}{4Z_2}\right)$$

$$Z'_1 = \frac{Z_1 Z_2 \cdot 4m}{Z_1(1 - m^2) + 4Z_2} \tag{13.105}$$

or

$$Z_1' = \frac{mZ_1 \cdot \left(\frac{4m}{1-m^2}\right)Z_2}{mZ_1 + \left(\frac{4m}{1-m^2}\right)Z_2}.$$ (13.106)

It shows that the series arm of the m-derived π-network is a parallel combination of mZ_1 and $\left(\frac{4m}{1-m^2}\right)Z_2$. The m-derived π-network is shown in Figure 13.25.

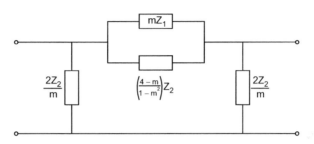

FIGURE 13.25 m-derived π-network.

13.9.3 m-derived Low-pass Filters

Figure 13.26 shows the m-derived low-pass T-section filter. This is obtained from that shown in Figure 13.23 by putting $Z_1 = j\omega L$ and $Z_2 = \frac{1}{j\omega C}$.

Therefore, each series arm consists of an inductance of value $\frac{mL}{2}$ and the shunt arm consists of a series combination of a capacitance mC and an inductance $\left(\frac{1-m^2}{4m}\right)L$.

It may be readily seen that the shunt arm is a series LC circuit which offers almost zero impedance at the frequency of its series resonance. At this resonant frequency, the transmission through the filter is zero (*i.e.*, the attenuation is infinite). Let this frequency of infinite attenuation be denoted f_∞.

Hence,

$$\frac{1}{\omega_\infty mC} = \left(\frac{1-m^2}{4m}\right)\omega_\infty L$$

or

$$\omega_\infty = \frac{2}{\sqrt{1-m^2}\sqrt{LC}}$$ (13.107)

or

$$f_\infty = \frac{1}{\pi\sqrt{LC}\sqrt{1-m^2}}.$$ (13.108)

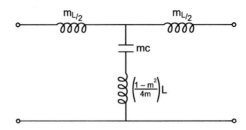

FIGURE 13.26 *m-derived low-pass T-section filter.*

Since the cut-off frequency for the constant-k low-pass filter is

$$f_C = \frac{1}{\pi\sqrt{LC}} \text{ or } \omega_C = \frac{2}{\sqrt{LC}},$$

Equation (13.107) reduces to

$$\omega_\infty = \frac{\omega_C}{\sqrt{1-m^2}}$$

or

$$m = \sqrt{1 - \left(\frac{f_C}{f_\infty}\right)^2}. \tag{13.109}$$

Since $f_\infty > f_C$, the range of m will be $0 < m < 1$.

Knowing f_C, R_0, and m, the values of the components of the m-derived low-pass filter may be calculated.

13.9.4 *m-derived π-network Low-pass Filter*

Figure 13.27 shows the m-derived low-pass π-section filter.

Again, for the low-pass filter $Z_1 = j\omega L$ and $Z_2 = \frac{1}{j\omega C}$.

Therefore, the series arm consists of a parallel combination of an inductance of value mL and a capacitance of value $(\frac{1-m^2}{4m})C$ and each shunt arm consists of a capacitance of value $\frac{mC}{2}$.

Now in this case the series arm is chosen so that it is resonant at the frequency f_∞ above cut-off frequency f_C.

Thus, at resonant frequency,

$$|mZ_1| = \left|\left(\frac{4m}{1-m^2}\right)Z_2\right| \tag{13.110}$$

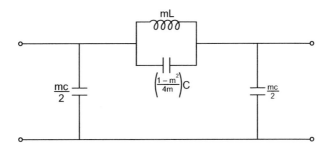

FIGURE 13.27 *m*-derived low-pass π-section filter.

or

$$\omega_\infty m\text{L} = \frac{1}{\omega_\infty \text{C}\left(\frac{1-m^2}{4m}\right)}$$

or

$$\omega_\infty = \frac{2}{\sqrt{\text{LC}}} \cdot \frac{1}{\sqrt{1-m^2}} \qquad (13.111)$$

or

$$f_\infty = \frac{1}{\pi\sqrt{\text{LC}}} \cdot \frac{1}{\sqrt{1-m^2}}. \qquad (13.112)$$

Since the cut-off frequencies for the constant-k low-pass filter are

$$f_C = \frac{1}{\pi\sqrt{\text{LC}}} \quad \text{or} \quad \omega_C = \frac{2}{\sqrt{\text{LC}}},$$

Equation (13.111) reduces to

$$\omega_\infty = \frac{\omega_C}{\sqrt{1-m^2}} \quad \text{or} \quad m = \sqrt{1-\left(\frac{f_C}{f_\infty}\right)^2}. \qquad (13.113)$$

Thus, in both T- and π-section m-derived filters, we get the same equation form.

Example 13.7. *Design an m-derived T- and π-networks low-pass filter with nominal characteristic impedance $R_0 = 900\,\Omega$, cut-off frequency $f_C = 0.85$ KHz, and infinite attenuation frequency $f_\infty = 1$ KHz.*

Solution: Given $R_0 = 900\,\Omega$, $f_C = 0.85\,\mathrm{KHz}$, $f_\infty = 1\,\mathrm{KHz}$, for the low-pass filter,

$$m = \sqrt{1 - \left(\frac{f_C}{f_\infty}\right)^2} = \sqrt{1 - (0.85)^2} = 0.526$$

$$L = \frac{R_0}{\pi f_C} = \frac{900}{\pi \times 0.85 \times 10^3} = 337\,\mathrm{mH}$$

$$C = \frac{1}{R_0 \pi f_C} = \frac{1}{900 \times \pi \times 0.85 \times 10^3} = 0.416\,\mu\mathrm{F}.$$

Now the value of the components for the T- and π-networks of the m-derived low-pass filter are

$$mC = 0.526 \times 0.416 = 0.219\,\mu\mathrm{F}$$

$$\frac{mC}{2} = 0.1094\,\mu\mathrm{F}$$

$$\left(\frac{1 - m^2}{4m}\right)C = 0.143\,\mu\mathrm{F}$$

$$mL = 0.526 \times 337 = 177.3\,\mathrm{mH}$$

$$\frac{mL}{2} = 88.63\,\mathrm{mH}$$

$$\left(\frac{1 - m^2}{4m}\right)L = 115.85\,\mathrm{mH}.$$

The m-derived T- and π-networks low-pass filter is as shown in Figure 13.28.

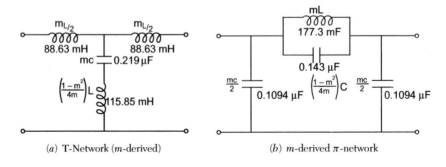

(a) T-Network (m-derived)   (b) m-derived π-network

FIGURE 13.28   T and π m-derived low-pass filter.

13.9.5 *m*-derived High-pass Filter (T-network)

Figure 13.29 shows a constant-*k* and *m*-derived high-pass filter.

$$Z_1 = \frac{1}{j\omega C} \quad \text{and} \quad Z_2 = j\omega L$$

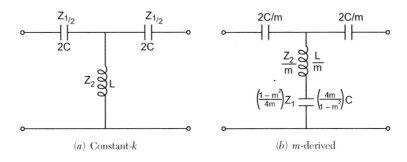

(a) Constant-*k*   (b) *m*-derived

FIGURE 13.29   High-pass constant-*k* and *m*-derived T-section filter.

Each series arm consists of a capacitance of value $\frac{2C}{m}$ and the shunt arm consists of a series combination of an inductance of value $\frac{L}{m}$ and a capacitance of value $(\frac{4m}{1-m^2})C$.

For the high-pass filter, the shunt arm is chosen so that it is resonant at resonant frequency f_∞,

$$\frac{\omega_\infty L}{m} = \frac{1}{\left(\frac{4m}{1-m^2}\right)\omega_\infty C}$$

or

$$\omega_\infty \equiv \frac{\sqrt{1-m^2}}{2\sqrt{LC}} \tag{13.114}$$

or

$$f_\infty = \frac{\sqrt{1-m^2}}{4\pi\sqrt{LC}}. \tag{13.115}$$

The cut-off frequency for the constant-*k* high-pass filter is

$$f_C = \frac{1}{4\pi\sqrt{LC}} \quad \text{or} \quad \omega_C = \frac{1}{2\sqrt{LC}}.$$

Now Equation (13.114) reduces to

$$\omega_\infty = \omega_C\sqrt{1-m^2} \text{ or } f_\infty = f_C\sqrt{1-m^2}$$

$$m = \sqrt{1-\left(\frac{f_\infty}{f_C}\right)^2}. \tag{13.116}$$

Since $f_\infty < f_C$, the value of m lies between 0 and 1 (i.e., $0 < m < 1$).

13.9.6 *m*-derived High-pass Filter (π-network)

Figure 13.30 shows a constant-k and m-derived π-section filter.

$$Z_1 = \frac{1}{j\omega C} \quad \text{and} \quad Z_2 = j\omega L$$

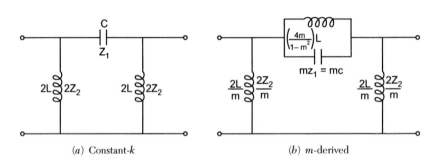

(a) Constant-k (b) *m*-derived

FIGURE 13.30   Constant-k and *m*-derived π-section high-pass filter.

Therefore, the series arm consists of a parallel combination of a capacitance of value $\frac{C}{m}$ and an inductance of value $\left(\frac{4m}{1-m^2}\right)L$. Each shunt arm consists of an inductance of value $\frac{2L}{m}$.

At resonant frequency f_∞, the reactance of capacitance $\frac{C}{m}$ equals the reactance of inductance $\left(\frac{4m}{1-m^2}\right)L$.

Thus,

$$\omega_\infty\left(\frac{4m}{1-m^2}\right)L = \frac{1}{\omega_\infty\frac{C}{m}}$$

or

$$\omega_\infty = \frac{\sqrt{1-m^2}}{2\sqrt{LC}} \tag{13.117}$$

or

$$f_\infty = \frac{\sqrt{1 - m^2}}{4\pi \sqrt{LC}}. \tag{13.118}$$

Since the cut-off frequency for the constant-k high-pass filter is

$$f_C = \frac{1}{4\pi \sqrt{LC}} \quad \text{or} \quad \omega_C = \frac{1}{2\sqrt{LC}},$$

Equation (13.117) will reduce to

$$\omega_\infty = \omega_C \sqrt{1 - m^2} \quad \text{or} \quad f_\infty = f_C \sqrt{1 - m^2} \tag{13.119}$$

or

$$m = \sqrt{1 - \left(\frac{f_\infty}{f_C}\right)^2}. \tag{13.120}$$

The value of f_∞ and m for the π-section is the same as for the T-section in the last section.

Example 13.8. *Design the T- and π-networks of an m-derived high-pass filter having nominal characteristic impedance $R_0 = 1$ KΩ, cut-off frequency $f_C = 2$ KHz, and infinite attenuation (or resonant) frequency f_∞ is 1.8 KHz.*

Solution: Given $R_0 = 1$ K$\Omega = 1000\,\Omega, f_C = 2$ KHz, $f_\infty = 1.8$ KHz, for the high-pass filter, we have

$$m = \sqrt{1 - \left(\frac{f_\infty}{f_C}\right)^2} = \sqrt{1 - (0.9)^2} = 0.436$$

and

$$C = \frac{1}{4\pi R_0 f_C} = \frac{1}{4\pi \times 10^3 \times 2 \times 10^3} = 0.0398\,\mu\text{F}$$

$$L = \frac{R_0}{4\pi f_C} = \frac{10^3}{4\pi \times 2 \times 10^3} = 0.0398\,\text{H} = 39.8\,\text{mH}.$$

Now the components values for the T- and π-section m-derived high-pass filter are

$$\frac{C}{m} = \frac{0.0398}{0.436}\,\mu\mathrm{F} = 0.0912\,\mu\mathrm{F}$$

$$\frac{2C}{m} = 0.1825\,\mu\mathrm{F}$$

$$\left(\frac{4m}{1-m^2}\right)C = 0.0857\,\mu\mathrm{F}$$

$$\frac{L}{m} = 91.28\ \mathrm{mH}$$

$$\frac{2L}{m} = 182.56\ \mathrm{mH}$$

$$\left(\frac{4m}{1-m^2}\right)L = \left[\frac{4\times 0.436}{1-(0.436)^2}\right] \times 39.8\ \mathrm{mH} = 85.7\ \mathrm{mH}.$$

The m-derived T- and π-networks high-pass filters are shown in Figures 13.31(a) and (b), respectively.

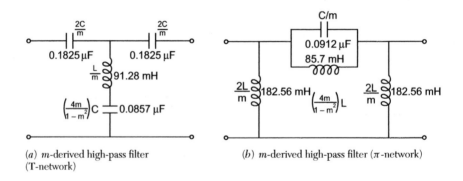

(a) m-derived high-pass filter
(T-network)

(b) m-derived high-pass filter (π-network)

FIGURE 13.31 m-derived T and π, high-pass filter.

13.10   m-DERIVED BAND-PASS FILTER

m-derived T-band-pass filter may be obtained by the usual procedure; namely:

(i) treating the entire series arm impedance as Z_1 and the shunt arm impedance as Z_2.

(*ii*) substituting these values of Z_1 and Z_2 in the *m*-derived T-section of Figure 13.23. The results are in the *m*-derived T-section band-pass filter of Figure 13.32.

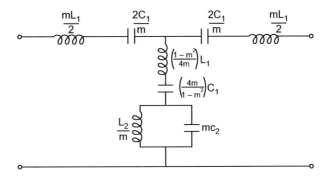

FIGURE 13.32   *m*-derived T-section band-pass filter.

Infinite attenuation in the circuit will result when the shunt arm impedance is zero at f_∞ or we can say, when the sum of impedance of the series combination of $(\frac{1-m^2}{4m})L_1$ and $(\frac{4m}{1-m^2})C_1$ and the impedance of the parallel combination of $\frac{L_2}{m}$ and mC_2 becomes zero. Therefore,

$$\left[j\omega_\infty \left(\frac{1-m^2}{4m} \right) L_1 + \frac{(1-m^2)}{j\omega_\infty 4mC_1} \right] + \frac{\frac{j\omega_\infty L_2}{m} \cdot \frac{1}{j\omega_\infty mC_2}}{\frac{j\omega_\infty L_2}{m} + \frac{1}{j\omega_\infty mC_2}} = 0$$

or

$$\frac{-\omega_\infty^2(1-m^2)L_1C_1 + (1-m^2)}{j\omega_\infty 4mC_1} + \frac{j\omega_\infty L_2}{-\omega_\infty^2 mL_2C_2 + m} = 0$$

or

$$[-\omega_\infty^2(1-m^2)L_1C_1 + (1-m^2)] \times [\omega_\infty^2 L_2C_2 - 1] = -\omega_\infty^2 \cdot 4L_2C_1$$

or

$$\left(\frac{1-m^2}{4} \right)(\omega_\infty^2 L_1C_1 - 1)(\omega_\infty^2 L_2C_2 - 1) = \omega_\infty^2 L_2C_1.$$

But for the band-pass filter,

$$L_1C_1 = L_2C_2 = \frac{1}{\omega_C^2}$$

$$\left(\frac{1-m^2}{4}\right)\left[\left(\frac{\omega_\infty}{\omega_C}\right)^2 - 1\right]\left[\left(\frac{\omega_\infty}{\omega_C}\right)^2 - 1\right] = 4\pi^2 f_\infty^2 L_2 C_1$$

or

$$\left(\frac{1-m^2}{4}\right)\left[\left(\frac{\omega_\infty}{\omega_C}\right)^2 - 1\right]^2 = 4\pi^2 f_\infty^2 L_2 C_1 \tag{13.121}$$

or

$$\left(\frac{1-m^2}{4}\right)\left[\left(\frac{f_\infty}{f_C}\right)^2 - 1\right]^2 = 4\pi^2 f_\infty^2 L_2 C_1. \tag{13.122}$$

For the band-pass filter by Equation (13.84),

$$f_C = \sqrt{f_1 f_2},\ L_2 = \left(\frac{f_2 - f_1}{4\pi f_1 f_2}\right) R_0 \text{ and } C_1 = \frac{(f_2 - f_1)}{4\pi R_0 f_1 f_2}.$$

Putting these values into Equation (13.122), we get

$$\left(\frac{1-m^2}{4}\right)\left|\frac{f_\infty^2 - f_1 f_2}{f_1 f_2}\right| = 4\pi^2 f_\infty^2 \left[\frac{(f_2 - f_1)^2}{16\pi^2 f_1 f_2}\right]$$

or

$$(1 - m^2)(f_\infty^2 - f_1 f_2)^2 = f_\infty^2 (f_2 - f_1)^2$$

or

$$f_\infty^2 - \frac{f_2 - f_1}{\sqrt{1 - m^2}} f_\infty - f_1 f_2 = 0. \tag{13.123}$$

Equation (13.123) is quadratic in f_∞, therefore,

$$f_\infty = \frac{\frac{f_2 - f_1}{\sqrt{1 - m^2}} \pm \sqrt{\left(\frac{f_2 - f_1}{\sqrt{1 - m^2}}\right)^2 + 4 f_1 f_2}}{2}$$

or

$$f_\infty = \frac{f_2 - f_1}{2\sqrt{1 - m^2}} \pm \frac{1}{2}\sqrt{\left(\frac{f_2 - f_1}{\sqrt{1 - m^2}}\right)^2 + 4f_1 f_2}. \qquad (13.124)$$

By Equation (13.124), it is clear that f_∞ has two values, hence, two frequencies of peak attenuation are given by

$$f_{\infty 1} = -\frac{f_2 - f_1}{2\sqrt{1 - m^2}} + \sqrt{\frac{(f_2 - f_1)^2}{4(1 - m^2)} + f_1 f_2} \qquad (13.125)$$

and

$$f_{\infty 2} = \frac{f_2 - f_1}{2\sqrt{1 - m^2}} + \sqrt{\frac{(f_2 - f_1)^2}{4(1 - m^2)} + f_1 f_2}. \qquad (13.126)$$

By Equations (13.125) and (13.126),

$$f_{\infty 2} - f_{\infty 1} = \frac{f_2 - f_1}{\sqrt{1 - m^2}}$$

or

$$m = \sqrt{1 - \left(\frac{f_2 - f_1}{f_{\infty 2} - f_{\infty 1}}\right)^2}.$$

From Equation (13.124), it is evident that the term under the radical sign is larger than the term $\frac{f_2 - f_1}{2\sqrt{1 - m^2}}$ and Equation (13.124) will yield one value of f_∞ as a negative frequency having no physical significance. Hence, the $\pm$ sign is prefixed to the quantity $\frac{f_2 - f_1}{2\sqrt{1 - m^2}}$ rather than to the quantity under the radical sign. Multiplying Equations (13.125) and (13.126), we get

$$f_{\infty 1} \cdot f_{\infty 2} = \left[\frac{(f_2 - f_1)^2}{4(1 - m^2)} + f_1 f_2\right] - \frac{(f_2 - f_1)^2}{4(1 - m^2)}$$

$$f_1 f_2 = f_C^2$$

or

$$f_C = \sqrt{f_1 f_2} = \sqrt{f_{\infty 1} \cdot f_{\infty 2}}. \qquad (13.127)$$

SOLVED PROBLEMS

Problem 13.1. *Design a low-pass filter (both T- and π-networks) having a cut-off frequency of 2 KHz and a nominal characteristic impedance of 500 Ω.*

Solution: Given, $f_C = 2$ KHz, $R_0 = 500$ Ω,

$$L = \frac{R_0}{\pi f_C} = \frac{500}{\pi \times 2 \times 10^3} = 79.57 \text{ mH}$$

$$C = \frac{1}{\pi f_C R_0} = \frac{1}{\pi \times 2 \times 10^3 \times 500} = 0.318 \,\mu\text{F}.$$

The T- and π-networks for the low-pass filter are shown in Figure 13.33.

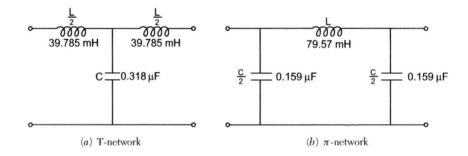

(a) T-network   (b) π-network

FIGURE 13.33   Constant-*k* low-pass filter.

Problem 13.2. *Design a constant-k high-pass filter (both T- and π-networks) having a cut-off frequency of 1 KHz and a nominal characteristic impedance of 500 Ω.*

Solution: Given, $f_C = 1$ KHz, $R_0 = 500$ Ω,
for the high-pass filter,

$$L = \frac{R_0}{4\pi f_C} = \frac{500}{4\pi \times 10^3} = 39.78 \text{ mH}$$

$$C = \frac{1}{4\pi f_C R_0} = \frac{1}{4\pi \times 10^3 \times 500} = 0.159 \,\mu\text{F}.$$

The constant-*k* T- and π-networks for the high-pass filter are shown in Figure 13.34.

Problem 13.3. *For the given T-section low-pass filter, determine the cut-off frequency and nominal characteristic impedance R_0.*

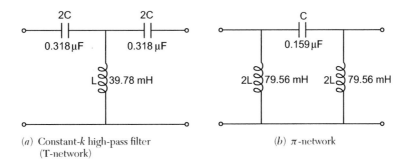

(a) Constant-k high-pass filter
    (T-network)

(b) π-network

FIGURE 13.34   **Constant-k T- and π-network for the high-pass filter.**

Solution: Total series inductance

$$L = 100 \text{ mH}$$
$$C = 0.02\,\mu\text{F}$$

Cut-off frequency

$$f_C = \frac{1}{\pi\sqrt{LC}}$$

$$= \frac{1}{\pi\sqrt{100 \times 10^{-3} \times 2 \times 10^{-8}}} = 0.711 \text{ KHz}$$

$$R_0 = \sqrt{\frac{L}{C}} = \sqrt{\frac{100 \times 10^{-3}}{2 \times 10^{-8}}} = 2.236 \text{ K}\Omega.$$

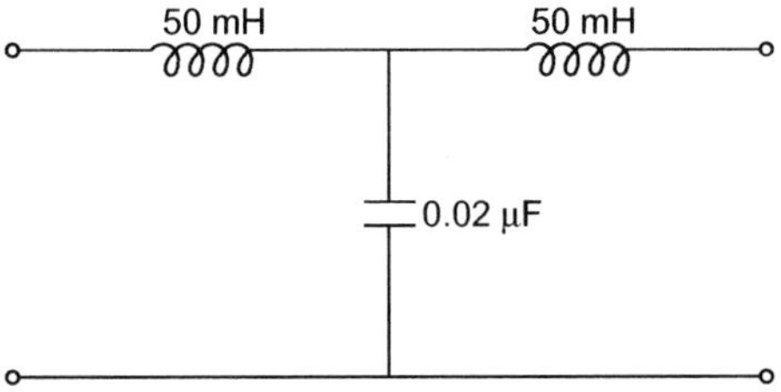

FIGURE 13.35

Problem 13.4. *For the given T-section high-pass filter, determine the cut-off frequency and nominal characteristic impedance* R_0.

Solution: Total series capacitance

$$C = 0.025\,\mu\text{F}$$
$$L = 80 \text{ mH}$$

$$f_C = \frac{1}{4\pi\sqrt{LC}} = \frac{1}{4\pi \times \sqrt{0.025 \times 10^{-6} \times 80 \times 10^{-3}}}$$

$$f_C = 1.779 \text{ KHz}$$

$$R_0 = \sqrt{\frac{L}{C}} = \sqrt{\frac{80 \times 10^{-3}}{0.05 \times 10^{-6}}} = 1.264 \text{ K}\Omega.$$

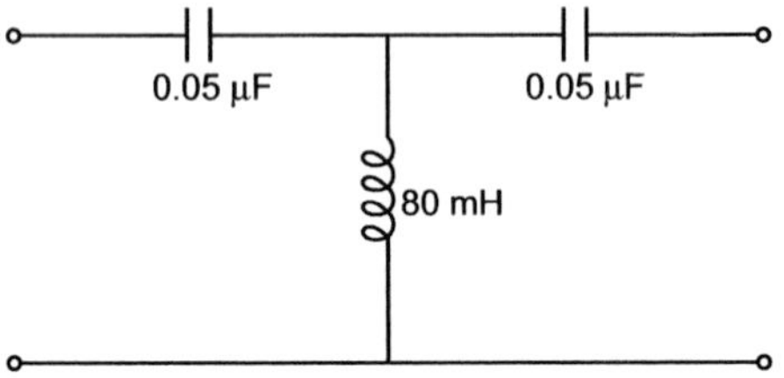

FIGURE 13.36

Problem 13.5. *Design m-derived T- and π-section low-pass filters for nominal characteristic impedance* $R_0 = 600 \ \Omega$. *Cut-off frequency* $= 1800$ Hz, *and infinite attenuation frequency* $f_\infty = 2$ KHz.

Solution: Given $R_0 = 600 \ \Omega, f_C = 1800$ Hz, $f_\infty = 2000$ Hz

$$m = \sqrt{1 - \left(\frac{f_C}{f_\infty}\right)^2} = \sqrt{1 - \left(\frac{1800}{2000}\right)^2} = 0.436.$$

Series arm inductance

$$L = \frac{R_0}{\pi f_C} = \frac{600}{\pi \times 1800} = 106.1 \text{ mH}$$

$$C = \frac{1}{R_0 \pi f_C} = \frac{1}{600 \times \pi \times 1800} = 0.2947 \ \mu\text{F}$$

$$\frac{mL}{2} = \frac{0.436 \times 106.1}{2} = 23.15 \text{ mH}$$

$$mC = 0.436 \times 0.2947 = 0.1285 \ \mu\text{F}$$

$$\left(\frac{1-m^2}{4m}\right)L = \frac{1 - (0.436)^2}{4 \times 0.436} \times 106.1 = 49.32 \text{ mH}$$

$$\frac{mC}{2} = \frac{0.1285}{2} = 0.0642 \ \mu\text{F}$$

$$mL = 23.15 \times 2 = 46.30 \text{ mH}$$

$$\left(\frac{1-m^2}{4m}\right)C = \frac{1 - (0.436)^2}{4 \times 0.436} \times 0.2947 = 0.1369 \ \mu\text{F}.$$

The m-derived T- and π-networks low-pass filter is as shown in Figure 13.37.

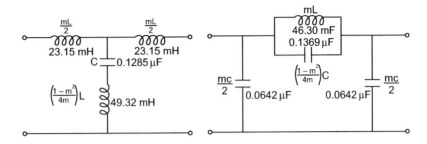

FIGURE 13.37

Problem 13.6. *Design the* T- *and* π-*sections of an m-derived high-pass filter having a nominal characteristic impedance of* 600 Ω, *cut-off frequency =* 3 *KHz, and an infinite attenuation at* 2.7 *KHz.*

Solution:

$$m = \sqrt{1 - \left(\frac{f_\infty}{f_C}\right)^2} = \sqrt{1 - \left(\frac{2.7}{3}\right)} = 0.436$$

Given

$$f_C = 3 \text{ KHz} \quad \text{and} \quad R_0 = 600 \ \Omega,$$

The series arm capacitance $C = \dfrac{1}{4\pi f_C R_0} = \dfrac{1}{4\pi \times 3\times 10^3 \times 600} = 0.0442 \ \mu\text{F},$
and the shunt arm inductance $L = \dfrac{R_0}{4\pi f_C} = \dfrac{600}{4\pi \times 3\times 10^3} = 15.9 \text{ mH}$

$$\frac{2C}{m} = 0.2027 \ \mu\text{F}$$

$$\frac{L}{m} = 36.5 \text{ mH}$$

$$\left(\frac{4m}{1-m^2}\right) C = 0.0951 \ \mu\text{F}$$

$$\frac{2L}{m} = 73 \text{ mH}$$

$$\frac{C}{m} = 0.1013 \ \mu\text{F}$$

$$\left(\frac{4m}{1-m^2}\right) L = 34.23 \text{ mH}.$$

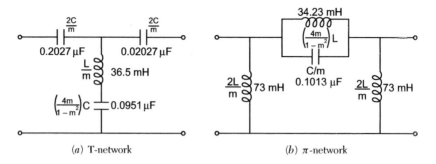

(a) T-network   (b) π-network

FIGURE 13.38

Problem 13.7. *Design a constant-k band-pass filter with cut-off frequencies of 2 KHz and 5 KHz and a nominal characteristic impedance of 600 Ω.*

Solution: Given $R_0 = 600\ \Omega$, $f_1 = 2$ KHz, $f_2 = 5$ KHz,

$$L_1 = \frac{R_0}{\pi(f_2 - f_1)} = \frac{600}{\pi \times 3000} = 63.66\ \text{mH}$$

$$C_1 = \frac{f_2 - f_1}{4\pi R_0 f_1 f_2} = \frac{3000}{4\pi \times 600 \times 2 \times 10^3 \times 5 \times 10^3} = 0.0397\ \mu\text{F}$$

$$C_2 = \frac{L_1}{R_0^2} = 0.1768\ \mu\text{F}$$

$$L_2 = \frac{R_0(f_2 - f_1)}{4\pi f_1 f_2} = \frac{600 \times 3 \times 10^3}{4\pi \times 10 \times 10^6} = 14.32\ \text{mH}.$$

Hence, the band-pass filter is as shown in Figure 13.39.

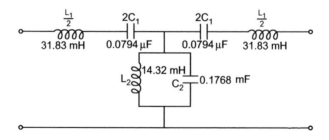

FIGURE 13.39

Problem 13.8. *Design a constant-k and stop filter with cut-off frequencies of 2 KHz and 5 KHz and nominal characteristic impedance of 600 Ω.*

Solution: Given $f_1 = 2$ KHz, $f_2 = 5$ KHz, $R_0 = 600\ \Omega$,

$$L_1 = \frac{R_0(f_2 - f_1)}{\pi f_1 f_2} = \frac{600 \times 3000}{\pi \times 2 \times 10^3 \times 5 \times 10^3} = 57.29\ \text{mH}$$

$$C_2 = \frac{(f_2 - f_1)}{R_0 \pi f_1 f_2} = \frac{L_1}{R_0^2} = 0.0159\ \mu\text{F}$$

$$C_1 = \frac{1}{4\pi R_0(f_2 - f_1)} = \frac{1}{4\pi \times 600 \times 3 \times 10^3} = 0.0442\ \mu\text{F}$$

$$L_2 = \frac{R_0}{4\pi (f_2 - f_1)} = C_1 R_0^2 = 81.43\ \text{mH}.$$

The band-stop filter is shown in Figure 13.40.

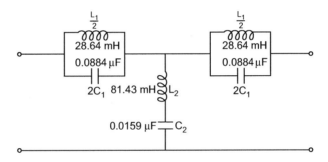

FIGURE 13.40

QUESTIONS FOR DISCUSSION

1. What is a low-pass filter ?
2. Define a high-pass filter ?
3. The value of a cut-off frequency depends on which factor ?
4. What is a band-pass filter ?
5. What is a band-stop filter ?
6. What is a filter ?
7. Can you design a filter (low- and high-pass) for cut-off frequency 50 Hz? If you can what is the value of the parameters ?
8. Give the formula for the cut-off frequency for a low- and high-pass filter.
9. What is nominal characteristic impedance ?
10. What is the relation between cut-off frequency and nominal characteristic for a low-pass filter (constant-k) ?

OBJECTIVE QUESTIONS

1. The value of the cut-off frequency for a low-pass filter is
   (a) $\dfrac{1}{2\pi\sqrt{LC}}$ (b) $\dfrac{1}{\pi\sqrt{LC}}$ (c) $\dfrac{1}{4\pi\sqrt{LC}}$ (d) $\dfrac{1}{3\pi\sqrt{LC}}$.

2. The value of the cut-off frequency for a high-pass filter is
   (a) $\dfrac{1}{2\pi\sqrt{LC}}$ (b) $\dfrac{1}{\pi\sqrt{LC}}$ (c) $\dfrac{1}{4\pi\sqrt{LC}}$ (d) $\dfrac{1}{3\pi\sqrt{LC}}$.

3. Two filters (low- and high-pass) with frequencies f_1 and f_2, respectively, are connected in series. The condition for the band-pass filter will be
   (a) $f_1 > f_2$ (b) $f_1 < f_2$ (c) $f_1 = f_2$ (d) $f_1 \leq f_2$.

4. Two filters (low- and high-pass) with frequencies f_1 and f_2, respectively, are connected in parallel. The condition for the band-stop filter will be
   (a) $f_1 > f_2$ (b) $f_1 < f_2$ (c) $f_1 = f_2$ (d) $f_1 \leq f_2$.

5. For a symmetric T-network having series impedances $\frac{Z_1}{2}$ and shunt impedance Z_2, the image impedance is
   (a) $\sqrt{Z_1 Z_2\left(1 + \dfrac{Z_1}{4Z_2}\right)}$ (b) $\dfrac{Z_1 Z_2}{Z_1 + Z_2}$

   (c) $\dfrac{Z_1 + Z_2}{Z_1 Z_2}$ (d) $\sqrt{\dfrac{Z_1 Z_2}{1 + \frac{Z_1}{4Z_2}}}$

6. For an π-network having series impedance Z_1 and shunt impedance $2Z_2$, the image impedance is
   (a) $\sqrt{Z_1 Z_2\left(1 + \dfrac{Z_1}{4Z_2}\right)}$ (b) $\dfrac{Z_1 Z_2}{Z_1 + Z_2}$ (c) $\dfrac{Z_1 + Z_2}{Z_1 Z_2}$ (d) $\sqrt{\dfrac{Z_1 Z_2}{1 + \frac{Z_1}{4Z_2}}}$.

7. The m-derived T- and π-sections low-pass filter has cut-off frequencies 1800 Hz and 2000 Hz (infinite attenuation frequency), the value of m is
   (a) 0.436   (b) 0.81   (c) 0.19   (d) 0.536.

UNSOLVED PROBLEMS

1. What is a low-pass filter ? Prove $f_C = \dfrac{1}{\pi\sqrt{LC}}$ for a constant-k low-pass filter.

2. What is a high-pass filter ? Prove $f_C = \frac{1}{4\pi\sqrt{LC}}$ for a constant-k high-pass filter (T-section).

3. What is a band-pass filter ? Prove that the resonant frequency of the shunt arm or series arm is the geometric mean of the cut-off frequencies.

4. Design a constant-k low-pass filter having a cut-off frequency of 2000 Hz and a nominal characteristic impedance $R_0 = 1$ KΩ.

5. Design a constant-k low-pass filter having a cut-off frequency of 2000 Hz and a nominal characteristic impedance $R_0 = 600$ Ω. Also find the frequency at which this filter offers attenuation of 19.1 dB.

6. Design a constant-k high-pass T- and π-section filter having a cut-off frequency of 6 KHz and a nominal characteristic resistance $R_0 = 600$ Ω.

7. Design a prototype band-pass filter section having cut-off frequencies of 2 KHz and 5 KHz and a nominal characteristic impedance of 600 Ω.

8. Determine the cut-off frequency and nominal characteristic impedance R_0 for the networks shown in Figure 13.41:

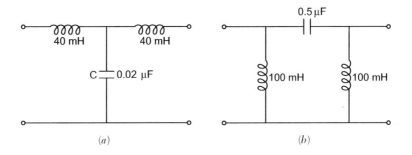

FIGURE 13.41

9. Determine the nominal characteristic impedance and cut-off frequency for the filter shown in Figure 13.42.

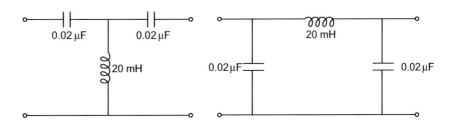

FIGURE 13.42

10. Design m-derived T- and π-sections low-pass filters for nominal characteristic impedance $R_0 = 500\ \Omega$, cut-off frequency $= 1.8$ KHz, and infinite attenuation frequency $f_\infty = 2$ KHz.

11. Design T- and π-sections of an m-derived high-pass filter having a nominal characteristic impedance of 500 Ω, cut-off frequency of 4 KHz, and infinite attenuation at 3.6 KHz.

12. Determine the cut-off frequency for a constant-k low-pass filter (T-section) as shown in Figure 13.43 and characteristic impedance.

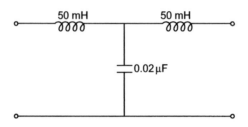

FIGURE 13.43

13. Prove for the m-derived band-pass filter,

$$m = \sqrt{1 - \left(\frac{f_2 - f_1}{f_{\infty 2} - f_{\infty 1}}\right)^2}.$$

14. Z_1 and Z_2 are series and shunt arm impedance of a T- and π-network. Prove that

$$Z_{0\pi} = \frac{Z_1 Z_2}{Z_{0T}},$$

where $Z_{0\pi}, Z_{0T}$ are characteristic impedances for π- and T-networks.

Chapter 14

FOURIER ANALYSIS

14.1 INTRODUCTION

The Fourier series technique was developed by the French mathematician Jean Baptiste Joseph Fourier (1768–1830). The Fourier series is a mathematical technique used for analyzing periodic functions by decomposing such a function into a weighted sum of simpler *sinusoidal* components.

In a simple way we can say, "A Fourier series is an expansion of a periodic function $f(t)$ in terms of an infinite sum of sines and cosines".

14.2 PERIODIC FUNCTIONS

A *periodic function* is that which repeats itself after regular intervals of time. Mathematically, we can explain a periodic function as

$f(t) = f(t+T)$ for every t where T is known as time period of the function.
$$f(t) = f(t \pm nT), \tag{14.1}$$

where n is an integer.

By Figure 14.1(a) and (b) it is clear,

$$f(t) = f(t+T)$$

at

$$t = 0$$
$$f(0) = f(T) = 0.$$

The value of function $f(t)$ is the same after time period "T."

The smallest value of T that satisfies Equation (14.1) is known as the time period.

The functions $\sin \omega t$ and $\cos \omega t$ are the most commonly used periodic functions and time period T equals $\frac{2\pi}{\omega}$, where ω is known as angular frequency.

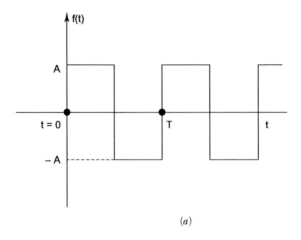

(a)

FIGURE 14.1

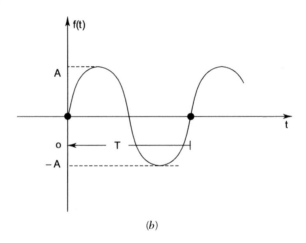

(b)

FIGURE 14.1

14.3   EVEN AND ODD FUNCTIONS

Let there be a function

$$f(t) = f(-t). \tag{14.2}$$

A function satisfying Equation (14.2) is said to be an *even function*. For example, $\cos \omega t, |\sin t|, t^2, t^n$ (for even values of n) are even functions.

Let there be a function

$$f(t) = -f(-t). \tag{14.3}$$

A function satisfying Equation (14.3) is said to be an *odd function*. For example, $\sin \omega t, t^n$ (for odd values of n).

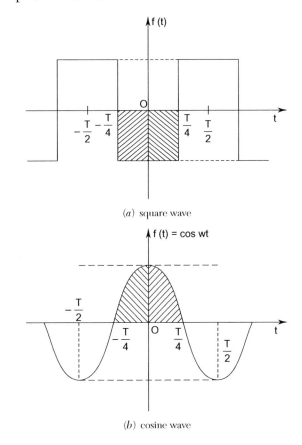

(a) square wave

(b) cosine wave

FIGURE 14.2   Examples of even functions.

14.3.1   Properties of Even and Odd Functions

Let the symbols e and o denote even and odd functions.

Any function $f(t)$ can be written in terms of its even and odd parts,

$$f(t) = f_e(t) + f_o(t). \tag{14.4}$$

If we replace t by $(-t)$ in Equation (14.4), we get

$$f(-t) = f_e(t) + f_o(-t) \tag{14.5}$$

and we know by Equations (14.2) and (14.3),

$$f_e(t) = f_e(-t) \text{ and } f_o(t) = -f_o(-t).$$

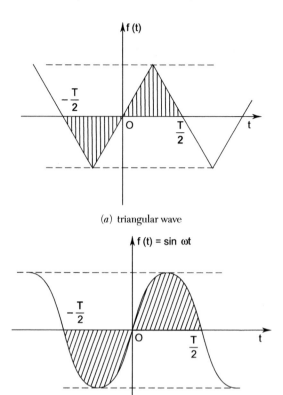

(a) triangular wave

(b) sine wave

FIGURE 14.3   Examples of odd functions.

Solving Equations (14.4) and (14.5), we get

$$f_e(t) = \frac{1}{2}[f(t) + f(-t)] \qquad (14.6)$$

$$f_o(t) = \frac{1}{2}[f(t) - f(-t)]. \qquad (14.7)$$

Thus, if $f(t)$ is given, we can find its even and odd parts. Even and odd functions possess the following properties:

The sum of any two or more even functions is an even function,

$$f_{e1}(t) + f_{e2}(t) + \cdots = f_e(t). \qquad (14.8(a))$$

The sum of any two or more odd functions is an odd function,

$$f_{o1} + f_{o2}(t_1 + \cdots = f_o(t). \qquad (14.8(b))$$

The multiplication of two or more even functions is an even function,

$$f_{e1}(t) \times f_{e2}(t) = f_e(t). \tag{14.9(a)}$$

The multiplication of two odd functions is an even function,

$$f_{o1}(t) \times f_{o2}(t) = f_e(t). \tag{14.9(b)}$$

The multiplication of an even and odd function is an odd function,

$$f_{e1}(t) \times f_{o2}(t) = f_o(t). \tag{14.10}$$

The multiplication of a constant with an even or odd function is an even or odd function simultaneously,

$$A \times f_{e1}(t) = f_e(t) \tag{14.11}$$

$$A \times f_{o1}(t) = f_o(t), \tag{14.12}$$

where A is a constant.

NOTE ▶ If an even function is added to an odd function, the result is neither even nor odd. Such a result is said to have even and an odd parts as given by Equation (14.4).

By integral properties for an even function $f_e(t)$,

$$\int_{-t}^{t} f_e(t)\, dt = 2 \int_{0}^{t} f_e(t)\, dt. \tag{14.13}$$

For an odd function $f_o(t)$,

$$\int_{-t}^{t} f_o(t)\, dt = 0. \tag{14.14}$$

14.4 FOURIER SERIES

A periodic function $f(t)$ of time period T satisfying Dirichlet conditions can be expanded into the following series:

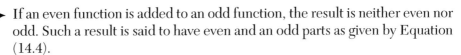

$$f(t) = a_0 + a_1 \cos \omega t + a_2 \cos 2\omega t + \cdots + a_n \cos n\omega t + b_1 \sin \omega t$$
$$+ b_2 \sin 2\omega t + \cdots b_n \sin n\omega t. \tag{14.15}$$

Equation (14.15) can be expressed in the trigonometric form,

$$f(t) = a_0 + \sum_{n=1}^{\infty} \left(a_n \cos n\omega t + b_n \sin n\omega t \right). \tag{14.16}$$

14.4.1   Dirichlet Conditions for Any Function f(t)

1. It is a single valued function.
2. The integral $\int_t^{t+T} |f(t)|\, dt$ exists and is finite for any arbitrary value of t.
3. If $f(t)$ is discontinuous there are a finite number of discontinuities in any one period.
4. The function $f(t)$ has a finite number of maxima and minima in any one period.

These conditions are called *Dirichlet conditions* and should be satisfied by a periodic function $f(t)$. These are necessary and sufficient conditions. Equation (14.16) shows that a periodic function consists of sine and cosine components of angular frequencies $0, \omega, 2\omega - n\omega$. The component with zero angular frequency is called the D.C. component. The number n represents the order of harmonies. For example, $n = 1$, the component $(a_1 \cos \omega t + b_1 \sin \omega t)$ is called the *first harmonic* or the fundamental component of the waveform. The term $(a_n \cos n\omega t + b_n \sin n\omega t)$ is called the nth harmonic component of the waveform. *Fourier analysis* is a process to determine the values of Fourier constants $(a_0, a_1, \ldots, a_n \text{ and } b_1, b_2, \ldots, b_n)$.

14.5   ALTERNATIVE FORMS OF FOURIER SERIES

Equation (14.16) is the trigonometric form of the Fourier series,

$$f(t) = a_0 + \sum_{n=1}^{\infty} (a_n \cos n\omega t + b_n \sin n\omega t).$$

We have $a_n \cos n\omega t + b_n \sin \omega t = c_n \sin (n\omega t + \phi)$, where

$$c_n = \sqrt{a_n^2 + b_n^2} \text{ and } \phi = \tan^{-1}\left(\frac{a_n}{b_n}\right).$$

Also, $a_n \cos n\omega t + b_n \sin n\omega t = c_n \cos (n\omega t + \phi)$, where

$$c_n = \sqrt{a_n^2 + b_n^2} \text{ and } \phi_n = \tan^{-1}\left(\frac{b_n}{a_n}\right).$$

Hence, the other form of Fourier series

$$f(t) = a_0 + \sum_{n=1}^{\infty} c_n(\sin (n\omega t + \phi)) \tag{14.17}$$

and

$$f(t) = a_0 + \sum_{n=1}^{\infty} c_n \cos{(n\omega t + \theta_n)} \qquad (14.18)$$

or

$$f(t) = a_0 + c_1 \cos{(\omega t + \theta_1)} + c_2 \cos{(2\omega t + \theta_2)} \ldots c_n \cos{(n\omega t + \theta_n)}. \qquad (14.19)$$

It may be seen that c_n and θ_n contain all the information needed to construct the Fourier Series of the form given by Equation (14.19).

14.6   USEFUL INTEGRALS IN FOURIER SERIES

The following integrals are useful to determine Fourier series coefficients:

$$\int_{t_1}^{t_1+T} \sin{n\omega t}\, dt = 0, \int_{0}^{2\pi} \sin{n\theta}\, d\theta = 0 \qquad (14.20)$$

$$\int_{t_1}^{t_1+T} \cos{n\omega t}\, dt = 0, \int_{0}^{2\pi} \cos{n\theta}\, d\theta = 0 \qquad (14.21)$$

$$\int_{t_1}^{t_1+T} \sin{m\omega t}\cos{n\omega t}\, dt = 0, \int_{0}^{2\pi} \sin{m\theta}\cos{n\theta}\, d\theta = 0 \qquad (14.22)$$

For all m and n, where m and n are integers,

$$\int_{t_1}^{t_1+T} \sin{m\omega t}\sin{n\omega t}\, dt = \begin{cases} \frac{T}{2} & m = n \neq 0 \\ 0 & m \neq n \\ 0 & m \text{ or } n = 0 \end{cases} \qquad (14.23)$$

$$\int_{t_1}^{t_1+T} \cos{m\omega t}\cos{n\omega t} = \begin{cases} \frac{T}{2} & m = n \neq 0 \\ 0 & m \neq n \\ T & m = n = 0 \end{cases} \qquad (14.24)$$

$$\int_{0}^{2\pi} \sin^2{n\theta}\, d\theta = \pi \qquad (14.25)$$

$$\int_{0}^{2\pi} \cos^2{n\theta}\, d\theta = \pi \qquad (14.26)$$

$$\int_{-\pi}^{\pi} \sin n\theta \cos m\theta\, d\theta = 0 \text{ for all } m \text{ and } n \qquad (14.27)$$

$$\int_{-\pi}^{\pi} \cos n\theta \cos m\theta\, d\theta = \begin{cases} 0 \text{ for } & n \neq m \\ \pi \text{ for } & n = m. \end{cases} \qquad (14.28)$$

14.7   EVALUATION OF FOURIER SERIES CONSTANTS

14.7.1   Evaluation of a_0

By Equation (14.16),

$$f(t) = a_0 + \sum_{n=1}^{\infty} (a_n \cos n\omega t + b_n \sin n\omega t).$$

Integrating both sides, the above Equation (14.16) over one period $(t_1, t_1 + T)$, where t is arbitrary, is

$$\int_{t_1}^{t_1+T} f(t)\, dt = a_0 \int_{t_1}^{t_1+T} dt + \int_{t_1}^{t_1+T} \sum_{n=1}^{\infty} (a_n \cos n\omega t + b_n \sin n\omega t)\, dt.$$

We write it as

$$\int_{t_1}^{t_1+T} f(t)\, dt = a_0 T + \sum_{n=1}^{\infty} \left[a_n \int_{t_1}^{t_1+T} \cos n\omega t\, dt + b_n \int_{t_1}^{t_1+T} \sin n\omega t\, dt \right].$$

By Equations (14.20) and (14.21),

$$\int_{t_1}^{t_1+T} \sin n\omega t\, dt = 0 \text{ and } \int_{t_1}^{t_1+T} \cos n\omega t\, dt = 0.$$

Hence,

$$\int_{t_1}^{t_1+T} f(t)\, dt = a_0 T$$

$$\boxed{a_0 = \tfrac{1}{T} \int_{t_1}^{t_1+T} f(t)\, dt}. \qquad (14.29)$$

We can say a_0 is the average value of $f(t)$ over a period.

14.7.2 Evaluation for a_n

To evaluate a_n, multiply Equation (14.16) by $\cos m\omega t$ and integrate over one period $(t_1, t_1 + T)$.

$$\int_{t_1}^{t_1+T} f(t)\cos m\omega t\, dt = a_0 \int_{t_1}^{t_1+T} \cos m\omega t\, dt$$
$$+ \int_{t_1}^{t_1+T} \sum_{n=1}^{\infty} (a_n \cos n\omega t + b_n \sin n\omega t)\cos m\omega t\, dt$$

or

$$\int_{t_1}^{t_1+T} f(t)\cos m\omega t\, dt = a_0 \int_{t_1}^{t_1+T} \cos m\omega t\, dt + \sum_{n=1}^{\infty} a_n \int_{t_1}^{t_1+T} \cos n\omega t$$
$$+ \cos m\omega t + \sum_{n=1}^{\infty} b_n \int_{t_1}^{t_1+T} \sin n\omega t \cos m\omega t\, dt$$

Now by Equations (14.21), (14.22), and (14.24), we get

$$\int_{t_1}^{t_1+T} f(t)\cos m\omega t\, dt = a_0 \times 0 + b_n \times 0 + a_n \times \frac{T}{2} \text{ (only when } m = n)$$

$$\boxed{a_n = \tfrac{2}{T} \int_{t_1}^{t_1+T} f(t)\cos m\omega t\, dt}. \tag{14.30}$$

Putting $n = m$ into Equation (14.30), we get

$$\boxed{a_n = \tfrac{2}{T} \int_{t_1}^{t_1+T} f(t)\cos n\omega t\, dt}. \tag{14.31}$$

14.7.3 Evaluation for b_n

To evaluate b_n, multiply Equation (14.16) by $\sin m\omega t$ and integrate over one period $(t_1, t_1 + T)$

$$\int_{t_1}^{t_1+T} f(t)\sin m\omega t\, dt = a_0 \int_{t_1}^{t_1+T} \sin m\omega t\, dt$$
$$+ \int_{t_1}^{t_1+T} \sum_{n=1}^{\infty} (a_n \cos n\omega t + b_n \sin n\omega t)\sin m\omega t\, dt$$

or

$$\int_{t_1}^{t_1+T} f(t) \sin m\omega t \, dt = a_0 \int_{t_1}^{t_1+T} \sin m\omega t \, dt$$

$$+ \sum_{n=1}^{\infty} a_n \int_{t_1}^{t_1+T} \cos n\omega t \sin m\omega t \, dt$$

$$+ \sum_{n=1}^{\infty} b_n \int_{t_1}^{t_1+T} \sin n\omega t \sin m\omega t \, dt.$$

Now by Equations (14.20), (14.22), and (14.23), we get

$$\int_{t_1}^{t_1+T} f(t) \sin m\omega t \, dt = a_0 \times 0 + a_n \times 0 + b_n \times \frac{T}{2} \text{ (only when } n = m).$$

$$\boxed{b_n = \frac{2}{T} \int_{t_1}^{t_1+T} f(t) \sin m\omega t \, dt} \qquad (14.32)$$

Putting $n = m$, we get

$$\boxed{b_n = \frac{2}{T} \int_{t_1}^{t_1+T} f(t) \sin n\omega t \, dt}. \qquad (14.33)$$

In summary, we can write

$$f(t) = a_0 + \sum_{n=1}^{\infty} (a_n \cos n\omega t + b_n \sin n\omega t), t_1 < t < t_1 + T,$$

where

$$\boxed{\begin{aligned} a_0 &= \frac{1}{T} \int_{t_1}^{t_1+T} f(t) \, dt \\ a_n &= \frac{2}{T} \int_{t_1}^{t_1+T} f(t) \cos n\omega t \, dt \\ b_n &= \frac{2}{T} \int_{t_1}^{t_1+T} f(t) \sin n\omega t \, dt \end{aligned}}.$$

These constants depend on n and the nature of the waveform of $f(t)$.

14.8   FOURIER SERIES CONSTANTS FOR PERIOD $(0 + 0.2\pi)$

Let $A(\theta)$ be a periodic function, with period 0 to 2π, thus

$$A(\theta) = A(\theta + 2\pi).$$

We know

$$\theta = \omega t = \frac{2\pi}{T}t$$

and let

$$A(\theta) = f(t)$$

$$f(t) = a_0 + \sum_{n=1}^{\infty} a_n \cos n\omega t + \sum_{n=1}^{\infty} b_n \sin n\omega t$$

$$g\theta = a_0 + \sum_{n=1}^{\infty} a_n \cos n\theta + \sum_{n=1}^{\infty} b_n \sin n\theta.$$

Since

$$\theta = \frac{2\pi}{T}t, d\theta = \frac{2\pi}{T} dt \text{ or } dt = \frac{T}{2\pi}d\theta.$$

By Equations (14.29), (14.31), and (14.33), putting $t_1 = 0$ and $T = 2\pi$,

$$a_0 = \frac{1}{T} \int_{t_1}^{t_1+T} f(t) \, dt = \frac{1}{T} \int_{0}^{T} A(\theta)\frac{T}{2\pi}d\theta = \frac{1}{2\pi} \int_{0}^{2\pi} A(\theta)\frac{T}{2\pi}d\theta$$

$$a_0 = \frac{1}{2\pi} \int_{0}^{2\pi} A(\theta)d\theta \tag{14.34}$$

$$a_n = \frac{2}{T} \int_{0}^{2\pi} A(\theta) \cos n\theta \frac{T}{2\pi}d\theta = \frac{1}{\pi} \int_{0}^{2\pi} A(\theta) \cos n\theta d\theta. \tag{14.35}$$

Similarly,

$$b_n = \frac{1}{\pi} \int_{0}^{2\pi} A(\theta) \sin n\theta d\theta. \tag{14.36}$$

14.9   LIMITS OF INTEGRATION PER TIME PERIOD

The general limits of integration for finding the Fourier Series coefficients are t_1 and $(t_1 + T)$. The limits can be any values of t per one time period.

Since t_1 is arbitrary it may be chosen according to convenience. Usually the limits of integration are chosen as follows: 0 to T, $-\frac{T}{2}$ to $\frac{T}{2}$, $-\frac{T}{4}$ to $\frac{3T}{4}$.

In this case, we can write

$$a_0 = \frac{1}{T} \int_{t_1}^{t_1+T} f(t)\,dt = \frac{1}{T} \int_{0}^{T} f(t)\,dt = \frac{1}{T} \int_{-\frac{T}{2}}^{\frac{T}{2}} f(t)\,dt \qquad (14.37)$$

$$a_n = \frac{2}{T} \int_{t_1}^{t_1+T} f(t) \cos n\omega t\,dt = \frac{2}{T} \int_{0}^{T} f(t) \cos n\omega t\,dt$$

$$= \frac{2}{T} \int_{-\frac{T}{2}}^{\frac{T}{2}} f(t) \cos n\omega t\,dt \qquad (14.38)$$

$$b_n = \frac{2}{T} \int_{t_1}^{t_1+T} f(t) \sin n\omega t\,dt = \frac{2}{T} \int_{0}^{T} f(t) \sin n\omega t\,dt$$

$$= \frac{2}{T} \int_{-\frac{T}{2}}^{\frac{T}{2}} f(t) \sin n\omega t\,dt. \qquad (14.39)$$

14.10 FOURIER CONSTANTS IN TERMS OF EVEN AND ODD FUNCTIONS

By Equation (14.4),

$$f(t) = f_e(t) + f_o(t).$$

Putting the value of $f(t)$ by Equation (14.4) into Equations (14.37), (14.38), and (14.39),

$$a_0 = \frac{1}{T} \int_{-\frac{T}{2}}^{\frac{T}{2}} f_e(t)\,dt + \frac{1}{T} \int_{-\frac{T}{2}}^{\frac{T}{2}} f_o(t)\,dt \qquad (14.40)$$

$$a_n = \frac{2}{T} \int_{-\frac{T}{2}}^{\frac{T}{2}} f_e(t) \cos n\omega t\,dt + \frac{2}{T} \int_{-\frac{T}{2}}^{\frac{T}{2}} f_o(t) \cos n\omega t\,dt \qquad (14.41)$$

$$b_n = \frac{2}{T} \int_{-\frac{T}{2}}^{\frac{T}{2}} f_e(t) \sin n\omega t\,dt + \frac{2}{T} \int_{-\frac{T}{2}}^{\frac{T}{2}} f_o(t) \sin n\omega t\,dt. \qquad (14.42)$$

14.11 WAVEFORM SYMMETRIES

The procedure of evaluating Fourier constants is straightforward, but when a signal exhibits symmetrical properties, the number of steps in evaluating constants can be reduced. A simple procedure can be adopted in Fourier

analysis. There are three types of symmetry:
(*a*) Even function symmetry,
(*b*) Odd function symmetry, and
(*c*) Half-wave or mirror symmetry.

14.11.1 Even Symmetry

A function is said to be even if

$$f(t) = f(-t).$$

Geometrically, an even function is symmetrical not only with respect to the vertical axis passing through the origin at $t = 0$ but also with respect to all vertical lines at $\frac{nT}{2}(n = \pm 1, \pm 2, \dots)$, since all periodic functions with period T also satisfy the condition $f(t) = f(t + T)$. Some examples of even symmetry are shown in Figure 14.4.

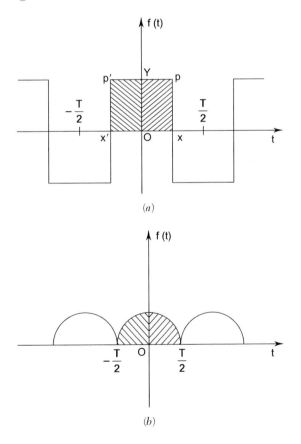

(*a*)

(*b*)

FIGURE 14.4 **Even waveforms possessing symmetry.**

NOTE ► For an even function, the area about the vertical axis is the same in magnitude (Area $oxpy$ = Area $ox'p'y$).

For even functions, $f_o(t) = 0$ and $f(t) = f_e(t)$.
Putting $f_o(t) = 0$ into Equations (14.40), (14.41), and (14.42)

$$a_0 = \frac{1}{T} \int_{-\frac{T}{2}}^{\frac{T}{2}} f_e(t)\, dt \tag{14.43}$$

$$a_n = \frac{2}{T} \int_{-\frac{T}{2}}^{\frac{T}{2}} f_e(t) \cos n\omega t\, dt \tag{14.44}$$

$$b_n = \frac{2}{T} \int_{-\frac{T}{2}}^{\frac{T}{2}} f_e(t) \sin n\omega t\, dt. \tag{14.45}$$

$f_e(t) \cos n\omega t$ is an even function and
$f_e(t) \sin n\omega t$ is an odd function.

Now by Equations (14.13) and (14.14), the property of integral

$$\int_{-t}^{t} f_e(t)\, dt = 2 \int_{0}^{t} f_e(t)\, dt \text{ and } \int_{-t}^{t} f_o(t)\, dt = 0.$$

Applying Equations (14.13) and (14.14) in Equations (14.43), (14.44), and (14.45), we get

$$a_0 = \frac{1}{T} \int_{-\frac{T}{2}}^{\frac{T}{2}} f_e(t)\, dt = \frac{2}{T} \int_{0}^{\frac{T}{2}} f_e(t)\, dt \tag{14.46}$$

$$a_n = \frac{4}{T} \int_{0}^{\frac{T}{2}} f_e(t) \cos n\omega t\, dt \tag{14.47}$$

$$b_n = 0. \tag{14.48}$$

Equation (14.48) shows that the coefficients of all sine terms are zero. Thus, the Fourier series of an even function contains only cosine terms and the constant term a_0. Thus, we see that for even function symmetry $[f(t) = f_e(t)]$.

$$a_0 = \frac{2}{T} \int_{0}^{\frac{T}{2}} f(t)\, dt$$

$$a_n = \frac{4}{T} \int_{0}^{\frac{T}{2}} f(t) \cos n\omega t\, dt$$

$$b_n = 0.$$

14.11.2 Odd Symmetry

A function $f(t)$ is said to be odd if

$$f(t) = -f(-t).$$

Geometrically, the odd function symmetry is a combination of symmetry about the origin and symmetry about the zero axis as shown in Figure 14.5.

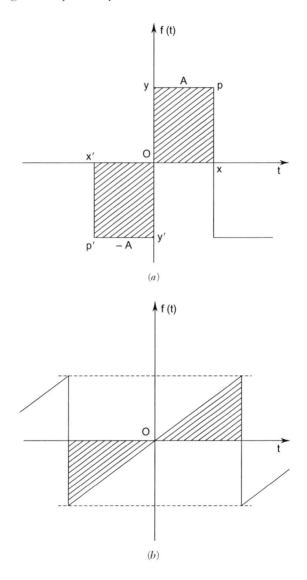

(a)

(b)

FIGURE 14.5 **Waveform possessing odd symmetry.**

For an odd function, the area about the horizontal axis is the same in magnitude (Area oxpy = Area $ox'p'y'$).

For an odd function,

$$f_e(t) = 0 \text{ and } f(t) = f_o(t).$$

Putting $f_e(t) = 0$ into Equations (14.40), (14.41), and (14.42)

$$a_0 = \frac{1}{T} \int_{-\frac{T}{2}}^{\frac{T}{2}} f_o(t)\, dt = \frac{1}{T} \int_{-\frac{T}{2}}^{\frac{T}{2}} f(t)\, dt \tag{14.49}$$

$$a_n = \frac{2}{T} \int_{-\frac{T}{2}}^{\frac{T}{2}} f_o(t) \cos n\omega t\, dt \tag{14.50}$$

$$b_n = \frac{2}{T} \int_{-\frac{T}{2}}^{\frac{T}{2}} f_o(t) \sin n\omega t\, dt \tag{14.51}$$

$f_o(t) \cos n\omega t$ is an odd function and
$f_o(t) \sin n\omega t$ is an even function.

Now by Equations (14.13) and (14.14), the property of integral,

$$\int_{-t}^{t} f_e(t)\, dt = 2 \int_{0}^{t} f_e(t)\, dt \text{ and } \int_{-t}^{t} f_o(t)\, dt = 0.$$

Applying Equations (14.13) and (14.14) in Equations (14.49), (14.50), and (14.51), we get

$$a_0 = 0 \tag{14.52}$$
$$a_n = 0 \tag{14.53}$$

$$b_n = \frac{4}{T} \int_{0}^{\frac{T}{2}} f(t) \sin n\omega t\, dt. \tag{14.54}$$

Thus, the Fourier series of an odd function contains only sine terms. Cosine terms, and constant terms are zero. Thus, we see that, for odd function symmetry $[f(t) = -f(-t)]$.

$$\boxed{\begin{aligned} a_0 &= 0 \\ a_n &= 0 \\ b_n &= \frac{4}{T} \int_{0}^{\frac{T}{2}} f(t) \sin n\omega t\, dt \end{aligned}}$$

14.11.3 Half-Wave or Mirror Symmetry

A function $f(t)$ is said to possess half-wave symmetry when

$$f(t) = -f\left(t \pm \frac{T}{2}\right). \tag{14.55}$$

Figure 14.6 shows functions possessing half-wave symmetry. In each case, the wave-form extendings for time. Interval, say $-\frac{T}{2}$ to 0, is negative of the waveform extending from 0 to $\frac{T}{2}$. This function is neither even nor odd and thus is formed by an even function and an odd function. A wave possesses half-wave symmetry when the negative portion of the wave is the mirror image of the positive portion of the wave.

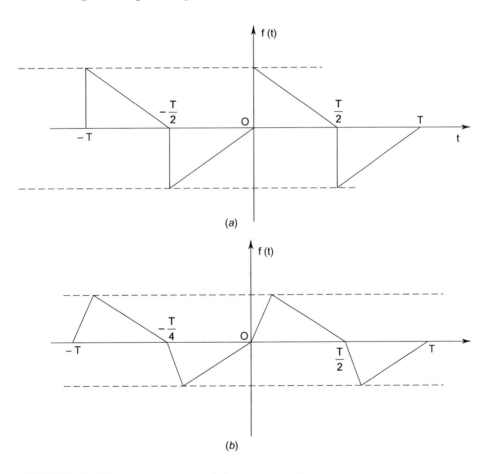

(a)

(b)

FIGURE 14.6   Waveforms possessing half-wave symmetry.

Now we can proceed to evaluate the Fourier coefficients. In the case of a function with half-wave symmetry, let us choose the periodic time T as extending from $-\frac{T}{2}$ to $+\frac{T}{2}$.

$$a_n = \frac{2}{T} \int_{-\frac{T}{2}}^{\frac{T}{2}} f(t) \cos n\omega t \, dt$$

$$= \frac{2}{T} \int_{-\frac{T}{2}}^{0} f(t) \cos n\omega t \, dt + \frac{2}{T} \int_{0}^{\frac{T}{2}} f(t) \cos n\omega t \, dt$$

$$= I_1 + I_2 \text{ (say)}$$

$$I_1 = \frac{2}{T} \int_{-\frac{T}{2}}^{0} f(t) \cos n\omega t \, dt \qquad (14.56)$$

Let us substitute $\left(t + \frac{T}{2}\right)$ for t in Equation (14.55) so that the new limits of integration will be 0 to $\frac{T}{2}$ instead of $-\frac{T}{2}$ to 0.

$$I_1 = \frac{2}{T} \int_{0}^{\frac{T}{2}} f\left(t + \frac{T}{2}\right) \cos n\omega \left(t + \frac{T}{2}\right) d\left(t + \frac{T}{2}\right)$$

$$= \frac{2}{T} \int_{0}^{\frac{T}{2}} f\left(t + \frac{T}{2}\right) \left(\cos n\omega t \cos \frac{n\omega T}{2} - \sin n\omega t \sin \frac{n\omega T}{2}\right) dt$$

But $\omega T = 2\pi$ and therefore, $\sin \frac{n\omega T}{2} = \sin n\pi = 0$.
And by Equation (14.55), $f(t) = -f\left(t + \frac{T}{2}\right)$ or

$$f\left(t + \frac{T}{2}\right) = -f(t).$$

$$I_1 = \frac{2}{T} \int_{0}^{\frac{T}{2}} -f(t) \cos n\omega t \cos n\pi \, dt$$

$$I_1 = -\frac{2}{T} \cos n\pi \int_{0}^{\frac{T}{2}} f(t) \cos n\omega t \, dt \qquad (14.57)$$

$$a_n = I_1 + I_2$$

$$= -\frac{2}{T} \cos n\pi \int_{0}^{\frac{T}{2}} f(t) \cos n\omega t \, dt + \frac{2}{T} \int_{0}^{\frac{T}{2}} f(t) \cos n\omega t \, dt$$

$$a_n = \frac{2}{T}(1 - \cos n\pi) \int_{0}^{\frac{T}{2}} f(t) \cos n\omega t \, dt. \qquad (14.58)$$

when n is even, $\cos n\pi = 1$ and $a_n = 0$

when n is odd, $\cos n\pi = -1$ and

$$a_n = \frac{4}{T} \int_0^{\frac{T}{2}} f(t) \cos n\omega t \, dt. \qquad (14.59)$$

Similarly,

$$b_n = 0, \text{ when n is even.}$$

When n is odd,

$$b_n = \frac{4}{T} \int_0^{\frac{T}{2}} f(t) \sin n\omega t \, dt.$$

NOTE ▶ The Fourier series expansion of a periodic function with half-wave symmetry contains only odd harmonies.

$$a_0 = 0$$

when n is even,

$$a_n = b_n = 0$$

when n is odd,

$$a_n = \frac{4}{T} \int_0^{\frac{T}{2}} f(t) \cos n\omega t \, dt$$

$$b_n = \frac{4}{T} \int_0^{\frac{T}{2}} f(t) \sin n\omega t \, dt$$

TABLE 14.1 Summary of Symmetry Conditions for Periodic Functions

Type of Symmetry	Conditions	Property	a_0	a_n	b_n
EVEN	$f(t) = -f(t)$	cosine terms only	$\frac{2}{T}\int_0^{\frac{T}{2}} f(t)\,dt$	$\frac{4}{T}\int_0^{\frac{T}{2}} f(t)\cos n\omega t\,dt$	0
ODD	$f(t) = -f(-t)$	sine terms only	0	0	$\frac{4}{T}\int_0^{\frac{T}{2}} f(t)\sin n\omega t\,dt$
Half wave	$f(t) = -f\left(t \pm \frac{T}{2}\right)$	odd n (harmonies) only	0	$\frac{4}{T}\int_0^{\frac{T}{2}} f(t)\cos n\omega t\,dt$	$\frac{4}{T}\int_0^{\frac{T}{2}} f(t)\sin n\omega t\,dt$

14.12   USE OF FOURIER SERIES

Fourier series is very useful in solving some of the differential equations that arise in mathematical physics such as the wave and the heat equations. Fourier series is also very useful in electrical engineering (especially in network analysis), vibration analysis, acoustics, optics, signal processing, and data compression.

SOLVED PROBLEMS

Problem 14.1. *Find the waveform symmetry of the following waveform (Figure 14.7).*

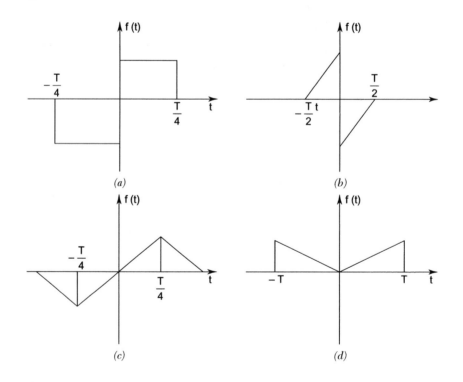

FIGURE 14.7

Solution: For an odd function, $f(t) = -f(-t)$.

For an even function, $f(t) = -f(t)$.

By inspection, Figure. 14.7(a) is an odd function.

Figure 14.7(*b*) is another odd function.
Figure 14.7(*c*) is also an odd function.
Figure 14.7(*d*) is an even function.

Problem 14.2. *Find the Fourier coefficients and the Fourier series of the square-wave function $f(t)$ defined by*

$$f(t) = \begin{cases} 0 & for -\pi \leq t < 0 \\ 1 & for\ 0 \leq t < \pi \end{cases} \quad and\ f(t + 2\pi) = f(t).$$

Solution: $f(t)$ is a periodic function with period 2π and its graph is shown in Figure 14.8.

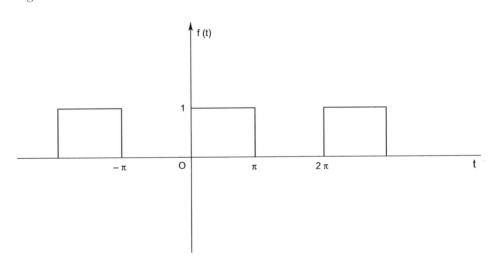

FIGURE 14.8

By using the formula,

$$a_0 = \frac{1}{2\pi} \int_{-\pi}^{\pi} f(t)\, dt = \frac{1}{2\pi} \int_{-\pi}^{0} 0\, dt + \frac{1}{2\pi} \int_{0}^{\pi} 1\, dt$$

$$a_0 = \frac{1}{2\pi} [t]_0^{\pi} = \frac{1}{2}$$

$$a_n = \frac{2}{T} \int_{-\pi}^{\pi} f(t) \cos n\omega t\, dt$$

$$= \frac{2}{2\pi} \int_{0}^{\pi} 1 \cos nt\, dt \left[\omega = \frac{2\pi}{T}\right] \text{ and } T = 2\pi$$

$$= \frac{1}{\pi} \left[\frac{\sin nt}{n} \right]_0^\pi = 0$$

$$b_n = \frac{2}{T} \int_{-\pi}^{\pi} f(t) \sin \omega t \, dt = \frac{2}{2\pi} \int_0^\pi 1 \sin nt \, dt$$

$$= \frac{1}{\pi} \left[-\frac{\cos nt}{n} \right]_0^\pi = -\frac{1}{n\pi} (\cos n\pi - \cos 0)$$

$$b_n = \begin{cases} 0 & \text{if } n \text{ is even} \\ \frac{2}{n\pi} & \text{if } n \text{ is odd.} \end{cases}$$

Therefore, the Fourier series of function $f(t)$ is

$$f(t) = a_0 + a_1 \cos t + a_2 \cos 2t + a_3 \cos 3t + \cdots + b_1 \sin t$$
$$+ b_2 \sin 2t + b_3 \sin 3t$$

$$f(t) = \frac{1}{2} + \frac{2}{\pi} \sin t + 0 \sin 2t + \frac{2}{3\pi} \sin 3t + 0 \sin 4t + \cdots$$

$$f(t) = \frac{1}{2} + \frac{2}{\pi} \sin t + \frac{2}{3\pi} \sin 3t + \cdots$$

Problem 14.3. *Find the amplitude of the 7th harmonic of the given waveform (Figure 14.8(a)) and $f(t + 2\pi) = f(t)$.*

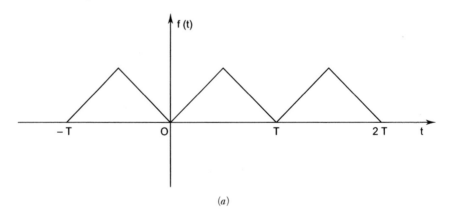

(a)

FIGURE 14.8

Solution: Since the given waveform is an even function,

$$b_n = 0$$

$$a_5 = \frac{4}{T} \int_0^{\frac{T}{2}} f(t) \cos 5\omega t \, dt$$

$$a_5 = \frac{4}{T} \int_0^{\frac{T}{2}} \frac{2t}{T} \cos\left(\frac{2\pi}{T} \times 5t\right) dt$$

$$= \frac{2}{49\pi^2}(\cos 7\pi - 1) = -\frac{4}{49\pi^2}.$$

Problem 14.4. *Find the Fourier series of the triangular wave function as shown in Figure 14.9 by $f(t) = |t|$ for $-1 \le t \le 1$ and $f(t+2) = f(t)$ for all t.*

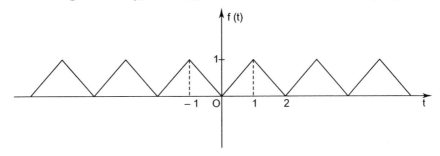

FIGURE 14.9

Solution: $f(t) = |t|$ is an even function. Hence, $b_n = 0$

$$a_0 = \frac{2}{T} \int_0^{\frac{T}{2}} f(t)\, dt \qquad\qquad [T = 2]$$

$$a_0 = \int_0^1 t\, dt = \left[\frac{t^2}{2}\right]_0^1 = \frac{1}{2}$$

$$a_n = \frac{4}{T} \int_0^{\frac{T}{2}} f(t) \cos n\omega t\, dt$$

$$a_n = 2 \int_0^1 t \cos n\pi t\, dt \qquad\qquad \left[\omega = \frac{2\pi}{T}\right]$$

$$a_n = \frac{2}{n^2\pi^2}(\cos n\pi - 1)$$

$$\cos n\pi = 1; \text{ if } n \text{ is even}$$
$$= -1; \text{ if } n \text{ is odd.}$$

$$a_n = \frac{2}{n^2\pi^2}(\cos n\pi - 1) = \begin{cases} 0 & \text{if } n \text{ is even} \\ -\frac{4}{n^2\pi^2} & \text{if } n \text{ is odd} \end{cases}.$$

Therefore, the Fourier series is

$$f(t) = \frac{1}{2} - \frac{4}{\pi^2}\cos \pi t - \frac{4}{9\pi^2}\cos 3\pi t - \frac{4}{25\pi^2}\cos 5\pi t - \cdots$$

$$f(t) = \frac{1}{2} - \sum_{k=1}^{\infty} \frac{4}{(2k-1)^2\pi^2} \cos(2k-1)\pi t \text{ for all } t.$$

Problem 14.5. *Find the Fourier series of the following function as shown in Figure 14.10.* $f(t) = t$ *from* $-\pi$ *to* π *and* $f(t + 2\pi) = f(t).$

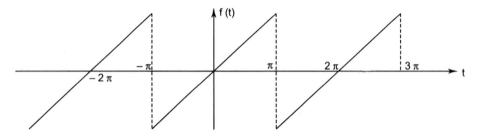

FIGURE 14.10

Solution: $f(t) = t$ is an odd function, hence, $a_n = 0$ and $a_0 = 0$,

$$b_n = \frac{4}{T} \int_0^{\frac{T}{2}} f(t) \sin n\omega t \, dt$$

$$b_n = \frac{4}{2\pi} \int_0^{\pi} t \sin nt \, dt$$

$$b_n = \frac{2}{\pi} \left(\left[-\frac{t \cos nt}{n} \right]_0^{\pi} + \left[\frac{\sin nt}{n^2} \right]_0^{\pi} \right)$$

$$b_n = \frac{2}{n}(-1)^{n+1}.$$

Hence, the Fourier series for this function is:

$$f(t) = 2 \sum_{n=1}^{\infty} \frac{(-1)^{n+1}}{n} \sin nt \text{ for } -\pi < t < \pi.$$

Problem 14.6. *Find the Fourier series for the square-wave voltage shown in Figure 14.11.*

Solution: The waveform may be defined over one period by

$$f(t) = \begin{cases} V_m & 0 < t < \pi \\ -V_m & \pi < t < 2\pi \end{cases}.$$

$$\omega = \frac{2\pi}{T} = 1 \qquad\qquad\qquad [T = 2\pi]$$

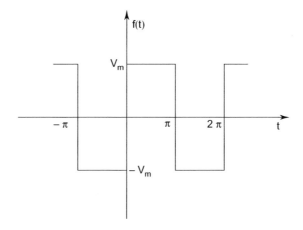

FIGURE 14.11

By Figure 14.11, it is clear, $f(t)$ is an odd function. So by using the odd symmetry formula,

$$a_0 = 0$$

$$a_n = 0$$

$$b_n = \frac{4}{T} \int_0^{\frac{T}{2}} f(t) \sin n\omega t \, dt$$

$$b_n = \frac{2}{\pi} \int_0^{\pi} V_m \sin nt \, dt$$

$$b_n = \frac{2V_m}{n\pi}(1 - \cos n\pi)$$

$$\cos n\pi = 1, \text{ when } n \text{ is even}$$
$$= -1, \text{ when } n \text{ is odd}$$

$$b_n = \frac{4V_m}{n\pi} \text{ for } n = 1, 3, 5, \dots.$$

Therefore, the Fourier series is

$$f(t) = \frac{4V_m}{\pi}\left(\sin t + \frac{1}{3}\sin 3t + \frac{1}{5}\sin 5t + \cdots\right).$$

Problem 14.7. . *Find the Fourier series of the following function as shown in Figure 14.12.*

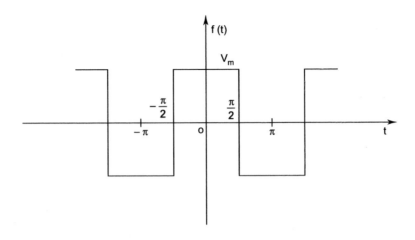

FIGURE 14.12

Solution: The wave may be defined over one period by

$$f(t) = \begin{cases} V_m & -\frac{\pi}{2} < t < \frac{\pi}{2} \\ -V_m & \frac{\pi}{2} < t < \frac{3\pi}{2}. \end{cases}$$

By Figure 14.12, it is clear, $f(t)$, is an even function, therefore, $b_n = 0$,

$$a_0 = \frac{1}{2\pi} \left[\int_{-\pi/2}^{\pi/2} V_m dt + \int_{\pi/2}^{3\pi/2} -V_m dt \right]$$

$$= \frac{1}{2\pi} [V_m \times \pi + (-V_m \times \pi)]$$

$$a_0 = 0.$$

We know a_0 is an average value of $f(t)$ over a period.
In one period, positive area = negative area, therefore, $a_0 = 0$.
By Equation (14.31),

$$a_n = \frac{2}{2\pi} \left[\int_{-\frac{\pi}{2}}^{\frac{\pi}{2}} V_m \cos nt \, dt + \int_{\frac{\pi}{2}}^{\frac{3\pi}{2}} -V_m \cos nt \, dt \right]$$

$$a_n = \frac{V_m}{\pi} \left\{ \left[\frac{\sin nt}{n} \right]_{-\frac{\pi}{2}}^{\frac{\pi}{2}} - \left[\frac{\sin nt}{n} \right]_{\frac{\pi}{2}}^{\frac{3\pi}{2}} \right\}$$

$$a_n = \frac{V_m}{n\pi} \left[3\frac{\sin n\pi}{2} - \sin \frac{3n\pi}{2} \right]$$

when n is even,

$$\sin\frac{n\pi}{2} = 0 \text{ and } \sin\frac{3n\pi}{2} = 0.$$

Therefore,

$$a_n = 0 \text{ for all even } n.$$

$$a_n = \frac{V_m}{n\pi}\left(3\sin\frac{n\pi}{2} - \sin\frac{3n\pi}{2}\right)$$

$$a_1 = \frac{4V_m}{\pi}, a_3 = -\frac{4V_m}{3\pi}, a_5 = \frac{4V_m}{5\pi}.$$

Therefore, the Fourier series for the given function is

$$f(t) = \frac{4V_m}{\pi}\left(\cos t - \frac{1}{3}\cos 3t + \frac{1}{5}\cos 5t + \cdots\right).$$

Problem 14.8. *Determine the Fourier series for the half-wave rectified voltage waveform shown in Figure 14.13.*

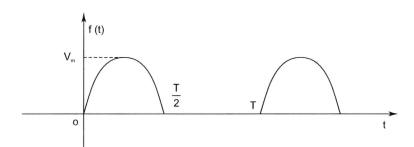

FIGURE 14.13

Solution:

$$T = 2\pi = \frac{2\pi}{\omega} \Rightarrow \omega = 1.$$

The above waveform is defined by

$$f(t) = \begin{cases} V_m\sin t & 0 < t < \pi \\ 0 & \pi < t < 2\pi \end{cases}$$

$$a_0 = \frac{1}{2\pi} \int_0^{2\pi} f(t)\, dt = \frac{1}{2\pi} \left[\int_0^{\pi} V_m \sin t\, dt + \int_{\pi}^{2\pi} 0 \cdot dt \right]$$

$$= \frac{V_m}{2\pi} [-\cos t]_0^{\pi} = \frac{V_m}{\pi}.$$

$$a_n = \frac{1}{\pi} \int_0^{2\pi} f(t) \cos nt\, dt = \frac{1}{\pi} \left[\int_0^{\pi} V_m \sin t \cos nt\, dt + \int_{\pi}^{2\pi} 0 \cos nt\, dt \right]$$

$$= \frac{V_m}{2\pi} \left[\int_0^{\pi} [\sin (n+1)t - \sin (n-1)t]\, dt \right]$$

$$= \frac{V_m}{2\pi} \left[-\frac{\cos (n+1)t}{n+1} + \frac{\cos (n-1)t}{n-1} \right]_0^{\pi}$$

$$= \frac{V_m}{2\pi} \left[-\frac{\cos (n+1)\pi}{n+1} + \frac{1}{n+1} + \frac{\cos (n-1)\pi}{n-1} - \frac{1}{n-1} \right].$$

When n is even, $\cos (n+1)\pi = -1, \cos (n-1)\pi = -1,$

$$a_n = \frac{V_m}{2\pi} \left[\frac{2}{n+1} - \frac{2}{n-1} \right]$$

$$a_n = -\frac{2V_m}{n(n^2-1)}.$$

When n is even, $n = 2, 4, 6, \ldots$.
When n is odd and not equal to 1, $\cos (n+1)\pi = 1, \cos (n-1)\pi = 1,$

$$a_n = \frac{V_m}{2\pi} \left[-\frac{1}{n+1} + \frac{1}{n+1} + \frac{1}{n-1} - \frac{1}{n-1} \right] = 0,$$

when $n = 3, 5, 7, \ldots$
When

$$n = 1,$$

$$a_1 = \frac{1}{\pi} \int_0^{2\pi} V_m \sin t \cos t\, dt$$

or

$$a_1 = \frac{V_m}{2\pi} \int_0^{\pi} \sin 2t\, dt = 0.$$

Therefore,

$$a_n = 0 \text{ for } n = 1, 3, 5, \ldots.$$

Now, the determination of b_n

$$b_n = \frac{1}{\pi} \int_0^{2\pi} f(t) \sin nt \, dt = \frac{1}{\pi} \int_0^{\pi} V_m \sin t \sin nt \, dt + \int_\pi^{2\pi} 0 \sin nt \, dt$$

$$= \frac{V_m}{\pi} \int_0^{\pi} \sin t \sin nt \, dt.$$

This integral is zero except for $n = 1$, therefore, $b_n = 0$ for $n = 2, 3, 4, \ldots$.

$$b_1 = \frac{V_m}{\pi} \int_0^{\pi} \sin^2 t \, dt = \frac{V_m}{\pi} \times \frac{\pi}{2} = \frac{V_m}{2}.$$

Hence, the Fourier series for the half-wave rectified voltage waveform is

$$f(t) = \frac{V_m}{\pi} + \frac{V_m}{2} \sin t - \frac{2V_m}{\pi} \sum_{n=2,4,6}^{\infty} \left(\frac{\cos nt}{n^2 - 1} \right).$$

Problem 14.9. *Find the Fourier series of the waveform shown in Figure 14.14.*

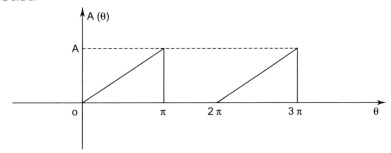

FIGURE 14.14

Solution: By inspection, it is observed that the given waveform is neither odd nor even. Thus, the series will contain both sine and cosine terms.

$$A(\theta) = \frac{A}{\pi} \theta \text{ for } 0 < \theta < \pi$$
$$= 0 \text{ for } \pi < \theta < 2\pi.$$

By Equations (14.34), (14.35), and (14.36)

$$a_0 = \frac{1}{2\pi} \int_0^{2\pi} A(\theta) d\theta = \frac{1}{2\pi} \left[\int_0^{\pi} \frac{A}{\pi} \theta d\theta + \int_\pi^{2\pi} 0 . d\theta \right]$$

$$= \frac{A}{2\pi^2} \left[\frac{\theta^2}{2} \right]_0^{\pi} = \frac{A}{4}$$

$$a_n = \frac{1}{\pi} \int_0^{2\pi} A(\theta) \cos n\theta d\theta$$

$$= \frac{1}{\pi} \left[\int_0^\pi \frac{A}{\pi}\theta \cos n\theta d\theta + \int_\pi^{2\pi} 0 \cos n\theta d\theta \right]$$

$$= \frac{A}{\pi^2 n^2}(\cos n\pi - 1)$$

$\cos n\pi = 1$, when n is even

$\qquad = -1$, when n is odd.

Therefore,

$$a_n = \frac{A}{\pi^2 n^2}(\cos n\pi - 1) = \begin{cases} 0 & \text{[when } n \text{ is even]} \\ -\frac{2A}{\pi^2 n^2} & \text{[when } n \text{ is odd]} \end{cases}$$

$$a_1 = -\frac{2A}{\pi^2}, a_3 = -\frac{2A}{9\pi^2}, a_5 = -\frac{2A}{25\pi^2} \text{ and so on.}$$

Now for b_n,

$$b_n = \frac{1}{\pi} \int_0^{2\pi} A(\theta) \sin n\theta d\theta$$

$$b_n = \frac{1}{\pi} \left[\int_0^\pi \frac{A}{\pi}\theta \sin n\theta d\theta + \int_\pi^{2\pi} 0 \sin n\theta d\theta \right]$$

$$b_n = \frac{A}{\pi^2} \int_0^\pi \theta \sin n\theta d\theta = -\frac{A}{n\pi} \cos n\pi$$

$\cos n\pi = -1$, when n is odd

$\qquad = 1$, when n is even.

$$b_n = -\frac{A}{n\pi} \cos n\pi = \begin{cases} \frac{A}{n\pi} & \text{[when } n \text{ is odd]} \\ -\frac{A}{n\pi} & \text{[when } n \text{ is even]} \end{cases}$$

Therefore, the Fourier series of function $A(\theta)$ is

$$A(\theta) = \frac{A}{4} - \frac{2A}{\pi^2} \cos\theta - \frac{2A}{9\pi^2} \cos 3\theta - \frac{2A}{25\pi^2} \cos 5\theta + \cdots$$
$$\frac{V}{\pi} \sin\theta - \frac{V}{2\pi} \sin 2\theta + \frac{V}{3\pi} \sin 3\theta - \cdots$$

Problem 14.10. *Find the Fourier series for the full-wave rectified voltage waveform shown in Figure 14.15.*

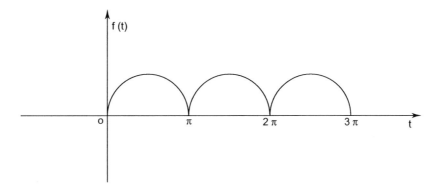

FIGURE 14.15

Solution: Here,

$$T = \pi \text{ and } \omega = \frac{2\pi}{T} = 2.$$

$$f(t) = |V_m \sin t| \text{ [function is positive only]}.$$

Therefore, its an even function $(b_n = 0)$,

$$a_0 = \frac{1}{T} \int_0^T f(t)\, dt$$

$$= \frac{1}{\pi} \int_0^\pi V_m \sin t\, dt = \frac{2V_m}{\pi}$$

$$a_n = \frac{2}{T} \int_0^T f(t) \cos n\omega t\, dt$$

$$a_n = \frac{2}{\pi} \int_0^\pi V_m \sin t \cos 2nt\, dt$$

$$a_n = \frac{V_m}{\pi} \int_0^\pi [\sin(t + 2nt) + \sin(t - 2nt)]\, dt$$

$$a_n = \frac{V_m}{\pi} \left[-\frac{\cos(1 + 2n)t}{1 + 2n} - \frac{\cos(1 - 2n)t}{1 - 2n} \right]_0^\pi$$

$$a_n = -\frac{V_m}{\pi} \left[\frac{\cos(1 + 2n)\pi - 1}{1 + 2n} + \frac{\cos(1 - 2n)\pi - 1}{1 - 2n} \right]$$

$$\cos(1 + 2n)\pi = -1 \text{ and}$$
$$\cos(1 - 2n)\pi = -1.$$

Therefore,

$$a_n = \frac{2V_m}{\pi}\left(\frac{1}{1+2n} + \frac{1}{1-2n}\right) = \frac{4V_m}{\pi}\left(\frac{1}{1-4n^2}\right).$$

Thus, the Fourier series is given by

$$f(t) = \frac{2V_m}{\pi} - \frac{4V_m}{\pi}\left(\frac{1}{3}\cos 2t + \frac{1}{15}\cos 4t + \frac{1}{35}\cos 6t + \cdots\right).$$

Problem 14.11. *Find the Fourier series coefficients of the waveform shown in Figure 14.16.*

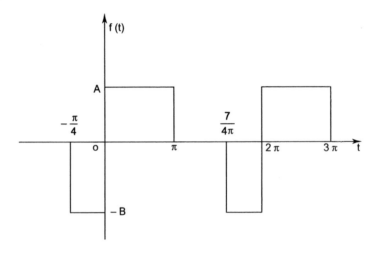

FIGURE 14.16

Solution: The function may be defined as

$$f(t) = \begin{cases} A; & 0 < t < \pi \\ 0; & \pi < t < \frac{7}{4}\pi \\ -B; & \frac{7}{4}\pi < t < 2\pi \end{cases}$$

$$a_0 = \frac{1}{T}\int_0^T f(t)\,dt$$

$$= \frac{1}{2\pi}\int_0^{2\pi} f(t)\,dt$$

$$= \frac{1}{2\pi}\left[\int_0^\pi A\,dt + \int_\pi^{\frac{7}{4}\pi} 0\,dt + \int_{\frac{7}{4}\pi}^{2\pi} -B\,dt\right]$$

$$a_0 = \frac{1}{2\pi} \left[A\pi - B \left(2\pi - \frac{7}{4}\pi \right) \right]$$

$$a_0 = \frac{1}{2} \left[A - \frac{B}{4} \right] = \frac{1}{8} [4A - B]$$

$$a_n = \frac{2}{T} \int_0^T f(t) \cos n\omega t \, dt \qquad\qquad\qquad [T = 2\pi]$$

$$a_n = \frac{1}{\pi} \left[\int_0^\pi A \cos nt \, dt + \int_{\frac{7}{4}\pi}^{2\pi} -B \cos nt \, dt \right] \qquad \left[\omega = \frac{2\pi}{T} = 1 \right]$$

$$a_n = \frac{1}{\pi} \left[\frac{A \sin nt}{n} \Big|_0^\pi + \frac{-B \sin nt}{n} \Big|_{\frac{7}{4}\pi}^{2\pi} \right]$$

$$a_n = \frac{A}{\pi n} \sin n\pi + \frac{-B}{n\pi} \sin 2n\pi + \frac{B}{n\pi} \sin n\frac{7\pi}{4}$$

$$\sin n\pi = \sin 2n\pi = 0, \text{ for } n = 1, 2, \ldots$$

$$a_n = \frac{B}{n\pi} \sin n\frac{7\pi}{4}.$$

Similarly,

$$b_n = \frac{2}{T} \int_0^T f(t) \sin n\omega t \, dt$$

$$b_n = \frac{1}{\pi} \int_0^\pi A \sin nt \, dt + \int_{\frac{7}{4}\pi}^{2\pi} -B \sin nt \, dt$$

$$b_n = \frac{1}{\pi} \left[-\frac{A \cos nt}{n} \Big|_0^\pi + \frac{B \cos nt}{n} \Big|_{\frac{7}{4}\pi}^{2\pi} \right]$$

$$b_n = \frac{1}{n\pi} \left[-A(\cos n\pi - 1) + B \left(1 - \cos n\frac{7}{4}\pi \right) \right]. \qquad [\cos n\pi = (-1)^n]$$

Problem 14.12. *Find the Fourier series of the function shown in Figure 14.17.*

Solution:

$$f(t) = \begin{cases} 0 & \text{for } 0 < t < \pi \\ V & \text{for } \pi < t < 2\pi \end{cases}$$

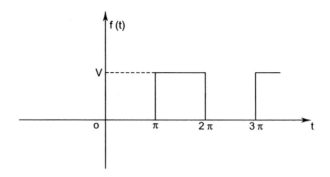

FIGURE 14.17

The determination of a_0,

$$a_0 = \frac{1}{T} \int_0^T f(t)\,dt$$

$$a_0 = \frac{1}{2\pi} \int_0^{2\pi} f(t)\,dt$$

$$= \frac{1}{2\pi}\left[\int_0^{\pi} 0\,dt + \int_{\pi}^{2\pi} V\,dt\right] = \frac{V}{2\pi}[t]_{\pi}^{2\pi} = \frac{V}{2}.$$

The determination of a_n,

$$a_n = \frac{2}{T} \int_0^T f(t)\cos n\omega t\,dt$$

$$a_n = \frac{1}{\pi}\left[\int_{\pi}^{2\pi} V\cos n\omega t\,dt\right] \qquad \left[\omega = \frac{2\pi}{T} = 1\right]$$

$$a_n = \frac{V}{\pi}\left[\frac{\sin n\omega t}{n\omega}\right]_{\pi}^{2\pi}$$

$$a_n = \frac{V}{n\pi}[\sin 2n\pi - \sin n\pi] = 0.$$

Now, the determination of b_n,

$$b_n = \frac{2}{T} \int_0^T f(t)\sin n\omega t\,dt$$

$$b_n = \frac{2}{2\pi} \int_0^{2\pi} f(t)\sin n\omega t\,dt$$

$$= \frac{1}{\pi}\left[\int_{\pi}^{2\pi} V\sin nt\,dt\right]$$

$$b_n = \frac{V}{\pi}\left[-\frac{\cos nt}{n}\right]_{\pi}^{2\pi} = \frac{V}{n\pi}[\cos n\pi - \cos 2n\pi]$$

$$b_n = \frac{V}{n\pi}[(-1)^n - 1] = \begin{cases} 0 & [\text{when } n \text{ is even}] \\ -\frac{2V}{n\pi} & [\text{when } n \text{ is odd}]. \end{cases}$$

Thus, the Fourier series of $f(t)$ is

$$f(t) = \frac{V}{2} - \frac{2A}{\pi}\left[\sin \omega t + \frac{1}{3}\sin 3\omega t + \frac{1}{5}\sin 5\omega t + \cdots\right].$$

Problem 14.13. *Find the Fourier series of the single rectangular pulse shown in Figure 14.18 and $f(t+4) = f(t)$.*

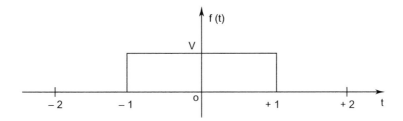

FIGURE 14.18

Solution: The above waveform may be defined over one period by

$$f(t) = \begin{cases} 0 & -2 < t < -1 \\ V & -1 < t < 1 \\ 0 & 1 < t < 2 \end{cases}.$$

The determination of a_0,

$$a_0 = \frac{1}{T}\int_0^T f(t)\,dt$$

$$a_0 = \frac{1}{4}\left[\int_0^4 f(t)\,dt\right]$$

$$= \frac{1}{4}\left[\int_{-2}^{-1} 0\,dt + \int_{-1}^{1} V\,dt + \int_{1}^{2} 0\,dt\right]$$

$$a_0 = \frac{1}{4} \times 2V = \frac{V}{2}.$$

The determination of a_n,

$$a_n = \frac{2}{T} \int_0^T f(t) \cos n\omega t \, dt$$

$$a_n = \frac{1}{2} \int_{-1}^1 V \cos n\omega t \, dt \qquad \text{[Function is zero for all other limits]}$$

$$a_n = \frac{V}{2} \times 2 \int_0^1 \cos \frac{n\pi}{2} t \, dt \qquad \left[\omega = \frac{2\pi}{T} = \frac{\pi}{2} \right]$$

$$a_n = \frac{2V}{n\pi} \sin \frac{n\pi}{2}.$$

The determination of b_n,

$$b_n = \frac{2}{T} \int_0^T f(t) \sin n\omega t \, dt$$

$$b_n = \frac{1}{2} \int_{-1}^1 V \sin n\omega t \, dt.$$

$b_n = 0$ by integral properties and Equation 14.14. Thus, the Fourier series of $f(t)$ is

$$f(t) = \frac{V}{2} + \frac{2V}{\pi} \left[\cos \frac{\pi}{2} t - \frac{1}{3} \cos \frac{3\pi}{2} t + \cdots \right].$$

Problem 14.14. *Find the Fourier series of the following defined function,*

$$f(t) = \begin{cases} -t & for - \pi < t < 0 \\ t & for \ 0 < t < \pi \end{cases} \quad an \ even \ function.$$

Solution: $f(t)$ is even, hence, $b_n = 0$.
The determination of a_0,

$$a_0 = \frac{1}{T} \int_0^T f(t) \, dt = \frac{1}{2\pi} \int_0^{2\pi} f(t) \, dt$$

$$= \frac{1}{2\pi} \left[\int_{-\pi}^0 -t \, dt + \int_0^\pi t \, dt \right]$$

$$a_0 = \frac{1}{2\pi} \left[-\frac{t^2}{2} \Big|_{-\pi}^0 + \frac{t^2}{2} \Big|_0^\pi \right]$$

$$a_0 = \frac{1}{2\pi} \left[+\frac{1}{2} \times \pi^2 + \frac{1}{2} \pi^2 \right] = \frac{\pi}{2}.$$

The determination of a_n,

$$a_n = \frac{2}{T} \int_0^T f(t) \cos n\omega t \, dt$$

$$a_n = \frac{1}{\pi} \int_0^{2\pi} f(t) \cos nt \, dt \qquad \left[\omega = \frac{2\pi}{T} = 1 \right]$$

$$= \frac{1}{\pi} \left[\int_{-\pi}^0 -t \cos nt \, dt + \int_0^\pi t \cos nt \, dt \right]$$

$$a_n = \frac{2}{\pi n^2} (\cos n\pi - 1) = \begin{cases} 0 & [\text{when } n \text{ is even}] \\ -\frac{4}{n^2 \pi} & [\text{when } n \text{ is odd}] \end{cases}$$

Thus, the Fourier series of given function $f(t)$ is

$$f(t) = \frac{\pi}{2} - \frac{4}{\pi} \left(\cos t + \frac{1}{9} \cos 3t + \frac{1}{25} \cos 5t + \cdots \right).$$

Problem 14.15. *Find the Fourier series of the function*

$$f(t) = \begin{cases} 0 & -\pi < t < 0 \\ \pi & 0 < t < \pi \end{cases}.$$

Solution: Follow problem 14.2.
In this case,

$$a_0 = \frac{\pi}{2}$$

$$a_n = 0$$

$$b_n = \frac{1}{n} [1 - (-1)^n] = \begin{cases} 0 & \text{if } n \text{ is even} \\ \frac{2}{n} & \text{if } n \text{ is odd} \end{cases}.$$

Therefore, the Fourier series of $f(t)$ is

$$f(t) = \frac{\pi}{2} + 2 \left(\sin t + \frac{1}{3} \sin 3t + \frac{1}{5} \sin 5t + \cdots \right).$$

Problem 14.16. *Find the Fourier series of the function,*

$$f(t) = \begin{cases} 0 & -2 \le t < 0 \\ t & 0 \le t \le 2 \end{cases}.$$

Solution: The determination of a_0,

$$a_0 = \frac{1}{T}\int_0^T f(t)\,dt = \frac{1}{4}\int_0^4 f(t)\,dt$$

$$= \frac{1}{4}\left[\int_{-2}^0 0\,dt + \int_0^2 t\,dt\right] = \frac{1}{4}\left[\frac{t^2}{2}\right]_0^2 = \frac{1}{2}.$$

The determination of a_n,

$$a_n = \frac{2}{T}\int_0^T f(t)\cos n\omega t\,dt \qquad\qquad \left[\omega = \frac{2\pi}{T} = \frac{\pi}{2}\right]$$

$$a_n = \frac{1}{2}\left[\int_0^2 t\cos\frac{n\pi}{2}t\,dt\right]$$

$$= \frac{2}{n^2\pi^2}(\cos n\pi - 1) = \frac{2}{n^2\pi^2}[(-1)^n - 1].$$

The determination of b_n,

$$b_n = \frac{2}{T}\int_0^T f(t)\sin n\omega t\,dt$$

$$b_n = \frac{1}{2}\left[\int_0^2 t\sin\frac{n\pi}{2}t\,dt\right]$$

$$= \frac{2}{n\pi}(-1)^{n+1}.$$

Therefore, the Fourier series of $f(t)$ is

$$f(t) = \frac{1}{2} + \sum_{n=1}^\infty\left[\frac{2}{n^2\pi^2}\{(-1)^n - 1\}\cos\frac{n\pi t}{2} + \frac{2}{n\pi}(-1)^{n+1}\sin\frac{n\pi t}{2}\right].$$

Problem 14.17. *Find the Fourier series of the waveform shown in Figure 14.19.*

Solution: By inspection, it's clear the given function has odd symmetry. Thus,

$$a_0 = 0 \text{ and } a_n = 0$$

$$T = 2\pi,\, \omega = \frac{2\pi}{T} = 1 \text{ and } \theta = \omega t \Rightarrow \theta = t.$$

It may be revealed that if the period 0 to 2π is considered, the function and integration are to be carried out in two parts (from 0 to π and 0 to 2π).

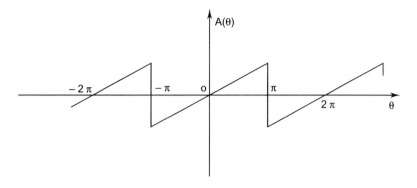

FIGURE 14.19

On the other hand, from $-\pi$ to π, the function and integration are to be carried out in one way only (from $-\pi$ to π, the waveform is a single straight line equation)

$$A(\theta) = \frac{A}{\pi}\theta \qquad \text{[Straight line equation } y = mc]$$

$$b_n = \frac{2}{T}\int_{-\pi}^{\pi} f(t)\sin nt\, dt$$

and

$$f(t) = A(\theta) = \frac{A}{\pi}\theta \text{ and } \theta = t \Rightarrow d\theta = dt$$

$$b_n = \frac{1}{\pi}\int_{-\pi}^{\pi}\frac{A}{\pi}\theta \sin n\theta\, d\theta$$

$$b_n = \frac{A}{\pi^2}\left[\frac{1}{n^2}\sin n\theta - \frac{\theta}{n}\cos n\theta\right]_{-\pi}^{\pi}$$

$$b_n = \frac{-2A}{n\pi}\cos n\pi$$

$$\cos n\pi = \begin{cases} -1 & \text{when } n \text{ is odd} \\ 1 & \text{when } n \text{ is even} \end{cases}.$$

Therefore, the Fourier series of the given function is

$$f(t) = \frac{2A}{\pi}\left[\sin \omega t - \frac{1}{2}\sin 2\omega t + \frac{1}{3}\sin 3\omega t - \frac{1}{4}\sin 4\omega t + \cdots\right].$$

Problem 14.18. *Find the Fourier series for the function given by*

$$f(t) = (t + \pi), \text{ when } -\pi < t < \pi \text{ and } f(t + 2\pi) = f(t).$$

Solution: The given function can be represented as shown in Figure 14.20.

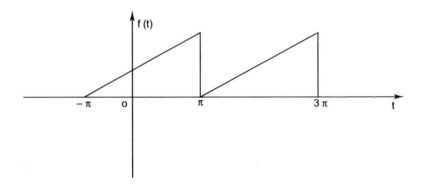

FIGURE 14.20

The determination of a_0,

$$a_0 = \frac{1}{T} \int_{t_1}^{t_1+T} f(t)\, dt = \frac{1}{2\pi} \int_{\pi}^{2\pi} f(t)\, dt \qquad \begin{bmatrix} t_1 & = -\pi \\ t_1 + T = & \pi \end{bmatrix}$$

$$= \frac{1}{2\pi} \int_{-\pi}^{\pi} (t + \pi)\, dt = \frac{1}{2\pi} \left[\frac{t^2}{2} + \pi t \right]_{-\pi}^{\pi} = \pi.$$

The determination of a_n using Equation (14.30),

$$a_n = \frac{2}{T} \int_{t_1}^{t_1+T} f(t) \cos n\omega t\, dt \qquad \left[\omega = \frac{2\pi}{T} = 1 \right]$$

$$= \frac{1}{\pi} \int_{-\pi}^{\pi} (t + \pi) \cos nt\, dt.$$

$$a_n = 0.$$

The determination of b_n using Equation (14.33),

$$b_n = \frac{2}{T} \int_{t_1}^{t_1+T} f(t) \sin n\omega t\, dt$$

$$= \frac{1}{\pi} \int_{-\pi}^{\pi} (t + \pi) \sin nt\, dt$$

$$b_n = -\frac{2}{n} \cos n\pi = -\frac{2}{n}(-1)^n = \frac{2}{n}(-1)^{n+1}.$$

The Fourier series of the given function is

$$f(t) = 2\left[\sin t - \frac{1}{2}\sin 2t + \frac{1}{3}\sin 3t - \frac{1}{4}\sin 4t + \cdots\right].$$

Problem 14.19. *Determine the Fourier coefficients of the waveform shown in Figure 14.21.*

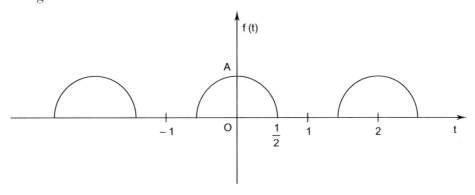

FIGURE 14.21

Solution: The waveform possesses even symmetry, hence, $b_n = 0$,

$$f(t) = A\cos\frac{2\pi}{T}t \text{ for } -0.5 \text{ to } +0.5.$$

$f(t)$ is zero for the other time interval $(-1$ to $1)$.
   The determination of a_0 using Equation (14.29),

$$a_0 = \frac{1}{T}\int_{t_1}^{t_1+T} f(t)\,dt = \frac{1}{2}\int_{-1}^{1} f(t)\,dt$$

$$= \frac{1}{2}\int_{-0.5}^{+0.5} A\cos\pi t\,dt \qquad\qquad [T = 2]$$

$$= \frac{A}{2}\left[\frac{\sin\pi t}{\pi}\right]_{-0.5}^{+0.5} = \frac{A}{\pi}.$$

The determination of a_n using Equation (14.31),

$$a_n = \frac{2}{T}\int_{t_1}^{t_1+T} f(t)\cos n\omega t\,dt \qquad\qquad \left[\omega = \frac{2\pi}{T} = \pi\right]$$

$$= \int_{-0.5}^{+0.5} A\cos\pi t\cos n\pi t\,dt$$

$$a_n = \frac{A}{2\pi}\left[\frac{\sin(1-n)\pi t}{(1-n)} + \frac{\sin(1+n)\pi t}{1+n}\right]_{-0.5}^{+0.5}$$

$$a_n = \frac{A}{\pi}\left[\frac{\sin(1-n)\frac{\pi}{2}}{1-n} + \frac{\sin(1+n)\frac{\pi}{2}}{1+n}\right].$$

Problem 14.20. *Determine the Fourier series coefficients of the function shown in Figure 14.22.*

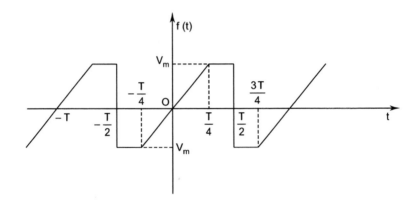

FIGURE 14.22

Solution: The given function may be defined as

$$f(t) = \begin{cases} \frac{4V_m}{T}t & \text{for } -\frac{T}{4} \text{ to } \frac{T}{4} \\ V_m & \text{for } \frac{T}{4} \text{ to } \frac{T}{2} \\ -V_m & \text{for } \frac{T}{2} \text{ to } \frac{3T}{4} \end{cases}.$$

By inspection, it is clear that the given function possesses odd symmetry. Therefore, $a_0 = 0$ and $a_n = 0$,

$$b_n = \frac{2}{T}\int_{t_1}^{t_1+T} f(t)\sin n\omega t\, dt$$

$$b_n = \frac{2}{T}\left[\int_{-\frac{T}{4}}^{\frac{T}{4}} \frac{4V_m t}{T}\sin n\omega t\, dt + \int_{\frac{T}{4}}^{\frac{T}{2}} V_m \sin n\omega t\, dt + \int_{\frac{T}{2}}^{\frac{3T}{4}} -V_m \sin n\omega t\, dt\right]$$

$$b_n = \frac{16V_m}{T^2}\int_0^{\frac{T}{4}} t\sin n\omega t\, dt + \frac{2V_m}{T}\int_{\frac{T}{4}}^{\frac{T}{2}} \sin n\omega t\, dt - \frac{2V_m}{T}\int_{\frac{T}{2}}^{\frac{3T}{4}} \sin n\omega t\, dt$$

$$b_n = \frac{16V_m}{T^2}\left[-\frac{1}{n\omega}\cos n\omega t + \frac{1}{n^2\omega^2}\sin n\omega t\right]_0^{\frac{T}{4}}$$

$$+ \frac{2V_m}{T}\left[-\frac{\cos n\omega t}{n\omega}\right]_{\frac{T}{4}}^{\frac{T}{2}} - \frac{2V_m}{T}\left[-\frac{\cos n\omega t}{n\omega}\right]_{\frac{T}{2}}^{\frac{3T}{4}}$$

$$\omega = \frac{2\pi}{T}, \frac{\omega T}{2} = \pi \text{ and } \frac{\omega T}{4} = \frac{\pi}{2}$$

$$b_n = \frac{16V_m}{T^2 n\omega}\left[1 - \cos\frac{n\pi}{2} + \frac{1}{n\omega}\sin\frac{n\pi}{2}\right]$$

$$+ \frac{2V_m}{Tn\omega}\left[\cos\frac{n\pi}{2} - \cos n\pi\right] + \frac{2V_m}{Tn\omega}\left[\cos\frac{3n\pi}{2} - \cos n\pi\right]$$

$$\omega = \frac{2\pi}{T}$$

$$b = \frac{8V_m}{n\pi T}\left[1 - \cos\frac{n\pi}{2} + \frac{T}{2\pi n}\sin\frac{n\pi}{2}\right] + \frac{V_m}{n\pi}\left[\cos\frac{n\pi}{2} - (-1)^n\right]$$

$$+ \frac{V_m}{n\pi}\left[\cos\frac{3n\pi}{2} - (-1)^n\right].$$

QUESTIONS FOR DISCUSSION

1. What is a periodic function ?
2. What is an even and odd function ?
3. Define even and odd functions and provide examples of each.
4. Is a sine wave an even or odd function ?
5. Is the multiplication of two odd functions even or odd ?
6. What is the formula for Fourier coefficients a_0 ?
7. Why does $b_n = 0$ for even symmetry ?
8. Why does $a_n = 0$ for odd symmetry ?
9. Explain even and odd symmetry.
10. What is half-wave symmetry ?
11. What is the use of Fourier series ?

OBJECTIVE QUESTIONS

1. For even symmetry, which coefficients are zero
   (a) a_0 (b) a_n (c) b_n (d) a_0 and a_n?

2. For odd symmetry, which coefficients are zero
   (a) a_0 (b) a_n (c) b_n (d) a_0 and a_n?

3. For half-wave symmetry, which coefficients are zero
   (a) a_0 (b) a_n (c) b_n (d) a_0 and a_n?

4. The function $f(t) = t + A$. Which of the Fourier series coefficients is/are zero
   (a) a_0 (b) a_n (c) b_n (d) a_0 and a_n.

5. The multiplication of sine wave and cos wave results in an
   (a) Even function   (b) Odd Function
   (c) Neither even or odd   (d) Either even or odd.

UNSOLVED PROBLEMS

1. Find the Fourier series of the function.

$$f(t) = \begin{cases} 1 & \text{if } |t| < 1 \\ 0 & \text{if } 1 \le |t| < 2 \end{cases} \quad \text{and} \quad f(t+4) = f(t).$$

2. Find the Fourier series of the function.

$$f(t) = \begin{cases} 0 & \text{if } -2 \le t < 0 \\ 1 & \text{if } 0 \le t < 1 \\ 0 & \text{if } 1 \le t < 2 \end{cases} \quad \text{and} \quad f(t+4) = f(t).$$

3. Find the Fourier series for the voltage waveform shown in Figure 14.23.

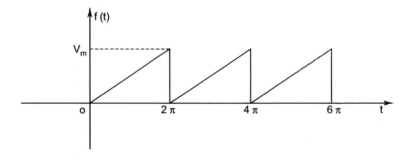

FIGURE 14.23

4. Find the Fourier series of the function

$$f(t) = \begin{cases} -t & \text{if } -4 \le t < 0 \\ 0 & \text{if } 0 \le t < 4 \end{cases} \quad \text{and } f(t+8) = f(t).$$

5. Find the Fourier series of the function

$$f(t) = (1-t) \text{ for } -1 \le t \le 1 \text{ and } f(t+2) = f(t).$$

6. Find the Fourier series of the function

$$f(t) = \begin{cases} 0 & \text{if } -\pi < t < 0 \\ \cos t & \text{if } 0 \le t < \pi \end{cases}.$$

7. Find the Fourier series for the voltage waveform shown in Figure 14.24.

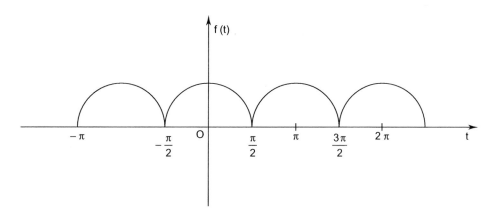

FIGURE 14.24

8. Find the Fourier series of the function

$$f(t) = \begin{cases} -1 & \text{if } -4 \le t < 0 \\ 3 & \text{if } 0 \le t < 4 \end{cases}.$$

9. Find the Fourier series of the function

$$f(t) = \begin{cases} t & \text{if } -1 \le t < 0 \\ 1-t & \text{if } 0 \le t < 1 \end{cases}.$$

10. Find the Fourier series of the function

$$f(t) = e^t, -2 \le t < 2.$$

11.   (a) Show that, if $-1 \le t \le 1$, then

$$t^2 = \frac{1}{3} + \sum_{n=1}^{\infty} (-1)^n \frac{4}{n^2 \pi^2} \cos n\pi t.$$

(b) By substituting a specific value of t, show that

$$\sum_{n=1}^{\infty} \frac{1}{n^2} = \frac{\pi^2}{6}.$$

12. Find the Fourier series of the waveform shown in Figure 14.25.

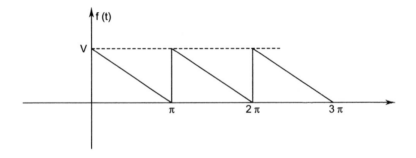

FIGURE 14.25

13. Find the trigonometric Fourier series for the waveform shown in Figure 14.26.

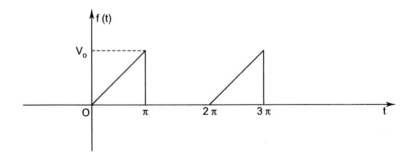

FIGURE 14.26

14. Determine the Fourier series for the waveform shown in Figure 14.27.

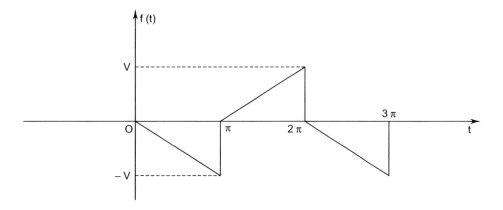

FIGURE 14.27

15. Find the Fourier Series for the waveform shown in Figure 14.28.

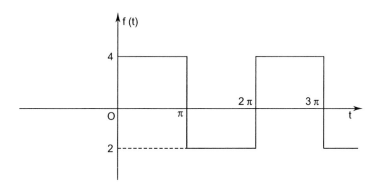

FIGURE 14.28

Appendix A

SELECTED ANSWERS TO EXERCISES

CHAPTER 1

1. (*b*)   3. (*c*)   5. (*c*)   7. (*c*)   9. (*c*)
11. (*a*)   13. (*b*)   15. (*d*)   17. (*d*)   19. (*d*)

CHAPTER 2

1. (*c*)

CHAPTER 3

1. (*a*)   3. (*b*)   5. (*d*)   7. (*b*)   9. (*b*)   11. (*c*)

CHAPTER 4

1. (*d*)   3. (*d*)   5. (*c*)   7. (*b*)   9. (*a*)

CHAPTER 5

1. (*c*)   3. (*c*)   5. (*d*)   7. (*a*)   9. (*d*)   11. (*a*)

CHAPTER 6

1. (*b*)   3. (*b*)   5. (*a*)   7. (*a*)   9. (*b*)

CHAPTER 7

1. (*d*)   3. (*b*)   5. (*a*)   7. (*c*)   9. (*b*)   11. (*d*)

CHAPTER 8

1. (*b*)   3. (*b*)   5. (*c*)   7. (*c*)   9. (*c*)   11. (*c*)

CHAPTER 9

1. (*c*)   3. (*b*)

CHAPTER 10

1. (*c*)   3. (*b*)   5. (*d*)   7. (*d*)   9. (*d*)
11. (*b*)   13. (*c*)   15. (*a*)   17. (*c*)

CHAPTER 11

1. (*b*)   3. (*c*)   5. (*d*)   7. (*b*)

CHAPTER 12

1. (*d*)   3. (*d*)   5. (*a*)

CHAPTER 13

1. (*b*)   3. (*a*)   5. (*a*)   7. (*a*)

CHAPTER 14

1. (*c*)   3. (*a*)   5. (*b*)

Appendix B

ABOUT THE CD-ROM

Included on the CD-ROM are simulations and other files related to network analysis and circuits topics

See the "README" files for any specific information/system requirements related to each file folder, but most files will run on Windows XP or higher

INDEX